UNIVERSITY OF STRATHCLYDE

30125 00749898 2

D1761091

Consumer-led food product development

ANDERSONIAN LIBRARY
★
WITHDRAWN FROM LIBRARY STOCK
★
UNIVERSITY OF STRATHCLYDE

Related titles:

Understanding consumers of food products
(ISBN 978-1-84569-009-0)
It is very important for food businesses, scientists and policy-makers to understand consumers of food products: in the case of businesses to develop successful products and in the case of policy-makers to gain and retain consumer confidence. Consumers' requirements and desires are affected by issues such as culture, age and gender while issues that are important to consumers nowadays such as diet and health or GM foods will not always be so significant. Therefore food businesses and policy-makers need to understand consumers' attitudes and influences upon them in order to respond effectively. Edited by two distinguished experts, this book is an essential guide for food businesses, food scientists and policy-makers.

Food product development
(ISBN 978-1-85573-468-5)
Product development, from refining an established product range to developing completely new products, is the lifeblood of the food industry. It is, however, a process fraught with risk which often ends in failure. This book explains the means to making product development a success. Filled with examples and practical suggestions, and written by a distinguished team with unrivalled academic and industry expertise, *Food product development* is an essential guide for research and product development staff, and for managers throughout the food industry concerned with this key issue.

Details of these books and a complete list of Woodhead titles can be obtained by:

- visiting our web site at www.woodheadpublishing.com
- contacting Customer Services (email: sales@woodhead-publishing.com; fax: +44 (0) 1223 893694; tel.: +44 (0) 1223 891358 ext. 130; address: Woodhead Publishing Limited, Abington Hall, Abington, Cambridge CB21 6AH, England)

Consumer-led food product development

Edited by
Hal MacFie

CRC Press
Boca Raton Boston New York Washington, DC

WOODHEAD PUBLISHING LIMITED
Cambridge England

Published by Woodhead Publishing Limited, Abington Hall, Abington,
Cambridge CB21 6AH, England
www.woodheadpublishing.com

Published in North America by CRC Press LLC, 6000 Broken Sound Parkway, NW,
Suite 300, Boca Raton, FL 33487, USA

First published 2007, Woodhead Publishing Limited and CRC Press LLC
© 2007, Woodhead Publishing Limited except Chapter 12, © 2006 mmr Research
Worldwide Limited
The authors have asserted their moral rights.

This book contains information obtained from authentic and highly regarded sources.
Reprinted material is quoted with permission, and sources are indicated. Reasonable
efforts have been made to publish reliable data and information, but the authors and
the publishers cannot assume responsibility for the validity of all materials. Neither the
authors nor the publishers, nor anyone else associated with this publication, shall be
liable for any loss, damage or liability directly or indirectly caused or alleged to be
caused by this book.

Neither this book nor any part may be reproduced or transmitted in any form or by
any means, electronic or mechanical, including photocopying, microfilming and
recording, or by any information storage or retrieval system, without permission in
writing from Woodhead Publishing Limited.

The consent of Woodhead Publishing Limited does not extend to copying for
general distribution, for promotion, for creating new works, or for resale. Specific
permission must be obtained in writing from Woodhead Publishing Limited for such
copying.

Trademark notice: Product or corporate names may be trademarks or registered
trademarks, and are used only for identification and explanation, without intent to
infringe.

British Library Cataloguing in Publication Data
A catalogue record for this book is available from the British Library.

Library of Congress Cataloging in Publication Data
A catalog record for this book is available from the Library of Congress.

Woodhead Publishing Limited ISBN 978-1-84569-072-4 (book)
Woodhead Publishing Limited ISBN 978-1-84569-338-1 (e-book)
CRC Press ISBN 978-1-4200-4399-0
CRC Press order number: WP4399

The publishers' policy is to use permanent paper from mills that operate a sustainable
forestry policy, and which has been manufactured from pulp which is processed using
acid-free and elementary chlorine-free practices. Furthermore, the publishers ensure that
the text paper and cover board used have met acceptable environmental accreditation
standards.

Project managed by Macfarlane Production Services, Dunstable, Bedfordshire, England
(e-mail: macfarl@aol.com)
Typeset by Godiva Publishing Services Limited, Coventry, West Midlands, England
Printed by TJ International Limited, Padstow, Cornwall, England

O
664·67
CON

2 8 JUN 2007

UNIVERSITY OF STRATHCLYDE
UNIVERSITY LIBRARY

Contents

Contributor contact details

(* = main contact)

Editor

Hal MacFie
Hal MacFie Training Services
43 Manor Road
Keynsham
Bristol BS31 1RB
UK
E-mail: hal@halmacfie.com

Chapter 1

H. L. Meiselman
Natick Soldier Center
Natick, MA 01760-5020
USA
E-mail:
    Herbert.L.Meiselman@us.army.mil

Chapter 2

H. Tuorila
Department of Food Technology
P.O. Box 66
FI-00014
University of Helsinki
Finland
E-mail: hely.tuorila@helsinki.fi

Chapter 3

P. Rozin
Department of Psychology
University of Pennsylvania
3720 Walnut St.
Philadelphia, PA 19104-6241
USA
E-mail: rozin@psych.upenn.edu

Chapter 4

M. R. Yeomans
Reader in Experimental Psychology
Department of Psychology
University of Sussex
Brighton BN1 9QH
UK
E-mail: martin@sussex.ac.uk

Chapter 5

A. Saba
Istituto Nazionale di Ricerca per gli
    Alimenti e la Nutrizione
Via Ardeatina 546
00178 Rome
Italy
E-mail: saba@inran.it

Chapter 6

M. Siegrist
ETH Zürich
Institute for Environmental Decisions
    (ED)
Consumer Behaviour
Universitätsstrasse 22, CHN J75.1
CH-8092 Zürich
Switzerland
E-mail: michael.siegrist@env.ethz.ch

Chapter 7

D. Buck
Product Perceptions Ltd
Windsor Place
Faraday Road
Crawley RH10 9TF
UK
E-mail: dominic.buck@
    productperceptions.com

Chapter 8

A. Krystallis
National Agricultural Research
    Foundation (NAGREF)
Institute of Agricultural Economics
    and Policy (IGEKE)
5 Parthenonos str.
141 21 N. Hrakleio
Athens
Greece
E-mail: Krystallis.igeke@nagref.gr

Chapter 9

K. Brunsø* and K. G. Grunert
MAPP
Department of Marketing and
    Statistics
Aarhus School of Business
University of Aarhus
Haslegaardsvej 10
8210 Aarhus V
Denmark
E-mail: Kab@asb.dk, Klg@asb.dk

Chapter 10

A. V. Cardello
Senior Research Scientist
Consumer Behavior and Product
    Performance
US Army Natick Soldier, R, D & E
    Center
Natick, MA 01760-5020
USA
E-mail:
    armand.cardello@natick.army.mil

Chapter 11

E. P. Köster*
Wildforsterweg 4A
3881 NJ Putten
The Netherlands
E-mail: ep.koster@wxs.nl

J. Mojet
Wageningen UR-CICS
Building no. 118
Bornsesteeg 59
6708 PD Wageningen
The Netherlands
E-mail: Jos.Mojet@wur.nl

Chapter 12

D. Thomson
Chairman
mmr Research Worldwide Ltd
(Visiting Professor, University of
    Reading)
Wallingford House
46 High Street
Wallingford OX10 0DB
UK
E-mail:
    d.thomson@mmr-research.com

Chapter 13

H. Stone* and J. L. Sidel
Tragon Corporation
365 Convention Way
Redwood City, CA 94063
USA
E-mail: hstone@tragon.com

Chapter 14

E. van Kleef* and H. C. M. van Trijp
Wageningen University
Marketing and Consumer Behaviour
    Group
Hollandseweg 1
6706 KN Wageningen
The Netherlands
E-mail: Ellen.vanKleef@wur.nl

Chapter 15

H. Moskowitz
Moskowitz Jacobs Inc
1025 Westchester Ave
White Plains, NY 10604
USA
E-mail: mjihrm@sprynet.com

Chapter 16

R. Popper* and J. J. Kroll
Vice President, R&D
Peryam & Kroll Research Corporation
3033 W. Parker Road, Ste. 217
Plano, TX 75023
USA
E-mail:
    richard.popper@pk-research.com

Chapter 17

L. Rothman
Kraft Foods
1 Kraft Court
Mail Code GV 653
Glenview, IL 60025
USA
E-mail: lrothman@kraft.com

Chapter 18

M. O'Mahony
Department of Food Science and
   Technology
University of California
Davis, CA 95616
USA
E-mail: maomahony@ucdavis.edu

Chapter 19

J. F. Delwiche
The Ohio State University
2015 Fyffe Road
Columbus, OH 43210
USA
E-mail: Delwiche.1@osu.edu

Chapter 20

Y. Lohéac*
ESC Bretagne Brest
2 avenue de Provence
CS 23812
29238 Brest cedex 3
France
E-mail: youenn.loheac@
   esc-bretagne-brest.com

S. Issanchou
INRA
UMR FLAVIC INRA-ENESAD
17 rue Sully BP 86510
21065 Dijon cedex
France
E-mail:
   sylvie.issanchou@dijon.inra.fr

Chapter 21

M. Martens* and H. Martens
Norwegian Food Research Institute
   (Matforsk)
Osloveien 1
N-1430 Aas
Norway
E-mail: Magni.martens@matforsk.no
   mma@kvl.dk

Michel Tenenhaus
HEC School of Management
   (GRECHEC)
Rue de la Liberation
78351 Jouy-en-Josas
France

Vincenzo Esposito Vinzi
ESSEC Business School
Department of Information and
   Decision Systems
Avenue Bernard Hirsch B.P. 50105
95021 Cergy-Pontoise
France

Chapter 22

J. Bogue* and D. Sorenson
Lecturer
Department of Food Business and
   Development
University College Cork
Cork
Ireland
E-mail: j.bogue@ucc.ie

Chapter 23

Hal MacFie
Hal MacFie Training Services
43 Manor Road
Keynsham
Bristol BS31 1RB
UK
E-mail: hal@halmacfie.com

Preface

New product development (NPD) is a ubiquitous activity in the food and beverages and personal products industry. Many companies see it as the source of their future expansion plans, some view it more realistically as the means by which they will stay afloat in a fast changing world. An army of consultants, qualitative and quantitative market researchers has grown up to support this activity from outside the companies. Within the company, marketing and product development personnel battle and collaborate to come up with the great new product or line extension. Sometimes the product is a result of some new breakthrough technology such as freeze drying or new ingredient such as artificial sweeteners or, increasingly, due to new understanding in other areas such as medicine (plant sterol products). However, by far the majority of new products still arise by making small changes to a current line, adding a new flavour, improving the current product by reformulation.

Looking at this frenzied beehive of NPD activity, do we need another book on product development? I believe there are three reasons why this book is relevant.

1. The current level of successful launches of products is still very low and clearly there must be something else that needs to be done to reduce the level of failure.
2. The cost of launching new products with healthy ingredients or new technologies is moving towards that of introducing pharmaceuticals and so for these products there can be no question of launching a dud in terms of consumer acceptance.
3. The 21st century consumer in affluent countries is a sophisticated animal who is ready to spend good money on products that cater to his or her

particular needs. Food and beverage companies have to go where the appliance and automotive companies have gone before and understand what makes these consumers tick, foresee their future needs, and start to develop products several years upstream of launch.

The inevitable conclusion is that we must turn our attention to the consumer with an intensity that has not been seen before. We must make sure that our short-term methods to assess liking and purchase intent are accurate and predictive. We must try to understand what causes changes in taste preference with age and among different groupings. We must understand how consumers shop, how their lives are changing, how their views of new and old production methods are going to change. Above all we must ensure that we gather knowledge and understanding that enables us to make products that people want to buy, that they enjoy consuming and that exceed their expectations so they want to buy them again. Sounds easy doesn't it! Hmm!

For these reasons I was therefore honoured when Sarah Whitworth of Woodhead Publishing invited me to edit a text that would focus on consumer-led NPD. I have been surprised and delighted by the enthusiasm and commitment which the chapter authors, all very busy people, have shown in writing and revising the chapters in their spare time to a tight time schedule. Herb Meiselman provides an excellent introduction and overview of the chapters in Chapter 1 so I will not repeat that here. Suffice it to say that it is larger, more diverse, and better than I ever imagined it would be when we started to write to authors.

I will say a word about who this book is aimed at and how it should be used. I see it as a gateway to enter the large amount of excellent consumer science that has already been conducted. For people working out there in the field, we hope that some of the chapters will catch your attention and stimulate you to read more about the science that underlies consumer behaviour or the new methods that are being used. All chapters give sources of further information and advice and these should be accessible to the general reader. For the graduate and undergraduate students who are interested in joining the great endeavour of feeding, stimulating, and entertaining the global consumer population, the science-based chapters should give you a firm background and good springboard for research projects. However we hope that you find the more practical chapters useful, and many of you will be surprised at how little the consumer is considered in many small companies that you join.

A final vote of thanks to the Woodhead staff led by Sarah. Their understanding and good natured persistence in 'helping' us authors meet our timescales was unremitting and ensured a timely text.

Enjoy the book and may your research be productive and your products successful!

Hal MacFie

Part I

Understanding consumer food choice and acceptance

1

Integrating consumer responses to food products

H. L. Meiselman, Natick Soldier Center, USA

1.1   Introduction

The food business, from agriculture to food service is probably the largest business in the world. Eating is a necessity for everyone and a pleasure for many. Therefore considerable efforts in both the research world and non-scientific world have explored the factors that contribute to eating. The 'single variables approach' has resulted in generally simplistic models of eating and of attempts to control eating. Control of eating is involved in weight maintenance, weight loss, or weight gain, all of which are health goals under varying circumstances. While research in the Western world has recently focused on eating too much, certain populations in the West eat too little, and many populations around the world are undernourished.

Eating research can be divided into three classes of variables (Meiselman, 2003). Much eating research has focused on characteristics of the product, that is *food variables*, such as palatability, appearance, and flavor. Others have focused on characteristics of the eater; these *people variables* include responsiveness to food cues (Rodin *et al.*, 1976), restrained eating (Westenhoefer *et al.*, 1999), and people's expectations (Cardello, 1994). Finally, others have focused on the environment in which we eat, considering *environmental variables* such as physical, social and economic factors (Bell and Meiselman, 1995; Meiselman, 1996; Herman *et al.*, 2003; Wansink, 2004; Stroebele and de Castro, 2004). From many studies of eating, we know that food choice, consumption, and liking are controlled by a very large number of variables. But these variables are rarely considered together, perhaps because they are overwhelming together or because the researchers and writers tend to specialize and want to focus on smaller classes of variables. Also, traditional experimental method has taught us to vary one thing while keeping other things constant. For these reasons, we know a lot about each category of eating variables (people, food, environment), but we know relatively less about how these classes interact. For example, we know a lot about people's expectations of foods, but very little about how expectations vary across different foods and different environments. And we know a lot about the relationship of food constituents and food flavors to palatability, but very little about how these food variables vary across people and environments. One of the major challenges in food-related research is integration of variables – we know relatively little about how the many variables integrate to produce our overall behavior towards food. This is one of the great challenges for the statisticians who analyze human eating data and human food product data – they need to help us integrate the data from different areas of research. And researchers in general need to develop and explore integrating research designs.

Authors have attempted to conceptualize eating more broadly, describing the interaction of the classes of variables that they consider important. Some include just one or two of these three main classes of variables, while other models consider all three classes of food, people, and environmental variables. Different authors give different weights to these three classes of variables. It is not the purpose of this chapter to review all models, but a selective presentation of some of them will hopefully aid the reader in integrating the material in this book, and the huge area of food and eating.

Thus the purpose of this chapter is to try to organize and integrate food variables, person variables, and environmental variables. This will be done in the context of models of food choice and food acceptance, and these models will be illustrated by the material in this book. It is hoped that this introductory integrative chapter will help put each of the later chapters into a broader perspective. It is also hoped that this chapter will encourage others to consider not only what variables contribute to food choice and foods acceptance, but how those variables interact to produce eating as we know it in the real world.

This element of realism adds another aspect to this chapter. Much of what has been studied related to human eating has taken place in the laboratory, with questionable relevance to natural eating (Meiselman, 1992) – see also Chapter 11. In presenting various variables and models of their action and interaction, I will attempt to put things into a more realistic eating perspective. This requires distinguishing variables that account for very small amounts of variance from variables that account for larger amounts. Weak variables might have little or no effect in the real world unless they cumulate to produce bigger effects. One of the weaknesses of the traditional laboratory research is to present any statistically significant effects; research in the real world is more geared to produce and present larger effects which are strong enough to overcome the noise in the real world.

Further, I will also try to analyze how the variables and models presented relate to food product development. The goal of a more realistic orientation and a product orientation go together because real products must survive in the real world. It is important to keep in mind that different researchers have different goals in their basic and applied food-related research. Product development is just one of the foci of research and, necessarily, it puts great emphasis on the product. Other variables (people, environment) are seen as they relate to the product, and food developers will rarely put environmental or people variables in higher importance than the food products themselves. But another large group of researchers is interested in health issues, including obesity, childhood food patterns, elderly food patterns, and the role of diet in various diseases. These researchers are less interested in products as products, and might use artificial food products to mimic foods, and to better control important food constituents. Health-oriented researchers might be more interested in people variables and environmental variables if they believe they contribute to healthy/unhealthy eating. Finally, some (academic) researchers are free to conduct research free of the constraints and biases of either product or health, and these people might concentrate on variables depending on their interests, their funding sources and other factors. Thus we cannot expect everyone to have the same interests, the same biases, and therefore the same model of how eating works.

I will begin with material that focuses on the food product. Since almost all of the material in this book is consumer-oriented, even the product chapters will contain some consumer orientation. The key distinction is that in the food-focused chapters, the object of study is the product, while in the person-oriented chapters, the object of study is the consumer.

1.2   Focus on the product

A number of earlier models of food acceptance focused on food variables; these variables are often referred to as physicochemical properties. Harper's model (Harper, 1981) is basically a sensory model linking physical, sensory and affective variables (Fig. 1.1). Physicochemical factors are the beginning stage of Land's model of food acceptance (Land, 1983) (Fig. 1.2) and Cardello's model of food acceptance (Cardello, 1996) (Fig. 1.3). These two models emphasize food variables, and especially emphasize their sensory properties. Both models include other non-food factors. Land includes a stage of 'central integration' where things such as 'experience, expectation, availability, advertising, price, occasion, etc.' are added. Thus, Land adds in both person variables (for example, expectation) and environmental variables (for example, occasion). But the main emphasis in these two models of food acceptance is the food/sensory basis. Many earlier conferences and books had this sensory-product focus (Solms and Hall, 1981; Thomson, 1988).

Stone and Sidel (Chapter 13) introduce product development, emphasizing the multidisciplinary challenge from a team of contributors including the following: purchasing, marketing, marketing research (consumer insights), production, quality control, packaging, logistics, sales, corporate, legal, purchasing, chefs, and the product development group itself. Added to this internal group are external suppliers or partners. Other chapters in this book, including Bogue and

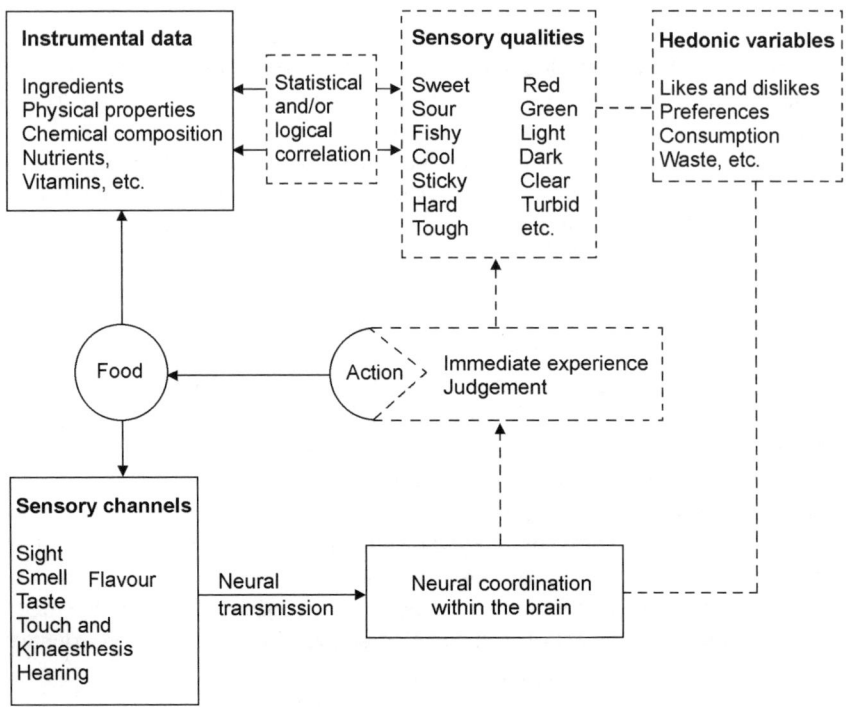

Fig. 1.1   Harper's sensory model of food acceptance (Harper, 1981).

Sorenson (Chapter 22) emphasize the team approach to product development. Stone and Sidel present the overall scheme of product development, and the frequent exceptions to the scheme. This scheme of product development is the basis for much of the work reported in this book, and the chapter provides a good introduction to the product development process and the roles of sensory professionals for those who are not familiar with this commercial process.

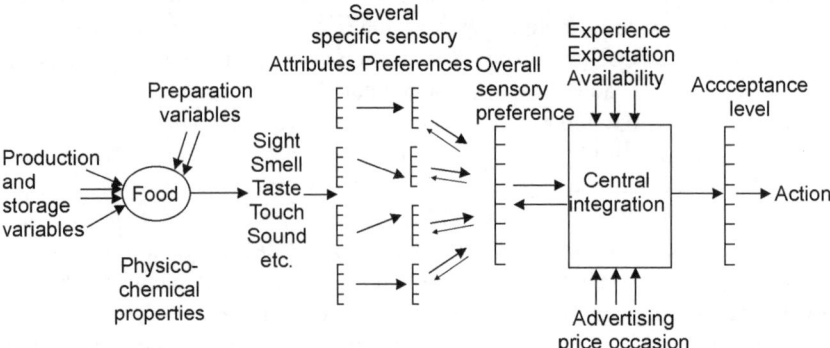

Fig. 1.2   Land's sensory model of food acceptance with an added step for 'central integration' (Land, 1983).

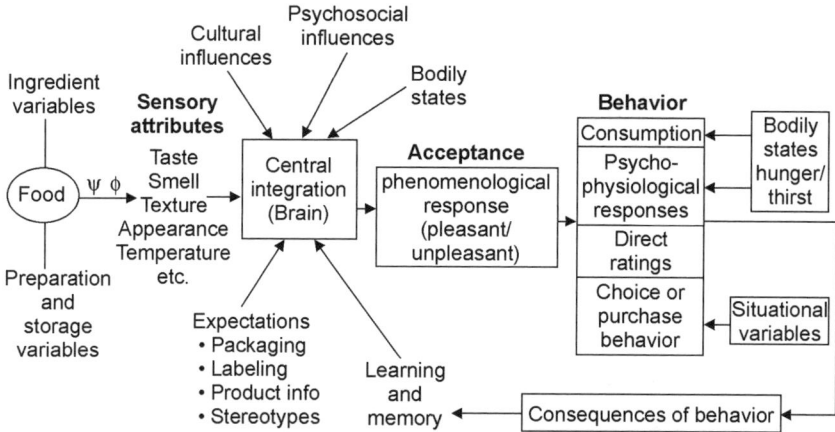

Fig. 1.3   Cardello's model of food-related behaviors. The initial step in the model is sensory (Cardello, 1996).

Stone and Sidel emphasize the role of sensory, stating 'Sensory should have a key role in the product development effort, beginning when the new product team is organized.' They emphasize both the strategic and practical outcomes of sensory involvement: the strategic defines how consumers perceive the product in relation to the marketing plan and other existing products, while the practical measures and defines the product.

Stone and Sidel support using separate groups for sensory or attitudinal information. They present advice on selection of subjects, and the design and analysis of tests. Stone and Sidel's preference for subjects is not the random approach of selecting consumers – they prefer to target product 'qualified subjects', whom they define as frequent product users who exhibit good discrimination performance. For tests, they distinguish descriptive analysis and preference testing. They define descriptive analysis as 'a methodology that provides word descriptions (attributes) of products that also includes the intensities, the strengths for each of those attributes.' They distinguish two types of preference testing, paired preference testing and the direct measurement of liking.

An emphasis on food variables characterizes Tuorila's treatment of food acceptance (Chapter 2), which begins with the statement that 'the success of foods greatly depends on the extent to which their sensory quality appeals to the target population.' Tuorila's model (Fig. 1.4) of the relationship between food and acceptance (affective or other holistic response) is a version of Cardello's model (Fig. 1.3). Interposed between the sensory attributes of food and behavior are cognitive factors such as expectation and attitude. However the connection between sensory attributes and these cognitive variables is not clarified by the chapter. Many cognitive variables will be discussed below under person variables ('focus on the person').

Tuorila reviews the individual food senses, and then goes on to connect sensory perception to affective response (acceptance). Tuorila claims that one

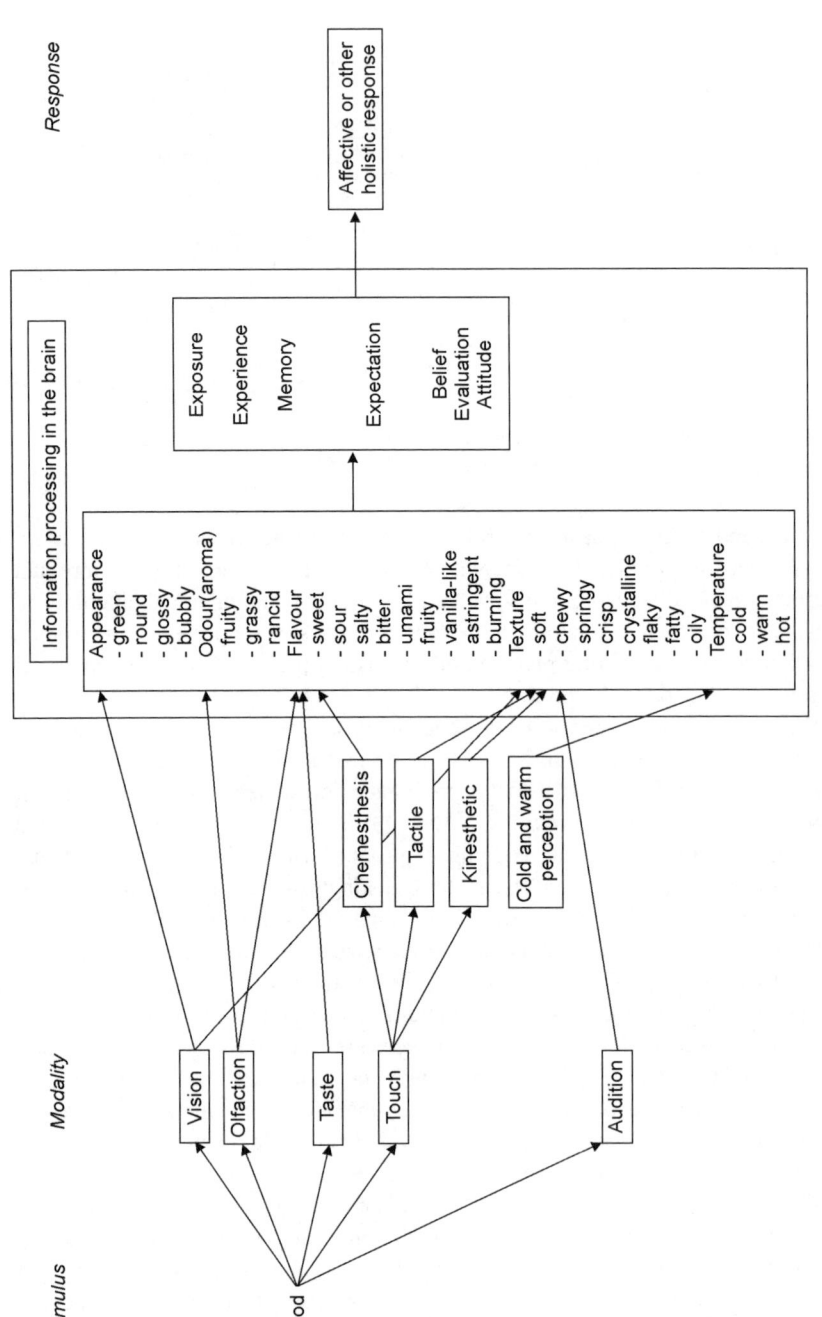

Fig. 1.4 Tuorila's model.

has to taste a food to give an hedonic impression, but we need to keep in mind the primary rule of culture – culture determines what we like and dislike, and we do not need to taste ants in order to say we do not like them. The material on olfaction is interesting in that food aroma is composed of many odorous compounds that are not detected individually, and also food aroma is mainly detected retronasally rather than orthonasally. These two results suggest that individual odour compounds are artificially studied, and sniff studies are artificial also. To the product developer or health researcher it might be difficult to relate simple sensory methods and phenomena to complex human behavior.

Tuorila goes on to demonstrate the importance of cognitive variables such as branding, familiarity, novelty and source. Tuorila presents a very helpful chart of methods/techniques for measuring affective response/acceptance (Chapter 2, Table 2.2) and presents some useful rules of measurement, such as measuring reactions to food names on a questionnaire is different from measuring reactions to actual foods.

I noted above the question of how much sensory factors contribute to food choice, acceptance, and consumption. After all of the sensory research and sensory evaluation of the past half century, one of the unknowns in the food industry is the degree to which sensory factors contribute to overall product success or failure when compared with factors such as culture, price, availability, etc. While we do not have a definitive answer to that question, Tuorila discusses the correlations between liking and consumption – most likely $r = 0.5$–0.7, accounting for 25–50% of variance. Tuorila points out the lack of high correlation between acceptance and consumption in single laboratory sessions. This underscores the real challenge for sensory research – to quantify how much of final consumer behavior is accounted for by sensory factors. Popper and Kroll (Chapter 16) raise the same question when they note: 'whether a difference in sensory perception is of a magnitude to result in a difference in food acceptability is a separate, and usually more critical, question.'

Continuing with the emphasis on sensory variables, O'Mahony presents (Chapter 18) information on difference and preference testing. O'Mahony refers to preference/acceptance testing as '... the bottom line. Will the product succeed ...?' O'Mahony focuses on laboratory testing for those without access to the marketplace (market, home). One might question whether the laboratory will provide the answer to the 'bottom line' question. Another test strategy is analytical sensory testing where one seeks to use the human senses as instruments. Descriptive analysis focuses on products, not people, and most concerns that would be applied to consumers (instructions, settings, etc.) are not thought to be relevant. A final test strategy is consumer perception, using actual consumers. The task here is much simpler than for the trained panelist in analytical testing, but many complexities are introduced when dealing with untrained consumers. While the goal is to 'reproduce ordinary eating conditions,' O'Mahony claims that we do as well as we can. Meiselman (1992) has suggested that we need to do better in choosing test situations to better predict the real world. O'Mahony presents a detailed analysis of test selection for each the above types of test, based mainly on statistical criteria.

Product development involves getting the right ingredients at the right levels, and this aspect of product development is the aim of just-about-right scales, called JAR scales. This is covered by Rothman (Chapter 17) who points out that acceptability/liking information or behavioral choice information do not yield information to directly adjust product attributes. JAR scales are bipolar scales with a center point labeled 'just about right' or 'just right.' Thus, JAR scales assume an ideal point against which consumers rate products. The interested reader is referred to a forthcoming ASTM manual, and to a Workshop held at the 2003 Pangborn Sensory Science Meeting (Popper, 2004).

JAR scales can be line scales or category scales, the latter with from three to nine points. Rothman has an excellent discussion of strengths and weaknesses of JAR scales. Weaknesses include: the use of language for scale end-points, the choice of end-points for negative attributes, the cognitively complex task of ideal points, the use of never enough attributes, the impact of JAR scales on other scales, and the interdependence of attributes. Rothman raises the interesting question of whether JAR scales are a sensory task or a cognitive task; it would seem that asking an ideal point is a cognitive task, reflecting the complex nature of this seemingly simple task. Rothman emphasizes that the distribution of JAR scores is the goal, not an average score. In addition, the goal for any product is to achieve a maximum of just about right ratings, indicating no need for change. Rothman introduces penalty analysis as a technique to relate JAR scales and hedonic scales, pointing out that penalty analysis helps prioritize which attributes need to be changed.

Delwiche (Chapter 19) presents another scaling topic in new product development, probabilistic or Thurstonian modeling. Delwiche emphasizes that these approaches are based on the view that perception is not constant, even with a constant physical stimulus. Thurstonian models are only applied to confusable stimuli, not to easily or completely distinguishable stimuli. Delwiche points out the different models applied to the observation that people often give different ratings to the same products. The deterministic view attributes this to the judge's responses, since the physical stimulus does not change. The probabilistic view attributes this to both variations in input to the judge, and variation in the judge's responses. Probabilistic models can estimate perceptual distance between products by both rating and discrimination procedures. One of the methods of probabilistic modeling is Thurstonian ideal point modeling which is applied to JAR data, as Rothman also notes. Delwiche presents other probabilistic methods and notes that they have yet to show widespread use perhaps because they are more difficult to use.

One of the most popular product classes of recent years is the class of healthy food products. These include reduced calorie foods, functional foods, low-fat foods, etc. The 2005 Pangborn Symposium included a workshop on healthy food products (Kilcast, 2006). Bogue and Sorenson (Chapter 22) begin their discussion of low-calorie foods by defining reduced calorie foods as 'a food product with lower amounts of calories, through either calorie or fat reduction.' They then state the challenge of 'whether consumers are willing to make trade-offs in terms of the sensory characteristics.' Thus they take a product orientation to low-calorie foods. Bogue and Sorenson identify a number of challenges for product developers, including proper labeling and descriptors on labels and poor consumer expectations of sensory trade-offs. A key question is whether to reduce/remove fat or select products naturally lower in fat.

Bogue and Sorenson present a specific example of development of a reduced calorie beverage in three stages showing the following task and the method:

- Concept ideation – focus groups.
- Concept screening – sensory analysis of four products:
  - descriptive sensory analysis,
  - consumer preference.
- Conjoint analysis.

Concept ideation is also discussed in this volume by van Kleef and van Trijp (Chapter 14) and Moskowitz (Chapter 15).

Research that focuses on the product is often supported by food product companies or agricultural interests. One of the main challenges in this type of research is to connect product attributes to the consumer in a way which is actionable, that is, a way which makes a concrete recommendation for a product characteristic at a quantitative level. Actionability is one of the attractions of JAR scales discussed above. Van Kleef and van Trijp include actionability in their analysis of new product methods (see Chapter 14). One way of making product characteristics more actionable is to link the product characteristic to the consumer's needs and wants, and these approaches are presented below in the next section. This is not to suggest that research with a product orientation does not consider the consumer. The models of food choice and acceptance presented by Tuorila, Cardello and others focus on the product and its sensory attributes but also include other person variables.

The basis of product design is a product concept, which Moskowitz (Chapter 15) defines as statements about products, although emphasizing that concepts themselves are not well defined. Moskowitz notes that concept research has been an applied aspect of marketing research, generally without specific scientific training. It has been viewed as an art, not a science. He notes that food technology deals with the physical aspects of products, while concept science (if it existed) would deal with what features of concepts work in consumers' minds. He notes that traditional consumer science deals with decision making, not products. The new concept science would begin with identifying consumer needs, and incorporating them into product concepts. Concept tests have dealt with predicting success rates of products and diagnosing features which drive acceptance; the latter is the topic of Moskowitz's chapter. Moskowitz notes that consumers are surprisingly good at identifying what elements of products are important to them. Perhaps this is sometimes overlooked in the sophisticated methods that have grown up around marketing research and consumer research.

To help identify specific statements in the product concept or positioning concept, Moskowitz has turned to conjoint analysis. In addition to identifying conjoint analysis as a tool for concept development, Moskowitz has identified the Internet as the tool of the future. The main reason for this is cost; marketing research, including conjoint analysis, can be time-intensive. While sample representativeness was an early issue for the Internet, it is no longer viewed as a major issue in the 21st century for many countries. Internet testing of concepts has been applied to benefit screening, full concept testing, and experimental design using concepts. Moskowitz methodically lists steps for Internet-based conjoint analysis studies, and presents several examples/applications.

Köster and Mojet (Chapter 11) deal with both product success and with boredom and product failure. This is one of the few chapters in this book specifically addressing why products fail, and how our methods contribute to that failure rate. Chapter 14 by van Kleef and van Trijp also address the dual nature of product failure and success. In product development research, the goal is product success, but perhaps we need to understand more about product failures, and dislike in general, to understand our methods better. Dislike has been addressed by Land (1988) and especially Rozin (1996).

One reason why products fail may be product boredom. Köster and Mojet address why some products have a short life cycle. They relate this short life cycle to intrinsic properties of the food as well as the adequacy of long-term acceptance testing. They go on to claim that product testing that involves novelty and surprise will evoke the wrong type of consumer curiosity. They suggest that as products become more familiar, their appeal changes, with some less complex products changing from more liked to less liked.

Köster and Mojet also address the issues of choice and testing duration. They note that 'the subjects simply have no choice to answer other questions and feel obliged to give an answer.' I have argued that providing consumers with no product choice in testing is the same type of problem; once consumers volunteer for a study, they usually have no choice of rejecting a product, and no choice of not giving an opinion. Köster and Mojet go on to clarify that 'diversive exploration with its attention for the pleasurable and aesthetic aspects of the product' sets in after exploration is completed. They raise the interesting methodological question of whether there should be a trade-off of number of subjects versus length of consumer testing, rather than focusing only on the former.

Köster and Mojet recommend (longer) home use tests with the following characteristics:

- at least seven repetitions of the product;
- the normal frequency of product use;
- use of a diary of product use;
- unexpected home visits to validate use data;
- questionnaires at first use (hedonic and frequency data) and last use (hedonic and JAR data);
- follow-up interview one week later.

Köster and Mojet also present a quick central location screening test ('boredom test') when two or three new product versions are compared with the market leader, and an extended boredom test which combines the central location and home use tests. They go on to discuss methods for products that produce aversion after a time.

Köster and Mojet raise the interesting question whether one can assume that people's response behaviors differ only in degree and not in essence. They question the reliance of many methods of averaging consumer data – are many of these responses on different underlying dimensions? Are we tapping this information with current methods? Are we becoming too dependent on the 'rather flimsy correlations that make up preference maps' and other statistical techniques that do not show a proper appreciation for the human consumer?

1.3   Focus on the person

This is probably the longest section of this volume and of this chapter. In attempts to develop successful products, companies have increasingly turned to studying the consumer who will purchase and hopefully repurchase their products. This has yielded a number of different approaches to studying consumers. Some methods are more person-focused, others try to link the person and the product, and a few try to link the person, the product, and the situation. First, some human-focused models are presented.

Researchers at Cornell University (Furst *et al.*, 1996; Connors *et al.*, 2001) have developed a model of food choice (Fig. 1.5) based on qualitative research on populations near the university in upstate New York. While the limited sample is a major drawback, this model draws from people's events and experiences, focusing on ideals, personal factors, resources, social relationships, and food contexts. Common values include health, taste, cost, convenience, and managing relationships. We will see some of these values throughout consumer based models (for example, means–end chains and lifestyle-related eating, Chapter 9). Most of these also show up in the Food Choice Questionnaire on what factors are important in eating (Steptoe *et al.*, 1995). The task for a consumer confronted with a food choice situation is to select among competing values, often based on the context of the choice. The role of tradition and novelty are important with respect to values; people cling to their traditional values, but are capable of changing values based on new information, new situations, new relationships and new environments.

Buck (Chapter 7) presents an overview of qualitative and quantitative consumer methods aimed at uncovering attitudes and motivations. The qualitative methods are often used at the beginning of consumer research. Qualitative research is not supposed to be statistically representative of the population; the goal is to develop exploratory hypotheses that can be confirmed in more quantitative research. Buck points out that one should not choose qualitative or quantitative approaches; the best results often come from using both methods in an appropriate mix.

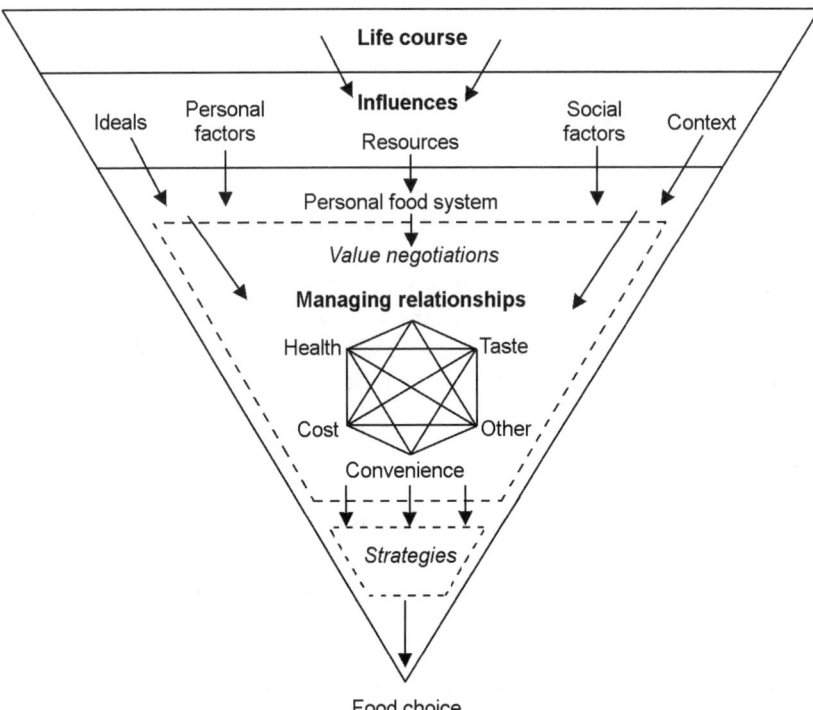

Fig. 1.5   Cornell model (Furst *et al.*, 1996; Connors *et al.*, 2001).

Buck presents the details of one-on-one depth interviews, which last 30–60 minutes, and aim at uncovering individual information without peers present. He discusses laddering, which is also covered in this book in Chapter 8. Repertory grid or the method of triads require that respondents split three stimuli/products into a pair and a single, and describe why the pair are similar and the single different to yield constructs. This method helps to uncover how consumers group or differentiate products. It helps us position new products with existing products, or existing products with competitor products. All of the depth interview techniques are time consuming in both data collection and analysis and therefore costly.

Group data collection tries to reduce the time/cost of individual approaches and introduce the effects of peer presence. Buck points out that the group might be especially selected to cover a range of consumer types, although some specialists recommend that groups be kept more homogeneous, and that additional groups be run for heterogeneity of response. There is also a question of whether the group moderator should be specially trained and experienced in group work. This of course raises the cost. Additional time is needed to work out the carefully designed guide for the questions and topics, which proceed from the general to the more specific and conclude with a summary of the group's findings. Buck notes that groups might be less intimidating to people than one-on-one interviews, but peer pressure might work against some people being open or having an exaggerated interest in a product. Products can be introduced to groups with prior or contemporary product placement, and products can be mapped for similarity/difference during the group meeting.

Buck also discusses quantitative consumer methods. Usage and attitude studies are conducted by a trained interviewer on a large number of consumers. There is a more complete traditional approach, or an event-centered approach. The large samples in such studies are useful for segmentation based on behaviors, attitudes, or needs.

Buck also presents some interesting possibilities on where consumer research is going. He notes the development of automated software to count words, but cautions that the same word can vary in meaning from person to person or from occasion to occasion. Developments in neuroscience might someday help consumer researchers to validate qualitative information with 'hard data' from associated neural activity. Further developments in measuring the impact of products on mood and even physical and cognitive behavior are also encouraging.

Thomson (Chapter 12) explores the relationship between product sensory attributes and emotional responses, which he calls SensoEmotional optimization and distinguishes it from more typical SensoHedonic optimization. Thomson suggests that 'featuring emotionally active sensory characteristics in brand architecture and brand communications ... can cue emotions that are fundamental to the way ... consumers ... perceive and interpret their brands'. Thomson lists the following conditions for SensoEmotional optimization:

- the presence of a relevant, uplifting emotion;
- a credible emotion within the product category;
- reasonable to build the emotion into the brand;
- the emotion must complement the sensory attributes of the product, and vice versa;
- optimization of the sensory component for maximum effectiveness.

Thomson points out that optimizing on emotions and optimizing on hedonics (liking) can lead to different outcomes, and he questions whether liking should always be the appropriate criterion, especially in a branded consumption test. He suggests going beyond liking or emotion to a more holistic construct such as 'enjoyment of the total product experience'. Enjoyment is like the concept of satisfaction, which began in the service industry and has now spread to the product industry. One definition of satisfaction compares what is expected to what is received (expectation *vs* experience). Thomson suggests getting consumers to focus more on brands when evaluating products. This might be quite different from the sensory researchers who want to focus on the product without the interference of factors such as brand or packaging. However, Thomson claims that 'much of the pleasure people get from consuming foods and drinks derives from their sensory characteristics' which brings us back to the position of Tuorila, Cardello and the other sensory-based researchers with whom we began this chapter. Thomson does not offer any real evidence other than anecdotes that sensory variables control pleasure or control emotions. But he does challenge the 'conventional wisdom' of linking liking and purchase, sensory characteristics and liking, the validity of central location tests and other 'out of context' tests. In this regard, his concerns are similar to those of Köster and Mojet noted above.

Thomson recommends a lengthy moderator-led group process to conduct the optimization. He does not recommend direct questioning to elicit emotional responses, but rather comparative questions (e.g. milk chocolate compared with white chocolate), and the use of imagery (pictorial images). The goal is to define emotional characteristics of products and brands in words and/or images. At a later stage, Thomson recommends validation groups of experienced testers. Interestingly, throughout the chapter he focuses on chocolate which perhaps makes for a good SensoEmotional optimization example.

1.3.1   Connecting product attributes to consumers

A popular consumer research method which emphasizes values is means–end chains and laddering. This topic is covered by Krystallis (Chapter 8), who defines means–end chains as follows: 'Means–end chains (MECs) are hierarchical cognitive structures that model the basis for personal relevance by relating consumers' product knowledge to their self-knowledge.' These are relationships between product attributes and product consequences. The product attributes can be concrete (sweet) or abstract (healthy). The product consequences fall along a continuum as follows: functional, psychological, instrumental values, or terminal values. This sequence of *attributes–consequences–values* is the heart of the means–end chain, and it has been suggested that the bottom half of the chain is product knowledge and the top half of the chain is self-knowledge. Two of the underlying assumptions of this approach are that 'consumers choose actions that produce desired consequences' and that values 'play a dominant role in guiding choice patterns.' Interestingly, the assumptions of means–end chains and laddering do not include the notion that people can freely verbalize these product associations. Krystallis includes in her very thorough review references that question the validity of means–end chains, as well as question the validity of the laddering technique. Researchers have focused on whether the technique is a measure of memory or of motivation.

Laddering is the technique associated with means–end chains, and is covered in detail in Chapter 8. Laddering consists of direct probing questions in one-on-one interviews, asking questions such as 'why?' or 'why is this important to you?' This seemingly simple approach can take an hour or more. The interview data are analyzed by content analysis procedures, keeping in mind the attribute–consequence–value dimension. A summary table (showing the frequency of connections) and hierarchical value map are then produced. These can help to understand 'why an attribute or consequence is important' and also represents the path from product attribute to personal value. Krystallis presents a number of cautions: (1) ladders must be in the consumer's own language and cognitive structures, (2) ladders should result from genuine retrieval of information, (3) ladders should be based on generally agreed upon cognitive categories, and (4) laddering data reduction should be theory based. Means–end chains derived from ladders are shown in aggregate, with the assumption that one can characterize a group. Investigators often condense the overall data set by removing redundancy and establishing cut-off levels. For the interested reader Chapter 8 (Section 8.4) contains an exceptional list of organized references on means–end chain uses including food and non-food products.

Means–end chains are also covered by Brunsø and Grunert (Chapter 9) who focus on two consumer methods: means–end chains and food-related lifestyles. They begin by identifying means–end chains as a more product-oriented method, while food-related lifestyles is based on a more abstract view of consumers. By linking product attributes to consequences of consumption and ultimately to values, means–end chains are of potentially high value to producers who want to understand what motivates consumers. Again, Brunsø and Grunert note that these attribute–values connections are assumed to exist. To my knowledge, there is no evidence for these connections. Brunsø and Grunert give a number of examples of mean–end chains applications, and provide useful conclusions on the common factors seen in all food examples: hedonic-related attitudes, health-related attitudes, naturalness-related attitudes, and convenience-related attitudes. (Of course, these are some of the same factors found in food choice questionnaires such as Steptoe et al., 1995.) The suggestion that health and naturalness are as important as hedonics and convenience is questionable, especially across different cultures. And one could question whether these results reflect mainly European views of eating (see Saba, Chapter 5, on risk and naturalness).

Brunsø and Grunert present the food-related lifestyle method, in which lifestyle is defined and delineated from concepts such as values. They state that 'food-related lifestyle represent a "means–end chain approach" to lifestyle' linking food attitudes to food consequences. The food-related lifestyles instrument contains 69 items on 23 dimensions, rated on a seven-point scale. The 23 dimensions cover five main areas:

1. Ways of shopping.
2. Cooking methods.
3. Quality aspects.
4. Consumption situations.
5. Purchasing motives.

The food-related lifestyles instrument has identified a number of food consumer segments from research in several European countries:

- The uninvolved food consumer – interest in food quality limited to convenience; young, urban.
- The careless food consumer – like the uninvolved, except they are interested in quality; young, urban.

- The conservative food consumer – traditional meal patterns; taste and health; older, least educated, rural.
- The rational food consumer – look at product information and prices; interested in all aspects of food quality; plan meals; women with families, large percentage do not work.
- The adventurous food consumer – preparation of meals; quality; younger; high education and income; urban.

Particular countries show additional segments. All of the consumer types find taste and health attitudes very important, while there is more variation in attitudes towards naturalness and convenience.

Brunsø and Grunert recommend that means–end chains and food-related lifestyle measurement be combined in applications to food products. First select a consumer segment based on the food-related lifestyles method; then use means–end chains to relate product attributes to consumer values. For further discussion of laddering, the reader is referred to van Kleef and van Trijp (Chapter 14) who analyze a number of consumer research methods.

One of the most frequent factors included in models of food choice is expectation, which is covered in this volume by Cardello (Chapter 10). Cardello emphasizes that both theoretical models and practical measurement tools are available to measure expectations. This includes what consumers expect from products, how product characteristics produce expectations, how well the products meet expectations, and effects of failing to meet expectations. Cardello links his treatment of expectations to liking, 'Ultimately, product success depends on the consumer's liking or disliking of the product,' and this is similar to the role of liking (acceptance) in the treatment of sensory affects by Tuorila. Cardello emphasizes these intrinsic properties, as well as extrinsic or cognitive factors such as brand, package, and label. Cardello presents expectations as an important link between extrinsic and intrinsic factors. He presents a thorough historical treatment of expectations; one of the early applications in food was in food service not in product development where it was finally applied in the 1980s by Cardello himself. Cardello also presents an exhaustive presentation of the different models of expectations.

Cardello summarizes methodological advice for incorporating expectations into product research. He notes that sensory expectations have rarely been studied in favor of affect-based expectations. Sensory expectations are beliefs that products will contain certain sensory attributes at certain levels. Affect expectations are beliefs that products will be liked or disliked to a certain degree. Sensory expectations are more product-linked, and affect expectations are more marketing linked to brand, packaging, advertising, etc. Cardello argues that consumer quality judgments are a type of expectation. He also mentions ideational expectations (beliefs that products will elicit certain cognitions) and conative expectations (beliefs about purchase) as future directions for expectations research.

Cardello raises the interesting question of whether like–dislike ratings of food names are in fact expectation ratings. This is a form of measuring expected liking. Expectations exist only for known foods. Cardello advises using measures of belief strength or confidence because both high and low expectations can have high and low confidence. It is risky to assume that well-known brands automatically yield high expectations with high confidence.

Expectations are easily manipulated using verbal and non-verbal (contextual) information. This area has been reviewed in a number of articles. The dependent variables in most studies are liking/disliking or sensory attribute ratings; purchase intent and other behavioral intentions are not as common but might be more sensitive to expectations. Purchase intent, in terms of willingness to pay, is discussed by Lohéac and Issanchou (Chapter 20). Expectations exhibit individual differences probably based on personality differences (reviewed by Deliza and MacFie, 1996), suggesting that evaluation of expectation effects needs to be done at the level of the individual, not group means.

Yeomans (Chapter 4) moves between the psychology and biology of the consumer in his considerations. He focuses on two key concepts: need states, which maintain biochemical parameters within a range, and hedonically driven eating. Do we eat to maintain a balance in the body, or for pleasure?

Yeomans notes that flavour–consequence learning is the most widely held theory of how appetite is involved in developing food preferences. Yeomans gives examples of flavour–consequence learning: (1) the relationship between energy density of foods and preference, presumably based on the association of hunger reduction and flavour, and (2) the relationship between flavours and the associated effects of pharmacological components such as caffeine. Rozin also addresses flavour–consequence learning (Chapter 3). Another form of conditioning is the flavour–flavour model of evaluative conditioning, which supports the observation that adding something liked, such as sugar, increases preference while adding something aversive decreases preference.

Yeomans discusses motivational influences on food preferences, when the nutritional needs of the consumer affect learning and expression of flavour preferences. Flavours are liked more when hungry than when sated. Studies have been carried out with single foods and single food constituents (e.g. caffeine). Yeomans comments on the need for, but the complexity of, studying complex food choices. Does manipulation of motivational state alter food choice – that is not clear but it seems that they might. While Yeomans supports the rule that 'the primary psychobiological influence on food choice is sensory hedonics, eating for pleasure,' he also suggests that people might change their hedonic preference based on energy content, and that this might vary with hunger level.

Yeomans calls for research in natural settings such as the home, and for longer-term research studies, which is also recommended by Köster and Mojet (Chapter 11). And while few chapters in this volume explicitly deal with meals, Yeomans tackles meal initiation and meal size. The control of meal initiation remains elusive, with blood sugar and hormones possibly involved. The control of meal size is affected by a set of variables which act through negative feedback to reduce the desire to eat. Meal size is also affected by palatability variables which encourage eating. This brings us back to sensory hedonics or eating for pleasure. Yeomans catalogues the evidence that, under laboratory conditions, there is a linear relationship between rated pleasantness and subsequent voluntary intake. The well-known de Castro studies (de Castro *et al.*, 1990) also find that overall palatability predicts meal size independent of hunger. Can palatability override satiety and produce over-consumption? This is an important health question. Yeomans suggests that one way to provide acceptable food and still combat overeating is to provide food that is more satiating. Yeomans argues that actually changing product composition has an effect on intake but changing labeling only does not. While there is relatively little work in this area, it is clear that there is a major discrepancy about the effect of information on liking and the effect of information on intake.

Finally, Yeomans argues that tests must be conducted realistically because liking ratings are sensitive to current body needs. Products designed for hungry consumers need to be tested with hungry consumers. And, 'product development has to include measures of intake, and the impact of repeated intake on product acceptability.' The main messages from this chapter are the importance of learning, and the constant interaction between psychology and biology in the control of eating. Yeomans provides advice to the product developer, focusing on the finding that consumers learn to associate flavours and consequences.

MacFie (Chapter 23) presents a detailed chapter on preference mapping, putting this important technique within the overall context of this book. MacFie points out that preference is used for incremental product changes or reformulations rather than the entirely new product developments to which van Kleef and van Trijp refer. MacFie begins with the relatively short history of preference mapping, and its relationship to central location testing. Central location testing (CLT) along with home use tests (HUT) are the two cornerstones of consumer testing, although Meiselman has suggested use of institutional settings as well (Meiselman, 2003). In this volume, Köster and Mojet suggest greater use of improved home use testing to take advantage of both the long-term and the natural aspects of HUTs.

MacFie begins with an outline of CLT, which is not covered elsewhere in this book. He notes that 'CLTs are conducted as marketplace screens, for competitive benchmarking and to compare the performance of a number of prototypes with the current product.' He identifies a number of issues of CLTs (summarized in Table 23.2) including: sequential monadic testing (*vs* monadic), the number of stimuli/products (6–8–12+?), the range of stimuli, the number of panelists (120+?), the number of subject clusters, the actual location, controlled preparation and presentation of samples, warm-up samples, efficiency of testing (often 1 day), restriction on the number of questions, and avoidance of JAR scales. This section is an excellent introduction to commercial CLT for those not familiar with this important part of product development.

MacFie then goes on to describe cluster analysis, internal preference mapping, and external preference mapping. Cluster analysis is a prior step to preference mapping, although MacFie notes that 'Cluster analysis may be thought of as a preliminary to preference mapping, although I prefer to view it as the lowest level of preference mapping in terms of assumptions about how the population is behaving.' Thus MacFie presents cluster analysis and internal and external preference mapping as a sequence or hierarchy: External preference mapping is level 3 of the hierarchy of preference mapping methods.

Level 1 – cluster analysis – assumes only that the respondents fall into different segments.
Level 2 – internal preference mapping – does not necessarily assume different groups but does assume a shared perceptual space.
Level 3 – external preference mapping – assumes a shared perceptual space but additionally that we have an external definition of that perceptual space.

MacFie lays out the three techniques with detailed examples and practical advice.

1.3.2 Identifying consumer needs

Van Kleef and van Trijp (Chapter 14) represent a consumer orientation in the identification of consumer needs for new product development. They include the history of new product development from its traditional linear technological approach in the 1960s to the interface of marketing and product technology today. And they underscore the poor scorecard for new product development – the success rate is low, and is not improving. This point is also emphasized by Köster and Mojet (Chapter 11), Thomson (Chapter 12), and Stone and Sidel (Chapter 13). This is the challenge for all of us interested in product design and product testing. What is our value-added? Does sensory/consumer research eliminate the losing products and retain the winning products? Van Kleef and van Trijp argue that the former (which they refer to as Type-1) is under control; current consumer methods can eliminate 'ideas, concepts, prototypes, and as-marketed products prior to their market launch.' Of course that assumes every company uses these methods! The real challenge, according to van Kleef and van Trijp, is identifying new product ideas (Type-2). Interestingly they also point out that confirmatory research is the predominant research in new product development.

Identifying new product ideas involves understanding consumers, 'how they perceive products, how their needs are formed and influenced, and how they make product choices based on them.' This must be done without an actual new product, which is yet to be developed. However, van Kleef and van Trijp argue that this early work is critical to later product success. And, they point out, early consumer research in product development usually relies on 'focus groups, surveys and the study of demographic data.' Van Kleef et al. (2005), developed a taxonomy of ten methods for consumer research in new product development (Fig. 1.6). This taxonomy depends on two variables: actionability and abstract-ness of the information provided, i.e. how easy or difficult it is to act on the information, and the newness of the products considered from very new products

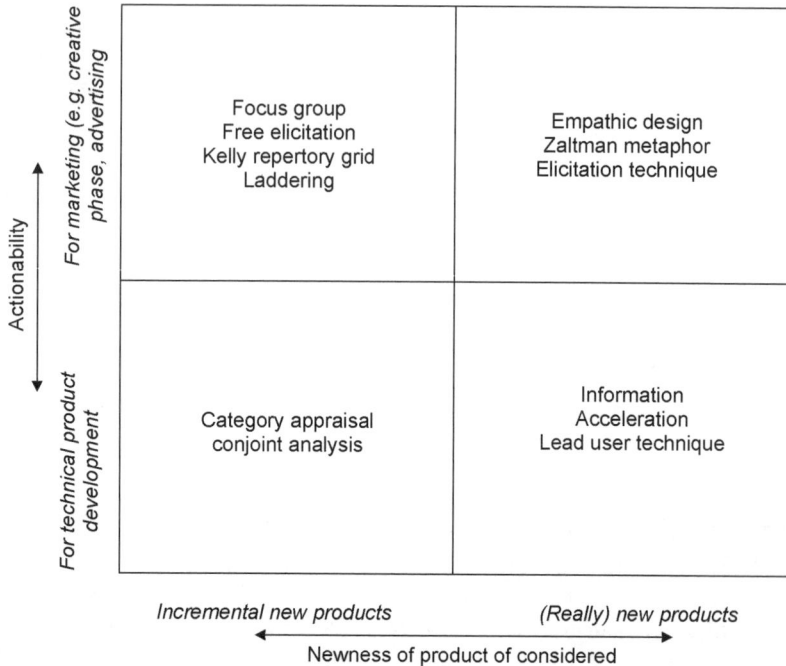

Fig. 1.6 A categorization of new product development techniques based on actionability and product newness (Van Kleef *et al.*, 2005).

to incrementally new products. The latter are variations on existing products (new flavor, etc.) and account for much of new product development.

The analysis of ten consumer methods led the authors to rate certain methods appropriate for really new products (empathic design, Zaltman metaphor elicitation technique, information acceleration, lead user technique). Other techniques are appropriate for incremental new products (focus group, free elicitation, Kelly repertory grid, laddering, category appraisal, and conjoint analysis). Readers of sensory/consumer research will no doubt recognize that the latter techniques are more often the methods of published research, and some of them are chapters in this book; the former techniques are found in marketing research and few of them are in this book. This is a big challenge to the field of sensory/consumer research.

We have dealt with consumer variables of increasing complexity. One of the most complex aspects of humans' relationship with food is the subject of risk beliefs and ethics which is covered by Saba (Chapter 5). Risk perception is viewed as 'a form of an attitude towards a specific object, such as a potential hazard.' Risk includes risk to oneself, to the environment, and to future genera-tions. People overestimate and underestimate risks. Uncontrollable and unnatural are more threatening; risks to oneself are less threatening than those to others. This 'optimistic bias' is especially important in food-related health risks such as fat and heart disease, which people think they can control. Saba emphasizes that the underestimation of these risks is difficult to change, suggesting that other strategies be used. Many risks subject to optimistic bias possess a degree of free choice (how much fat is consumed). Another bias is that attractive outcomes are seen as less risky. Although not much emphasis is placed on individual differences, there is evidence of gender and geographical effects.

A growing issue for both product consumers and product producers is ethics. Ethical values are deeply held beliefs about right and wrong. Ethical considerations are sometimes more important than perceived risks. Part of the ethical issue is that foods raised under different conditions are often indistinguishable. Saba points out that while all consumers react to 'traditional food quality aspects such as freshness and taste (traditional traits)' some consumers are also influenced by 'reflection traits' which reflect ethical concerns. These concerns extend beyond production, processing, and handling to issues of globalization. Saba points, however, that 'Naturalness ... is a vague concept that is notoriously difficult to define.' This leaves both the product designer and the consumer researcher without a firm basis for addressing many concerns about foods. Nevertheless, Saba argues that consumers' judgments are often understandable if studied. Such studies might use some of the consumer research methods presented in this volume.

One class of products which is receiving more attention is the class of products aimed at children. Popper and Kroll (Chapter 16) address the measurement of acceptability of products for 8–12-year-olds. They argue that very young children require more observational methods, and children over 12 can use adult methods. Popper and Kroll review the research on sensory differences between children and adults, and conclude that children might be less capable of detecting slight product differences. Popper and Kroll also explore the important role of familiarity and novelty in children, and the role of exposure in overcoming neophobia. Popper and Kroll note the potential impact of testing novel products with more familiar ones – for example, a novel dip with a more familiar chip. Testing novel foods in the laboratory is one of the themes throughout this book. Popper and Kroll question whether testing which allows children to become used to products is needed – a scenario that is difficult in most brief laboratory testing.

Popper and Kroll review some of the problems children have in testing situations:

- difficulty with concept formation and classification;
- limited attention span;
- difficulty ranking things in order of magnitude;
- memory limitations, affecting a series of stimuli;
- difficulty attending to multidimensional stimuli;
- large variation in children of different ages (especially changing stages).

Reviews of children's testing abilities can be found in Guinard (2001) and an ASTM publication (2003). Popper and Kroll discuss hedonic scaling with children, drawing the two important conclusions that children can use verbal scales as well as face scales, and that longer scales are as good as shorter scales. They caution on possible confusions using face scales. With rating scales, children often rate higher than adults, sometimes using the highest scale point frequently. Children also seem very sensitive to scale presentation on the page, although research shows that children interpret the meanings of the hedonic scale the same as adults. Children also seem to be able to use sensory scales and JAR scales, with children often changing the five-point scale to a three-point scale. For both sensory and JAR scales, Popper and Kroll raise the question of whether children understand all of the sensory terms. Judgments of appearance, size, amount, basic tastes, and simple texture appear to be possible.

Popper and Kroll note that children aged 5–7 can be tested either one on one or in a classroom situation where a test moderator explains the task. They cite a ranking procedure that can be used with 3–5-year-olds who successively select their favorite from a group of stimuli. Popper and Kroll also cite research showing better discrimination using mothers as interviewers for ages 3–4, but not older. Obviously, one needs to balance the risk of bias from the mother with better discrimination by the young child.

Popper and Kroll note the need for more research on many of these issues, and also on the possible need for research on age relevance for research. They also raise the interesting question of whether children are more sensitive to other factors in the 'product concept fit', 'how it handles, its play value, the image it projects, and how successfully all these aspects are integrated.' Thus Popper and Kroll move within this chapter from consideration of basic sensory and hedonic issues – acceptance – to consideration of more subtle contextual factors in testing (for example, presence of mother and time of day), and raise this integration question at the end. Every part of the field of sensory/consumer research should be asking these integration questions. And every part of the field needs to pay more attention to the importance of culture – most of the information presented on age comes from the United States, and the rest from other Western countries. Would these same principles hold for China, India, or Japan? Are there universal principles for children's testing?

In the task of connecting products to consumers, a great deal of data can be generated by the methods outlined in this book. A number of statistical approaches to handling the data have been proposed. Martens and her co-authors (Chapter 21) properly describe 'Food product development is a multivariate challenge ...'. This is a point I have been trying to make for years; simple descriptions and simple models of human eating and the food products that feed human eating are generally inappropriate. Eating is controlled by literally hundreds, if not thousands, of variables, and what is generally missing in research and in models of data is a fuller and more comprehensive view of eating. Advanced statistical approaches offer the promise of being able to handle the large amounts of data appropriate to such multivariate challenges. Martens *et al.* present partial least squares (PLS) as describing these 'multivariate observations in terms of their patterns of co-variation.' They further point out the two goals of PLS methods: the exploratory discovery of predictive relationships in data, and the confirmatory testing of theoretical predictions. The chapter provides historical and theoretical background for PLS methods, followed by a guide and examples.

1.4   Focus on social, economic and physical context

Land (1983, 1988) emphasizes the importance of context, stating 'acceptability is meaningless without context, and contextual incongruity can have a marked negative effect' (Land, 1988, p. 482). Thomson, in this volume, echoes the concern with 'out of context tests'. Land's model (Fig. 1.2), while emphasizing sensory factors, includes context. A prior model (Harper, 1981) did not include context (Fig. 1.2). Marshall (1995) includes context in the food provisioning process (Fig. 1.7, Table 1.1) in which food moves through the following sequence: production, distribution, preparation, consumption, and disposal. Social context as well as meal context are implied in some of these stages. Marshall suggests that studies of eating must include more than just the one stage of consumption. This might be the broadest definition of context.

Other models have ranged widely incorporating variables contributing to food-related behavior. Khan (1981) included the following categories of factors:

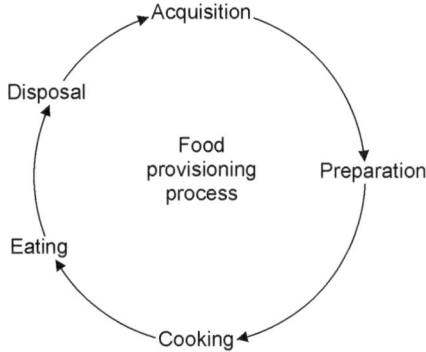

Fig. 1.7   Marshall model: the food provisioning process from food production to food disposal, and appropriate locations for each step (Marshall, 1995).

Table 1.1   The food provisioning process and appropriate locations (from Marshall, 1995)

Phase	Locus
Production	Farm
Distribution	Market
Preparation	Kitchen
Consumption	Table
Disposal	Scullery

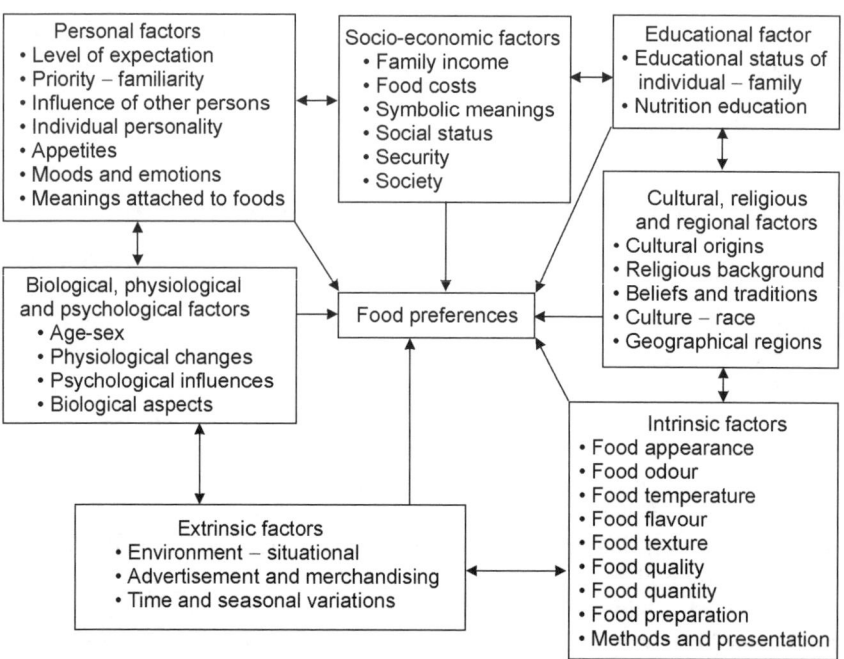

Fig. 1.8   Khan's seven factor model of food preferences (Khan, 1981).

personal factors (e.g. level of expectation, socio-economic factors (e.g. family income), educational factor (e.g. educational status), cultural, religious and regional factors (e.g. culture–race), intrinsic factors (e.g. food quality), extrinsic factors (e.g. environment), biological, physiological, and psychological factors (e.g. age–sex) (Fig. 1.8). At the time of Khan's paper in 1981, this was perhaps the broadest collection of factors presented in one model, with some categories limited ('food costs') and others expansive ('cultural origins'). Recently, Köster and Mojet (2007) have presented a very broad model of factors involved in eating and drinking behavior (Fig. 1.9). Mojet's model is based on six main factors: extrinsic product characteristics, intrinsic product characteristics, biological and physiological, psychological, situational, sociocultural. The model being used in this paper (product, individual, environment) contains two of Mojet's factors in each main factor as follows: product (extrinsic product characteristics, intrinsic product characteristics), individual (biological and physiological, psychological), and environment (situational, sociocultural). Examination of Fig. 1.9 highlights the complexity of eating and drinking, and also highlights the complexity of producing successful food products and designing successful solutions for food-related health problems.

While not a model of choice or food acceptance, Meiselman's 'measurement scheme for institutional products' (Meiselman, 1994) proposed different contexts for increasingly more complex testing goals and environments from basic sensory research in the laboratory to field testing as follows:

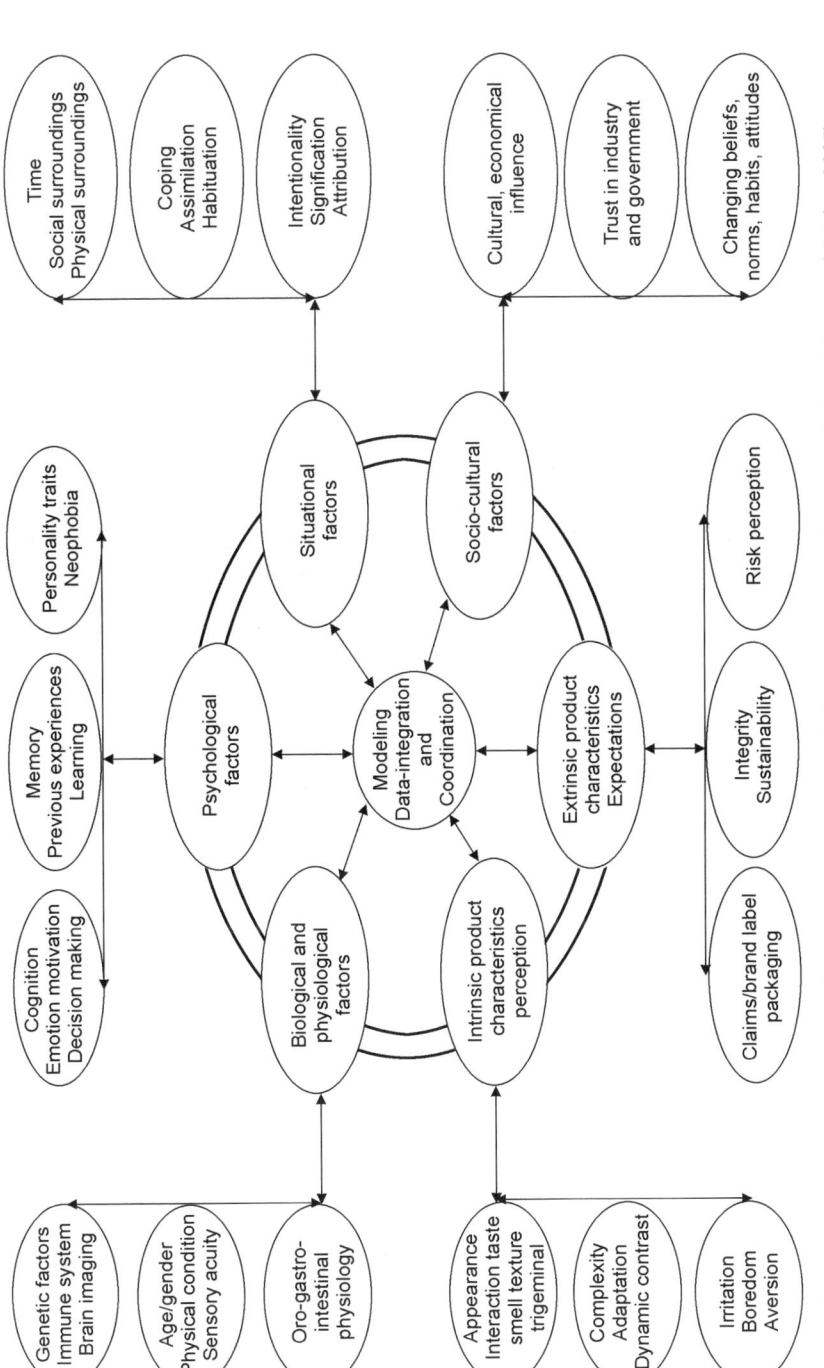

Fig. 1.9 Köster and Mojet's six factor model of eating and drinking behavior and food choice (Köster and Mojet, 2007).

- Phase 1 Consumer marketing.
- Phase 2 Individual item sensory testing.
- Phase 3 Consumer meal testing – laboratory.
- Phase 4 Consumer meal testing – field.
- Phase 5 Prototype testing.
- Phase 6 Extended ration use validation.
- Phase 7 Quality control testing.

Within this volume, Köster and Mojet (Chapter 11) call for greater use of home use testing (HUT) to address the problem of novelty in laboratory testing of new products. HUT provides an opportunity for subjects to experience a product over an extended time in a natural setting, thus increasing familiarity and decreasing novelty. Boutrolle and colleagues (2005, 2007) have recently studied home use testing, and provide a review of previous research comparing home use tests with central location and laboratory tests. A key future trend in consumer-oriented product development will be a better understanding of how different testing locations, especially the home, affect product appreciation. King *et al.* (2004, 2007) have studied which variables in laboratory testing might be critical in providing more suitable contexts. Results show that providing consumers with a choice, and providing a meal context might be two critical factors.

Two chapters in this book (20 and 3) specifically address broader contextual issues: economic context and cultural context. Economic context is addressed in Chapter 20 by Lohéac and Issanchou. This is generally an understudied area relating to product quality, especially among researchers more focused on intrinsic product attributes such as sensory attributes. Lohéac and Issanchou note that purchase intent is increasingly used as a dependent variable, since it is not enough to know that people like a product. We need to know whether they like it at a particular price, so the most frequent method is asking respondents whether or not they would buy a product at a stated price. This 'hypothetical method' does not involve any real purchase and therefore no financial consequences, and results in a positive bias in which consumers state a higher hypothetical price then found with other methods. Lohéac and Issanchou distinguish the hypothetical method from the 'reservation price (i.e. the maximum price a consumer is ready to pay)', and they recommend the auction method to reveal this amount.

Most research auctions are done through the Vickrey method and the Becker–DeGroot–Marschack (BDM) method, both of which are designed to elicit the maximum price one would pay for an item. With the Vickrey method, each consumer submits a sealed bid, and the winner pays the lower price below his/her bid, so this is often called a second-price auction. The auctioneer announces all of the bids; this is an interactive, competitive environment. The same approach can be extended to nth price auctions to produce more winners.

The BDM method compares the reservation price of each bidder to a random price previously defined. There is no competition; and one could do this procedure with one consumer. Lohéac and Issanchou state that 'the two methods induce an experimental market environment', although the BDM approach is not competitive based on auctioneer feedback. The BDM approach does not require that the bidders actually want the item, while the Vickrey method requires that you first ask consumers whether they want the item. Another issue for Vickrey auctions is the need for a large number of bidders.

Lohéac and Issanchou note that hedonic and auction methods produce the same ranking of products, and the authors discuss reasons why both methods are valuable. Auctions are usually used for existing products, although methods exist for testing novel items with proposed characteristics.

Rozin (Chapter 3) addresses the important issue of culture, and claims that culture is the most important variable in determining what we eat and what we do not eat. Rozin adds a number of interesting and important issues to his introductory material: (1) the existing data on human food choices are not only based almost entirely on the West, but a biased sample of the affluent, educated West; (2) people eat for nutrition and pleasure, but also for complex social reasons, and for moral reasons; (3) the mixing of cultures shows a 'conservatism of cuisine', food habits are one of the last things to change in the direction of the host country in immigrant population; (4) there is often more variation explained within culture than between cultures; (5) observation of lower classes in non-Western cultures is more likely to demonstrate strong cultural differences.

One way in which culture affects eating is in the foods that we eat. Rozin notes the approach of Elizabeth Rozin, who developed the concept of flavor principles to denote the characteristic flavorings of cuisines. This should be very important to those who want to translate one food to another cuisine – the secret is in the mix of flavorings, because 'It is the flavoring, more than any other feature, that imparts the cultural distinctiveness to a food.'

Rozin points out two of the major influences of culture: the distinction between individualism and collectivism (the role of the individual), and the role of the elderly. Rozin points out the rather startling conclusion that 'we do not know how food preferences are created', and that neither genetic differences or family influences are a key factor. Important factors include (1) exposure, (2) evaluative conditioning, and (3) social influence. Rozin applies these cultural concerns to a range of products and issues.

1.5   Conclusions and future trends

This is an impressive book on consumer-led food product development. It is impressive on the range of topics, the detail with which topics are presented, indicating the depth of the fields, and the relative newness of most of the supporting fields and research. While many books are updates of earlier presentations of a field, this text could not have been produced 10 years ago. Many of the variables and methods barely existed 20 years ago. Much of the research has been developed during the 1980s and 1990s, and continues to mature today. It is hoped that the future of these fields will be as rich as the recent past.

This book is about product development, and a number of chapters give views of overall product development, the sequence of product development, and the choices made along the way as to strategy and methods. What is emphasized in this book? There are two strong emphases – consumers, and properties of products including sensory properties. There are many examples of attempts to connect consumers to products in keeping with the book's consumer orientation. A variety of variables and methods are reported that try to measure human perceptions, cognitions, and behavior and relate these to products.

However, the analysis of these variables and methods by van Kleef and van Trijp suggest that the methods being used do not tap all of the critical human variables, because it remains difficult to predict market success of products, especially for novel products. The concern with current consumer research methods is echoed by Köster and Mojet, and also by Thomson, who question the validity of many of the laboratory methods, and emphasize the problems of testing the novel products to which van Kleef and van Trijp refer. This issue is a critical goal for consumer research in product development: how to measure long-term liking and acceptance of novel products when consumers respond to novelty in a manner that does not predict long-term acceptance. Along with long-term liking, van Kleef and van Trijp raise the issue of actionability, linking consumer needs and wants to product characteristics. The methods of consumer-led product development must continue to develop in a direction which serves two goals: (1) serving practical roles in product development, (2) while also contributing to research to yield a better understanding of consumer behavior. One direction which this might take is a further move towards, and integration with, marketing, as was envisaged by Howard Moskowitz at the first Pangborn Sensory Science Symposium.

This book focuses on food products, not eating per se. The models of food choice and food quality included in this chapter generally acknowledge the importance of eating context. Another future goal for consumer oriented product development is to place foods within the context of meals. The interested reader is referred to edited volumes on meals such as the interdisciplinary volume by Meiselman (2000), the more anthropological volume by Walker (2002), and the 2004 Pangborn Workshop by Pliner. Future research needs to further explore how to improve the predictive value of consumer testing of foods by serving foods in a valid meal context. At present, research has not uncovered which aspects of complex meal situations are critical, but consideration should be given to appropriate meal accompaniments, serving times, serving amounts, and meal duration.

Further use of meal context is compatible with Köster and Mojet's call for increased use of appropriate in-home long-term testing. Meal testing seems to be at odds with traditional central location tests of single products, but perhaps future developments will bring together these approaches.

Most of this book is directed toward achieving successful products. This book also contains consideration of product failure, which is welcome. It is widely acknowledged that product failures are common in the food industry, and we need as much attention to why our methods fail as to why some products succeed. My own view is that product acceptance is a very complex topic, involving many variables relating to products, people, and environments. We generally need more complex methods that will include and integrate these many variables.

Integration of variables will lead to consideration of concepts such as overall satisfaction, which is used more in the service industry than the product industry. Perhaps more product research should take place in service settings (restaurants, institutions) instead of the two choices of central location tests and home use tests. In this book, Thomson uses the term 'enjoyment' in the same broad approach as satisfaction, and questions whether 'liking' should be the sole product development criterion. Another possibility is that we need to more fully address issues such as values and emotions. Thomson's chapter is an interesting look at a new approach to dealing with emotional aspects of products, and several chapters address values through means–end chains.

Another way of looking at the integration of variables is to include the behavior which represents that integration – namely eating (or intake, or consumption). Yeomans (Chapter 4) represents a broad consideration of many biological variables, and further integration of sensory and consumer science with nutrition and biology are clearly in the future.

Several chapters question the tendency to report consumer data as overall averages. Most of us are now aware of the need to segment consumer data. Increasingly, specific groups will be dealt with separately. Popper and Kroll's chapter on children is a good example of this trend, and it is hoped that future volumes will address designing products for the growing elderly population as well.

Finally, as food and food research continue to globalize, there will be increased emphasis on the role of culture. Several chapters in this book focus on critical cultural variables such as values, ethics, and risks. However, the current dominance of Western data and Western food choice models need to be balanced by consideration of data and models from other parts of the world. This raises the interesting question of whether there are universals in food-related behavior. How will Western models fit in other parts of the world? *The Meal* by Walker (2002) presents some fascinating views of eating around the world, strongly suggesting that the image of the Western meal does not hold everywhere, and further that meals continue to change over history. We eat very differently today from how we did in the 19th or even the early and mid-20th centuries. Those of us interested in how people choose and consume foods need to look forward to how people will eat 30, 50, and 100 years from now. History tells us that food products and the meals composed with them will be very different.

1.6   References

ASTM (2003) *ASTM Standard Guide for Sensory Evaluation of Products by Children*, E-2299-03. West Conshohocken, PA: ASTM International.

BELL, R. and MEISELMAN, H.L. (1995) The role of eating environment environments in determining food choice. In Marshall, D. (Ed), *Food Choice and the Consumer*. Glasgow: Blackie Academic and Professional, 292–310.

BOUTROLLE, I., ARRANZ, D., ROGEAUX, M. and DELARUE, J. (2005) Comparing central location test and home use test results: application of a new criterion. *Food Quality and Preference*, **16**, 704–713.

BOUTROLLE, I., DELARIE, J., ARRANZ, D., ROGEAUX, M. and KÖSTER, E.P. (2007) Central location test vs. home use test: contrasting results depending on product type. *Food Quality and Preference*, **18**, 490–499.

CARDELLO, A.V. (1994) Consumer expectations and their role in food acceptance. In MacFie, H.J. and Thomson, D.M.H. (Eds), *Measurement of Food Preferences*. London: Blackie Academic, 253–297.

CARDELLO, A.V. (1996) The role of the human senses in food acceptance. In Meiselman, H.L. and MacFie, H.J.H. (Eds), *Food Choice Acceptance and Consumption*. Glasgow: Blackie Academic and Professional, 1–82.

CONNORS, M., BISOGNI, C., SOBAL, J. and DEVINE, C. (2001) Managing values in personal food systems. *Appetite*, **36**, 189–200.

DE CASTRO, J.M., BREWER, E.M., ELMORE, D.K. and OROZOCO, S. (1990) Social facilitation of the spontaneous meal size of humans occurs regardless of time, place, alcohol and snacks. *Appetite*, **15**, 89–101.

DELIZA, R. and MACFIE, H.J.H. (1996) The generation of sensory expectation by sensory cues and its effect on sensory perception and hedonic ratings: a review. *Journal of Sensory Studies*, **11**, 103–128.

FURST, T., CONNORS, M., BISOGNI, C.A., SOBAL, J. and FALK, L.W. (1996) Food choice: a conceptual model of the process. *Appetite*, **26**, 247–266.

GUINARD, J.-X. (2001) Sensory and consumer testing with children. *Trends in Food Science and Technology*, **11**, 273–283.

HARPER, R. (1981) The nature and importance of individual differences. In Solms, J. and Hall, R.L. (Eds), *Criteria of Food Acceptance: How Man Chooses What He Eats*. Zurich: Forester, 220–237.

HERMAN, C.P., ROTH, D.A. and POLIVY, J. (2003) Effects of the presence of others on food intake: a normative interpretation. *Psychological Bulletin*, **129**(6), 873–886.

KHAN, M.A. (1981) Evaluation of food selection patterns and preferences. *CRC Critical Reviews in Food Science and Nutrition*, **15**, 129–153.

KILCAST, D. (2006) Workshop summary: overview of developments in healthy eating and challenges to sensory professionals. *Food Quality and Preference*, **17**, 629–634.

KING, S.C., WEBER, A.J., MEISELMAN, H.L. and LV, N. (2004), The effect of meal situation, social interaction, physical environment and choice on food acceptability. *Food Quality and Preference*, **15**, 645–654.

KING, S.C., MEISELMAN, H.L., HOTTENSTEIN, A.W., WORK, T.M. and CRONK, V. (2007), The effect of contextual variables on food acceptability: a confirmatory study. *Food Quality and Preference*, **18**, 58–65.

KÖSTER, E.P. and MOJET, J. (2007) Diversity in the determinants of food choice: a psychological perspective. *Food Quality and Preference*, submitted.

LAND, D.G. (1983) What is sensory quality? In Williams, A.A. and Atkin, R.K. (Eds), *Sensory Quality in Foods and Beverages: Definition, Measurement and Control*. Chichester: Ellis Horwood.

LAND, D.G. (1988) Negative influences on acceptability and their control. In Thomson, D.M.H. (Ed), *Food Acceptability*. London: Elsevier Applied Science, 77–88.

MARSHALL, D. (1995) Introduction: food choice, the food consumer and food provisioning. In Marshall, D. (Ed), *Food Choice and the Consumer*. Glasgow: Blackie Academic and Professional, 3–17.

MEISELMAN, H.L. (1992) Methodology and theory in human eating research. *Appetite*, **19**, 49–55.

MEISELMAN, H.L. (1994) A measurement scheme for developing institutional products. In MacFie, H.J.H and Thomson, D.M.H. (Eds), *Measurement of Food Preferences*. London: Blackie Academic and Professional, 1–24.

MEISELMAN, H.L. (1996) The contextual basis for food acceptance, food choice, and food consumption: the food, the situation and the individual. In Meiselman, H.L. and MacFie, H.J.H. (Eds), *Food Choice Acceptance and Consumption*. Glasgow: Blackie Academic and Professional, 239–263.

MEISELMAN, H.L. (Ed.) (2000) *Dimensions of the Meal*. Gaithersburg, MD: Aspen.

MEISELMAN, H.L. (2003) A three factor approach to understanding food quality: the product, the person and the environment. *Food Service Technology*, **3**, 99–105.

PLINER, P. (2004) Workshop Summary: What to Eat. A multi-discipline view of meals. *Food Quality and Preference*, **15**, 901–905.

POPPER, R. (2004) Workshop Summary: Data Analysis Workshop: Getting the most out of just-about-right data. *Food Quality and Preference*, **15**, 891–899.

RODIN, J., MOSKOWITZ, H.R. and BRAY, G.A. (1976) Relationship between obesity, weight loss, and taste responsiveness. *Physiology and Behavior*, **17**, 591–597.

ROZIN, P. (1996) The socio-cultural context of eating and food choice. In Meiselman, H.L. and MacFie, H.J.H. (1996) *Food Choice Acceptance and Consumption*. Glasgow: Blackie Academic and Professional, 83–104.

SOLMS, J. and HALL, R.L. (Eds) (1981) *Criteria of Food Acceptance: How Man Chooses What He Eats*. Zurich: Forester.

STEPTOE, A., POLLARD, T.M. and WARDEL, J. (1995) Development of a measure of the motives underlying the selection of food: the Food Choice Questionnaire. *Appetite*, **25**, 267–284.

STROEBELE, N. and DE CASTRO, J.M. (2004) Effect of ambience on food intake and food choice. *Nutrition*, **20**, 821–838.

THOMSON, D.M.H. (Ed.) (1988) *Food Acceptability*. London: Elsevier Applied Science.

VAN KLEEF, E., VAN TRIJP, H.C.M. and LUNNING, P. (2005) Consumer research in the early stages of new product development: a critical review of methods and techniques. *Food Quality and Preference*, **16**, 181–202.

WALKER, H. (Ed.) (2002) *The Meal*, Totnes, UK: Prospect Books, 113–122.

WANSINK, B. (2004), Environmental factors that increase the food intake and consumption volume of unknowing consumers, *Annual Reviews in Nutrition*, **24**, 455–479.

WESTENHOEFER, J., STUNKARD, A.J. and PUDEL, V. (1999) Validation of the flexible and rigid control dimensions of dietary restraint. *International Journal of Eating Disorders*, **26**(1), 53–64.

2

Sensory perception as a basis of food acceptance and consumption

H. Tuorila, University of Helsinki, Finland

2.1 Introduction

Human senses mediate information from the environment, this being essential for survival. An important function of senses is to help identify edible material and prevent ingestion of material that may have an adverse effect on health and well-being. We are protective of taking foreign particles into our body (Rozin *et al.*, 1995), but starting and continuing eating are inevitable and need to be supported by appropriate mechanisms. The motivation to eat is organised by the homeostatic (hunger/satiety) and hedonic (rewarding) systems (Saper *et al.*, 2002). Sensory stimulation is connected to ingestion via cephalic phase responses that act as a preparatory phase for the utilisation of ingested food (Mattes, 1997a).

Given the wide supply of food products and intense competition in the current market, the success of foods greatly depends on the extent to which their sensory quality appeals to the target population. Other benefits, such as health aspects, may enhance the perceived value of food, but they are useless if the sensory quality fails to attract consumers (Tuorila and Cardello, 2002). Sensory analysis of food and data on consumer responses are therefore fundamental information when the success of a product is to be predicted.

This chapter deals with the role of sensory perception in the consequent acceptance and consumption of foods and reflects on the implications that these mechanisms may have for the product developer setting out to improve or design new foods and beverages. The phrase food acceptance is defined and measured in the literature in numerous ways, but here it is used to refer to the broad category of affective responses to food. Prior to reviewing the relationships among perception, affection and consumption it is necessary to describe the nature and function of each of them. Comprehensive reviews are available on the functions of our senses in food acceptance (Cardello, 1996). Therefore the following review gives a general basis for understanding the perception of food and the measurement of affection and consumption, and reviews some recent research areas such as the integration of perceptions.

2.2   The sensory system

All sensory modalities are involved with the perception of food (Fig. 2.1); however, the senses play an important role at different phases. Visual perception or olfactory signals are often the first to provide information of the qualities of food. Other sensory modalities come into play when the food is touched, tasted or eaten, and later inputs complement or revise the early visual and olfactory information. Although senses are categorised into five modalities presented in the figure, they act in a multimodal manner. Thus, several modalities participate in the observation or identification of a food, supporting each other's perceptual outcome. Multimodality is also evident in the human categorisation of food attributes: texture and flavour, two major categories of food quality, are perceived in a complementary or integrative manner via different sense modalities – flavour by chemical senses taste, smell and chemesthesis, and texture by visual, auditory and somesthetic systems.

Multimodality stresses the importance of both initial and entire impression of a product. For a product to be acceptable, the initial impression creates an expectation that either has to be confirmed or in a successful, positive, way disconfirmed.

2.2.1   Vision

The human eye responds to a certain part of electromagnetic radiation of the light, resulting as the perception of colour and other appearance attributes. The latter are a broad group of properties, such as transparency, turbidity and shine, and properties related to visually perceived texture, such as smooth, lumpy, rough, flaky, crystalline and viscose.

The major impact of colour on the identification of flavour of foods has been well demonstrated: congruent colour helps to identify the corresponding flavour, while no colour or, worse, an incongruent colour results in false alarms, i.e. misled associations in flavour identification tasks (e.g. DuBose et al., 1980). Using wine as an example, recent research confirms the steering role of colour in flavour perception: enology students (Morrot et al., 2001) and, to a lesser extent, trained wine experts (Parr et al., 2003) were misled by red colouring of white wine samples so that odour and flavour attributes typically characterising red wines were reported to be present in coloured white wines.

The appearance of foods and beverages used to be considered an important part of the quality, but this recent research helps to understand the true power of

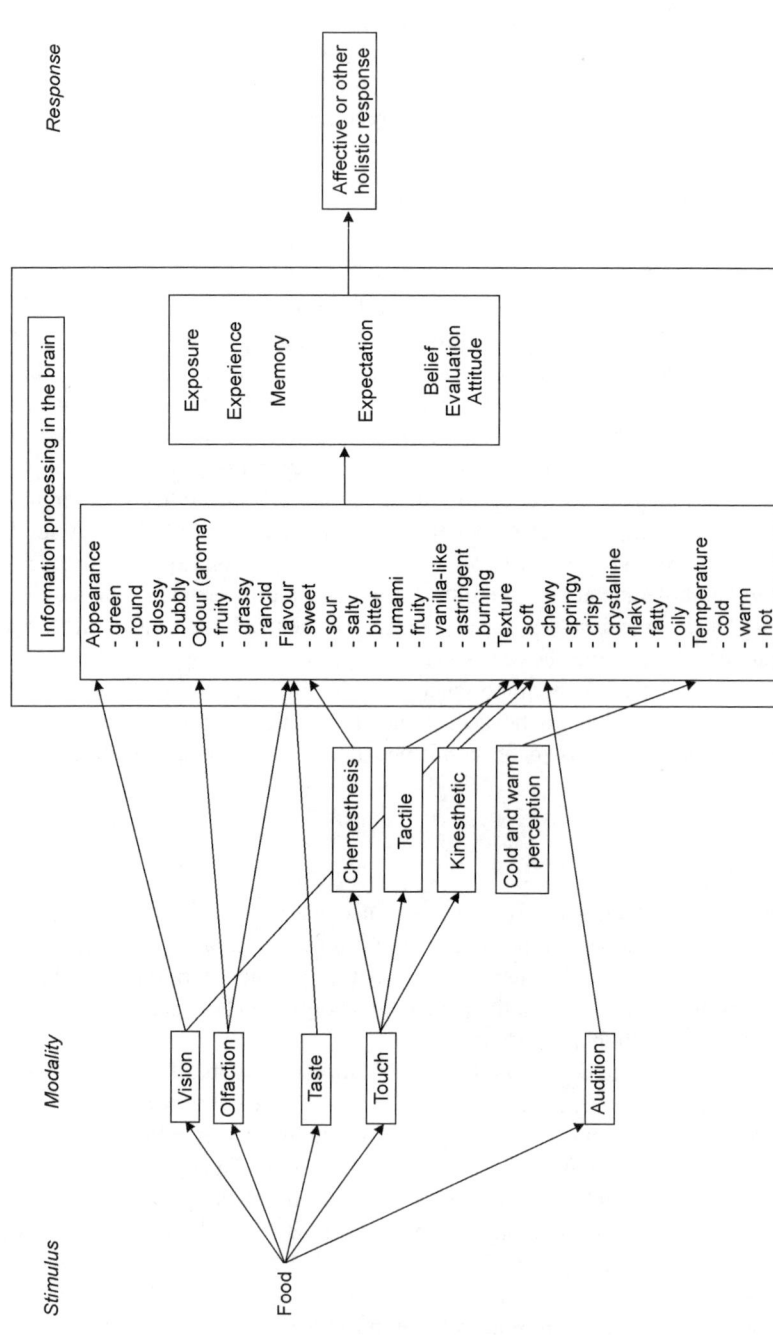

Fig. 2.1* Perception of food via different modalities and the integration and interpretation of the perceptions. Attributes listed in the third column are examples of the attributes in this category. * The left side of the graph was adapted from Fisher and Scott (1997).

visual cues. Whether effects such as those observed with wine tasters will last in long-term testing remains to be investigated.

2.2.2 Olfaction

Breathing in through the nostrils brings to the olfactory epithelium evaporated odorous chemical compounds triggering the perception of orthonasal odour. During mastication and swallowing, further odorous compounds pass through nasopharynx channel to the olfactory epithelium, thus causing the perception of retronasal odour. Traditionally, most research attention has been directed to orthonasal perception of odour (see, e.g., Firestein, 2001), yet the retronasal function is of utmost importance for food flavour (e.g. Small and Prescott, 2005).

A wide variety of odorous compounds can be perceived by a human. Buck and Axel (1991) found the gene family of about 1000 genes encoding odour receptors in the mouse. In humans, the corresponding number of genes is 650, yet almost half of these are pseudogenes (Malnic *et al.*, 2004). Taken that each gene expresses one receptor, there are 340 different receptors. However, receptors can respond to more than one odorous molecule, and odorous molecules can activate several receptors. For further complexity, food aroma is composed of tens or hundreds of odorous compounds. This diversity translates into rich and individually varying patterns of odour perception. Individual capabilities of differentiating among odours vary by exposures and training (e.g. Larsson, 2002).

The challenge of understanding the differences in olfactory perceptions is also due to the variability and vagueness of the vocabulary used. Frequently, the descriptions refer to the impact (calming, arousing, disgusting) or to the source (rose, pen, perfume) of an odour (Dubois and Rouby, 2002). Several odour classifications have been developed during the past centuries (see examples in Table 2.1), but such classifications do not have a deeper scientific basis. For an untrained layperson, naming any particular odour is difficult, but hedonic description of an odour is natural. Thus, the affective dimension can be considered fundamental for odours (Rouby and Mustafi, 2002). However, odour description can be remarkably aided by the provision of descriptive words and by training (Cain, 1982; de Wijk and Cain, 1994).

The human olfactory system does not perceive odours as individual compounds, but rather as patterns based on a group of odorants. The ability to detect a single odour compound from a mixture is not substantially improved by training: odour compounds are blended in a manner that cannot be broken down by the human odour detection and recognition devices (Livermore and Laing, 1998). The olfactory cortex interprets signals deriving from mixtures of compounds as unitary (Zou and Buck, 2006).

The retronasal odour is perceived via olfactory epithelium in the nose, similarly to orthonasal smell, but an individual locates the perception of retronasal odour in his/her mouth, where the source of smell is during tasting (see

Table 2.1   Examples of odour classifications (adapted from Rouby and Bensafi, 2002, and Amoore, 1970)

Author	Category	Example of the source of odour
Linneaus	Aromatic	Laurel
(1765)	Fragrant	Lime, jasmine blossom
	Ambrosia	Amber, musk
	Onion-like	Garlic
	Goat	Goat, sweat
	'Narcotic odours'	Solanaceous plants
	Nauseating	Foul, rotten, putrid
Zwaardemaker	Ethereal	Acetone, chloroform
(1895)	Aromatic	Camphor, benzaldehyde
	Floral and balsamic	Jasmin, vanilla
	Ambrosiac	Amber, musk
	Onion-like	Mercaptans, garlic
	Empyreumatic	Tar, smoke
	Caprylic	Fatty acids, cheese
	Repelling	Coriander, bugs
	Nauseating	Rotten meat, skatole
Amoore	Ethereal	Ethylene dichloride
(1970)	Camphor	1,8-Cineol
	Musk	Pentadecanolactone
	Floral	Phenylethylmethylethyl carbinole
	Mint	Mentone
	Pungent	Formic acid
	Putrid	Dimethyl disulphide

Small and Prescott, 2005). Thereby retronasal smell, along with taste and chemesthetic sensations, becomes an important part of food flavour.

Proper and precise description of odours requires systematic, empirical learning based on stimuli representing different odours. The fact that even trained sensory panelists perceive patterns based on a bunch of odorous compounds and not separate compounds, as demonstrated by Livermore and Laing (1998), stresses the importance of using humans as instruments in odour and flavour analyses. Research on retronasal odour perception is new and there is a clear need to understand the similarities and differences of ortho- and retronasal perception, to better understand the overall flavour perception.

2.2.3   Taste

Taste perception is initiated on the tongue, on which taste buds are located in three types of taste papillae. Taste buds contain receptor cells with microvilli which have active reception sites. Transduction of taste stimulation to nerve impulses takes place with different mechanisms, typically via regulation of ion channels of receptor cells (primarily sour and salty tastes that are induced by H^+ of organic acids and Na^+ of sodium chloride) or via a G-protein coupled cascade of reactions (primarily sweet, bitter and umami substances) (see Lindemann, 2001).

Five distinct tastes are currently postulated: sweet, sour, salty, bitter and umami (brothy). Tastes can be elicited by a range of substances (Table 2.2), but the most typical sweet, salty and sour perceptions are created by a few substances only. For instance, sweet substances other than sucrose may have a different time intensity profile, they may leave a bitter aftertaste or elicit extra coolness in the mouth (Grenby, 1996). Sourness stems from acids which, in addition to sour taste, also have minor 'off-tastes' of salty, bitter and sweet (Settle *et al.*, 1986) and qualities referring to vinegar and astringency (Hartwig and McDaniel, 1995). Thus, some acids also stimulate chemesthetic perceptions (see 'Chemesthesis' in Section 2.2.4).

In food, saltiness is primarily due to the presence of sodium chloride. Many other sodium salts and mixtures of salts have been investigated for their taste

Table 2.2   Chemical compounds eliciting perceptions of taste

Taste quality	Compounds	Reference(s)
Sweet	Sucrose* Other carbohydrates (*fructose, lactose, mannose*) Sugar alcoholes (*sorbitol, xylitol*) Intensive sweeteners (*saccharin, acesulfame* K, trehalose, etc.) Aspartame	Spillane (1996)
Sour	Citric acid* Tartaric acid* Lactic acid	Settle *et al.* (1986) Hartwig and McDaniel (1995)
Salty	Sodium chloride* Other sodium salts (*potassium chloride*, etc.) Other salts	Mattes (1997b) Breslin and Beauchamp (1996)
Bitter	Caffeine* Quinine sulphate* or chloride* Sucrose octaacetate (SOA) Naringin Limonin Isohumulones Peptides Phenyl thiocarbamide (PTC) and *n*-propylthiouracil (PROP)	Maga (1990) Guinard *et al.* (1994) Prescott and Tepper (2004) Cubero-Castillo and Noble (2004)
Umami	Sodium glutamate* Inosine monophosphate (IMP) Guanine monophosphate (GMP)	Bellisle (1999)

* Typical model substance, used for the training of sensory panellists, recommended by ISO and ASTM standards.

properties. Sodium salts other than NaCl have less salty taste (Schiffman *et al.*, 1980; van der Klaauw and Smith, 1995), while salts can have mutual perceptual interactions that can facilitate the use of their combinations (Breslin and Beauchamp, 1995). With respect to the search of salt substitutes, no major breakthrough has until now been reported.

Bitter substances are probably the largest group of taste-eliciting substances. Sensitivities to various bitter substances are not correlated (Cubero-Castillo and Noble, 2004). This group is further enriched by the special case of PTC (phenyl-thiocarbamide) or PROP (*n*-propylthiouracil). Parts of populations do not taste PTC and PROP as bitter, while others taste them as very or extremely bitter. A gene regulating the ability to taste the bitterness of PTC has been located (Kim *et al.*, 2003), and the ability to taste PROP – that chemically and perceptually resembles PTC and is now used as its substitute – has been suggested to predict other chemosensory capabilities in humans (Prescott and Tepper, 2004). The current evidence does not uniformly support this view, and further research will show to what extent this special phenotype has such wider implications.

The taste of umami is typically elicited by sodium glutamate, either alone or combined with ribonucleotides that enhance the intensity. A common amino acid, glutamic acid is thought to represent proteins and thus to have evolutionarily justified function (Lindemann, 2001). Umami gained credibility as a distinct taste, when specific glutamate receptors were identified on the human tongue (see review by Scott, 2005). There is wide variability, including non-tasters, in the sensitivity to the most common umami substance, monosodium glutamate (Lugaz *et al.*, 2002).

Although taste perceptions are not many, each taste quality has a salient role in certain foods or beverages. Through multimodal perceptions, tastes also support the impressions obtained from other sensory modalities.

2.2.4   Touch (somesthesis)

Somatosensory perceptions consist of feelings of cutaneous and deep touch, muscle tension and joint position (see Cardello, 1996). Feelings of warmth, cold and irritation belong to the cutaneous perceptions. They occur in all parts of body and are critical for the perceptions of texture and temperature. In addition to physical stimulation, chemical stimulation mediated via free nerve endings belongs to this system (see Fig. 2.1). Perceptions related to chemical stimulation are called chemesthesis, while those related to physical stimulation are referred to as tactile and kinaesthetic perceptions.

Chemesthesis
Chemesthesis, in earlier literature often referred to as trigeminal stimulation, is based on chemical stimulation of free nerve endings of mainly trigeminal, but also other, nerves in the mouth, nose and eyes (Green, 2002). Oral chemesthetic sensations are described as burning, stinging, tingling, feeling of numbness, hot, warm, cool or cold, and pain. Typical food products causing these sensations

include a range of spices (e.g. mustard, chilli, cinnamon, pepper and cloves contain compounds capable of irritation), alcoholic beverages (containing the irritant ethanol) and cooling agents such as eucalyptus and menthol (Prescott and Stevenson, 1995). The sensation often emerges slowly and dissipates even slower. The irritation is either based on direct stimulation of the free nerve ending or occurs via enzymatic reaction of an irritant (such as carbon dioxide) in the epithelium (Alimohammadi and Silver, 2002).

Tactile and kinaesthetic perceptions
Tactile perception is mainly based on mechanoreceptors that are densely located in fingertips and in the area of face, including the mouth. Tactile perception in the mouth is called mouthfeel (Lawless and Heymann, 1999), and typical examples of mouthfeel are astringency, oral viscosity and oiliness. The mechanoreceptors react to pressure, stretching and vibration. The kinaesthetic perceptions are related to joint positions and using force, thus textural characteristics as defined by Szczesniak (1963), such as hardness, adhesiveness, cohesiveness and elasticity, are perceived via kinaesthetic system. Astringency is important for the quality of wine, tea and some other beverages. The complex perception is mainly based on phenolic compounds precipitating on mouth surfaces, but also taste receptors may be involved (Monteleone *et al.*, 2004).

In contrast to odours, there are plenty of words to describe tactile and kinaesthetic perceptions (e.g. Drake, 1989) which, together with well-developed multivariate research techniques, give a good basis for the analysis of perceived textural differences in products (see, e.g., Roininen *et al.*, 2004).

Food texture is traditionally considered less interesting or less important than flavour. Texture is noticed when flavour is mild or when texture does not correspond to expectation (Szczesniak 1963). However, texture can contribute to perceptions in a powerful way. For example, the flavour of benzaldehyde was perceived differently from iso-sweet solutions that were either aqueous or thickened with a hydrocolloid (Cook *et al.*, 2003). Such multimodal effects deserve careful consideration during the design of products.

2.2.5 Audition
Auditory perceptions, in the context of food, are usually related to food texture. Disintegration of a food (or any other material) when a force is exerted on it, may result in sounds. This occurs mainly if a food is crisp, crunchy or hard – typically fruit, vegetables, cereal products and sweets can create a sound while in the mouth. The sounds during a breakage of food vary and are typical of some textures, for which reason it has been suggested that auditory perception would in these cases be the primary source of information of the type of a texture.

2.2.6 Interpretation of perceptions in the brain
The sensory system consists of receptors in periphery, as described above, of nerves mediating impulses, and of brain areas which receive the impulses and

interpret them. The information reaches the brain as impulses, and only the brain is capable of making sense of the message. Thus, the brain can be thought of as the most crucial sense organ. Specialised brain regions receive and interpret the information from different sense modalities, and the orbitofrontal cortex has a central role in integration of the information (Rolls, 2005). The multimodality of the information means that anatomically separated units provide a holistic perception. Recent research is rapidly accumulating information on the brain processes related to the perception of food.

Flavour perception is based on chemical stimulation during the time food is in the mouth, whether this stimulation derives from taste, retronasal odour or chemesthesis. In a comprehensive review of food flavour, Small and Prescott (2005) discuss the integration as a result of brain networking via learning. For example, the brain learns to combine vanilla odour with sweet taste, and therefore the brain responses to this combination occur more strongly and even in a different region of the brain, compared with the brain response to an unfamiliar and thus, incongruent combination of vanilla with salty taste. Although flavour of food consists of impulses that reach the brain via different senses, the learning process unites the perception into a coherent, seemingly single, experience from which the parts are difficult to separate again.

Texture is another multimodally perceived food attribute, perception being based on visual, tactile, kinaesthetic and auditory stimulation. The interplay among the sense modalities is well agreed on, but the neural integration has received less research interest than flavour. Until now, research on brain responses to texture has focused on the perception of fat and oral viscosity (Rolls, 2005), both mainly tactile perceptions.

Whether multimodal processes unify perceptions of flavour, texture or entire foods, once they have been formed, they facilitate the perception process by producing 'missing links'. For example, individual thresholds for the odour of benzaldehyde (bitter almond) were lower when a person tasted sweet solution while smelling, compared with a situation in which no sweet solution was in the mouth (Dalton et al., 2000). During learning about foods, our brain builds up networks that help us to take cues across modalities. When a person smelling vanillin simultaneously views a picture of ice cream, he or she detects the odour faster and more accurately than in a situation in which a picture displaying an incongruent object is present (Gottfried and Dolan, 2003); and when a wine-taster perceives attributes of red wine in a white wine that was coloured red (Morrot et al. 2001), he or she performs the evaluation task in a logical way that was programmed into his or her brain, even that the outcome may be disappointing for those who would like to see an analytically correct performance rather than a demonstration of 'perceptual illusion'.

2.2.7   From perception to affection
If we ask our consumer panel to rate how much they would like a food based on its name, appearance, aroma or any combination of these pieces of information,

we will no doubt get the ratings we asked for. However, ratings not involving tasting and consumption are tentative, because they are based on expectations and are subject to alteration, when more sensory information accumulates (e.g. Tuorila *et al.*, 1994; Arvola *et al.*, 1999). To provide a realistic estimate of affection, one has to taste or preferably consume a food: perceptions during consumption are at the heart of affection.

Affection is processed in the brain separately from perceived intensity of a stimulus (Small *et al.*, 2003). However, cognitions are capable of modifying affections, and the effect can be observed as brain activity. For example, presenting isovaleric acid to respondents as the odour of cheddar cheese vs body odour turned out to activate different regions in the brain (de Ajauro *et al.*, 2005). New brain research also corroborates the effect of cognition at a brand level: two cola drinks were presented unlabelled and labelled by their brand names to their regular users. In the unlabelled condition, the preferred drink caused a neural response of different magnitude than the non-preferred drink, and when labelled with brand names, the responses were reported to be dramatically different (McClure *et al.*, 2004). Thus, brain research proves many earlier findings based on subjective reports of affection to foods, although the interpretation of increases in brain activity may sometimes be difficult.

2.3   Prediction of consumption from sensory-affective responses

Liking for food and food preferences predict food consumption. However, both variables can be measured in a number of ways that reflect different aspects of the concept. The outcome of prediction depends on the way the concepts are operationalised.

2.3.1   Measuring affective responses

Rating procedures
Self-reports of affection towards a food stimulus are typically measured using procedures listed in Table 2.3. The task is basically simple, as each respondent is asked to express his or her personal impression. Responses should reveal the best liked sample(s), preference order or distances among samples on a continuum reflecting affection.

The scales differ from each other by the level of measurement. Consequently, different statistical procedures are available for data analysis (parametric or non-parametric). Paired comparison and ranking represent ordinal scale; category and graphic ratings are, broadly, interval scales; and labelled affective magnitude (LAM) scales and *ad libitum* mixing are ratio scales. Data obtained using an ordinal scale are rough, but the evaluation task is easy even for special groups, such as children. In specific instances, this procedure may therefore be the best option. The category and graphic scales are most commonly used, as

Table 2.3 Procedures for measuring affective responses to foods: advantages and limitations

Task (verbal anchors)	Advantages	Limitations
Paired comparison: Which one do you like better?	Simple task. Suitable for any age group.	Good for two samples only. Does not indicate how much better one sample is over the other, and why it is better.
*Ranking (*n *samples):* Rank samples according to your preference from 1 to *n*	Extension of paired comparison. Simple task.	With increasing number of samples, the task becomes complex. Does not indicate the relative preference differences among the samples.
Category scale 1: Bipolar, verbally anchored 9-point dislike/ like scale (dislike extremely– like extremely)	Verbally anchored categories are easy to use. Gives degree of like/dislike. Validated and widely used.	Often only positive end from neutral to like extremely (5–9) is useful. Validated in English only, translation is a challenge (*cf.* the word 'dislike').
Category scale 2: Unipolar 7- or 9-point liking, verbal anchoring at the ends (do not like at all– like very much)	Allows an affective response without the 'useless' negative (dislike) end of the scale.	Verbal anchoring of intermediate points difficult. Not validated.
Category scale 3: Bipolar, verbally anchored 5-, 7-, or 9-point pleasantness (unpleasant–pleasant)	Easy to use verbally anchored categories. Gives degree of pleasantness.	Useful, if ratings of unpleasantness are expected. Not clear to what extent ratings correspond to ratings of like/dislike.
Category scale 4: Bipolar 5-, 7- or 9-point scale anchored with distressed *vs.* happy faces ('smiley face' scales)	Especially designed help children to understand the rating. Gives degree of like/dislike.	Young children do not understand.
Visual analogue scales (VAS) for • pleasantness • like/dislike • liking (usually a line scale)	Graphic appearance of the scale helps to discern the degrees of affection and mutual distances.	Verbal anchoring only at the ends (for bipolar, also the centre). Laborious to manually decode the ratings.
LAM (labelled affective magnitude scale) (highest imaginable unpleasant – highest imaginable pleasant sensation)	Graphic appearance of the scale helpful. Good for rating 'extreme' perceptions. Avoids 'ceiling effect'.	Discriminates among stimuli only if extreme stimuli are present.

Table 2.3   Continued

Task (verbal anchors)	Advantages	Limitations
Just-right scale Category or graphic scales to measure a particular attribute (middle point 'just right', the ends 'too much' and 'too little')	Allows an affective response to a specific attribute.	Careful data analysis necessary to avoid misleading information (e.g. mean values!)
Ad libitum *mixing* (two stimulus strengths are mixed to the optimum strength of the observer)	Allows determination of individual optimum of a specific attribute.	Suitable to limited cases only (optimum needs chemical or physical determination).

they provide reasonably precise information on the degree of affection and, at the same time, they are illustrative and easy to use for most respondents. LAM is a newly developed and validated scale (Schutz and Cardello, 2001; Cardello and Schutz, 2004), with the idea derived from a corresponding intensity scale LMS (labelled magnitude scale) (Green *et al.*, 1996). LAM has been shown to correspond to category scales, but as a special bonus, the quasi-logarithmic scale (with end anchors referring to highest imaginable perception) allows the expression of extreme responses. This aspect is important when extreme responses are to be expected and therefore, a risk of a ceiling effect on ratings exists.

Most of the affective scales can be used for the collection of overall affective or attribute-specific responses. For example, respondents can be asked to rate the overall pleasantness of juice or pleasantness of sweetness (of a juice). Just-right scales and *ad libitum* mixing target specifically to one attribute, for which an optimal strength is sought. When analysed carefully, separating from each other the emerging subgroups with possibly variable optima, the latter methods are capable of yielding important information of differences in preference among consumer segments.

The scales are structured with different verbal expressions. Verbal anchors defining numerical scale points make the task easier, but varying interpretation of verbal expressions are a potential source of bias, so careful work is needed for the definition of scale points (e.g. Jones *et al.*, 1955). Unstructured and graphic scales avoid this potential bias, but may result in a larger variability in the data.

The choice of the most appropriate rating procedure for a specific study depends on a number of factors. If the clarity and concreteness are the crucial issues (e.g. a busy testing situation or attentional limitations of very young or old respondents) then simple tasks (paired comparison, ranking, or scales with few verbally labelled categories) may be the best choice. Also smiley faces (Ellis, 1967) can add to the clarity in such cases. If all samples to be presented are known to be pleasant and well liked, one has to be careful to allow differentiation among positive affective responses. On the other hand, information of negative responses to samples, even among a small segment of respondents, might be crucial, and then the bipolar scale is the only choice. If meaningful comparisons of ratings from time to time are an issue, then one should carefully choose a rating scale that is most appropriate and stick to its use as much as possible.

Stimuli

The stimuli to be rated can be actual food samples that are tasted, but they can also be names of foods. It is important to note that ratings assigned to samples vs those assigned to food names correlate weakly, if at all. A tendency to use more extreme ratings for food names than for actual stimuli has been shown with both between-subjects data sets (Cardello and Maller, 1982) and within-subject designs (e.g. Tuorila, 1987). Ratings given to food names reflect attitudes to the food concerned, while the actual food may not be considered as great or as bad as the positive or negative attitude to it. It has been suggested that rating of a food name is based on the memory of the best or worst example (see Cardello and Maller, 1982). Thus, collecting ratings of food names (in a background questionnaire) can be very useful for checking how well the present (tasted) food stimulus performs compared with the memory of product type or category to which it belongs.

Affective ratings are sensitive to context factors, such as other samples in the testing situation or the verbal definition of the respective product category. The actual samples rated within a session affect the other ratings, by forming a frame of reference for the evaluation. The literature shows many examples of how the range, in terms of sensory (e.g. sweetness level) or affective dimension, is capable of shifting ratings (see Lawless and Heymann, 1999). The name or description of samples provides a category and therefore, a context, for the evaluation (see Cardello *et al.*, 1985; Tuorila *et al.*, 1994). Also the product category matters: when a speciality beer or coffee was considered a member of the regular coffee or beer category, tendency for a higher contrast between the special vs regular was observed, compared with the situation in which respondents had subcategorised the products into separate regular vs gourmet groups (Zellner *et al.*, 2002).

Since it is well known that respondents take cues from context factors, such as described above, the potential effects of cues need to be carefully considered when planning trials. Only the necessary samples and the right product category should be included, unless the effect of these variables on ratings is tested.

Respondents

Ordinary consumers can cope with any type of rating tasks related to affection, whereas specific subgroups of consumers need special procedures. For example, infant responses are mainly measured by observing their behaviour, e.g. amount of liquid consumed or suckling behaviour. Young children need special attention, often individual interviews and adapted evaluation procedures (Guinard, 2001). The same is often true of elderly respondents, although their capabilities vary strongly, depending on their age, health and overall well-being. Also, the elderly may be reluctant to express negative opinion, and their lenient or too high rating behaviour may result in misinterpretation of their affection. Hedonic ranking that forces the elderly to express their preference may result in better discrimination among the samples than hedonic scales (Barylko-Pikielna *et al.*, 2004).

2.3.2   Food consumption operationalised

Methods of collecting information on food consumption have been primarily developed and validated in nutritional sciences, usually with a goal to estimate intakes of nutrients at an individual or population level. Typical methods include 24 h food recall, dietary history and food frequency questionnaires (FFQ) (Patterson and Pietinen, 2004). Individually reported food intakes are, via data banks estimates of nutrients, translated into nutrient intakes.

In consumer sciences, the interest seldom extends beyond the purchase (intention), choice or consumption of a single food or a food category. Hence, the consumer scientist needs instruments reliably reflecting or representing intended or actual choice or consumption of a certain food or food category. Basically these instruments are similar to those used in nutrition research, but the scope is more limited. Table 2.4 shows a range of possible instruments.

2.3.3   How affective responses and food consumption relate to each other
Affective ratings vs use frequencies

A common estimate of the relationship between affective response to a food and the consumption is that affection predicts 25–50% of consumption (see Cardello and Maller, 1982). Such a prediction stems from correlation coefficients within the range 0.5 to 0.7 between the two measures, usually operationalised as self-reported ratings. In a study with Finnish women ($n = 136$), correlations between ratings of liking and use frequency of sweet foods ranged from 0.57 to 0.73 (Lähteenmäki and Tuorila, 1994). Our recent survey ($n = 669$), based on ten food names as stimuli, showed correlations 0.33 to 0.67, mean 0.52 (pleasantness ratings vs. use frequencies), and 0.38 to 0.61, mean 0.53 (ratings of liking vs. use frequencies), roughly corroborating the estimates in the literature (Tuorila *et al.*, 2007). In a survey with 87 college-age US women and a large number of foods, the correlations ranged from -0.04 to 0.62 (median 0.40) (Drewnowski and Hann, 1999). As pointed out by the authors, the partly low correlations in this study were due to an imbalance between the two variables, when a food was either very well liked, but very rarely used, or vice versa.

Technically, the correlations between affection and consumption should be highest in studies such as the ones cited above, in which both ratings were assumingly based on the same mental image of a food item, and both variables were rated with a paper-and-pencil (or corresponding) technique. The use frequency scales are rough and respondents want to be consistent, both favouring a

Table 2.4 Procedures for measuring food use or consumption tendencies: advantages and limitations

Task	Advantages	Limitations
Actual choice Which one do you choose? Which one do you want to eat now?	Simple, concrete task. Suitable for any age group.	Laborious data collection. One choice may not be representative. Momentary factors affect the choice.
Intended choice Which one would you prefer to taste/eat?	Simple, fairly concrete task.	Subject to social desirability bias.
Frequency of current or future (intended) use Verbally anchored response options, e.g. from never to daily	Easy to collect and respond.	Subject to social desirability bias.
Purchase interest Bipolar, verbally anchored 7 or 9-point scale (not at all–extremely interested)	Easy to collect and respond.	Subject to social desirability bias.
Likelihood of buying/using Unipolar 7 or 9-point scale, verbally anchored at the ends (unlikely–likely)	Easy to collect and respond.	Subject to social desirability bias.
Consumed quantity Estimated by observation or from waste (or even self-reports)	Concrete.	Momentary factors affect the consumed amount. One measurement may not be representative.

tendency to correlations higher than in the reality. For more valid comparisons, separating between the sets of instruments spatially (not to ask the questions in a single shot) or temporarily (different sets of ratings presented in separate questionnaires or sessions) is advisable when possible.

Affective ratings vs food choices

Consumers often have a choice among various foods. Self-selected foods are usually rated as pleasant and they are eaten in larger quantities (de Castro *et al.*, 2000); thus the choice is likely to reflect affection. Accordingly, de Graaf *et al.* (2005) found that in the course of decreasing hedonic ratings (9-point hedonic scale) of foods, the chances to choose a product for the second time decreased from 52 (rating 9) to 8% (rating 1). Hence, a clear correspondence was found between the actual choice and the hedonic rating, but one might expect even a stronger prediction: why did the patterns of choice at the second time not run from 100 to 0%?

Table 2.5   Choice of an apple ($n = 92$) vs. a chocolate bar ($n = 190$) as a snack: reasons given for the choice of each option as percentages of all reasons (Roininen and Tuorila, 1999)

Reason given	Apple (%)	Chocolate bar (%)
Health/nutrition related		
Healthfulness	29.3	3.7
Light	10.9	–
Obtain energy	1.1	9.0
Allergic to the alternative	6.5	10.5
Pleasure related		
Good taste	25.0	31.6
Momentary desire	25.0	19.5
Availability		
Alternative available at home	2.2	9.5
Give to other family members	–	4.7
Convenience	–	3.7
Is worth more money	–	7.8
Total	**100**	**100**

A single choice is a tricky measure: simple itself, but complex by the background. Overt behaviour is influenced by a large number of mutually related reasons, some obvious and others more hidden. Personal and situational factors often steer the choice away from the affection based choice: for example, variety seeking (Ratner *et al.*, 1999) as well as tendency for dieting and perceived hunger (Tuorila *et al.*, 2001) are capable of shifting a choice. The multitude of reasons of a single choice was demonstrated when an apple vs chocolate bar was given to respondents as a reward for participation in a survey (Table 2.5; Roininen and Tuorila, 1999). The choice of apple was typical of those who customarily ate apples often and considered them more pleasant than chocolate; those who chose the chocolate bar rated the pleasantness and use frequencies of both items as fairly similar. The sophisticated laboratory experiment of Lange *et al.* (1999) on purchase behaviour of juice similarly demonstrated that factors other than preference guide choice. The authors proposed that variety seeking, needs of other family members, and other products' good value for money and similar financial aspects all played a role.

One should note that processes related to the first vs. subsequent choices are different. The first choice is based on expectation due to product image, appearance and other information. Lack of alternatives may also play a role. Tasting brings the chemosensory and somesthetic experience into play whereby the subjects may completely revise their earlier impression (Arvola *et al.*, 1999). In subsequent choices, the overall sensory experience is incorporated among the reasons of choice. For the food producer, seeking for profitability, the subsequent choices are crucial, thus the overall sensory quality is an essential cornerstone for repeated choices.

Affective ratings vs reported purchase interests
Information of purchase prospects is crucial for the prediction of the success of a new product, whereby the association between affective ratings and purchase interests are relevant. Similarly to affection, purchase intentions have been rated in a number of ways. Guinard *et al.* (1996) and Bower *et al.* (2003) had their respondents rate their purchase interest in regular and reduced-fat products (Guinard) and yellow spreads (Bower) from definitely would not vs definitely would buy; Kähkönen and Tuorila (1999) collected ratings of buying probability of reduced-fat and regular-fat products *within the next month* (probably vs probably not); Tuorila *et al.* (1998) had yogurt-type snacks rated for purchase interest (not at all–very interested). In all these studies, a health or fat-related claim was part of the design.

Although affection and purchase interest are related, they are different issues. Label information, given in all studies cited above, is a strong element in purchase decision. For example, Guinard *et al.* (1996) found that their older adults preferred buying reduced-fat products, although these were not preferred in hedonic tests. Others found that affective responses remained unaffected by information, but the purchase interest was affected by it. Thus, the information does not guide the affection, but it does guide the action. This finding is also supported by actual choices made by respondents in a study by Stein *et al.* (2003).

A somewhat positive affection is apparently needed to prompt subsequent choices even when purchase is supported by motives other than affection. The affection will then gradually shift to reach a balance with individual use patterns. Thus, purchase interest cannot be maintained without positive affection, but most of all, it reflects the overall utilitarian value of the product to a consumer.

Affective ratings vs consumed quantities
Several short-term comparisons have been conducted between affective ratings and consumed quantities in laboratory settings. This line of research was started by Lucas and Bellisle (1987), who challenged the validity of hedonic tests in the prediction of product success by showing that *ad libitum* consumption suggested lower optimal sweetness of yogurt than taste-and-spit hedonic ratings. Thus, the subjects indicated the highest liking for a sweet yogurt, but ate a larger volume of a less sweet yogurt. Shepherd *et al.* (1991) found that regardless of varying affective responses to saltiness levels (0.1–0.6% Na^+) of tomato soups, the consumed quantities were similar in laboratory sessions; the outcome suggests that saltiness (in the range used) was not a salient predictor of consumption. Daillant and Issanchou (1991) found that not all tested consumers of yogurt responded to the varying level of sweetness by changing quantities of consumption. Hellemann and Tuorila (1991) found that the pleasantness and consumption of bread in laboratory conditions correlated $r = 0.63$ when pleasantness was based on a taste-and-spit test and $r = 0.82$, based on post-consumption ratings. Thus, post-consumption ratings were more reliable predictors of consumption, probably because of the more realistic exposure.

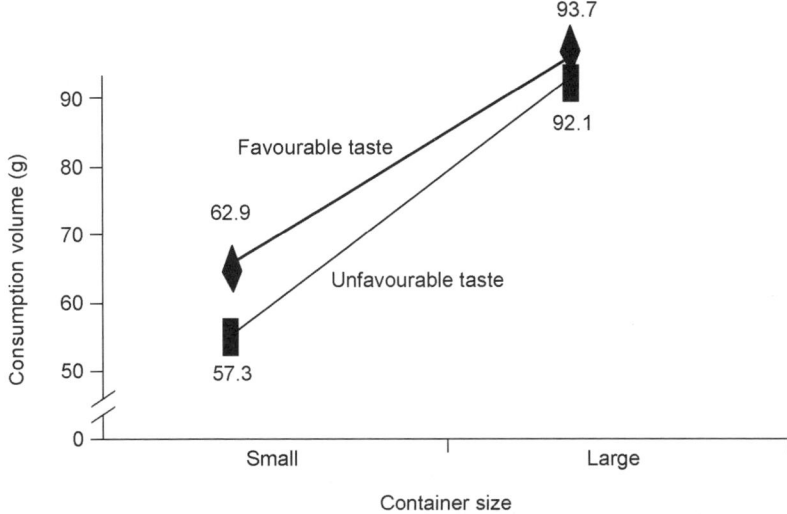

Fig. 2.2   The impact of favourability of taste and container size on consumption of popcorn. The data of 'favourable' group is based on 54 and those of 'unfavourable' taste on 67 subjects (Wansink and Park, 2001).

In a study by Wansink and Park (2001), popcorn consumption at the movies varied largely based on whether the available amount was 'small' (120 g) or 'large' (240 g). This underlines again the context dependence of consumption. Although the respondents rated the statement 'this popcorn tasted better than most movie popcorn I've had' strikingly different (9-point agree/disagree, favorable taste, mean = 7.5, unfavourable taste, mean = 2.5), these varying impressions had only a slight effect on consumption (Fig. 2.2).

Affective ratings, choice information and rough estimates of consumption were collected for a large group of food items in nine settings in an army field study during 4–11 days (de Graaf *et al.*, 2005). The choices and consumed quantities were recorded, and the measures of consumption were correlated with affective ratings given to each food (based on tasting), resulting in $r = 0.22$ to 0.62 (mean 0.45) for main dishes and $r = 0.13$ to 0.56 (mean 0.35) for snacks. Notably, items with top hedonic ratings 8 or 9 (9-point scale) were all consumed while, with lower ratings, the consumed proportion of the available supply decreased. However, once chosen, 46% of the portions of the even most disliked items were consumed, maybe because of the field conditions and the lack of alternatives.

Two lessons emerge from these studies. First, laboratory consumption data warn that taste-and-spit tests may not allow enough exposure to a sample for reliable prediction of product performance. The concern is relevant, although the outcome is not obvious in all cited data. It should be noted that consumption in the laboratory tends to be artificial, and setting the stage for the collection of such data is a challenge. The field study overcomes the problem of taste-and-spit exposure and laboratory conditions, and proves the association between affection and choice. It also provides the second lesson: the consumption of the entire portion is likely when the product is very well liked, but otherwise leftovers are to be expected. This implication is important for catering industries.

Example: Familiarity and novelty as barriers to change

Familiar foods, as opposed to their unfamiliar counterparts, are generally preferred in any population (e.g. Tuorila *et al.*, 1994, 2001). This is natural because of the long history of exposures and the range of mechanisms available to integrate a person to one's own culture (Rozin, 1982; Rozin and Vollmecke, 1986). Along with suspicion to a novel product itself, consumers may have doubts about the production technologies (Cardello, 2003). In the course of exposures, i.e. in the course of familiarization, the responses to new foods often become more favourable (Pliner, 1982; Birch, 1999). Neophobic individuals (food neophobia determined using FNS, Table 2.6, Pliner and Hobden, 1992) tend to be more negative about unfamiliar foods than less neophobic individuals.

Information (name of the food or any other description reducing the uncertainty of the identity of the product) is capable of enhancing the affective ratings of unfamiliar foods to some extent (Pelchat and Pliner, 1995; Cardello *et al.*, 1985; Tuorila *et al.*, 1994) (Fig. 2.3). The information of beneficial properties of a new product helped to enhance hedonic responses if introduced at the very initial stage of familiarization, while the favourable information given later, after one-week home use, did not enhance later hedonic responses (Kähkönen *et al.*, 1996).

Besides for entire foods, the power of familiarity is true for stimuli of separate sense modalities. The rated familiarity of odour correlated positively with hedonic and intensity ratings in a cross-cultural study (Distel *et al.*, 1999), and information provided on the source of odour increased pleasantness ratings of pleasant odours (Distel and Hudson, 2001).

Responses to novelty can be examined using different frames of reference, one being the social representation theory dealing with perceptions of everyday phenomena shared by the population. The theory states that familiarization with a novelty involves anchoring it to a familiar element of an individual's world (Moscovici, 1984). Supporting this view, Tuorila *et al.* (1994) found that associating an unfamiliar food to a familiar and well-liked food enhances chances for the former to be perceived as acceptable. Bäckström *et al.* (2004) developed a questionnaire to describe dimensions of consumer responses to novelty in foods and based on survey data, using social representation theory as a point of departure,

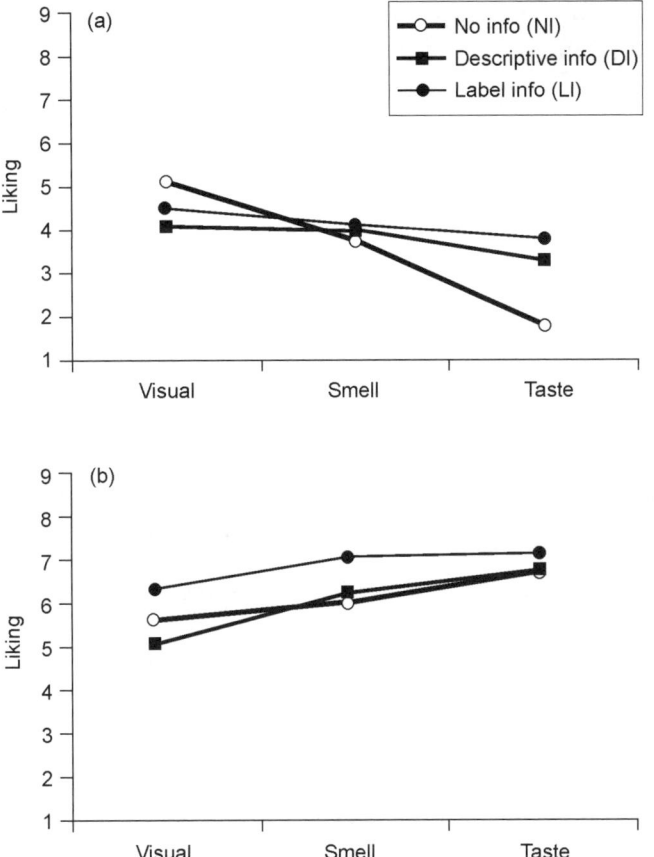

Fig. 2.3 Mean expected (visual and smell conditions) and actual liking (taste condition) for (a) a new (Finnish alcoholic beer) and (b) a familiar (root beer) beverage among US respondents. Number of subjects in each of the three verbal experimental conditions: no information, $n = 42$; descriptive information, $n = 40$; label information, $n = 39$ (Tuorila *et al.*, 1994).

Huotilainen and Tuorila (2005) found that trust and suspicion are the competing forces central for the perception of new foods.

If a new product is difficult to categorise, 'truly new' (Loken, 2006), the resistance to novelty may become an obstacle and biased responses may accrue during the first exposures. The roles of information, associations and individual traits are worth taking into account when the product is to be launched. Exposures may have to be repeated before reliable ratings can be expected. Home use experiments enable repeated exposures and therefore serve a purpose, although they suffer from the lack of control in testing conditions.

2.4 Individual factors modulating affective responses and consumption

Liking for foods is largely culturally learned; thus culture as a whole and, more specifically, the family and closer social surroundings are important when children are repeatedly exposed to the available food supply (Rozin and Vollmecke, 1986; Zellner, 1991; Birch, 1999). The outcome of the socialisation, at an individual level, is a reasonable balance between affection and consumption, leaving however plenty of space for individual preferences and food use patterns. Individual perceptual or sociocognitive differences can affect consumption directly or via sensory-affective path, or both. Of such individual effects, the perceptual capabilities and mental dispositions will be discussed.

2.4.1 Sensory capabilities

Sensitivity to chemosensory and other food-related stimuli varies widely across individuals and is often suspected of impacting the affections. Possible impacts of sensitivities fall at least into three categories.

The less sensitive a person is to a certain stimulus, the higher intensity s/he needs to reach the optimal strength

Since 1990s, this assumption has prompted research on the acceptance of foods among the elderly whose impaired chemosensory capabilities have been assumed to be a reason for reduced appetites. Although the elderly may prefer higher optima for aroma substances, compared with the young (de Graaf *et al.*, 1996), the intensification of aroma did not improve the acceptance of foods (measured as affection and consumption) among those elderly whose sensory capabilities were impaired (Koskinen *et al.*, 2003, 2005).

In the search of reasons for unhealthy eating habits, sensory capabilities have also been considered as a potential modulator, and thus the perception and intakes of nutritionally adverse food components have also been investigated. Shepherd and Farleigh (1989) reviewed the literature for individual sensitivities to sodium chloride and fat vs the intake of salty or high-fat foods. In the great majority of studies, no relationship was found between sensitivities and consumption. Further support for the lack of association between sensitivity, affection and consumption has been provided by Guinard *et al.* (1999).

Overall, the evidence of the effect of sensory sensitivity on the preference and use of foods is not convincing.

A common underlying trait impacts perceptions and food consumption

Extensive research of the genetically determined tasting status for PROP has prompted this line of inference. The non-tasters, tasters and supertasters of PROP are assumed to have different food consumption habits (for a review, see Duffy *et al.*, 2004), but other authors consider such influences unlikely (Drewnowski, 2004). Further evidence for proving or disproving this case is likely to emerge in the future.

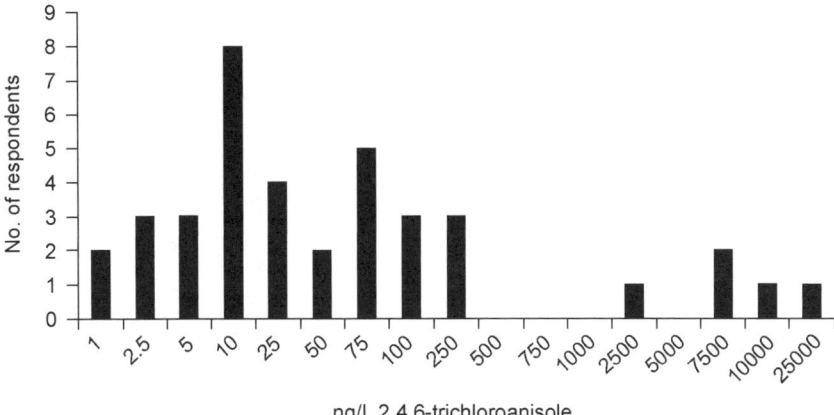

Fig. 2.4   Distribution of individual odour thresholds of TCA (2,4,6-trichloroanisole) in wine (Sauvignon Blanc) among 38 respondents (Suprenant and Butzke, 1996).

Individual sensitivity to compounds with adverse sensory effects restricts the consumption of the concerned foods
Off-flavours are a distinct example of adverse effects of a single compound on acceptance and consumption. Sensitivity to cork taint in wine, due to the compound 2,4,6-trichoroanisole (TCA), illustrates the case. In a population of 38 respondents, the threshold for TCA varied between 1 and 25 000 ng per litre (Fig. 2.4) (Suprenant and Butzke, 1996). The high sensitivity (low threshold) clearly limits the use of tainted wine (Prescott *et al.*, 2005) and thus, the presence of off-flavour has an implication for the acceptance vs rejection of wine.

Another example of this category is the sensitivity to bitter taste that has long been deemed to lead to disliking for foods with a bitter flavour note (Fischer *et al.*, 1961), although in another study, sensitivity to PROP appeared to be a better predictor of a choice of 'strong-tasting' foods than the sensitivity to quinine sulphate (Glanville and Kaplan, 1965). Tanimura and Mattes (1993) found that sensitivities to bitter substances (other than PROP) were related to non-use of certain bitter foods.

Thus, it seems fair to assume that chemosensory sensitivity to certain disliked substances can lead to the rejection of a food or beverage.

2.4.2   Attitudes, traits and other dispositions
Individual beliefs, attitudes and expectations as well as personality traits have plenty of potential to interact with the acceptance of foods (e.g. Cantin and Dube, 1999). Their importance grows in the wealthy Western countries, in which the supply of food is large and abundant product information is available. Thus, consumers are segmented based on their values, attitudes, personality traits and other mental constructs.

An increasing number of verbal scales for the characterisation of consumer segments is available in the literature, e.g. food neophobia scale, food choice

Table 2.6 Examples of instruments to measure food-related traits, attitudes and other dispositions

Instrument	Subscales	Aim	Items	Reference
Food neophobia scale (FNS)	–	To quantify the tendency to relate to new foods	10 statements, to be rated on a 7-point scale (disagree–agree)	Pliner and Hobden (1992)
Food Choice Questionnaire	Health Mood Convenience Sensory appeal Natural content Price Weight control Familiarity Ethical concern	To characterise motives related to food choice	Total of 36 items rated for importance on 4-point scale (not at all important–very important)	Steptoe et al. (1995)
Health and Taste Attitude Scales (HTAS)	General health interest Light product interest Natural product interest Craving for sweet foods Using food as a reward Pleasure	To characterise health and pleasure related food attitudes	Total of 38 items rated on a 7-point scale (disagree–agree)	Roininen et al. (1999)
Three-factor Eating Questionnaire (TFEQ)	Cognitive restraint Disinhibition Perception of hunger	To characterise coping with dieting	Total of 51 items rated on different scales (true–false, seldom–always, etc.)	Stunkard and Messick (1985)
Children's Eating Behaviour (CEBQ)	Responsiveness to food Enjoyment of food Satiety responsiveness Slowness in eating Fussiness Emotional overeating Emotional undereating Desire for drinks	To characterise children's eating styles	Total of 35 items to be rated by parents on 5-point scale (never–always)	Wardle et al. (2001)

questionnaire, health and taste attitude scales, various forms of restrained eating questionnaires, and children's eating behaviour (rated by parents). These scales are briefly characterised in Table 2.6, and more scales are available elsewhere (e.g. compilation by Diehl, 2006). Typically, verbal scales comprise several statements or questions around the issue of concern that are rated using categories from disagree to agree, not important to important, or the like. Based on the ratings, a composite score is computed to each respondent, reflecting his or her placement on the measured dimension.

Four major issues should be taken into account when considering the use of composite verbal attitude or trait scales:

1. The researcher needs to understand the validity status of the scale – without documented validity, there is no proof that a scale measures what it is claimed to measure.
2. The researcher needs to understand the target of the scale and evaluate the suitability to the assumed respondents.
3. The researcher needs to understand the cultural context when using scales developed and validated in a foreign (or past) cultural environment, as the meaning of verbal expressions may differ from the original.
4. Translation of scales to other languages needs care to preserve the original meaning.

After these reservations have been carefully considered, scales quantifying consumer traits and attitudes can be used to reveal important aspects of preferences of modern consumers.

2.5 When sensory perceptions are ignored

The food producer strives to maintain the sensory attributes of commercial food products within the documented limits (specification). For the consumer this means that the product should consistently correspond to his or her sensory and hedonic expectations. A deviation from the specification is subject to consumer complaints. However, there are reports showing that a change in sensory quality does not always result in adverse consumer responses.

An interesting series of experiments describes the consumption of normal and slightly adulterated (oxidised off-flavour) milk in dormitory and laboratory environments (Vickers et al., 1999). On the 9-point hedonic scale, the adulterated milk was one point lower than the control milk, but no effect on consumption was found. As possible reasons for this, the authors proposed the following: (1) the limited choice in the consumption situation, (2) subjects may have earlier exposures to oxidised flavour in milk, and thus the slightly impaired taste fell within a normal range, (3) milk was so essential for the meal context that it was consumed regardless the perceived change, and (4) the recruited subjects were milk likers who enjoyed the milk in spite of quality problem. The explanations reflect well the circumstances in which regular consumers taste and use the foods and beverages. The non-responsiveness of a consumer panel to flavour differences is furthermore supported by Shepherd *et al.* (1991) whose subjects, in a laboratory experiment, ate the same amount of tomato soup regardless of saltiness variations. It is also in line with the study of Wansink and Park (2001) whose subjects were driven by the supply of popcorn, rather than by the perceived taste.

In another study on consumer responses to deviation, colour of orange juice was adulterated by added greenish hue (Tepper, 1993). Respondents ($n = 342$) rated the adulterated and control sample on just-right scales for colour, sweetness, flavour and overall liking. Although colour was rated dramatically atypical, the overall liking remained almost unaffected. The conclusion was that, although colour is important for the first impression of a juice, its impact gets less when the sample is tasted. Thus, the latter sensory impression revises the initial impression. The result was also in line with other studies indicating that chemosensory properties, rather than appearance, dominate perception of the overall quality of beverages. However, in a real consumption situation the consumer does not necessarily engage tasting, if the initial impression of the product is not appealing or not as expected.

A third example of disregard of sensory change comes from the study on cork taint (Prescott *et al.*, 2005) showing that, in general, the rejection of tainted wine did not switch on at the threshold of the perceived off-flavour, but at a higher concentration. This result is consistent with other studies suggesting that the tolerated changes in products are generally larger than the perceived changes (Conner *et al.*, 1988; Laurila *et al.*, 1996). More strikingly, 10% of respondents did not reject even a heavily tainted wine. This may be due either to individual insensitivity to TCA (Fig. 2.4; Suprenant and Butzke, 1996) or to the respondents' interpretation that the off-flavour was part of the wine flavour. The latter explanation parallels the proposition by Vickers *et al.* (1999) that the off-flavour of milk was regarded as being within the range of normal variation.

More generally, the lenience observed in the cases described above can also be explained by expectation theory (see Chapter 10), and especially by the assimilation model. This model states that when the actual product performance disconfirms the expectation, the evaluation of the deviant product shifts towards expectation. Assimilation is most likely when the difference between expectation and performance is not too large.

2.6   Future trends

During the past few years, research on brain responses to food stimuli has made immense progress, and multimodal nature of perceptions is understood at a new depth. More research can be expected on the unresolved issues of understanding how texture and flavour perceptions and preferences are processed in the brain. Brain research on preferences emphasises the learning aspect in the acquisition of food preferences: this will probably be an important issue in future research, in view of the crucial implications of food preferences for nutrition.

Because of the intense competition in the food market and Western consumers' expectations of quality, sensory quality of marketed foods and beverages is more important than ever. Affective responses of consumers need to be considered seriously, and advanced tools for the analysis of consumer responses and prediction of consumer preferences are needed. Since the mid-1980s, tools such as preference mapping and the LAM scale have been developed and brought into use, and it would be not surprising to see more of specific tools emerging. In particular, improved instruments to predict long-term acceptance of products would be useful.

Individual attitudes, traits and dispositions will probably continue growing in importance. Tools are needed for proper consumer segmentation. The tricky aspect of verbal segmentation instruments is their validity, as they tend to be local and culture specific. The statements used as items of instruments should reflect the thoughts of and discriminate among modern consumers. At the same time, they should be general and relatively timeless to last reliably in research use.

New foods and beverages are developed in a speedy rhythm and they enter the market thick and fast. Most of them are not true novelties, but rather modifications of the familiar products. Whatever is the level of novelty, knowledge of crucial success factors of different types of products among different consumer groups is extremely important. For better prediction of food market, future research should continue identifying fundamental success factors of new products.

2.7   Sources of further information and advice

DOTY R L (2003), *Handbook of Olfaction and Gustation*, New York, Marcel Dekker.

LAWLESS H T and HEYMANN H (1999), *Sensory Evaluation of Food: Principles and Practices*, Gaithersburg, MD, Aspen.

MEISELMAN H L and MACFIE H J H (1996), *Food Choice, Acceptance, and Consumption*, London, Blackie.

ROUBY C, SCHAAL B, DUBOIS D, GERVAIS R and HOLLEY A (2002), *Olfaction, Taste, and Cognition*, Cambridge, Cambridge University Press.

2.8   References and further reading

ALIMOHAMMADI H and SILVER W L (2002), 'Chemesthesis: hot and cold mechanisms', *Chemosense*, **4**(2), 1–9.

AMOORE J E (1970), *Molecular Basis of Odor*, Springfield, Charles C Thomas.

ARVOLA A, LÄHTEENMÄKI L and TUORILA H (1999), 'Predicting the intent to purchase unfamiliar and familar cheeses: the effects of attitudes, expected liking and food neophobia', *Appetite*, **32**, 113–126.

BÄCKSTRÖM A, PIRTTILÄ-BACKMAN A-M and TUORILA H (2004), 'Willingness to try new foods as predicted by social representations and attitude and trait scales', *Appetite*, **43**, 75–83.

BARYLKO-PIKIELNA N, MATUSZEWSKA I, JERUSZKA M, KOZLOWSKA K, BRZOZOWSKA A and ROSZKOWSKI W (2004), 'Discriminability and appropriateness of category scaling versus ranking methods to study sensory preferences in elderly', *Food Qual Pref*, **15**, 167–175.

BELLISLE F (1999), 'Glutamate and the UMAMI taste: sensory, metabolic, nutritional and behavioural considerations. A review of the literature published in the last 10 years', *Neurosci Biobehav Rev*, **23**, 423–438.

BIRCH L L (1999), 'Development of food preferences', *Annu Rev Nutr*, **19**, 41–62.

BOWER J A, SAADAT M A and WHITTEN C (2003), 'Effect of liking, information and consumer characteristics on purchase intention and willingness to pay more for a fat spread with a proven health benefit', *Food Qual Pref*, **14**, 65–74.

BRESLIN P A S and BEAUCHAMP G K (1995), 'Suppression of bitterness by sodium: variation among bitter taste stimuli', *Chem Senses*, **20**, 609–623.

BUCK L and AXEL R (1991), 'A novel multigene family may encode odorant receptors: a molecular basis for odor recognition', *Cell*, **65**, 175–187.

CAIN W S (1982), 'Odor identification by males and females: predictions vs. performance', *Chem Senses*, **7**, 129–142.

CANTIN I and DUBE L (1999), 'Attitudinal moderation of correlations between food liking and consumption', *Appetite*, **32**, 367–381.

CARDELLO A V (1996), 'The role of human senses in food acceptance', in Meiselman H L and MacFie H J H, *Food Choice, Acceptance, and Consumption*, London, Blackie, 1–82.

CARDELLO A V (2003), 'Consumer concerns and expectations about novel food processing technologies: effects on product liking', *Appetite*, **40**, 217–233.

CARDELLO A V and MALLER O (1982), 'Relationships between food preferences and food acceptance ratings', *J Food Sci*, **47**, 1553–1557, 1561.

CARDELLO A V and SCHUTZ H G (2004), 'Numerical scale-point locations for constructing the LAM (labeled affective magnitude) scale', *J Sens Stud*, **19**, 341–346.

CARDELLO A V, MALLER O, MASOR H B, DUBOSE C and EDELMAN B (1985), 'Role of consumer expectancies in the acceptance of novel foods', *J Food Sci*, **50**, 1707–1718.

CARDELLO A V, BELL R and KRAMER M F (1996), 'Attitudes of consumers toward military and other institutional foods', *Food Qual Pref*, **7**, 7–20.

CHRISTENSEN C M (1984), 'Food texture perception', *Adv Food Res*, **29**, 159–199.

CONNER M T, BOOTH D A, CLIFTON V J and GRIFFITHS R P (1988), 'Individualized optimization of the salt content of white bread for acceptability', *J Food Sci*, **53**, 549–554.

COOK D J, HOLLOWOOD T A, LINFORTH R S T and TAYLOR A J (2003), 'Oral shear stress predicts flavour perception in viscous solutions', *Chem Senses*, **28**, 11–23.

CUBERO-CASTILLO E and NOBLE A C (2004), 'Relationship of 6-*n*-propylthiouracil status to bitterness sensitivity', in Prescott J and Tepper B J, *Genetic Variation in Taste Sensitivity*, New York, Marcel Dekker, 105–116.

DAILLANT B and ISSANCHOU S (1991), 'Most preferred level of sugar: rapid measure and consumption test', *J Sens Stud*, **6**, 131–144.

DALTON P, DOOLITTLE N, NAGATA H and BRESLIN P A S (2000), 'The merging of the senses: integration of subthreshold taste and smell', *Nature Neurosci*, **3**, 431–432.

DE ARAUJO I E, ROLLS E T, VELAZCO M I, MARGOT C and CAYEUX I (2005), 'Cognitive modulation of olfactory processing', *Neuron*, **46**, 671–679.

DE CASTRO J M, BELLISLE F, DALIX A-M and PEARCEY S M (2000), 'Palatability and intake relationships in free-living humans: characterization and independence of influence in North Americans', *Physiol Behav*, **70**, 343–350.

DE GRAAF C, VAN STAVEREN W and BUREMA J (1996), 'Psychophysical and psychohedonic functions of four common food flavours in elderly subjects', *Chem Senses*, **21**, 293–302.

DE GRAAF C, KRAMER F M, MEISELMAN H L, LESHER L L, BAKER-FULCO C, HIRSCH E S and WARBER J (2005), 'Food acceptability in field studies with US army men and women: relationship with food intake and food choice after repeated exposures', *Appetite*, **44**, 23–31.

DE WIJK R A and CAIN W S (1994), 'Odor quality: discrimination versus free and cued identification', *Percept Psychophys*, **56**, 12–18.

DIEHL J M (2006), *Fragebögen zur Erfassung ernährungs- und gewichtsbezogener Einstellungen und Verhaltensweisen (Questionnaires for the assessment of eating- and weight-related attitudes and behaviors)*, Giessen, Department of Psychology, University of Giessen.

DISTEL H and HUDSON R (2001), 'Judgement of odor intensity is influenced by subjects' knowledge of the odor source', *Chem Senses*, **26**, 247–251.

DISTEL H, AYABE-KANAMURA S, MARTINEZ-GOMEZ M, SCHICKER I, KOBAYAKAWA T, SAITO S and HUDSON R (1999), 'Perception of everyday odors – correlation between intensity, familiarity and strength of hedonic judgement', *Chem Senses*, **24**, 191–199.

DRAKE B (1989), 'Sensory texture/rheological properties: a polyglot list', *J Texture Stud*, **20**, 1–27.

DREWNOWSKI A (2004), '6-*n*-Propylthiouracil sensitivity, food choices, and food consumption', in Prescott J and Tepper B, *Genetic Variation in Taste Sensitivity*, New York, Marcel Dekker, 179–193.

DREWNOWSKI A and HANN C (1999), 'Food preferences and reported frequencies of food consumption as predictors of current diet in young women', *Am J Clin Nutr*, **70**, 28–36.

DUBOIS D and ROUBY C (2002), 'Names and categories for odours: the veridical label', in Rouby C, Schaal B, Dubois D, Gervais R and Holley A, *Olfaction, Taste, and Cognition*, Cambridge, Cambridge University Press, 47–66.

DUBOSE C N, CARDELLO A V and MALLER O (1980), 'Effects of colorants and flavorants on identification, perceived flavor intensity, and hedonic quality of fruit-flavored beverages and cake', *J Food Sci*, **45**, 1393–1415.

DUFFY V B, LUCCHINA L A and BARTOSHUK L M (2004), 'Genetic variation in taste: potential biomarker for cardiovascular disease risk?' in Prescott J and Tepper B, *Genetic Variation in Taste Sensitivity*, New York, Marcel Dekker, 195–228.

ELLIS B H (1967), 'Preference testing methodology – part 1', *Food Technol*, **22**(5), 49–56.

FIRESTEIN S (2001), 'How the olfactory system makes sense of scents', *Nature*, **413**, 211–218.

FISCHER R, GRIFFIN F, ENGLAND S and GARN S M (1961), 'Taste thresholds and food dislikes', *Nature*, **191**, 1328.

FISHER C and SCOTT T R (1997), *Food Flavours: Biology and Chemistry*, Cambridge, The Royal Society of Chemistry.

GLANVILLE E V and KAPLAN A R (1965), 'Food preference and sensitivity of taste for bitter compounds', *Nature*, **205**, 851–853.

GOTTFRIED J A and DOLAN R J (2003), 'The nose smells what the eye sees: crossmodal visual facilitation of human olfactory perception', *Neuron*, **39**, 375–386.

GREEN B G (2002), 'Psychophysical measurement of oral chemesthesis', in Simon S A and Nicolelis M A L, *Methods in Chemosensory Research*, Boca Raton, FL, CRC Press, 3–19.

GREEN B G, DALTON P, COWART B, SHAFFER G, RANKIN K and HIGGINS J (1996), 'Evaluating the "Labeled Magnitude Scale" for measuring sensations of taste and smell', *Chem Senses*, **21**, 323–334.

GRENBY H, ed. (1996), *Advances in Sweeteners*, London, Blackie.

GUINARD J X (2001), 'Sensory and consumer testing with children', *Trends Food Sci Technol*, **11**, 273–283.

GUINARD J-X, HONG D Y, ZOUMAS-MORSE C, BUDWIG C and RUSSELL G F (1994), 'Chemo-reception and perception of the bitterness of isohumulones', *Physol Behav*, **56**, 1257–1263.

GUINARD J X, SMICIKLAS-WRIGHT H, MARTY C, ABU-SABHA R, SOUCY I, TAYLOR- DAVIS S and WRIGHT C (1996), 'Acceptability of fat-modified foods in a population of older adults: contrast between sensory preference and purchase intent', *Food Qual Pref*, **7**, 21–28

GUINARD J X, SECHEVICH P J, MEAKER K, JONNALAGADDA S S and KRIS-ETHERTON P (1999), 'Sensory responses to fat are not affected by varying dietary energy intake from fat and saturated fat over ranges common in the American diet', *J Am Diet Assoc*, **99**, 690–696.

HARTWIG P and McDANIEL M R (1995), 'Flavor characteristics of lactic, malic, citric, and acetic acids at various pH levels', *J Food Sci*, **60**, 384–388.

HELLEMANN U and TUORILA H (1991), 'Pleasantness ratings and consumption of open sandwiches with varying NaCl and acid contents', *Appetite*, **17**, 229–238.

HUOTILAINEN A and TUORILA H (2005), 'Social representation of new foods has a stable structure based on suspicion and trust', *Food Qual Pref*, **16**, 565–572.

JONES L V, PERYAM D R and THURSTONE L L (1955), 'Development of a scale for measuring soldiers' food preferences', *Food Res*, **20**, 512–520

KÄHKÖNEN P and TUORILA H (1999), 'Consumer responses to reduced and regular fat content in different products: effects of gender, involvement and health concern', *Food Qual Pref*, **10**, 83–91.

KÄHKÖNEN P, TUORILA H and RITA H (1996), 'How information enhances acceptability of a low-fat spread', *Food Qual Pref*, **7**, 87–94.

KIM U-K, JORGENSON E, COON H, LEPPERT M, RISCH N and DRAYNA D (2003), 'Positional cloning of the human quantitative trait locus underlying taste sensitivity to phenylthiocarbamide', *Science*, **299**, 1221–1225.

KOSKINEN, S, KÄLVIÄINEN N and TUORILA H (2003), 'Flavor enchancement as a tool for increasing pleasantness and intake of a snack product among the elderly', *Appetite*, **41**, 87–96.

KOSKINEN S, NENONEN A and TUORILA H (2005), 'Intakes of cold cuts in the elderly are predicted by olfaction and mood, but not by flavor type or intensity of the products', *Physiol Behav*, **85**, 314–323.

LÄHTEENMÄKI L and TUORILA H (1994), 'Attitudes towards sweetness as predictors of liking and use of various sweet foods', *Ecol Food Nutr*, **31**, 161–170.

LANGE C, ROUSSEAU F and ISSANCHOU S (1999), 'Expectation, liking and purchase behavior under economical constraint', *Food Qual Pref*, **10**, 31–39.

LARSSON M (2002), 'Odor memory: a memory systems approach', in Rouby C, Schaal B, Dubois D, Gervais R and Holley A, *Olfaction, Taste, and Cognition*, Cambridge, Cambridge University Press, 231–245.

LAURILA E, LÄHTEENMÄKI L, RITA H and TUORILA H (1996), 'Pleasantness in relation to difference threshold of NaCl in mashed potato', *Food Qual Pref*, **7**, 225–228.

LAWLESS H T and HEYMANN H (1999) *Sensory Evaluation of Food: Principles and*

Practices, Gaithersburg, MD, Aspen.

LINDEMANN B (2001) 'Receptors and transduction in taste', *Nature*, **413**, 219–225.

LIVERMORE A and LAING D G (1998), 'The influence of odor type on the discrimination and identification of odorants in multicomponent odor mixtures', *Physiol Behav*, **65**, 311–320.

LOKEN B (2006), 'Consumer psychology: categorization, inference, affect, and persuasion', *Annu Rev Psychol*, **57**, 453–485.

LUCAS F and BELLISLE F (1987), 'The measurement of food preferences in humans – do taste-and-spit tests predict consumption?', *Physiol Behav*, **39**, 739–743.

LUGAZ O, PILLIAS A-M and FAURION A (2002), 'A new specific ageusia: some humans cannot taste L-glutamate', *Chem Senses*, **27**, 105–115.

MAGA J A (1990), 'Compound structure versus bitter taste', in Rouseff R L, *Bitterness in Foods and Beverages*, New York, Elsevier, 35–48.

MALNIC B, GODFREY P A and BUCK L B (2004), 'The human olfactory receptor gene family', *Proc Natl Acad Sci USA*, **101**, 2584–2589.

MATTES R D (1997a), 'Physiologic responses to sensory stimulation by food: nutritional implications', *J Am Diet Assoc*, **97**, 406–413.

MATTES R D (1997b), 'The taste for salt in humans', *Am J Clin Nutr*, **65**(Suppl), 692S–697S.

MCCLURE S M, LI J, TOMLIN D, CYPERT K S, MONTAGUE L M and MONTAGUE P R (2004), 'Neural correlates of behavioral preference for culturally familiar drinks', *Neuron*, **44**, 379–387.

MONTELEONE E, CONDELLI N, DINNELLA C and BERTUCCIOLI M (2004), 'Prediction of perceived astringency induced by phenolic compounds', *Food Qual Pref*, **15**, 761–769.

MORROT G, BROCHET F and DUBOURDIEU D (2001), 'The color of odors', *Brain and Language*, **79**, 309–320.

MOSCOVICI S (1984), 'The phenomenon of social representations', in R M Farr and S Moscovici, *Social Representations*, Cambridge, Cambridge University Press, 3–69.

PARR W, WHITE G and HEATHERBELL D A (2003), 'The nose knows: influence of colour on perception of wine aroma', *J Wine Res*, **14**, 79–101.

PATTERSON R E and PIETINEN P (2004), 'Assessment of nutritional status in individuals and populations', in Gibney M J, Margetts B M, Kearney J M and Arab L, *Public Health Nutrition*, Oxford, Blackwell, 66–82.

PELCHAT M L and PLINER P (1995), ' "Try it.You'll like it." Effects of information on willingness to try novel foods', *Appetite*, **24**, 153–166.

PLINER P (1982), 'The effects of mere exposure on liking for edible substances', *Appetite: J Intake Res*, **3**, 283–290.

PLINER P and HOBDEN K (1992), 'Development of a scale to measure the trait of food neophobia for edible substances', *Appetite*, **19**, 105–120.

PRESCOTT J and STEVENSON R J (1995), 'Pungency in food perception and preference', *Food Rev Int*, **11**, 665–698.

PRESCOTT J and TEPPER B J, eds (2004), *Genetic Variation in Taste Sensitivity*, New York, Marcel Dekker.

PRESCOTT J, NORRIS L, KUNST M and KIM S (2005), 'Estimating a "consumer rejection threshold" for cork taint in white wine', *Food Qual Pref*, **16**, 345–349.

RATNER R K, KAHN B E and KAHNEMAN D (1999), 'Choosing less-preferred experiences for the sake of variety', *J Cons Res*, **26**, 1–15.

ROININEN K and TUORILA H (1999), 'Health and taste attitudes in the prediction of use frequency and choice between less healthy and more healthy snacks', *Food Qual Pref*, **10**, 357–365.

ROININEN, K, LÄHTEENMÄKI L and TUORILA H (1999), 'Quantification of consumer attitudes to health and hedonic characteristics of foods', *Appetite*, **33**, 71–88.

ROININEN K, FILLION L, KILCAST D and LÄHTEENMÄKI L (2004), 'Exploring difficult textural properties of fruit and vegetables for the elderly in Finland and the United Kingdom', *Food Qual Pref*, **15**, 517–530.

ROLLS E T (2005), 'Taste, olfactory, and food texture processing in the brain, and the control of food intake', *Physiol Behav*, **85**, 45–56.

ROUBY C and BENSAFI M (2002), 'Is there a hedonic dimension to odors?', in Rouby, C, Schaal B, Dubois D, Gervais R and Holley A, *Olfaction, Taste, and Cognition*, Cambridge, Cambridge University Press, 140–159.

ROZIN P (1982), 'Human food selection: the interaction of biology, culture and individual experience', in L M Barker, *The Psychobiology of Human Food Selection*, Chichester, Ellis Horwood, 225–254.

ROZIN P and VOLLMECKE T A (1986), 'Food likes and dislikes', *Annu Rev Nutr*, **6**, 433–456.

ROZIN P, NEMEROFF C, HOROWITZ M, GORDON B and VOET W (1995), 'The borders of the self: contamination sensititivity and potency of the body apertures and other body parts', *J Res Pers*, **29**, 318–340.

SAPER C B, CHOU T and ELMQUIST J K (2002), 'The need to feed: homeostatic and hedonic control of eating', *Neuron*, **36**, 199–211.

SCHIFFMAN S S, MCELROY A E and ERICKSON R P (1980), 'The range of taste quality of sodium salts', *Physiol Behav*, **24**, 217–224.

SCHUTZ H G and CARDELLO A V (2001), 'A labeled affective magnitude (LAM) scale for assessing food liking/disliking', *J Sens Stud*, **16**, 117–159.

SCOTT K (2005), 'Taste recognition: food for thought', *Neuron*, **48**, 455–464.

SETTLE R G, MEEHAN K, WILLIAMS G R, DOTY R L and SISLEY A C (1986), 'Chemosensory properties of sour tastants', *Physiol Behav*, **36**, 619–623.

SHEPHERD R and FARLEIGH C A (1989), 'Sensory assessment of foods and the role of sensory attributes in determining food choice', in Shepherd R, *Handbook of the Psychophysiology of Human Eating*, New York, Wiley, 25–56.

SHEPHERD R, FARLEIGH C A and WHARF S G (1991), 'Effect of quantity consumed on measures of liking for salt concentrations in soup', *J Sens Stud*, **6**, 227–238.

SMALL D M and PRESCOTT J (2005), 'Odor/taste integration and the perception of flavor', *Exp Brain Res*, **166**, 345–357.

SMALL D M, GREGORY M D, MAK Y E, GITELMAN D, MESULAM M M and PARRISH T (2003), 'Dissociation of neural representation of intensity and affective valuation in human gustation', *Neuron*, **39**, 701–711.

SPILLANE W J (1996), 'Molecular structure and sweet taste', in T H Grenby, *Advances in Sweeteners*, London, Blackie, 1–25.

STEIN L J, NAGAI H, NAKAGAWA M and BEAUCHAMP G K (2003), 'Effects of repeated exposure and health-related information on hedonic evaluation and acceptance of a bitter beverage', *Appetite*, **40**, 119–129.

STEPTOE A, POLLARD T M and WARDLE J (1995), 'Development of a measure of the motives underlying the selection of food: the food choice questionnaire', *Appetite*, **25**, 267–284.

STUNKARD A J and MESSICK S (1985), 'The three-factor eating questionnaire to measure dietary restraint, disinhibition, and hunger', *J Psychonom Res*, **29**, 71–83.

SUPRENANT A and BUTZKE C E (1996), 'Implications of odor threshold variations on sensory quality control of cork stoppers', *Proc 4th Int Symp Cool Clim Vitic Enol*, 70–74.

SZCZESNIAK A S (1963), 'Classification of textural characteristics', *J Food Sci*, **28**, 385–389.

TANIMURA S and MATTES R D (1993), 'Relationships between bitter taste sensitivity and consumption of bitter substances', *J Sens Stud*, **8**, 31–41.

TEPPER B J (1993), 'Effects of slight color variation on consumer acceptance of orange juice', *J Sens Stud*, **8**, 145–154.

TUORILA H (1987) 'Selection of milks with varying fat contents and related overall liking, attitudes, norms and intentions', *Appetite*, **8**, 1–14.

TUORILA H and CARDELLO A V (2002), 'Consumer responses to an off-flavour in juice in the presence of specific health claims', *Food Qual Pref*, **13**, 561–569.

TUORILA H, MEISELMAN H, BELL R, CARDELLO A V and JOHNSON W (1994) 'Role of sensory and cognitive information in the enhancement of certainty and liking for novel and familiar foods', *Appetite*, **23**, 231–246.

TUORILA H, ANDERSSON Å, MARTIKAINEN A and SALOVAARA H (1998), 'Effect of product formula, information and consumer characteristics on the acceptance of a new snack food', *Food Qual Pref*, **9**, 313–320.

TUORILA H, KRAMER F M and ENGELL D (2001), 'The choice of fat-free vs. regular-fat fudge: the effects of liking for the alternative and the restraint status', *Appetite*, **37**, 27–32.

TUORILA H, HUOTILAINEN A, LÄHTEENMÄKI L, OLLILA S, TUOMI-NURMI S and URALA N (2007), 'Comparison of affective rating scales and their relationship to variables reflecting food consumption', *Food Qual Pref*, in press.

VAN DER KLAAUW N J and SMITH D V (1995), 'Taste quality profiles for fifteen organic and inorganic salts', *Physiol Behav*, **58**, 295–306.

VICKERS Z, MULLAN L and HOLTON E (1999), 'Impact of differences in taste test ratings on the consumption of milk in both laboratory and food service setting', *J Sens Stud*, **14**, 249–262.

WANSINK B and PARK S B (2001), 'At the movies: how external cues and perceived taste impact consumption volume', *Food Qual Pref*, **12**, 69–74.

WARDLE J, GUTHRIE C A, SANDERSON S and RAPOPORT L (2001), 'Development of the children's eating behaviour questionnaire', *J Child Psychol Psychiatr*, **42**, 963–970.

ZELLNER D A (1991), 'How foods get be liked: some general mechanisms and some special cases', in R C Bolles, *The Hedonics of Taste*, Hillsdale, Laurence Erlbaum, 199–217.

ZELLNER D A, KERN B B and PARKER S (2002), 'Prediction for the good: subcategorization reduces hedonic contrast', *Appetite*, **38**, 175–180.

ZOU Z and BUCK L B (2006), 'Combinatorial effects of olfactory mixes in olfactory cortex', *Science*, **311**, 1477–1481.

3

How does culture affect choice of foods?

P. Rozin, University of Pennsylvania, USA

3.1 Food and culture in historical perspective

Although I cannot cite any quantitative evidence for or against the claim, I believe that culture is the single biggest determinant of food choice. That is, if one wanted to predict what foods an unknown adult person chose to eat, actually ate, or liked to eat, the single most informative feature of that person would be his or her culture. Age and gender do not account for much variance in food choice. More surprisingly, factoring out general cultural influences, the preferences of parents account for little variance in their adult children's food preferences (e.g. Rozin, 1991). Note that the low correlation between parents and their children in food preferences argues against a major role for either genetic differences or family-specific food experiences. Differences in beliefs and values, within culture (including attitude to overweight and to natural foods) surely explain some differences in food choice within culture. But these effects are small, in comparison to the culture-based differences between, say, rural Asian Indians, Mexicans, French and Americans.

Prior to the 20th century, there was very little cultural exchange with respect to foods. The major exception is the exchange of foods between the Western and Eastern Hemisphere consequent upon the 'discovery' of the Americas by Western European explorers (Crosby, 1972). Colonization and exploration in the 17th to 20th centuries produced some exchange of foods. Often, because of restricted amounts and high cost, only the upper classes in Europe experienced these new foods. However, the availability of cheap sugar imported from the Americas (Mintz, 1985) had a major effect on food consumption and choice in Western Europe. This made sugar available to all, and allowed for widespread use of sugar as an additive to foods like coffee that might otherwise be unpalatable.

By the late 20th century, massive improvements in transportation, refrigeration, and travel had led to the availability of many new foods in supermarkets in much of the developed world. I believe that there is a greater variety of foods in my local supermarket, in the United States, than was available to anyone in the world even 50 years ago. Globalization has come to the world of food very rapidly, for the more affluent peoples of the world, and some aspects of globalization, such as cola beverages, have come to virtually everyone. Differences between what foods are sought or consumed in different parts of the world are decreasing. Almost everyone loves things that are sweet and creamy, and through technology humans are finding more and more effective ways to produce such foods, and do it cheaply.

There are a few fundamental issues that must be foregrounded before any serious discussion of food, culture, and the consumer.

The most important point is that almost all research on the psychology of food choice, and probably most of the consumer research in industry, has been carried out in the Western developed world. This means Western Europe, the United States and Canada, Australia, and New Zealand. The populations of all of these countries constitute less than 15% of the world's population. Either India or China alone have more living human beings than all of the Western developed world. The people on whom our knowledge is based are relatively wealthy, eat a rather varied and moderately high-protein diet, and have a lot of exposure to many of the cuisines of the world. They spend a small minority of their total income on food (no more than 12% for Americans in 1990; *Economist*, 1990) compared with around 50% in the less developed countries. Much of their food could be thought to fall in the luxury category, and they have abundant choices. For most of the people in the world, what is eaten is what has always been eaten; tradition rather than choice dominates the food world. However, with increasing wealth and globalization, the world is slowly becoming more like the Western developed world, so that what we learn about that world will be more and more applicable.

The second critical point is that food is much more than nutrition and pleasure. For almost everyone in the world, it has those two aspects or functions. But food is deeply related to social life. In the Western developed world, the evening meal is often the only family get-together of the day, and food is the focus of many celebrations. Dates are often scheduled at restaurants or cafes. Chocolate is often a gift showing affection. Much of the conflict in families with children has to do with eating habits of the children. There is perhaps a universal tendency to avoid and dislike the food of one's enemies. The recent anointment of 'freedom fries' in some parts of the United States as a protest against France, and their French fries, is one of many examples. Throughout the world, developed or developing, food is deeply intertwined with social life.

There is an important moral aspect to food, as well. Some of this may be mediated by the belief that 'You are what you eat.' We have shown that this belief is held, implicitly, even by educated people in the developed world (Nemeroff and Rozin, 1989), and is often explicitly believed in traditional cultures. You are what you eat enhances the potential that food has as a moral vehicle. You can take on the moral attributes of your food, including in some cases, the people who have made your food. The food–moral link is often muted in the Western developed world, but it is clear that, for example, cigarette smoking in the United States has taken on negative moral implications. There is now evidence for Americans that consumption of foods such as meat and chocolate causes at least some individuals to have a lower moral opinion of a person (Stein and Nemeroff, 1995). In some cultures, the food–moral link is powerful and explicit. For some 800 million Hindu Indians, food is a 'bio-moral' substance (Appadurai, 1981). The type of food you eat, including who prepared it, embodies it with important moral properties. By eating food prepared by someone of a lower caste, one's moral purity is compromised. As Marriott (1968) has pointed out, one can reconstruct many of the important social relations in Hindu India by understanding the food rules. Food serves as a homogenizing function by the act of sharing, cementing, and confirming the relationship between individuals within the family. At the same time, refusal to eat the food of others has a heterogenizing function, confirming a separation between non-sharing parties (Appaduria, 1981).

A third critical point is that food, like all other cultural entities and practices, is influenced when cultures mix. The availability of food from other cultures, especially in the developed world, has had enormous effects on food intake. It is hard, while walking the streets of the United States, to realize that pizza is an Italian food. But there is something special about food, perhaps fostered by the particular intimacy of eating, of putting things into one's body. It is widely believed, but not well documented, that there is a 'conservatism of cuisine.' That is to say, food habits are one of the last things to change in the direction of the host country in immigrant populations. This seems to be the case. Some evidence comes from Italian Americans, who three generations after immigration, with their clothing and language totally American, still regularly consume Italian food many times a week, and almost always for the special Sunday meal (Goode et al., 1984).

A fourth point to bear in mind is that although there are very important cultural differences, in many domains including food, there is a great deal of variation within culture. On many psychological variables that have been explored, there is more variation explained within culture than between culture, even though there is much variation between cultures (Rozin, 2003). In multi-ethnic cultures, such as the United States, there is a great deal of between culture variation that is encompassed within the American designation; is this within or between culture variation?

A final point that is often forgotten: social structure within any society accounts for much variation. Generally speaking, in non-Western countries, the more affluent upper classes have much more contact with Western culture, and the financial means to participate in it. If one is interested in cultural variation, it is the lower social classes that probably embody these differences best. And, of course, there are many more mouths in the lower classes.

3.2   The ways culture affects food and food in life

The most obvious food–culture link is in the particular foods that are consumed. Culinary practices have been analyzed by Elisabeth Rozin (Rozin, 1982) into three important features: the basic foods, the basic preparation techniques, and the characteristic flavors placed on the traditional foods. For example, for Chinese cuisine, rice, pork, and a number of other foods are basic, the most common preparation technique is stir-frying, and the traditional flavorings (what Rozin, 1982, calls flavor principles) are a mixture of soy sauce, ginger-root, and rice wine. It is the flavoring, more than any other feature, that imparts the cultural distinctiveness to a food. Potatoes flavored with the Chinese flavor principle will taste Chinese, even though potatoes are not a major part of most Chinese cuisine.

There are other important culture-specific aspects to particular cuisines, such as the manner of eating (e.g. fork vs chopsticks), the ordering of foods in the meal, the etiquette of eating, the social rules about eating (for example, children eating at table with adults, women and men eating at same table), and the particular foods appropriate for particular individuals or particular times. The idea of a restaurant, where food is created for consumption by people who do not 'know' the consumers is relatively new. Eating away from home is becoming a more and more important part of eating in the developed world, with the time pressures of dual-income families. Eating 'out' is not a very common occurrence for most people in the world, though it will no doubt become more common everywhere.

One of the most powerful influences of culture on food is much more indirect, and has to do with the environments and technologies created by cultures, which provide opportunities for different types of food-ways. For example, wealthier and more technologically advanced cultures can ship foods from further away and keep them fresh for longer. Conveyances such as vending machines can bring foods into many environments in which they would not ordinarily be.

3.3   The big sense of culture and its relation to food and eating

When a person travels to experience another culture, what he or she usually does is to see the cultural environment – the buildings, streets, scenery, and, yes, the food. There is usually little opportunity to explore the 'minds' of the people in the other culture. And yet, what is of greatest interest to cultural psychologists is these minds. Much of what we know about culture differences has to do with cognitions and motivations. A number of psychologists have attempted to taxonomize major cultural differences across the globe. Two somewhat systematic attempts have been made by Hofstede (1982) and Schwartz (1992). Hofstede, on the basis of questionnaires completed in many different cultures, identifies four major dimensions of psychological difference by culture. These are uncertainty avoidance, power distance, individualism vs collectivism and masculinity vs femininity. Schwartz identifies 11 dimensions, which can be reduced, to some degree, to two, which he calls openness to change vs conservatism, and self-enhancement vs self-transcendence.

Much of the more recent research in psychology has focused on the contrast between individualistic and collectivist cultures (e.g. Markus and Kitayama, 1991; Triandis, 1995). In collectivist cultures, there is a stronger family influence, and a greater emphasis on conformity and social responsibilities. The individual is seen as interdependent with others, rather than dependent. Nisbett (2003) has described a major difference in pattern of thinking between Americans and East Asians. Perhaps the most important aspect of this difference, related to individualism vs collectivism, is an analytic attention to detail in the Western cultures, and a more holistic, integrative, context-sensitive outlook in the Eastern cultures. Those interested in consumers might find these differences in perspectives of interest.

In terms of food purchases, the organization of collective cultures implies that purchase decisions are more likely to be made on behalf of the family. The American tendency to cater to each family member's particular tastes is probably less prominent in more traditional and more collective cultures. In those cases, generally one eats what one has always eaten, and relishes it. One hundred flavors of ice cream are not so appealing in the traditional setting.

One major and important distinction between cultures has to do with attention and honor bestowed on people as a function of age. In the United States and many other Western-developed cultures, it is the middle-aged people who have the most power and respect. Children are adored but must know their place, and old people hopefully go away quietly, often in a home for the aged. In Japan and many other countries, by contrast, the small child rules; no punishment is meted out to him or her. And old people are respected and revered. The later years of life are not as reliably accompanied by decline in social importance, as in the United States. This pattern, in more traditional cultures, is reinforced by the fact that older people usually live in the same compound as their children and grandchildren, and play a significant role in the rearing of their grandchildren. Older people probably have more influence on food selection in more traditional cultures.

3.4   Culture and acquisition of food preferences

Put simply, we do not know how food preferences are created. Within culture, there is much variation in preference for particular dishes, and for particular basic foods. We do not know why. It does not seem to be due primarily to either genetic differences or family influences, since family resemblance in food preferences is low (Rozin, 1991). Common-sense, informal observation of the world suggest a number of mechanisms for preference creation, for each of which there is an important cultural influence.

The principal effect of culture on food choice is probably mediated by exposure. People generally choose foods they have experienced, and there is a generally positive relationship between exposure and liking (e.g. Zajonc, 1968). This holds for foods (e.g. Pliner, 1982). Culture is the principal determinant of what foods one is exposed to. In traditional settings, one is exposed only to locally produced foods, and those are a subset of all the potentially available foods in the local environment (e.g. excluding some plants, insects, and most vertebrates in most cultures). Exposure depends on availability and cost. When something is present, and affordable, there is the opportunity to develop a taste for it.

A second known mechanism for creation of preferences has to do with Pavlovian conditioning. There is no doubt that under certain conditions (and we do not know what those limits are), when a new flavor or food is contingently followed or accompanied by an already liked (or disliked) flavor, or by certain types of gastrointestinal and postabsorptive events (positive or negative), the liking for the new food can be changed. This is called evaluative conditioning (deHouwer *et al.*, 2001). This is probably how people get to like dark unsweetened coffee. They initially experience it with a distinct sweet taste, and the sweet taste pairing probably causes an increased liking for the coffee flavor, and perhaps even the bitterness. Cultures play an important role, because cultures program what foods will be eaten together. Thus, the jelly on a peanut butter and jelly sandwich may serve to increase the liking for both the peanut butter and bread.

Mere exposure and evaluative conditioning probably do not account for most food preferences. These can be attributed, rather vaguely, to social influence. We do know that evaluative conditioning can occur when the critical event (the unconditioned stimulus, in Pavlovian conditioning) is a positive or negative facial expression of a person who is consuming the food in question (Baeyens *et al.*, 1996). But much cultural influence probably manifests itself outside the conditioning paradigm. Desires to be adult, to be like admired people, and influences of this sort, operate in sometimes subtle ways to create likes or dislikes. And they also often fail to operate, as parents can testify. Birch *et al.* (1996) have ably reviewed much of this literature. One of the striking findings in this field is that explicit rewards for eating a food may increase the intake of the food, while the reinforcement is in force, but they do not create liking for the food.

3.5   Some examples of specific issues in product marketing and development in a cultural context

3.5.1   Chocolate

Edible chocolate is a human creation, made from an unlikely source; a very bitter bean that does not have a particularly appealing texture. After much innovation and technological advance over hundreds of years of Western history, chocolate has evolved into one of the most appealing foods on earth. It has a concentration of calories that ranks it very high among all foods; two highly appealing innate properties, sweetness and fattiness; an extremely attractive aroma, and the specially appealing property that it melts at body temperature, producing an oral sensation that is exquisite for most who sample it. From its Mexican origins, it arrived in Europe and became one of the most favored foods in the Western-developed world, and surely the most craved food in that part of the world.

Chocolate raises two fascinating cultural questions. There is no ready answer for either. First is the question of the geographical distribution of chocolate use. Although it grows only in tropical regions, it is widely consumed and adored almost entirely in temperate climes, particularly Western-developed countries. Why is chocolate not prized in South America, Africa, and Asia, the places where it grows? We do not know, but it is surely a temptation for chocolate manufacturers to spread it from the 15% of the world's population that loves it to the remaining 85%. And this could have the side result of increasing the pleasure of consumption for billions of people. Here are some possible reasons for the focus of chocolate consumption in Western developed countries, all pure speculation:

- It does not keep well at tropical temperatures, particularly since it melts at body temperature.
- It is just too expensive.
- The quality of the chocolate available for consumption in most tropical countries is poor.
- It does not fit into the relevant cuisines or cultural conceptions of what a food is. Chocolate may be seen in some cultures as a children's food. Also, chocolate reaches its pinnacle in desserts that are often not a highly developed part of traditional tropical cuisines, so it may not easily achieve its niche in the food cuisines of many cultures.

Given the success of cola beverages and many other Western products in the developing world, it is at least possible that the main problem may be simply exposure to good quality chocolate.

The second question has to do with the rather narrow distribution of this super food in terms of its culinary uses within the cultures that adore it. Chocolate is almost invariably consumed as part of a confection or sweet snack, or as a beverage (e.g. hot chocolate). It is rarely used on savory foods, and has not been a success as a flavoring for yogurt or soda, though it has scored immensely well as an ice cream flavor. Why, even in the parts of the world that love its flavor, texture, and aroma, is it used in such a narrow context? Why don't we put it on vegetables? (The same high popularity but narrow culinary distribution holds for the two other extremely popular xanthine food/beverages, tea and coffee.) Perhaps if we could answer this question, we would know more about how to make chocolate more acceptable in the developing world.

3.5.2 Dairy products

Dairy products constitute a major part of the food intake of people in the Western-developed world. They are a much more modest part of the diet of most of the rest of the world, and are virtually absent in the food of China, the most populous country in the world. It is true that the great majority of the people who are not in the Western-developed world are lactose intolerant, and so would experience gastrointestinal distress on consumption of moderate amounts of milk or other non-fermented dairy products. But this is not a sufficient account of the situation. Fermented milk products have low lactose levels and are usually well tolerated by lactose-intolerant people. The substantial intake of yogurt in India is one clear example of this. Surely the Chinese, among the most inventive people on earth in the domain of cuisine, could have figured out that fermented milk was easily digestible. They had extensive exposure to it by virtue of the long Mongol occupation of China, since the Mongols were major dairy consumers and consumed fermented milk products. A reasonably complete account of dairy rejection in China needs an important sociocultural component; lactose intolerance may be a contributing factor, but it is not the only one. One possibility is that hostility to the invading Mongols included hostility to their foods. This may have been enhanced by the fact that milk, as a body product, is potentially disgusting. And, on account of their odor, fermented products have a strong tendency to be disgusting. Yes, the Chinese consume and love some fermented products, such as 'thousand year old' eggs, and soy sauce. But even though most cultures love some fermented products, these same cultures are usually repelled by others. Many Americans and Europeans, for example, are fond of fermented grapes (wine) and milk (cheese and yogurt), but are deeply offended by decaying meat. For some in the world, some forms of decaying meat are much preferred to decaying dairy products. Therefore, part of the acceptance of dairy products has to do with traditions about what is food, what is not food, and what is disgusting.

3.5.3 Variety

One of the consequences of modern food technology and the free market is an enormous variety of potential foods. Much of this variation can be described as 'micro-variety': large families of very similar products, such as the 100 flavors of ice cream offered in some stores, or the wide variety of coffee, yogurt, or olive oil flavours and types. These must have some appeal to some consumers, because they flood the modern Western supermarket. It is very possible, based on recent research in psychology (Inyegar and Lepper, 2000; Schwartz, 2004), that under many situations, even Americans find a high micro-variety of choices somewhat aversive, and discouraging of consumption. In accord with this finding, the majority of specific products in a typical American supermarket sells fewer than one case a week (Kahn and McAllister, 1997). Most critically, it is clear that interest in micro-variety, and interest in new things, including food, is a cultural variable. Traditional cuisines highlight particular foods, and particular versions of those foods. Just as Americans are offended by chocolate string beans, which seems to violate their culinary sensibility, micro-varieties, as an alternative to a culturally proto-typical food, may violate the culinary sensibilities of people in some cultures. Americans are, perhaps, at one extreme here, though ironically, the recent data on the limits of the appeal of variety and choice come from Americans. My local supermarket has 150 different types of yogurt and 100 different types of antacids.

Recent comparative studies of Europeans and Americans suggests a notably higher preference for micro-variety in Americans (Rozin *et al.*, 2006). One extreme of this is the American diner, with literally hundreds of food choices. If one orders steak, one has the choice of any of a set of vegetables as one side order, and then French fries, home fried, baked or mashed potatoes. And then, the individual can season his or her food at the table with ketchup, mustard, hot sauce, salt, and pepper, at a minimum. The idea here seems to be that each individual has a unique preference function, and food should be individually tailored to it. The idea of individualized food may have some connection to the Protestant tradition that focuses on individualism. The more traditional idea is that there are best combinations, such as steak and French fries, in France, and that when one goes to at least a good restaurant, one wants the chef to offer just a few choices, and arrange them to his or her liking.

We know very little about attitudes to variety for most human beings, those in traditional, developing cultures. There is the classic opposition between the appeal of familiarity and potential boredom from lack of variety. Openness to new experiences has been identified as a major culture-difference variable (e.g. Schwartz, 1992), and it may have particular manifestations in the domain of food. In more collective cultures, a consensus about what is the best form of X may have much more influence than it does in more individualistic cultures.

3.5.4   Natural and genetically modified foods

At least in the Western-developed world, 'natural' is an attribute that makes a food more appealing (and often more expensive!). Interestingly, naturalness for medicines is much less appealing (Rozin *et al.*, 2004). There is a fascinating conflict between love for the natural, and desire to be protected from it by technology, as when facing death or natural disasters. There have been few studies in this area, but there seems to be little basis for the common belief that natural foods are both healthier and better tasting (Ames *et al.*, 1987; Schutz and Lorenz, 1976). Even for educated Westerners, when the natural preference is analysed, it seems to be less reasoned or rational, and more a commitment to the superiority of nature, per se (Rozin *et al.*, 2004). We do not know if this is true for the less developed world, but from what we know about the hyper-rationality of Westerners, it seems likely that if they revert rather quickly to an 'ideational' justification for natural preference, this would be all the more common in people from more traditional cultures. What this means, of course, is that evidence for risks of consuming natural foods, or advantages of processed foods in some instances, will probably fall on deaf ears.

Although the preference for natural is very high in Americans, it is striking that opposition to genetically modified foods is notably higher in Europeans than in Americans (Gaskell *et al.*, 1999). We currently have no account of this difference, but it is clear, from experiences with mad cow disease as well as genetic engineering, that complex political factors, general attitudes to technology, and trust in institutions all play a role in attitudes to genetic engineering (Frewer and Salter, 2003; Siegrist, 1999).

In recent work, it has been noted that lay American conceptions of natural are at some variance with 'expert' conceptions. Thus, lay Americans rate a cocker spaniel as more natural than a genetically modified plant or animal with a single gene insertion (Rozin, 2005a). It appears that judgments of naturalness have more to do with process than content. For example, Americans rate natural water with some chemicals removed from it as more natural than that same water, after the removed chemicals have been replaced. The first and third exemplars are chemically identical, but because they have been directly altered by people twice, they are less natural than the water with just the removal, which has been altered only once (Rozin, 2006).

As genetically modified foods become more common, it will be very important to understand the basis for their rejection, especially as they become less expensive and, perhaps, more flavorful and shelf stable. It is likely that increasing acceptability of such foods may involve different strategies in different cultures.

3.6 Looking directly at a cultural comparison: the food world of French and Americans

Claude Fischler and I, over the past few years, have been engaged in studies comparing French and American attitudes to food and food-related behaviors. Our motivation for this was the observation that Americans had become very ambivalent to food, with the pleasures of eating tempered by worries about obesity, attractiveness, and health. This concern seemed to be counterproductive, as evidenced by the facts that it is very difficult to maintain weight loss and that Americans are getting heavier. Furthermore, even though the French eat a diet higher in fat, eat fewer foods modified to reduce fat and salt, and worry less about the food–health link, they are at least as healthy as Americans. Our findings suggest a range of differences between the cultures (e.g. Rozin *et al.*, 1999, 2003; reviewed in Rozin, 2005b).

- The French tend to think of food as something to eat and experience, while Americans tend to think of food as chemicals entering the body and affecting it.
- The French prize food as a great pleasure, in comparison to some other pleasures, such as having a large and very comfortable house or hotel room.
- While the French modify their diet to 'improve' health less than Americans, they actually think their diet is healthier than Americans think their American diet is.

- The French eat slightly less than Americans, and the outcome of this is that they are less overweight.
- One reason the French eat less is that their portion sizes are smaller.
- A second reason is that they are less inclined to snack between meals.
- It is probably true, though good evidence is still lacking, that the French get more exercise, primarily through walking and bicycling as part of their normal life, than do Americans.

For the French, as opposed to Americans, the theme of eating brings forth ideas of moderation, as opposed to the abundance theme for Americans (think of stuffing oneself at an American Thanksgiving dinner). The French opt more for quality than quantity in food, more for taste than for shelf-life. They savor food experiences more, and are less bothered by the physiological consequences of eating. And they may reject the American tendency to think of food and drugs as on a continuum (thus neutraceuticals), and rather think of them as very separate entities, sold in different stores. The result of all this is a special reverence for food in the French, and an enjoyment of it, in moderation, less conflicted by health concerns. What this means in terms of products and marketing, is presumably that health appeals might be more effective in the United States than in France.

Of course, nothing is that simple. The French are major consumers of beef, and yet there was a substantial drop in beef intake after the 'mad cow' disease scare, and they are more concerned about genetic engineering than Americans. Even these two very similar cultures have important subtle differences that we are just beginning to understand (see Stearns, 1997, for an extended discussion of the food and body worlds of French and Americans).

3.7   Understanding cultural dimensions in food choice for food product development

There are important food universals, such as the preference for sweet tastes and fatty textures. There are also important ecological facts that have a major influence on food consumption and choice; some products grow in only some areas of the Earth. There are some universal realities, such as the fact that one cannot eat something unless one has access to it, and can afford it. Availability is itself largely a cultural variable, and so is preference. We do not know how cultures establish food preferences in the process of development, and how, on a larger scale, particular foods or dishes or attitudes to food and eating become part of a culture. It is striking and puzzling that hot chili peppers spread throughout the tropical and semi-tropical Eastern Hemisphere from their origins in the Western Hemisphere, but that tomatoes, potatoes, and corn, also from the Western Hemisphere, and with more initially acceptable tastes and great nutritional value, were accepted as human food in fewer areas. It is probably true that the upper classes around the world, with their strong orientation to Western ways, are more susceptible to change and more accepting of new Western technological advances in food processing. It is probably also true that some cultures are more receptive to novelty than others. One reason for this may be the importance of their own food traditions. France is dedicated to its cuisine. It is not even really clear what American cuisine is, and its breadth and multi-ethnicity might well promote a greater acceptance of new foods.

3.8   Future trends

There is no reason to think human proclivities to like fat and sweet, and to try to obtain food with as little effort as possible (part of our biological heritage) will change. There is every reason to believe that the food industry will find ways to make foods more appealing in flavor and texture, and less expensive. An insight into what might be possible can be gained, with great pleasure, by a meal at El Bulli, the incredibly innovative and marvellous restaurant in Catalonia. In the long run, easy food that tastes great will take over, but it is largely cultural variables that may determine how quickly, and in what ways this happens. Globalization is here and growing, and superior products, defined in terms of certain dimensions such as textural distinctiveness and variety, and certain flavors, will prevail for most people. Globalization will guarantee that more people will have the opportunity to try more foods, as will increasing wealth. However, for the near term, it is likely that there will be resistance, more in some cultures than others. And there will probably be, in many cases, a tenacious continuance of traditional food ways. Perhaps the work by Goode *et al.* (1984) is generally illustrative of what will happen: Italian Americans accept many modern and many American foods, but Sunday dinner is still Italian, after three generations. Modernity and tradition can live side by side, and they probably always will.

3.9   Sources of further information and advice

There are a few recent compendia that treat food in a cultural context in some detail. Two are the *Cambridge World History of Food* (Kiple and Ornelas, eds, 2000) and the *Encyclopedia of Food* (Katz, 2004). Excellent books that treat food in a cultural context include Leon Kass's (1994) *The Hungry Soul*, Michael Pollan's (2006) *The Omnivore's Dilemma*, Jared Diamond's *Guns, Germs and Steel*, and Claude Fischler's (1990) *L'Homnivore*. Books that provide a useful background about human food choice and some of the cultural influence on it include Barker (1982), Booth (1994), Meiselman and MacFie (1996), and Shepherd and Raats (2006), and a useful and broad, culturally sensitive model of food choice is presented by Sobal *et al.* (2006). The growing subfield of cultural psychology has much to contribute to our understanding of how people interact with their worlds, and food in particular, in a cultural context. The new *Handbook of Cultural Psychology* (Rozin, 2007) provides a broad perspective on this, with a particular chapter on the cultural psychology of food (Kitayama and Cohen, 2007). And then there is the magnificent array of ethnic cookbooks available in any substantial book store.

3.10   References

AMES, B., MAGAW, R. and GOLD, L. (1987). Ranking possible carcinogenic hazards. *Science*, **236**, 272–80.

APPADURAI, A. (1981). Gastro-politics in Hindu South Asia. *American Ethnologist*, **8**, 494–511.

BAEYENS, F., KAES, B., EELEN, P. and SILVERANS, P. (1996). Observational evaluative conditioning of an embedded stimulus element. *European Journal of Social Psychology*, **26**, 15–28.

BARKER, L. M. (Ed.) (1982). *The Psychobiology of Human Food Selection*. Bridgeport, CT: AVI.

BIRCH, L. L., FISHER, J. O. and GRIMM-THOMAS, K. (1996). The development of children's eating habits. In H. L. Meiselman & H. J. H. MacFie (eds.) *Food Choice, Acceptance and Consumption* (pp. 161–206). London: Blackie Academic and Professional.

BOOTH, D. A. (1994). *Psychology of Nutrition*. London: Taylor & Francis.

CROSBY, A. W. Jr. (1972). *The Columbian Exchange. Biological and Cultural Consequences of 1492*. Westport, CT: Greenwood Press.

DEHOUWER, J., THOMAS, S. and BAEYENS, F. (2001). Associative learning of likes and dislikes. A review of 25 years of research on human evaluative conditioning. *Psychological Bulletin*, **127**, 853–869

DIAMOND, J. (1997). *Guns, Germs, and Steel. The Fates of Human Societies*. New York: WW Norton.

Economist (1990). *The Economist Book of Vital World Statistics*. New York: Random House.

FISCHLER, C. (1990). *L'Homnivore*. Paris: Odile Jacob

FREWER, L.J. and SALTER, B. (2003). The changing governance of biotechnology: The politics of public trust in the agri-food sector. *Applied Biotechnology, Food Science and Policy*, **1** (4), 199–211.

GASKELL, G., BAUER, M. W., DURANT, J. and ALLUM, N. C. (1999). Worlds apart? The reception of genetically modified foods in Europe and the US. *Science*, **285**, 384–387.

GOODE, J. G., CURTIS, K. and THEOPHANO, J. (1984). Meal formats, meal cycles, and menu negotiation in the maintenance of an Italian-American community. In M. Dougles (ed.), *Food in the Social Order. Studies of Food and Festivities in Three American Communities* (pp. 143–218). New York: Russell Sage.

HOFSTEDE, G. (1982). Dimensions of national cultures. In R. Rath, H. S. Asthana, D. Sinha & J. B. P. Sinha (eds.), *Diversity and Unity in Cross-cultural Psychology* (pp. 173–187). Lisse, Netherlands: Swets and Zeitlinger B.V.

IYENGAR, S. and LEPPER, M. (2000). When choice is demotivating: Can one desire too much of a good thing. *Journal of Personality & Social Psychology*, **79**, 995–1006.

KAHN, B. E. and McALISTER, L. (1997). *Grocery Revolution. The new focus on the consumer*. Reading, MA. Addison-Wesley

KASS, L. (1994). *The Hungry Soul*. New York: The Free Press.

KATZ, S. (ed.) (2004). *Encyclopedia of Food*. New York: Scribner.

KIPLE, K. F. and ORNELAS, K. C. (eds.) *Cambridge World History of Food*. Cambridge, UK: Cambridge University Press.

KITAYAMA, S. and COHEN, D. (2007). *Handbook of Cultural Psychology*. New York: Guilford Press.

MARKUS, H. R. and KITAYAMA, S. (1991). Culture and the self: Implications for cognition, emotion, and motivation. *Psychological Review*, **98**, 224–253.

MARRIOTT, M. (1968). Caste ranking and food transactions: a matrix analysis. In: M. Singer and B. S. Cohn (eds.), *Structure and Change in Indian Society* (pp. 133–171). Chicago: Aldine.

MEISELMAN, H. L. and MACFIE, H. L. H. (Eds.) (1996). *Food Choice, Acceptance and Consumption*. London: Blackie.

MINTZ, S. (1985). *Sweetness and Power. The Place of Sugar in Modern History*. New York: Viking.

NEMEROFF, C. and ROZIN, P. (1989). 'You are what you eat:' Applying the demand-free 'impressions' technique to an unacknowledged belief. *Ethos. The Journal of Psychological Anthropology*, **17**, 50–69.

NISBETT, R. E. (2003). *The Geography of Thought. How Asians and Westerners Think Differently ... and Why*. New York: The Free Press.

PLINER, P. (1982). The effects of mere exposure on liking for edible substances. *Appetite*, **3**, 283–290.

POLLAN, M. (2006). *The Omnivore's Dilemma. A Natural History of Four Meals*. New York: Penguin Press

ROZIN, E. (1982). The structure of cuisine. In L. M. Barker (Ed.), *The Psychobiology of Human Food Selection* (pp. 189–203). Westport, CT: AVI.

ROZIN, P. (1991). Family resemblance in food and other domains: The family paradox and the role of parental congruence. *Appetite*, **16**, 93–102.

ROZIN, P. (2003). Five potential principles for understanding cultural differences in relation to individual differences. *Journal of Research in Personality*, **37**, 273–283.

ROZIN, P. (2005a). The meaning of 'natural': Process more important than content. *Psychological Science*, **16**, 652–658.

ROZIN, P. (2005b). The meaning of food in our lives: a cross-cultural perspective on eating and well-being. *Journal of Nutrition Education and Behavior*, **37**, S107–S112.

ROZIN, P. (2006). Naturalness judgments by lay Americans: process dominates content in judgments of food or water acceptability and naturalness. *Judgment and Decision Making*, 1 (2), 91–97.

ROZIN, P. (2007). Food and eating. In: S. Kitayama and D. Cohen (eds.), *Handbook of Cultural Psychology*. New York: Guilford Press.

ROZIN, P., FISCHLER, C., IMADA, S., SARUBIN, A. and WRZESNIEWSKI, A. (1999). Attitudes to food and the role of food in life: comparisons of Flemish Belgium, France, Japan and the United States. *Appetite*, **33**, 163–180.

ROZIN, P., KABNICK, K., PETE, E., FISCHLER, C. and SHIELDS, C. (2003). The ecology of eating: part of the French paradox results from lower food intake in French than Americans, because of smaller portion sizes. *Psychological Science*, **14**, 450–454.

ROZIN, P., SPRANCA, M., KRIEGER, Z., NEUHAUS, R., SURILLO, D., SWERDLIN, A. and WOOD, K. (2004). Natural preference: instrumental and ideational/moral motivations, and the contrast between foods and medicines. *Appetite*, **43**, 147–154.

ROZIN, P., FISCHLER, C., SHIELDS, C. and MASSON, E. (2006). Attitudes towards large numbers of choices in the food domain: a cross-cultural study of five countries in Europe and the USA. *Appetite*, **46**, 304–308.

SCHUTZ, H. G. and LORENZ, O. A. (1976). Consumer preferences for vegetables grown under 'commercial' and 'organic' conditions. *Journal of Food Science*, **41**, 70–73.

SCHWARTZ, B. (2004). *The Paradox of Choice. Why More is Less*. New York: HarperCollins.

SCHWARTZ, S. H. (1992). Universals in the conent and structure of values. Theoretical advances and empirical tests in 20 countries. *Advances in Experimental Social Psychology*, **25**, 1–65.

SHEPHERD, R. and RAATS, M. (eds.) (2006). *The Psychology of Food Choice*. Wallingford, UK: CABI Press.

SIEGRIST, M. (1999). A causal model explaining the perception and acceptance of gene technology. *Journal of Applied Social Psychology*, **29** (10), 1093–2106.

SOBAL, J., BISOGNI, C. A., DEVINE, C. M. and JASTRAN, M. (2006). A conceptual model of the food choice process over the life course. In: R. Shepherd and M. Raats (eds.) *The Psychology of Food Choice*. Wallingford, UK: CABI Press.

STEARNS, P. N. (1997). *Fat History. Bodies and Beauty in the Modern West*. New York: New York University Press.

STEIN, R. I. and NEMEROFF, C. J. (1995). Moral overtones of food: Judgments of others based on what they eat. *Personality & Social Psychology Bulletin*, **21**, 480–490.

TRIANDIS, H. C. (1995). *Individualism and Collectivism*. Boulder, CO: Westview Press.

ZAJONC, R. B. (1968). Attitudinal effects of mere exposure. *Journal of Personality and Social Psychology*, **9**, 1–27.

4

Psychobiological mechanisms in food choice

M. R. Yeomans, University of Sussex, UK

4.1 The importance of understanding psychobiological mechanisms in food choice

Humans evolved an appetite control system that was designed to protect the body from nutrient shortages and to allow us to exploit food supplies which were scarce. The modern-day consumer no longer faces the everyday pressure of searching out the rare resource of nutritional food which occupied humans during our evolutionary history. However, the modern consumer retains a highly complex appetite control system which is predisposed to allow us to identify and consume safe and nutritious foods. The environment consumers now live in is very different from that which shaped our appetite control systems, and this mismatch has been suggested as a contributory factor to the worldwide increase in incidence of obesity and disordered eating (Zimmet and Thomas, 2003). A key factor in modern food development must be a recognition that our ability to control intake can be compromised by factors such as disguised energy content. This chapter reviews our current understanding of food choice and preference from a psychobiological perspective, highlighting the relationship between food selection and preference and the appetite control system.

4.2 Need-states and hedonic rewards in eating

Two key concepts need to be understood. The first is the concept of need-state. This phrase originated in physiological models of controls of eating and drinking, and reflected a general approach to the physiology of motivation which was based on analogies between the way the body may regulate its internal environment (homeostasis) and the way physical control systems operated (e.g. Toates, 1986). This system's approach to motivation is still influential today, as evidenced by the continued use of phrases such as deprivation-induced eating. The basic notion of need-state is that in order for our bodies to function optimally, various biochemical parameters need to be maintained within a narrow physiological range. Examples include temperature, fluid levels and, in relation to food choice, the level of nutrients such as glucose. The basic control theory model of appetite suggests that the body detects perturbations from the ideal level of key physiological variables, and that our appetite is then driven by the need to restore homeostasis. These concepts are critical to food choice, since they make clear predictions about the relationship between the physical state of the body and the likely choices consumers will make.

A second critical concept is that of hedonically driven eating: eating for the pleasure of the sensory experience. Could stimulation of appetite by sensory cues underlie the current obesity crisis by causing over-eating? Or are our hedonic responses to food in part a reflection of our need-states? For example, does food taste more pleasant when we are hungry? A full analysis of these questions is beyond the scope of this chapter, and has been attempted elsewhere (Yeomans et al., 2004). However, the current state of our understanding of these issues is summarised in the following discussions, and the need-state versus pleasure debate underlies each of the topics explored in this chapter.

The chapter starts with the question of why we like foods, before exploring more explicit psychobiological influences on food choice. To understand these issues fully some knowledge of the mechanisms underlying appetite control is needed, and our current understanding of these complex mechanisms is summarised. The chapter ends by discussing the relevance of these ideas to food production, and how future findings may create new challenges for the food industry.

4.3   Psychobiological influences on acquisition and expression of food preferences

The only unequivocal innate flavour preference is for sweet taste, alongside an innate aversion to bitterness (Steiner et al., 2001). However, we have the potential to develop limitless flavour preferences, in the process reversing some of our instinctive reactions such as our aversion to bitterness. Although the complex learning processes that underlie the development of flavour preference are far from fully understood, experimental studies of flavour preference development in humans and other animals has helped identify many of the key processes. The two most important theories are highlighted here, while a more detailed analysis is provided by several reviews (Brunstrom, 2004; Yeomans, 2006; Zellner, 1991).

4.3.1   Flavour–consequence learning (FCL)

If we eat a novel food and subsequently become ill, we develop a strong aversion to the flavour of the food even if we know this was not the cause of the illness. Originally described in rats as conditioned taste aversion (CTA: Garcia and Koelling, 1966), this form of learning is now known to be based on similar principles to the conditioning process originally described by Pavlov in his classic work on salivation in dogs. Most consumers are likely to have developed some flavour aversions in this way. However, the principles underlying development of CTA have been used to offer an explanation for how we acquire food preferences as well as aversions.

Essentially, one way in which we conceive flavour preferences and aversions is through learned associations between the flavour of the food we are consuming and the effects the food then has on our body once ingested. CTA represents learning where the post-ingestive effect is aversive. However, if the effects are positive (for example, increased energy, reduced hunger or a specific pharmacological benefit such as the effects of alcohol or caffeine), then a flavour preference should develop. This broadening of CTA to a general psychobiological model of preference development through flavour–consequence learning (FCL) remains the most widely cited theory of how our appetite system is coupled with learning to generate food preferences.

The ideas behind FCL are heavily influenced by broader concepts in associative learning, based on the original learning principles set out in Pavlov's Nobel Prize winning work on appetitive responses to cues predicting the arrival of food to dogs. Pavlov identified a general principle of learning where responses to a biologically important stimulus (e.g. food, drink: defined as the unconditioned stimulus or UCS) could transfer to a second, neutral stimulus (the conditioned stimulus or CS). Accordingly, in FCL the primary association is thought to be between the perceived sensory characteristics of the ingested food or drink (acting as CS) and the post-ingestive effects of the food or drink (UCS). One way of conceptualising FCL is shown in Fig. 4.1a. Thus, as with other forms of learning, it is predicted that in most situations, changes in preferences generated by FCL will proceed progressively, with repeated experiences of flavour and consequences strengthening the change in preference. But what features of foods are most likely to support FCL?

4.3.2   FCL based on nutrients

It is clear that our appetite control system predisposes us to develop a preference for nutritional food while learning to avoid harmful substances and ignore those that have no benefit, such as grass. Most important are flavours that signal sources of energy, since the primary function of eating is to maintain the supply of energy to our bodies. There is a clear relationship between the energy density of foods and self-reported flavour preference (Drewnowski, 1998). Similar data are seen when considering preferences for fruit and vegetables by children, with children preferring those fruit that have the highest energy density (Gibson and

(a)   (b)

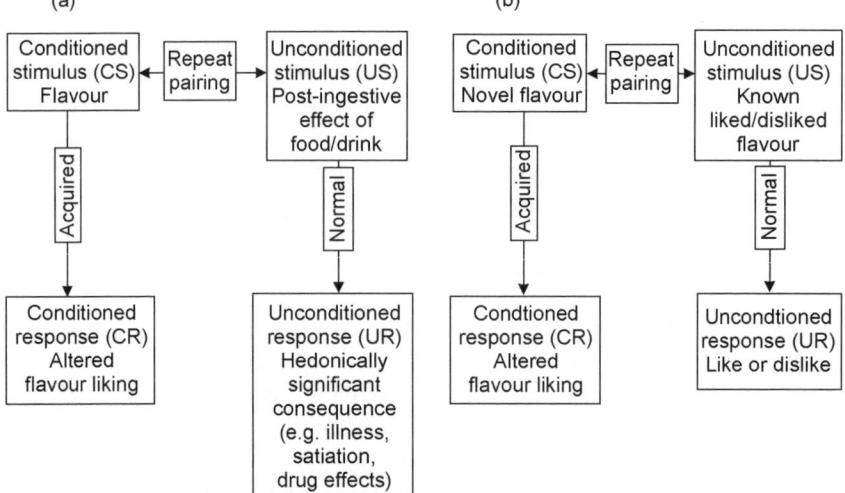

Fig. 4.1   A Pavlovian model of the nature of the associations in (a) flavour–consequence learning and (b) flavour–flavour learning.

Wardle, 2003). These observations are consistent with the predictions of FCL: past associations between flavours and the ability of foods to reduce hunger (which will depend on the amount of energy we consume) shape our preferences.

The strongest evidence for the importance of nutrients in generating FCL comes from studies in animals, where neutral flavours are selectively paired with the post-ingestive delivery of energy (Capaldi, 1992; Sclafani, 1999). A large number of energy-containing nutrients have been shown to induce FCL in rats: sucrose (Capaldi *et al.*, 1994; Fedorchak and Bolles, 1987; Harris *et al.*, 2000; Sclafani, 2002), glucose (Myers and Sclafani, 2001a,b), starch (Elizalde and Sclafani, 1988; Ramirez, 1994; Sclafani and Nissenbaum, 1988), fats (Lucas and Sclafani, 1989) and alcohol (Ackroff and Sclafani, 2001, 2002, 2003). The strength of the acquired preference and consistency of findings both point to FCL as a key mechanism in the development of flavour preferences by animals, consistent with the idea that FCL evolved as a solution to the 'omnivore's paradox': how to select nutritious and safe food from the huge variety of potential foods encountered in nature (Rozin and Vollmecke, 1986).

FCL has been discussed as an important component of the development of liking for foods with high energy density in humans (Stubbs and Whybrow, 2004). The most convincing laboratory-based studies have been with children, with the reported increase in preference for high-energy (carbohydrate) containing drinks relative to low-energy drinks (Birch *et al.*, 1990) and for high-fat relative to low-fat yoghurt (Johnson *et al.*, 1991; Kern *et al.*, 1993). In adults, consumption of flavours associated with high or low levels of protein at breakfast resulted in a preference for the high-protein version (Gibson *et al.*, 1995), while preference for the flavour of a soup with added starch increased relative to a soup with no starch (Booth *et al.*, 1982). Thus nutrient-based FCL appears to be a component of flavour preference development in humans.

4.3.3   FCL based on pharmacological components of foods and drinks

Some of the most consistent acquired flavour preferences are for drinks that contain substances with psychoactive consequences, such as alcohol and caffeine. Journey to another part of the world and the cuisine may be unfamiliar, but coffee or tea and familiar soft drinks typically dominate the drinks being consumed. Indeed, in these instances liking seems to be contrary to our natural instinct to dislike bitter tastes. FCL provides an obvious framework through which to explain this acquired liking: the specific flavour of the drink becomes a reliable and contingent predictor of the positive post-ingestive effect of the drink. The clearest examples in the laboratory use caffeine as the consequence, and have repeatedly shown acquired preferences for the flavours of drinks paired with caffeine (Richardson *et al.*, 1996; Rogers *et al.*, 1995; Tinley *et al.*, 2003; Yeomans *et al.*, 1998). More recently, the same approach has demonstrated caffeine-based FCL under more naturalistic conditions (Mobini *et al.*, 2005). Here, consumers evaluated their liking for two novel-flavoured iced tea drinks, and then they consumed these at home either at breakfast, at night or whenever they wished. Unbeknown to the participants, one drink contained caffeine and one did not. As can be seen (Fig. 4.2), regular caffeine consumers who drank

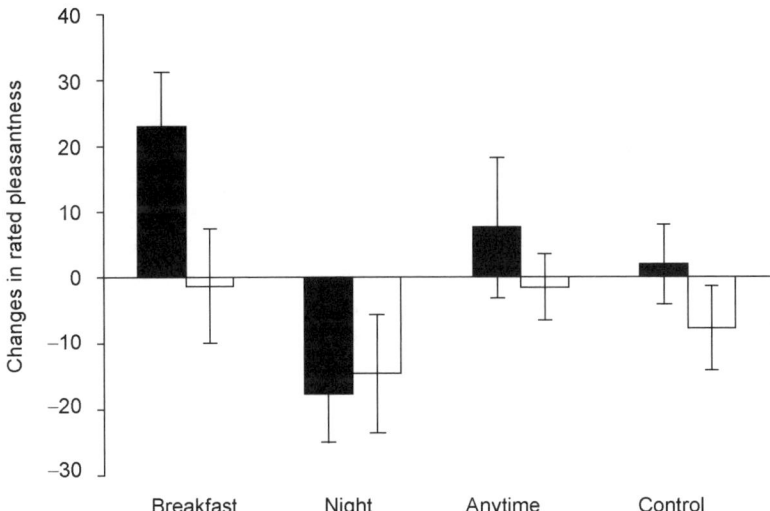

Fig. 4.2   Changes in rated pleasantness of drink flavours associated with either caffeine (■) or the absence of caffeine (□) after consuming these drinks at home for four weeks either as a breakfast drink, a night-time drink, consumed anytime or rated but not consumed (control). Adapted from Mobini *et al.* (2005) with permission.
(From *Food Quality and Preference*, 16, Mobini, Elliman and Yeomans, Conditioned liking for flavours paired with caffeine: a naturalistic study, 659–666, Copyright (2005), with permission from Elsevier.)

these drinks at breakfast developed a clear preference for the caffeine-paired flavour relative to the caffeine-free flavour.

4.3.4   Flavour–flavour models of evaluative conditioning (FFL)

It is common practice to add something we already like to a new food in order to increase overall acceptability. Thus we may start drinking coffee by adding milk and sugar, or try versions of alcohol where the bitter taste is disguised by a sweet mixer. Or a mother may encourage a toddler to try vegetables by smothering them in cheese sauce. These examples could all be explained by a second form of conditioned flavour association called evaluative conditioning (EC), which involves a change in evaluation of one stimulus by association with a second stimulus that is already liked or disliked (De Houwer et al., 2001; Field and Davey, 1999). Flavour-based EC, also known as flavour–flavour learning (FFL), involves pairing a neutral flavour CS with a liked or disliked flavour UCS. As with FCL, changes in liking are usually interpreted within an associative learning framework based on the principles of Pavlovian conditioning (Fig. 4.1b). For example, sweetness is innately liked, and the addition of sweetness to a wide variety of foods and drinks increases their immediate acceptability. FFL predicts that the simultaneous experience of a new flavour and sweetness results in increased liking for the sweet-associated flavour on its own. This is consistent with the development of liking for flavours as diverse as coffee, tea, beer and wine, yoghurt, etc., all of which are initially consumed in a sweetened form. This idea is supported by laboratory-based studies showing increased liking for sweet-paired food-odours and flavours (Yeomans et al., 2006; Zellner et al., 1983). Conversely, concurrent experience of a new flavour with a disliked flavour such as tween (Baeyens et al., 1990, 1995, 1996) or for a food-related odour with a bitter taste (Yeomans et al., 2006) reliably causes a flavour aversion to develop.

Overall, development of a dislike for flavour components consistently paired with an aversive flavour UCS is robust, and alongside CTA explain how human flavour aversions develop. Flavour preferences through FFL with sweet taste UCS may also be influential, particularly in consumers with strong sweet preferences. As with FCL, liking change with FFL requires relatively few pairings of CS and UCS flavours. FFL may therefore be important in human flavour preference development, although as with FCL more research is needed to determine the full scope and importance of these associations.

So what is the relevance of all these learning studies for developers of new food products? These studies have important considerations when considering the likely acceptability of new products. For example:

- That the reaction of consumers to products will change over time as they acquire associations with consequences through FCL. Thus product development that concentrates solely on first reactions may be a poor predictors of the likely success of a product.

- That initial responses to flavours can be modified by knowledge of how existing preferences can be modified through FFL. For example, if a product is developed that has a less than optimal flavour but that has the potential to become liked over time because of its post-ingestive effects, initial flavour acceptability could be enhanced by the addition of sweetness, which will act both to increase initial liking and promote liking for other elements of the overall product flavour through FFL.
- That product development must consider closely the expected effects consumers are looking for. A product to be used at mealtimes must be satiating, and this has to be true even of reduced-energy foods. In contrast, a snack aimed at boosting short-term energy needs to be formulated appropriately. In both cases, learning technology could be usefully applied to product development to ensure that the predicted effects deliver what the consumer wants in a way which enhances product acceptability over time.

4.4   Motivational influences on food preferences

As we have developed clear and well-supported models of flavour-preference development in the form of FFL and FCL, so it has also become clear that the nutritional needs of the consumer greatly affect both the ability to learn new flavour preferences and how and when these acquired preferences are subsequently expressed.

4.4.1   Motivational states and expression of flavour–consequence learning

The idea that the current motivational needs of a consumer influence their hedonic evaluation of a product makes intuitive sense. For instance, an acquired liking for a flavour that predicts that a food has a high energy content would be appropriate when the consumer was hungry, but responding to that same acquired liking when sated could lead to over-eating. Thus, it may be that the body has evolved a mechanism for not only acquiring food preferences, but also determining whether expression of these acquired preferences is an appropriate response. There is empirical support for this idea. In the study of FCL in children with drinks containing energy in the form of carbohydrate (Birch *et al.*, 1990), the acquired preference was much less when the children evaluated the drink when sated than when hungry. If this worked perfectly, and all flavour preferences were based on FCL, then it appears the appetite system underlying these types of preference has evolved in such a way as to guard against over-consumption.

The idea that preferences acquired through FCL are sensitive to acute need state are strongly supported by studies with caffeine as UCS (Yeomans *et al.*, 2000a,b). In this instance, state-dependency was found in two ways. Firstly, consumers had to be both caffeine-dependent (Rogers *et al.*, 1995; Tinley *et al.*, 2003, 2004) and acutely deprived of caffeine (Yeomans *et al.*, 1998) in order to develop liking for a novel flavour predictive of caffeine. Thus, the consumer had to be in an appropriate motivational state (both chronically dependent and acutely deprived of caffeine) in order for caffeine to be an effective reinforcer of FCL. Secondly, once caffeine-consumers had acquired a liking for a novel caffeine-paired drink, they expressed that acquired liking only when acutely in need of caffeine (Yeomans *et al.*, 2000a,b). Thus the body seems not only sensitive to the state needed to support the acquisition (learning) of FCL but also acutely sensitive to the relevance of these acquired preferences to the current need, in this instance for the effects of caffeine. Our recent study of acquired liking for caffeine-paired flavours in home consumers adds support to this conclusion. When consumers were allowed to choose when to consume these drinks themselves, those who consistently consumed the caffeinated drink after a period without caffeine developed a preference for the drink, whereas those who consumed their drinks within a couple of hours of consuming tea or coffee did not (Mobini *et al.*, 2005).

4.4.2   Motivational state and expression of flavour–flavour learning

While there is clear evidence that expression of flavour preferences developed through FCL are sensitive to the current needs of the consumer, whether this is so with FFL is much less clear. Studies in rats suggest that preferences generated by FFL are unaffected by hunger, whereas preferences based on FCL do. For example, rats were trained with simultaneous pairings of either an odour and sucrose or an odour with saccharin when hungry (Fedorchak and Bolles, 1987) and subsequently tested the expression of these preferences when food-deprived or sated. Hunger modified expression of the sucrose-, but not saccharin-, based preferences, implying that the flavour-energy component of the flavour–sucrose association was sensitive to current energetic needs. Likewise, preferences for odours associated with sucrose, but not saccharin, were greater if rats were trained hungry rather than sated (Capaldi *et al.*, 1994). However, a recent study in humans suggests that FFL is dependent on hunger-state at testing (Yeomans and Mobini, 2006). Here, liking for food-related odours acquired by pairing with a sweet taste, but not a bitter taste or water, was expressed when hungry, but not when sated (Fig. 4.3). This too implies that liking for flavours developed through FFL in humans should be regulated by needs.

   If both FFL and FCL-based preferences are sensitive to hunger, how then do we explain the over-eating that is partly responsible for the current increases in obesity? The answer probably lies in the accuracy of our appetite control system. As discussed earlier, it makes no sense for an animal to eat just sufficient of a nutritious food if the same food will have disappeared the next time the animal comes to find it. So a system which allows some degree of over-eating makes evolutionary sense.

So again, what do these findings mean for product development? The full implications are discussed later, but in brief the key implications are the following:

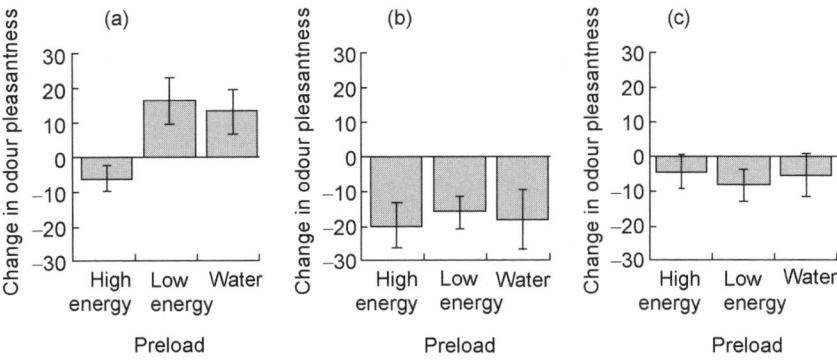

Fig. 4.3   The effects of high- and low-energy soup or water preloads on expression of acquired hedonic orthonasal evaluations of odours following repeated retronasal experience of the same odours paired with either 10% sucrose, 0.01% quinine hydrochloride or water: (a) sucrose-paired odours; (b) quinine-paired odour; (c) control odour. Adapted from Yeomans and Mobini (2006). (From *Food Quality and Preference*, 16, Mobini, Elliman and Yeomans, Conditioned liking for flavours paired with caffeine: a naturalistic study, 659–666, Copyright (2005), with permission from Elsevier.)

- A development process that tests potential new products without taking into account the relevant motivational state of the assessor is likely to fail. For example, if assessors are hungry at the point of testing, but consumers use the product in a low-hunger situation, this mismatch may lead to products failing. Conversely, developing a product with a panel tested at different times of day may be uninformative about acceptability for a product consumers will use in a consistent way: potential snack products need to be assessed at suitable snack times, and meal products at mealtimes.
- The development process has to think about likely changes in product acceptability with repeated consumption, consider what components of the product may facilitate these changes, and consider when consumers are likely to use the product. Thus liking may increase with repeated consumption for a product where the flavour incorporates novel aspects alongside a known liked quality (e.g. a sweet taste). But a product developed to enhance satiety, where liking change may be through FCL, needs to ensure the product is consumed primarily when hungry. This adds a role for how products are marketed to the development process.

4.5   Motivational influences on food choice

The previous section concentrated on how our motivational state altered our liking for foods, with clear evidence that flavours are liked more when hungry than when sated. But how does motivation relate to our selection of foods given a number of alternative options (food choice)? Psychobiological studies in this area are much more limited in scope because of the inherent complexity in what is being asked.

As we have seen, liking for the flavour of foods may increase by association with other food qualities (either in terms of flavour or nutrients). It is also known that liking is a very important determinant of food choice: people rarely consume food they do not like (De Castro, 2000), instead choosing foods for which they have acquired a liking. Thus the primary psychobiological influence on food choice is sensory hedonics, eating for pleasure. We also know a great deal about the biological basis of the pleasure experience from eating (Berridge, 2003). The main neural system underlying hedonic experience in general involves activation of opioid peptides (Kelley *et al.*, 2002), our own brain opiates. When opioid receptors are blocked, food tastes less pleasant (Yeomans and Gray, 2002), and both humans (Yeomans *et al.*, 1990) and animals (Cooper *et al.*, 1985) consequently are less likely to choose foods normally considered palatable over less palatable options. These data together show that hedonic experience is critical to food choice.

How then does manipulation of motivational state alter food choice? The answer here is less clear. One important study, however, showed clearly that hunger-state was an important factor in food choice independent of sensory hedonics (Tuorila *et al.*, 2001). In that study, women rated their liking for fat-free or regular energy hot fudge to be consumed on ice-cream which also contained either no fat or regular fat. They were then informed of their evaluations so that their choice was made with the explicit knowledge of which flavour they had rated as the more pleasant. They were then given a choice test between the two samples. Importantly, when tested, less hungry participants were more likely to choose the fat-free versions even though many of these participants had rated this as the less pleasant taste. The implication is that people may use information about energy content to guide food choice, contrary to their hedonic preference, but that hunger state can reverse this effect.

A further important influence on food choice relates to personal goals of the consumer. Thus people who have adopted a restrained style of eating either in order to reduce the risk of weight gain or in an attempt to lose weight may actively seek out foods that are less preferred over higher-energy, more preferred options. Although well intentioned, it is also clear that factors such as attempts to diet, especially where they have failed repeatedly (yo-yo dieting), may be counter-productive by disrupting more appropriate food choice behaviours (Mela, 2001). Indeed, the very act of restricting access to favourite foods may have the counter-productive effect of making these overvalued. For example, when women were required to carry a bag of chocolates with them for a day without eating them, they subsequently consumed twice as much when given free access to the same food relative to controls who had not had to deal with the temptation of free availability of a preferred food (Stirling and Yeomans, 2004). This effect was exacerbated in women with a history of restrained eating (Fig. 4.4), clearly showing that repeated attempts to avoid certain foods may lead to a reduced ability to resist these foods.

In summary, this section highlights the importance of anticipated pleasure in food choice, and how this may be modified by hunger state at the time when

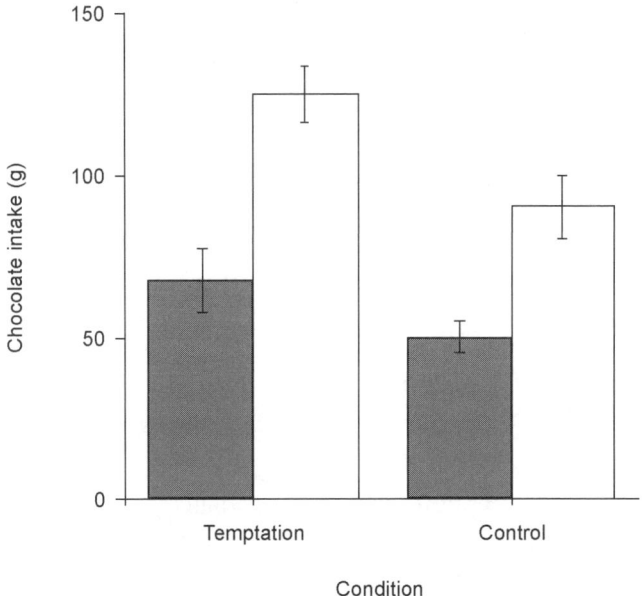

Fig. 4.4   The effects of self-denial on intake: chocolate consumed by high- (□) and low-restraint (■) women who had been required to carry but not consume chocolates for 24 h beforehand (temptation) and a control group with free access to chocolate. Adapted from Stirling and Yeomans (2004).

food is chosen. Longer-term self-restriction appears to heighten sensitivity to highly palatable foods, suggesting that, although we are able to direct our food choices based on what we consider healthy, over-rigid self-restriction is counter-productive. However, these conclusions have to be seen as tentative since the literature on motivational effects on food choice remains sketchy relative to the equivalent literature on acquired food preferences and on our understanding of the control of food intake. In order that future products can be developed that maximise our understanding of psychobiological influences on consumer behaviour, further research is urgently needed to help clarify these issues.

4.6   Motivational influences on food intake

Historically, the study of motivational influences on control of food intake has been the main focus of the psychobiological investigation of appetite, and a full discussion of factors involved in food intake control is beyond the scope of this section. Instead, the section attempts to summarise the key components of control of food intake, highlighting in particular the rapid advances made in recent years. The section also focuses on those aspects of control of food intake that may be most relevant to food product development, particularly with a view of considering what features of foods may contribute most strongly to prevention of over-consumption.

An important first question is whether meal-size is driven by hunger (i.e. energy and nutrient deficits generated by metabolism and nutrient use since the last meal) or satiety (the more we eat, the longer it is until we eat again). A classic way of discriminating these two options has been to explore the relationship between the size of meals and the period of time before and after each meal (De Castro, 1996). Based on extensive diary records, the size of voluntary meals shows a stronger hunger-related (preprandial) correlation than satiety-based (postprandial) relationship (De Castro and Elmore, 1988). This contrasts with rats that ate more frequently and showed a satiety-driven relationship, before switching to a hunger-driven relationship when the opportunity to eat was restricted to fewer meals per day (De Castro, 1988). Interestingly, when people were placed in an environment with no external cues (daylight, time, etc.) to help dictate mealtimes, their eating switched to a satiety-driven pattern (Bernstein et al., 1981). The overall implication is that our meal pattern is controlled mainly by habits: we associate certain times of day with eating, and these time cues act as critical stimuli for meal initiation.

4.6.1   Control of meal-size

How then is meal-size controlled? Intuitively, we might construct a simple model where the use of energy and nutrients by the body leads progressively to a state of hunger, while consumption reverses these deficits (satiety). However, meal duration is far too short for the body to assimilate the nutrients and use the signals generated by the nutrients to reverse hunger during each meal. Thus the processes that promote the initiation of eating (hunger) and those that determine when meals terminate have to be different, and here the main factors involved in both sets of processes are briefly summarised.

4.6.2   Psychobiological factors in meal initiation

The physiological basis of meal initiation has been long researched, but is still poorly understood. In recent years two different physiological cues have received particular attention. The first relates to a classic theory of hunger, the glucostatic model (Mayer, 1953), which was based on the general idea that feelings of hunger were generated by changes in the availability of blood glucose. Given that glucose is the primary energy source for the brain, ensuring a constant supply of blood glucose is clearly important. Although the simplistic idea that overall blood glucose levels equate with hunger does not fit with our current understanding of control of blood glucose levels (Campfield and Smith, 1990), an important observation was that spontaneous meal-taking by rats was preceded by a transient decline and then increase in blood glucose, with feeding starting during the ascending phase. Similar findings have since been reported in humans, with the spontaneous request for food by time-blinded participants again preceded by transient changes in glucose (Melanson et al., 1999). Thus transient changes in blood glucose levels may lead to an experience of hunger and so meal initiation. What remains unclear is whether the glucose changes are the key signal, or represent metabolic adjustments in anticipation of eating. We know that the rapid influx of nutrients generated by eating is a major challenge to homeostatic control (Woods, 1991), and so determining whether glucose changes are really a cause or consequence of the decision to eat remains uncertain.

For many years scientists have hypothesised that the body may have a specific hormone, the actions of which underlie our experience of hunger. However, until 1999 all appetite-related hormones were known to reduce appetite (i.e. were satiety signals). The discovery of a new hormone, called ghrelin, which stimulates food intake in animals (Nakazato *et al.*, 2001) and humans (Cummings *et al.*, 2001, 2004; Wren *et al.*, 2001), have led to the suggestion that it may act as a hunger hormone, a notion supported by the clear relationship between overall body-size and ghrelin levels, with abnormally high levels in patients with anorexia (Tanaka *et al.*, 2003), and low levels in obese patients (Kanumakala *et al.*, 2005; Shiiya *et al.*, 2002; Tschop *et al.*, 2001). The latter finding is also strong evidence that ghrelin is not a prerequisite for eating, since obese people clearly over-consume relative to their energy needs despite having low levels of ghrelin. Thus ghrelin appears to be a hormonal signal which works alongside leptin to signal the status of long-term energy stores (Hellstrom *et al.*, 2004).

4.6.3   Control of meal-size: palatability–satiety interactions

How then is meal-size controlled? It is now recognised that meal-size reflects the interaction of two different sorts of feedback systems (Smith, 2000). The first set of signals (satiation) relate to the processes underlying the development of satiety, and involve a complex sequence of cues incorporating learned, orosensory, gastric and post-gastric cues (the satiety cascade: Blundell and Tremblay, 1995). These processes act through negative feedback to reduce desire to eat and so lead to meal termination. In contrast, the second factor is seen as a feed-forward control, and relates to the ability of sensory characteristics of foods to stimulate short-term appetite (often referred to as palatability effects). Clear evidence for the importance of the latter processes is that the experience of hunger can increase in the early stages of a meal if the food is perceived as pleasant tasting (palatable) relative to bland or unpleasant tasting (Yeomans, 1996; Fig. 4.5). However, as eating progresses so hunger decreases and fullness increases until the meal ends. Once the meal has been ingested, post-absorptive satiety cues inhibit further feeding.

While a full discussion of the processes involved in satiation and satiety is beyond the scope of this chapter, a brief summary is warranted in order to evaluate the importance of the nature of the food consumed in determining the rate at which satiety develops. A discussion of the neural controls of appetite is not given here; the reader is referred to one of several recent reviews for summaries of the complex brain systems that underlie hunger and satiety (Berthoud, 2003; Schwartz *et al.*, 2000).

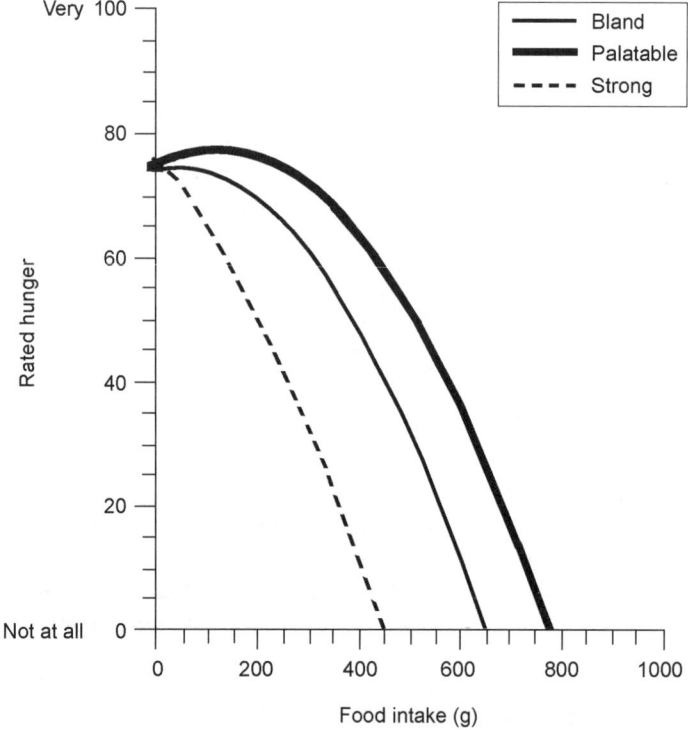

Fig. 4.5   Changes in rated hunger for normal weight men eating a palatable, bland or overly strong flavoured test meal. Redrawn from Yeomans (1996) with permission.

4.6.4   The role of volume and energy in determining meal-size

In the context of the current chapter, the key question is the extent to which the nutrient content of the ingested food controls intake. To what extent is meal-size a function of nutrient intake? The answer appears to be that actual nutrient content has little impact on intake until the consumer has been able to learn about the nutritional consequences of ingestion after the meal. The strongest evidence that this is so comes from studies looking at the short-term effects of nutrient supplementation or dilution. For example, on first exposure volunteers ate the same amount of porridge when this was served in low-energy or high-energy dense versions (Fig. 4.6), but ate more of the low-energy version once they had had a chance to learn the relationship between the porridge flavour and the post-ingestive consequence (Yeomans *et al.*, 2005). Similar conclusions have come from studies of energy density in relation to portion-size (Bell and Rolls, 2001; Kral and Rolls, 2004), where manipulated portion sizes have consistent effects on voluntary short-term intake (the more that was served, the more people consume), while manipulation of energy density had minimal effects. In contrast to the lack of evidence for effects of energy density or nutrients, there is strong evidence that people regulate the volume which they

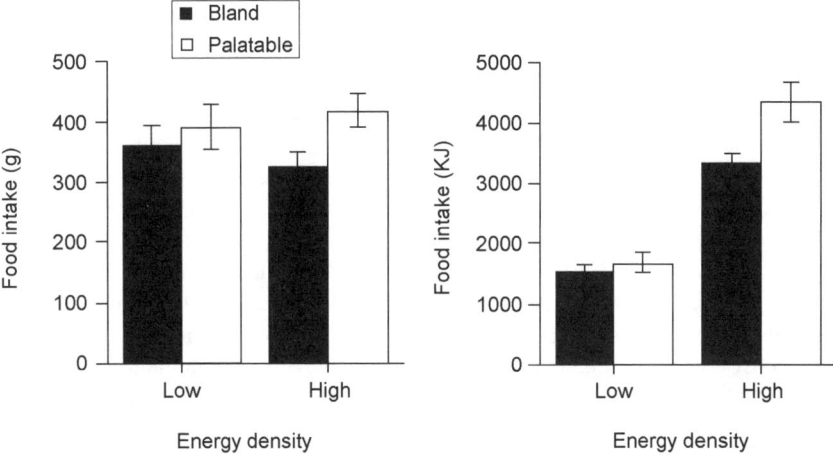

Fig. 4.6   Effects of high-energy (360 kJ) and low-energy (60 kJ) soup preloads on mass (left hand panel) and energy (right hand panel) of a lunch presented in bland or palatable forms. Data modified from Yeomans *et al.* (2001b), and reprinted with permission.

ingest (Rolls and Roe, 2002; Rolls *et al.*, 1998, 2000). Thus in the short term, how much food is served combined with orosensory and gastric metering of the volume ingested seem to be the strongest determinants of meal-size.

4.6.5   Gastric cues

What role does the stomach play in satiety? Is the progressive increase in feelings of fullness as a meal progresses a reflection of specific sensors measuring gastric filling? Current research identifies gastric stretch receptors as an element in satiety (Read *et al.*, 1994). However, although cues reflecting gastric distension play a role in controlling food intake, the effects of gastric stretch are a short-lived cue. It is also possible for people to discriminate the feelings of fullness generated by gastric stretch alone from the feelings of fullness generated by eating (French and Cecil, 2001). Thus the stomach may play an important role in control of meal-size only where people eat meals which approach gastric capacity.

4.6.6   Post-gastric cues: the satiety hormones

Food entering the small intestine is known to stimulate a number of hormones which are implicated in satiety. The best known of these is cholecystokinin (CCK), which appears to fit all the characteristics needed to be defined as a satiety hormone: (a) it is released in response to food cues, (b) CCK injection inhibits eating and can reduce feelings of hunger (Kissileff *et al.*, 1981; Muurahainen *et al.*, 1991) and (c) levels of circulating CCK correlate with feelings of fullness (French *et al.*, 1993). The CCK signalling system is now well characterised, and is clearly an important physiological component of satiety (Strader and Woods, 2005).

Two other hormones, polypeptide-Y 3-36 (PYY36) and glucagon-like peptide 1 (GLP-1) also appear to have satiety-like properties. In humans, most, but not all (Long *et al.*, 1999), studies report decreased food intake after administration of GLP-1, including studies in normal weight, diabetic and obese participants (Flint *et al.*, 1998, 2000; Gutzwiller *et al.*, 1999a,b; Näslund *et al.*, 1998, 1999b; Toft-Nielsen *et al.*, 1999). Reduced food intake is consistent with ratings of reduced hunger, and increased fullness following GLP-1 infusion (Gutzwiller *et al.*, 1999b; Näslund *et al.*, 1999a). PYY36 is also known to reduce food intake in animals, is released postprandially from the gastro-intestinal tract in proportion to the energy content of the ingested meal and induces satiety in humans (Batterham *et al.*, 2002). Infusions of PYY36 have been shown to reduce food intake in obese and lean subjects (Batterham *et al.*, 2003). Thus food arriving in the gut generates a sequence of signals which act to suppress appetite.

4.6.7   Palatability and the control of meal-size
As discussed earlier, hedonic motivation (eating for pleasure) is an important influence on short-term ingestion. In laboratory studies there is a linear relationship between the rated pleasantness of a food and subsequent voluntary food intake (Yeomans, 2007). Likewise, in diary-based studies of behaviour under naturalistic conditions, the overall palatability of a meal predicts voluntary meal-size independently of hunger state (De Castro *et al.*, 2000a,b). Indeed, in laboratory-based studies, enhancing the palatability of a test meal counteracts the ability of energy consumed beforehand to suppress intake (Yeomans *et al.*, 2001b), suggesting that palatability overrides satiety and can therefore lead to over-consumption. Palatability effects are also clear in animal studies (Davis, 1989), and as discussed briefly when evaluating influences on food choice, a great deal is now known about the neural basis of hedonic stimulation of appetite, with a clear role for opioid peptides in both animal (Bodnar, 2004) and human studies (Yeomans and Gray, 2002). This has led to increased interest in the hedonic component of eating as a possible contribution to the increased incidence of obesity, as well as parallels between the pleasures associated with eating and drug-based reward (Grigson, 2002; Kelley, 2004; Pelchat, 2002; Wang *et al.*, 2004).

4.6.8   Learning, satiation and satiety
This brief discussion of motivational controls of food intake ends by returning to considering further the role of learning in control of food intake. Despite the clear evidence for clear physiological influences on meal-size, these controls should be seen as influences rather than absolute controls. Learning to eat in response to cues in our environment can have dramatic impacts on food intake, and must override many of the physiological satiety signals discussed earlier. A

classic example came from a study in rats which were restricted to a single meal a day, signalled by a noise in their cage (Weingarten, 1983). The rats were then allowed free access to food. Despite their lack of food deprivation, hearing the same noise they had experienced as a meal cue when food deprived, reliably induced eating, with rats consuming as much as 20% of their daily intake in the period immediately after the noise. It is hard to construct similar tests in people, but another classic study did demonstrate that believing it was now their usual dinner-time was a sufficient cue for obese volunteers to eat more in a laboratory test (Schachter and Gross, 1968). A great deal of our daily eating may by similarly stimulated by such cues, and recent evidence in animals that the pattern of release of the appetite hormone ghrelin across the day itself depends on how many meals are habitually consumed (Sugino et al., 2002) suggests that habitual meal patterns may entrain the physiological controls of meal initiation.

4.7   Understanding psychobiological mechanisms in food choice for food product development

How might the large body of information of psychobiological influences on food preference, choice and appetite control help in the development of new food products? In truth, there are a myriad potential uses and influences of our enhanced understanding of the psychobiology of appetite that can be of value. This discussion focuses on three areas that may be of particular value.

4.7.1   Sensory impact and appetite

A simple response to the observation that enhanced palatability results in increased intake might be to suggest that future food products need to be less palatable. The widespread belief (with surprisingly little evidence) that the palatability of the foods available to consumers is higher today than in the past has been cited as an explanation for the increase in obesity. It is certainly well established that voluntary intake increases as a function of palatability (Yeomans, 2007). However, it is self-apparent that new food products will not be successful if consumers do not like the flavour of these products!

How then might foods be designed to utilise our knowledge of the factors involved in meal-size control to try to militate against over-eating as a consequence of palatability? The answer will lie in the ability to build into products elements that maximise the ability of the food to satiate so that the appetite-stimulating effects of flavour are counter-acted by enhanced satiety so leading to no net increase in intake. There are many ways this could be achieved. For example, macronutrients differ in the extent to which they lead to satiety, with fat recognised as the least satiating (Blundell and Burley, 1992) and protein the most satiating (Blundell et al., 1996; De Graaf et al., 1992; Vandewater and Vickers, 1996). Thus altering the macronutrient balance in products should alter the level of post-ingestive satiety.

4.7.2   The importance of learning

One obvious conclusion from the present review is that the relationship between a consumer and a food product is not a fixed one. The most important changes will occur over the first few times a product is consumed, as the consumer associates the flavour with other features of the product. It is important to remember that a great deal of this learning is not something consumers are aware of. For example, consumers develop a preference for the flavour of caffeinated drinks over drinks without caffeine without any awareness that caffeine was present (Rogers *et al.*, 1995). However, the observation that caffeinated drinks remain the most popular beverages worldwide is proof of the power of flavour–caffeine associations. Thus developing an appreciation that consumers should learn to associate flavours and after-effects will be an important element of successful product development, particularly since this learning will alter consumers' appreciation of food flavour.

Just as learned associations will alter consumers' liking for flavours, they may also promote learning of sensible consumption patterns by learning not to over-consume foods that make them feel more full (Booth, 1991). The important consideration here will be ensuring that products have a flavour that promotes learning: the product must stand out as different for this to happen. The prediction from this is that inconsistency between the relationship between flavours and consequences should lead to dysregulation of appetite, and so may lead to overeating (Stubbs and Whybrow, 2004). For example, if you consume a sugar-based soda one day and a diet version with the same (or very similar) flavour another, the body cannot detect a reliable pattern between flavour and consequence and so you cannot learn to moderate your intake of the energy-rich version. This implies that successful products need to have an optimal balance of macronutrients and carefully constructed flavours that facilitate flavour-nutrient learning.

4.7.3   Expectations and appetite control

The final issue is the extent to which information about a product may generate an explicit knowledge which itself may interact with the experience of a product once ingested and so direct consumer behaviour. The most obvious examples of this in the psychobiological literature have examined whether providing labels about nutritive content modify the ability of foods to satiate a consumer. Most evidence suggests that food labels have minimal effects on actual appetite regulation. Thus soup labels implying high fat or low energy had no impact on subsequent food intake, whereas actual energy in the soup modified lunch intake (Yeomans *et al.*, 2001a). Likewise, the rate at which pleasantness of an eaten food declines (sensory-specific satiety: SSS) is similarly insensitive to labelled nutrient information. Thus the labelled fat content of potato chips did not modify the rate at which SSS developed even after multiple-exposures to the snack (Miller *et al.*, 2000).

Do these results imply that providing clear nutritional advice on a label is of no value? The answer is of course no, since the label serves to help educate consumers. However, it would be wrong to assume that food labelling itself can help consumers control intake. The important factor remains the actual nutritional content of the food consumed regardless of the awareness of the consumer of that content.

4.8   Future trends

Unlike the trend for increased body size of consumers, which shows no sign of slowing, predicting how our psychobiological understanding of the relationship between consumer and food will develop is unclear. Much research effort at present is directed at the development of drug treatments to reduce body size in obese patients (Halford, 2006), but this increased understanding of these appetite suppressants will have knock-on effects for our broader understanding of appetite control. Thus a few years ago we knew that cannabis was associated with cravings for carbohydrate-rich foods (the munchies) but had no knowledge of the mechanism underlying this. Today we know a great deal about the role of our endogenous cannabinoids (Kirkham, 2005), leading to new treatments for obesity. The pace at which our understanding of the neural controls of feeding has developed has accelerated markedly, driven by pressures to help understand and treat obesity.

Areas where current understanding is incomplete include the role of learning in appetite control, the inability to respond in the short term to energy density and the realisation that not all consumers are the same. All of these areas will impact heavily on future product development. A critical issue will be finding ways to reduce energy density without impacting negatively on consumers' perception of flavour and satisfaction with the product. An ideal product would be an enjoyable eating experience which satisfies short-term appetite without promoting over-consumption, and only an appreciation of how the consumer experiences both the hedonic and sensory qualities of foods will allow development of products that meet this aim.

The final lesson from this brief review of psychobiological influences on food choice and appetite is that development of products has to be conducted in a way that is realistic to the situation where the product will be consumed. Most important is the data showing sensitivity of liking ratings to current needs. Developing the sensory qualities of products to taste good in a test situation that is different from the way in which consumers may use that product is unlikely to succeed. For example, developing a product using panels that are not hungry at the time of sampling may not be helpful in developing a product used by consumers as a snack to reduce hunger between meals. Likewise, product development has to include measures of intake, and the impact of repeated intake on product acceptability.

4.9   Sources of further information and advice

The reference list in Section 4.10 details all the papers cited in this review. However, there are many useful reviews that can be used to expand on the material from this chapter, and these are broken up by topic here.

Useful books include:

LOGUE, A.W. (2004) *The Psychology of Eating and Drinking*, 3rd edition. Hove: Brunner-Routledge.

HETHERINGTON, M.M. (2001) *Food Cravings and Addiction*. Leatherhead: Leatherhead Food RA.

MELA, D.J. and ROGERS, P.J. (1998) *Food, Eating and Obesity: The Psychobiological Basis of Appetite and Weight Control*. London: Chapman & Hall.

For information on how flavour preference develops, the following are accessible reviews:

ZELLNER, D. A. (1991). How foods get to be liked: some general mechanisms and some special cases. In R. C. Bolles (Ed.), *The Hedonics of Taste* (pp. 199–217). Hillsdale, NJ: Lawrence Erlbaum Associates.

YEOMANS, M. R. (2006). The role of learning in development of food preferences. In R. Shepherd and M. Raats (Eds.), *Psychology of Food Choice* (pp. 93–112). Wallingford: CABI.

The control of appetite is another area that has been extensively reviewed, with the following recommended:

BERTHOUD, H. R. (2003). Neural systems controlling food intake and energy balance in the modern world. *Current Opinion in Clinical Nutrition and Metabolic Care*, **6**(6), 615–620.

SMALL, C. J. and BLOOM, S. R. (2004). Gut hormones and the control of appetite. *Trends in Endocrinology and Metabolism*, **15**(6), 259–263.

4.10   References

ACKROFF, K. and SCLAFANI, A. (2001). Flavor preferences conditioned by intragastric infusion of ethanol in rats. *Pharmacology Biochemistry and Behavior*, **68**, 327–338.

ACKROFF, K. and SCLAFANI, A. (2002). Flavor quality and ethanol concentration affect ethanol-conditioned flavor preferences. *Pharmacology Biochemistry and Behavior*, **74**(1), 229–240.

ACKROFF, K. and SCLAFANI, A. (2003). Flavor preferences conditioned by intragastric ethanol with limited access training. *Pharmacology Biochemistry and Behavior*, **75**, 223–233.

BAEYENS, F., EELEN, P., VAN DEN BURGH, O. and CROMBEZ, G. (1990). Flavor-flavor and color–flavor conditioning in humans. *Learning and Motivation*, **21**, 434–455.

BAEYENS, F., CROMBEZ, G., HENDRICKX, H. and EELEN, P. (1995). Parameters of human evaluative flavor–flavor conditioning. *Learning and Motivation*, **26**, 141–160.

BAEYENS, F., CROMBEZ, G., DE HOUWER, J. and EELEN, P. (1996). No evidence for modulation of evaluative flavor–flavor associations in humans. *Learning and Motivation*, **27**, 200–241.

BATTERHAM, R. L., COHEN, M. A., ELLIS, S. M., LE ROUX, C. W., WITHERS, D. J., FROST, G. S., *et al.* (2003). Inhibition of food intake in obese subjects by peptide yy3-36. *New England Journal of Medicine*, **349**(10), 941–948.

BATTERHAM, R. L., COWLEY, M. A., SMALL, C. J., HERZOG, H., COHEN, M. A., DAKIN, C. L., *et al.* (2002). Gut hormone pyy(3-36) physiologically inhibits food intake. *Nature*, **418**, 650–654.

BELL, E. A. and ROLLS, B. J. (2001). Energy density of foods affects energy intake across multiple levels of fat content in lean and obese women. *American Journal of Clinical Nutrition*, **73**, 1010–1018.

BERNSTEIN, I. L., ZIMMERMAN, J. C., CZEISLER, C. A. and WEITZMAN, E. D. (1981). Meal patterns in 'free-running' humans. *Physiology and Behavior*, **27**(4), 621–623.

BERRIDGE, K. C. (2003). Pleasures of the brain. *Brain and Cognition*, **52**(1), 106–128.

BERTHOUD, H. R. (2003). Neural systems controlling food intake and energy balance in the modern world. *Current Opinion in Clinical Nutrition and Metabolic Care*, **6**(6), 615–620.

BIRCH, L. L., McPHEE, L., STEINBERG, L. and SULLIVAN, S. (1990). Conditioned flavor preferences in young children. *Physiology and Behavior*, **47**, 501–505.

BLUNDELL, J. E. and BURLEY, V. (1992). Evaluation of the satiating power of dietary fat in man. In Y. Oomura, S. Baba and T. Shimazu (Eds.), *Progress in Obesity Research* (pp. 543–457). London: Libbey.

BLUNDELL, J. E. and TREMBLAY, A. (1995). Appetite control and energy (fuel) balance. *Nutrition Research Reviews*, **8**, 225–242.

BLUNDELL, J. E., LAWTON, C. L., COTTON, J. R. and MACDIARMID, J. I. (1996). Control of human appetite – implications for the intake of dietary-fat. *Annual Review of Nutrition*, **16**, 285–319.

BODNAR, R. J. (2004). Endogenous opioids and feeding behavior: a 30-year historical perspective. *Peptides*, **25**(4), 697–725.

BOOTH, D. A. (1991). Learned ingestive motivation and the pleasures of the palate. In R. C. Bolles (Ed.), *The Hedonics of Taste*. Hillsdale, NJ: Lawrence Erlbaum Associates.

BOOTH, D. A., MATHER, P. and FULLER, J. (1982). Starch content of ordinary foods associatively conditions human appetite and satiation, indexed by intake and pleasantness of starch-paired flavours. *Appetite*, **3**, 163–184.

BRUNSTROM, J. M. (2004). Does dietary learning occur outside awareness? *Consciousness and Cognition*, **13**(3), 453–470.

CAMPFIELD, L. A. and SMITH, F. J. (1990). Systemic factors in the control of food intake. In E. M. Stricker (Ed.), *Handbook of Neurobiology: Volume 10, Neurobiology of Food and Fluid Intake* (pp. 183–206). New York: Plenum Press.

CAPALDI, E. D. (1992). Conditioned food preferences. *Psychology of Learning and Motivation*, **28**, 1–33.

CAPALDI, E. D., OWENS, J. and PALMER, K. A. (1994). Effects of food deprivation on learning and expression of flavor preferences conditioned by saccharin or sucrose. *Animal Learning and Behavior*, **22**, 173–180.

COOPER, S. J., JACKSON, A., MORGAN, R. and CARTER, R. (1985). Evidence for opiate receptor involvement in the consumption of a high palatability diet in nondeprived rats. *Neuropeptides*, **5**, 345–348.

CUMMINGS, D. E., PURNELL, J. Q., FRAYO, R. S., SCHMIDOVA, K., WISSE, B. E. and WEIGLE, D. S. (2001). A preprandial rise in plasma ghrelin levels suggests a role in meal initiation in humans. *Diabetes*, **50**(8), 1714–1719.

CUMMINGS, D. E., FRAYO, R. S., MARMONIER, C., AUBERT, R. and CHAPELOT, D. (2004). Plasma ghrelin levels and hunger scores in humans initiating meals voluntarily without time- and food-related cues. *American Journal of Physiology: Endocrinology and Metabolism*, **287**(2), E297–304.

DAVIS, J. D. (1989). The microstructure of ingestive behavior. *Annals of the New York Academy of Sciences*, **575**, 106–121.

DE CASTRO, J. M. (1988). The meal pattern of rats shifts from postprandial regulation to preprandial regulation when only five meals per day are scheduled. *Physiology and Behavior*, **43**, 739–746.

DE CASTRO, J. M. (1996). How can eating behavior be regulated in the complex environments of free-living humans? *Neuroscience and Biobehavioral Reviews*, **20**(1), 119–131.

DE CASTRO, J. M. (2000). Eating behaviour: lessons from the real world of humans. *Nutrition*, **16**, 800–813.

DE CASTRO, J. M. and ELMORE, D. K. (1988). Subjective hunger relationships with meal patterns in the spontaneous feeding behaviour of humans: evidence for a causal connection. *Physiology and Behavior*, **43**(2), 159–165.

DE CASTRO, J. M., BELLISLE, F. and DALIX, A.-M. (2000a). Palatability and intake relationships in free-living humans: measurement and characterisation in the French. *Physiology and Behavior*, **68**, 271–277.

DE CASTRO, J. M., BELLISLE, F., DALIX, A.-M. and PEARCEY, S. M. (2000b). Palatability and intake relationships in free-living humans: characterization and independence of influence in North Americans. *Physiology and Behavior*, **70**, 343–350.

DE GRAAF, C., HULSHOF, T., WESTSTRATE, J. A. and JAS, P. (1992). Short-term effects of different amounts of protein, fats, and carbohydrates on satiety. *American Journal of Clinical Nutrition*, **55**, 33–38.

DE HOUWER, J., THOMAS, S. and BAEYENS, F. (2001). Associative learning of likes and dislikes: a review of 25 years of research on human evaluative conditioning. *Psychological Bulletin*, **127**(6), 853–869.

DREWNOWSKI, A. (1998). Energy density, palatability and satiety: implications for weight control. *Nutrition Reviews*, **56**(12), 347–353.

ELIZALDE, G. and SCLAFANI, A. (1988). Starch-based conditioned flavor preferences in rats: influence of taste, calories and Cs–Us delay. *Appetite*, **11**, 179–200.

FEDORCHAK, P. M. and BOLLES, R. C. (1987). Hunger enhances the expression of calorie- but not taste-mediated conditioned flavour preferences. *Animal Behavior Processes*, **13**, 73–79.

FIELD, A. P. and DAVEY, G. C. L. (1999). Re-evaluating evaluative conditioning: an exemplar comparison model of evaluative conditioning effects. *Journal of Experimental Psychology: Animal Behavior Processes*, **25**, 211–224.

FLINT, A., RABEN, A., ASTRUP, A. and HOLST, J. J. (1998). Glucagon-like peptide-1 promotes satiety and suppresses energy intake in humans. *Journal of Clinical Investigation*, **101**, 515–520.

FLINT, A., RABEN, A., REHFELD, J. F., HOLST, J. J. and ASTRUP, A. (2000). The effect of glucagon-like peptide-1 on energy expenditure and substrate metabolism in humans. *International Journal of Obesity*, **24**, 288–298.

FRENCH, S. J. and CECIL, J. E. (2001). Oral, gastric and intestinal influences on human feeding. *Physiology and Behavior*, **74**, 729–734.

FRENCH, S. J., MURRAY, B., RUMSEY, R. D., SEPPLE, C. P. and READ, N. W. (1993). Is cholecystokinin a satiety hormone? Correlations of plasma cholecystokinin with hunger, satiety and gastric emptying in normal volunteers. *Appetite*, **21**(2), 95–104.

GARCIA, J. and KOELLING, R. A. (1966). Relation of cue to consequence in avoidance learning. *Psychonomic Science*, **4**, 123–124.

GIBSON, E. L. and WARDLE, J. (2003). Energy density predicts preferences for fruit and vegetables in 4-year-old children. *Appetite*, **41**(1), 97–98.

GIBSON, E. L., WAINWRIGHT, C. J. and BOOTH, D. A. (1995). Disguised protein in lunch after low-protein breakfast conditions food-flavor preferences dependent on recent lack of protein intake. *Physiology and Behavior*, **58**, 363–371.

GRIGSON, P. S. (2002). Like drugs for chocolate: separate rewards modulated by common mechanisms. *Physiology and Behavior*, **76**, 389–395.

GUTZWILLER, J. P., DREWE, J., GOKE, B., SCHMIDT, H., ROHRER, B., LAREIDA, J., et al. (1999a). Glucagon-like peptide-1 promotes satiety and reduces food intake in patients with diabetes mellitus type 2. *American Journal of Physiology*, **276**(5 Pt 2), R1541–1544.

GUTZWILLER, J. P., GOKE, B., DREWE, J., HILDEBRAND, P., KETTERER, S., HANDSCHIN, D., et al. (1999b). Glucagon-like peptide-1: a potent regulator of food intake in humans. *Gut*, **44**(1), 81–86.

HALFORD, J. C. (2006). Pharmacotherapy for obesity. *Appetite*, **46**(1), 6–10.

HARRIS, J. A., GORISSEN, M. C., BAILEY, G. K. and WESTBROOK, R. F. (2000). Motivational state regulates the content of learned flavor preferences. *Journal of Experimental Psychology: Animal Behavior Processes*, **26**(1), 15–30.

HELLSTROM, P. M., GELIEBTER, A., NASLUND, E., SCHMIDT, P. T., GLUCK, M., YAHAV, E., et al. (2004). Peripheral and central signals in control of eating in normal, obese and binge-eating human subjects. *British Journal of Nutrition*, **92**, S47–S57.

JOHNSON, S. L., McPHEE, L. and BIRCH, L. L. (1991). Conditioned preferences: young children prefer flavors associated with high dietary fat. *Physiology and Behavior*, **50**(6), 1245–1251.

KANUMAKALA, S., GREAVES, R., PEDREIRA, C. C., DONATH, S., WARNE, G. L., ZACHARIN, M. R., et al. (2005). Fasting ghrelin levels are not elevated in children with hypothalamic obesity. *Journal of Clinical Endocrinology and Metabolism*, **90**(5), 2691–2695.

KELLEY, A. E. (2004). Ventral striatal control of appetitive motivation: role in ingestive behavior and reward-related learning. *Neuroscience and Biobehavioral Reviews*, **27**(8), 765–776.

KELLEY, A. E., BAKSHI, V. P., HABER, S. N., STEININGER, T. L., WILL, M. J. and ZHANG, M. (2002). Opioid modulation of taste hedonics within the ventral striatum. *Physiology and Behavior*, **76**, 365–377.

KERN, D. L., MCPHEE, L., FISHER, J., JOHNSON, S. and BIRCH, L. L. (1993). The postingestive consequences of fat condition preferences for flavors associated with high dietary fat. *Physiology and Behavior*, **54**, 71–76.

KIRKHAM, T. C. (2005). Endocannabinoids in the regulation of appetite and body weight. *Behavioral Pharmacology*, **16**(5–6), 297–313.

KISSILEFF, H. R., PI-SUNYER, F. X., THORNTON, J. and SMITH, G. P. (1981). C-terminal octapeptide of cholecystokinin decreases food intake in man. *American Journal of Clinical Nutrition*, **34**, 154–160.

KRAL, T. V. and ROLLS, B. J. (2004). Energy density and portion size: their independent and combined effects on energy intake. *Physiology and Behavior*, **82**(1), 131–138.

LONG, S. J., SUTTON, J. A., AMAEE, W. B., GIOUVANOUDI, A., SPYROU, N. M., ROGERS, P. J., et al. (1999). No effect of glucagon-like peptide-1 on short-term satiety and energy intake in man. *British Journal of Nutrition*, **81**(4), 273–279.

LUCAS, F. and SCLAFANI, A. (1989). Flavor preferences conditioned by intragastric fat infusions in rats. *Physiology and Behavior*, **46**, 403–412.

MAYER, J. (1953). Genetic, traumatic and environmental factors in the etiology of obesity. *Physiological Reviews*, **33**, 472–508.

MELA, D. J. (2001). Determinants of food choice: relationships with obesity and weight control. *Obesity Research*, **9**(Supplement 4), 249S–255S.

MELANSON, K. J., WESTERTERP-PLANTENGA, M. S., SARIS, W. H. M., SMITH, F. J. and CAMPFIELD, L. A. (1999). Blood glucose patterns and appetite in time-blinded humans: carbohydrate versus fat. *American Journal of Physiology*, **277**, R337–R345.

MILLER, D. L., BELL, E. A., PELKMAN, C. L., PETERS, J. C. and ROLLS, B. J. (2000). Effects of dietary fat, nutrition labels, and repeated consumption on sensory specific satiety. *Physiology and Behavior*, **71**, 153–158.

MOBINI, S., ELLIMAN, T. D. and YEOMANS, M. R. (2005). Changes in the pleasantness of caffeine-associated flavours consumed at home. *Food Quality and Preference*, **16**, 659–666.

MUURAHAINEN, N. E., KISSILEFF, H. R., LACHAUSSEE, J. and PI-SUNYER, F. X. (1991). Effect of a soup preload on reduction of food intake by cholecystokinin in humans. *American Journal of Physiology*, **260**, R672–R680.

MYERS, K. P. and SCLAFANI, A. (2001a). Conditioned enhancement of flavor evaluation reinforced by intragastric glucose. *Physiology and Behavior*, **74**, 495–505.

MYERS, K. P. and SCLAFANI, A. (2001b). Conditioned enhancement of flavor evaluation reinforced by intragastric glucose. I. Intake acceptance and preference analysis. *Physiology and Behavior*, **74**, 481–493.

NAKAZATO, M., MURAKAMI, N., DATE, Y., KOJIMA, M., MATSUO, H., KANGAWA, K., *et al.* (2001). A role for ghrelin in the central regulation of feeding. *Nature*, **409**(6817), 194–198.

NÄSLUND, E., GUTNIAK, M., SKOGAR, S., ROSSNER, S. and HELLSTROM, P. M. (1998). Glucagon-like peptide 1 increases the period of postprandial satiety and slows gastric emptying in obese men. *American Journal of Clinical Nutrition*, **68**(3), 525–530.

NÄSLUND, E., BARKELING, B., KING, N., GUTNIAK, M., BLUNDELL, J. E., HOLST, J. J., *et al.* (1999a). Energy intake and appetite are suppressed by glucagon-like peptide-1 (glp-1) in obese men. *International Journal of Obesity*, **23**(3), 304–311.

NÄSLUND, E., BOGEFORS, J., SKOGAR, S., GRYBACK, P., JACOBSSON, H., HOLST, J. J., *et al.* (1999b). Glp-1 slows solid gastric emptying and inhibits insulin, glucagon, and pyy release in humans. *American Journal of Physiology*, **277**(3 Pt 2), R910–R916.

PELCHAT, M. L. (2002). Of human bondage: food craving, obsession, compulsion, and addiction. *Physiology and Behavior*, **76**, 347–352.

RAMIREZ, I. (1994). Flavor preferences conditioned with starch in rats. *Animal Learning and Behavior*, **22**, 181–187.

READ, N., FRENCH, S. A. and CUNNINGHAM, K. (1994). The role of the gut in regulating food intake in man. *Nutrition Reviews*, **52**, 1–10.

RICHARDSON, N. J., ROGERS, P. J. and ELLIMAN, N. A. (1996). Conditioned flavour preferences reinforced by caffeine consumed after lunch. *Physiology and Behavior*, **60**, 257–263.

ROGERS, P. J., RICHARDSON, N. J. and ELLIMAN, N. A. (1995). Overnight caffeine abstinence and negative reinforcement of preference for caffeine-containing drinks. *Psychopharmacology*, **120**, 457–462.

ROLLS, B. J. and ROE, L. S. (2002). Effect of the volume of liquid food infused intragastrically on satiety in women. *Physiology and Behavior*, **76**, 623–631.

ROLLS, B. J., CASTELLANOS, V. H., HALFORD, J. C., KILARA, A., PANYAM, D. and PELKMAN, C. L. (1998). Volume of food consumed affects satiety in man. *American Journal of Clinical Nutrition*, **67**, 1170–1177.

ROLLS, B. J., BELL, E. A. and WAUGH, B. A. (2000). Increasing the volume of a food by incorporating air affects satiety in man. *American Journal of Clinical Nutrition*, **72**, 361–368.

ROZIN, P. and VOLLMECKE, T. A. (1986). Food likes and dislikes. *Annual Review of Nutrition*, **6**, 433–456.

SCHACHTER, S. and GROSS, L. P. (1968). Manipulated time and eating behavior. *Journal of Personality and Social Psychology*, **10**, 98–106.

SCHWARTZ, M. W., WOODS, S. C., PORTE, D., JR., SEELEY, R. J. and BASKIN, D. G. (2000). Central nervous system control of food intake. *Nature*, **404**(6778), 661–671.

SCLAFANI, A. (1999). Macronutrient-conditioned flavor preferences. In H.-R. Berthoud and R. J. Seeley (Eds), *Neural Control of Macronutrient Selection* (pp. 93–106). Boca Raton, FL: CRC Press.

SCLAFANI, A. (2002). Flavor preferences conditioned by sucrose depend upon training and testing methods: two-bottle tests revisited. *Physiology and Behavior*, **76**, 633–644.

SCLAFANI, A. and NISSENBAUM, J. W. (1988). Robust conditioned flavor preference produced by intragastric starch infusions in rats. *American Journal of Physiology*, **255**, R672–R675.

SHIIYA, T., NAKAZATO, M., MIZUTA, M., DATE, Y., MONDAL, M. S., TANAKA, M., *et al.* (2002). Plasma ghrelin levels in lean and obese humans and the effect of glucose on ghrelin secretion. *Journal of Clinical Endocrinology and Metabolism*, **87**(1), 240–244.

SMITH, G. P. (2000). The controls of eating: a shift from nutritional homeostasis to behavioral neuroscience. *Nutrition*, **16**, 814–820.

STEINER, J. E., GLASER, D., HAWILO, M. E. and BERRIDGE, K. C. (2001). Comparative expression of hedonic impact: affective reactions to taste by human infants and other primates. *Neuroscience and Biobehavioral Reviews*, **25**, 53–74.

STIRLING, L. J. and YEOMANS, M. R. (2004). Effects of exposure to a forbidden food on eating in restrained and unrestrained women. *International Journal of Eating Disorders*, **35**, 59–68.

STRADER, A. D. and WOODS, S. C. (2005). Gastrointestinal hormones and food intake. *Gastroenterology*, **128**(1), 175–191.

STUBBS, R. J. and WHYBROW, S. (2004). Energy density, diet composition and palatability: influences on overall food energy intake in humans. *Physiology and Behavior*, **81**, 755–764.

SUGINO, T., YAMAURA, J., YAMAGISHI, M., OGURA, A., HAYASHI, R., KUROSE, Y., *et al.* (2002). A transient surge of ghrelin secretion before feeding is modified by different feeding regimens in sheep. *Biochemical and Biophysical Research Communications*, **298**(5), 785–788.

TANAKA, M., NARUO, T., YASUHARA, D., TATEBE, Y., NAGAI, N., SHIIYA, T., *et al.* (2003). Fasting plasma ghrelin levels in subtypes of anorexia nervosa. *Psychoneuroendocrinology*, **28**(7), 829–835.

TINLEY, E. M., YEOMANS, M. R. and DURLACH, P. J. (2003). Caffeine does not reinforce conditioned flavour liking in fully abstinent caffeine consumers. *Psychopharmacology*, **166**, 416–423.

TINLEY, E. M., DURLACH, P. J. and YEOMANS, M. R. (2004). How habitual caffeine consumption and dose influence flavour preference conditioning with caffeine. *Physiology and Behavior*, **82**(2/3), 317–324.

TOATES, F. M. (1986). *Motivational Systems*. Cambridge: Cambridge University Press.

TOFT-NIELSEN, M. B., MADSBAD, S. and HOLST, J. J. (1999). Continuous subcutaneous infusion of glucagon-like peptide-1 lowers plasma glucose and reduces appetite in type 2 diabetic patients. *Diabetes Care*, **22**, 1137–1143.

TSCHOP, M., WEYER, C., TATARANNI, P. A., DEVANARAYAN, V., RAVUSSIN, E. and HEIMAN, M. L. (2001). Circulating ghrelin levels are decreased in human obesity. *Diabetes*, **50**(4), 707–709.

TUORILA, H., KRAMER, F. M. and ENGELL, D. (2001). The choice of fat-free vs. regular-fat fudge: the effects of liking for the alternative and the restraint status. *Appetite*, **37**(1), 27–32.

VANDEWATER, K. and VICKERS, Z. (1996). Higher-protein foods produce greater sensory-specific satiety. *Physiology and Behavior*, **59**, 579–583.

WANG, G. J., VOLKOW, N. D., THANOS, P. K. and FOWLER, J. S. (2004). Similarity between obesity and drug addiction as assessed by neurofunctional imaging: a concept review. *Journal of Addictive Diseases*, **23**(3), 39–53.

WEINGARTEN, H. P. (1983). Conditioned cues elicit feeding in sated rats: a role of learning in meal initiation. *Science*, **220**, 431–433.

WOODS, S. C. (1991). The eating paradox: how we tolerate food. *Psychological Review*, **98**(4), 488–505.

WREN, A. M., SEAL, L. J., COHEN, M. A., BRYNES, A. E., FROST, G. S., MURPHY, K. G., *et al.* (2001). Ghrelin enhances appetite and increases food intake in humans. *Journal of Clinical Endocrinology and Metabolism*, **86**(12), 5992.

YEOMANS, M. R. (1996). Palatability and the microstructure of eating in humans: the appetiser effect. *Appetite*, **27**(2), 119–133.

YEOMANS, M. R. (2006). The role of learning in development of food preferences. In R. Shepherd and M. Raats (Eds.), *Psychology of Food Choice* (pp. 93–112). Wallingford, Oxford, UK: CABI.

YEOMANS, M. R. (2007). The role of palatability in control of food intake: Implications for understanding and treating obesity. In S. J. Cooper and T. C. Kirkham (Eds), *Progress in Brain Research: Appetite and Body Weight: Integrative Systems and the Development of Anti-obesity Drugs* (pp. 247–269). London: Elsevier.

YEOMANS, M. R. and GRAY, R. W. (2002). Opioids and human ingestive behaviour. *Neuroscience and Biobehavioral Reviews*, **26**, 713–728.

YEOMANS, M. R. and MOBINI, S. (2006). Hunger alters the expression of acquired hedonic but not sensory qualities of food-paired odors in humans. *Journal of Experimental Psychology: Animal Behavior Processes*, **32**, 460–466.

YEOMANS, M. R., WRIGHT, P., MACLEOD, H. A. and CRITCHLEY, J. A. J. H. (1990). Effects of nalmefene on feeding in humans: dissociation of hunger and palatability. *Psychopharmacology*, **100**(3), 426–432.

YEOMANS, M. R., SPETCH, H. and ROGERS, P. J. (1998). Conditioned flavour preference negatively reinforced by caffeine in human volunteers. *Psychopharmacology*, **137**(4), 401–409.

YEOMANS, M. R., JACKSON, A., LEE, M. D., NESIC, J. S. and DURLACH, P. J. (2000a). Expression of flavour preferences conditioned by caffeine is dependent on caffeine deprivation state. *Psychopharmacology*, **150**(2), 208–215.

YEOMANS, M. R., JACKSON, A., LEE, M. D., STEER, B., TINLEY, E. M., DURLACH, P. J., *et al.* (2000b). Acquisition and extinction of flavour preferences conditioned by caffeine in humans. *Appetite*, **35**, 131–141.

YEOMANS, M. R., LARTAMO, S., PROCTER, E. L., LEE, M. D. and GRAY, R. W. (2001a). The actual, but not labelled, fat content of a soup preload alters short-term appetite in healthy men. *Physiology and Behavior*, **73**(4), 533–540.

YEOMANS, M. R., LEE, M. D., GRAY, R. W. and FRENCH, S. J. (2001b). Effects of test-meal palatability on compensatory eating following disguised fat and carbohydrate preloads. *International Journal of Obesity*, **25**, 1215–1224.

YEOMANS, M. R., BLUNDELL, J. E. and LESHAM, M. (2004). Palatability: response to nutritional need or need-free stimulation of appetite? *British Journal of Nutrition*, **92**, S3–S14.

YEOMANS, M. R., WEINBERG, L. and JAMES, S. (2005). Effects of palatability and learned satiety on energy density influences on breakfast intake in humans. *Physiology and Behavior*, **86**, 487–499.

YEOMANS, M. R., MOBINI, S., ELLIMAN, T. D., WALKER, H. C. and STEVENSON, R. J. (2006). Hedonic and sensory characteristics of odors conditioned by pairing with tastants in humans. *Journal of Experimental Psychology: Animal Behavior Processes*, **32**, 215–228.

ZELLNER, D. A. (1991). How foods get to be liked: some general mechanisms and some special cases. In R. C. Bolles (Ed.), *The Hedonics of Taste* (pp. 199–217). Hillsdale, NJ: Lawrence Erlbaum Associates.

ZELLNER, D. A., ROZIN, P., ARON, M. and KULISH, C. (1983). Conditioned enhancement of human's liking for flavor by pairing with sweetness. *Learning and Motivation*, **14**, 338–350.

ZIMMET, P. and THOMAS, C. R. (2003). Genotype, obesity and cardiovascular disease – has technical and social advancement outstripped evolution? *Journal of Internal Medicine*, **254**(2), 114–125.

5

How do risk beliefs and ethics affect food choice?

A. Saba, Istituto Nazionale di Ricerca per gli Alimenti e la Nutrizione, Italy

5.1 Introduction

Food consumption has always been a matter subject to a complex network of cultural and individual factors. In the developed countries, consumers have developed more dynamic and complex demands. Industries and food processors have been claiming that food consumers have become more difficult to understand and to predict. On the one hand, technological innovation has given rise to all manner of food products of exceptional quality and safety. On the other hand, the complexity of the food chain leading to human consumption has increased. However, recent food emergencies (e.g. dioxin, BSE, *Salmonella* in eggs) have left the public with the impression that control systems do not have always the situation under control. The public has become more concerned about the risk from food hazards (Adam, 1999) also because of the decline of their trust in the regulation of food supply (Warren *et al.*, 1990). Public risk perceptions appear to drive beliefs about the acceptability of emerging technologies and food manufacturing, such as genetically modified foods or food irradiation and attitudes towards microbiological risks and food handling practices (Frewer *et al.*, 2004). Questions about food safety and security raise further ethical questions about responsibility and accountability and about acceptable or justifiable levels of risk (Serageldin, 1999; Thompson, 2001). Empirical studies support the claim that environmental attitudes regarding consumer behaviour are morally and ethically based (Guagnano *et al.*, 1995). The present chapter focuses on risk perception related to food selection and on the increasing role of moral and ethical considerations in the domain of food choice.

5.2 Consumer risk perception and food choice

Eating and drinking are, unfortunately, coupled with a multitude of potential health risks, but only a few have been studied from a consumer perspective. People frequently use a number of quick decision-making rules when dealing with uncertain situations. In many cases these so-called heuristics, or 'rules of thumb', allow people to function successfully in everyday life by making the most efficient use of limited cognitive abilities and time, and their inability to cope with vast amounts of information. Cognitive research suggests five situations in which people may be disposed to using such rules: when they are overloaded with information; when they do not have enough time; when the issues are not too important to them; when they have little knowledge or information on the topic; and when a specific shortcut comes to mind easily (Pratkanis and Aronson, 1992). Generally speaking, risk perception might be regarded as a form of attitude towards a specific object, such as a potential hazard. Risk may be conceptualised in terms of risk to human health, the environment, animal health and future generations (Miles and Frewer, 2001). People may overestimate some risks and underestimate others. Research by cognitive psychologists and social scientists has demonstrated that when laypersons make estimates of risk they do not merely calculate in accordance with statistical and probabilistic information.

While scientists consider public reactions to technical risk as reflecting 'ignorance' and 'irrationality', research has suggested that the public's reaction to risk is underpinned by hazard qualities not taken into account by experts (Slovic et al., 1980; Slovic, 1987, 1993). Slovic (1987) reported that two dimensions were important when laypeople judge risks: the first aspect being the extent to which a hazard is 'dreaded' (severe, likely, uncontrollable, involuntary, catastrophic) and the second being how 'known' the hazard is (known to science, new, has delayed effects). Involuntary/uncontrollable and unnatural risks are viewed as more threatening than those over which people perceive they have a choice, even if the probability of occurrence of the involuntary risk is very low (Slovic, 1987). However, a risk taken without the subject's knowledge or without the subject having decided to take it, especially if the risk stands to benefit someone other than the risk-taker, makes for a powerful outrage factor. This is particularly true when it comes to possible dangers associated with food.

Applying the psychometric paradigm approach to study food risk (Sparks and Shepherd, 1994; Siegrist 2003) has demonstrated that the main components responsible for variance in individuals' perception of risks were 'severity' and 'unknown risk'. Other research has shown that individuals tend to underestimate the risk to themselves relative to others from a variety of hazards and believe that, although negative events do occur, these events are relatively unlikely to harm them personally (Weinstein, 1987, 1989). People perceive that they are relatively invulnerable to a particular hazard but do not extend this invulnerability to others. This phenomenon is known as 'optimistic bias', or 'unrealistic optimism' (Weinstein, 1980). Many food-related hazards are associated with optimistic bias, such as heart attacks and heart disease, weight gain and obesity. Other health effects are related to consumption of a high-fat diet, drinking problems and other associated health effects, for example, diabetes and food poisoning. People seem to have a positive view of the risks to themselves from various hazards and also had a positive view of their own dietary behaviour, both in terms of intake of particular nutrients and of specific foods. Optimistic bias effects were also reflected in studies showing how some consumers with low fruit and vegetable intakes regarded themselves as 'high consumers' (Cox *et al.* 1998).

The implications of unrealistic optimism in health-related domain behaviour have long been of concern. Several authors suggested that an illusion of relative invulnerability to hazards might mean that people are less likely to adopt appropriate precautionary behaviour (van der Plight, 1994; Weinstein, 1989). Optimistic bias was found for the potential hazards included a high-fat diet (Gatenby 1996), along with microbiological and technological risks (e.g. genetically modified, GM, foods) (Sparks *et al.*, 1995; Frewer *et al.*, 1994). Generally speaking, people believe that they performed risky behaviours less often than other individuals and this optimistic bias may be a barrier to the initiation of health-protective dietary change. However, the effects of 'optimism' were larger for some of the hazards, being particularly marked for the lifestyle hazards of a high-fat diet and alcohol abuse, as well as for food poisoning from home-prepared foods.

Sparks *et al.* (1995) assessed respondents' beliefs about their susceptibility to heart disease, weight gain and feeling unwell due to high fat consumption. They found optimism for all three of these measures. In addition, there was evidence that individuals were optimistic about their standing on risk factors associated with dietary fat intake, as well as being optimistically biased about the likelihood of suffering any negative effects. Many people believe that a high-fat diet is not healthy and studies showed that health concerns are relevant in people's food choices and a high-fat diet is not considered to be healthy (Glanz *et al.*, 1998).

Other studies that have been carried out in the context of dietary fat intake have indicated that optimistic bias may be a barrier to the initiation of health-protective dietary change. Explanations have been put forward for such optimistic biases. There may be a need to deny risks in order to avoid anxiety, or people may not consider the likely actions taken by other people to avoid risks, thereby attaching too much weight to their own risk-avoiding behaviours (Weinstein, 1989). Whatever the cause of such biases, it is extremely difficult to reduce them (Weinstein and Klein, 1995), and therefore attempts to change bias may not be fruitful as a means of making dietary interventions more effective.

There is some evidence that individuals exhibit more optimistic bias about problems they believe they can control (Harris, 1996; Harris and Middleton, 1994; Lek and Bishop, 1995; Welkenhuysen *et al.*, 1996). For example, people are unaware that the home is a likely place for microbiological food risk, believing that the responsibility lies instead with food manufacturers or restaurants (Worsfold and Griffith, 1997). This distinction has been found in other studies (Woodburn and Raab, 1997), indicating that people believed that food eaten at home was at a lower risk of causing food poisoning than that eaten out, with particular mention of fast food restaurants. While the perceived risk from food poisoning caused by home-prepared food was the lowest of a set of hazards studied, the risk from food prepared by others was considerably higher (Frewer et al., 1994). The authors found that respondents felt that they had high control over the risks, perceiving low personal risk and high knowledge about food poisoning in the home. Then, controllability of the hazard was found to be an important factor influencing food risk perception. An optimistic bias and control of the risk was estimated to be high for example for such hazards as caffeine, high-sugar diet, alcohol, nutritional deficiencies and others (Sparks and Shepherd, 1994).

All these risks have something in common in that they seem to be the result of consumers' free choice and taking the risk is related mainly to the individual responsibility. Research has reported a lack of optimistic bias and a 'low-control' for other types of food risks, such as GM food or pesticides residues (Frewer et al., 1994; Sparks and Shepherd, 1994; Sparks et al., 1995). Generally speaking, although consumers may claim to be fairly confident about the safety of food in their supermarkets, they express high levels of concern when asked about specific food safety issues, such as pesticide residues or hormones in animals (Senauer, 1992). Consumer concern about the safety of food has been highlighted by a number of food scares in recent years. Specific issues of concern to consumers include bacteriological contamination, chemical residues, food irradiation and BSE. Although some of these factors pose very little actual risk to the community, it is the perceived risks that affect consumers' buying behaviour. Consumers who are concerned about food safety are mainly concerned with specific issues to do with food production and handling. Experience of personal harm from food in the form of allergic or intolerant reaction can also influence how people evaluate food risk. Allergic consumers are inclined to take more individual responsibility for risk protection than consumers without allergies, especially if they have been sensitised to food risk through adverse experiences (Gaivoronskaia and Hvinden, 2005).

Studies suggested that perceptions of risks and benefits play an important role in shaping consumer attitudes towards genetic engineering in food (Bredahl, 1999; Grunert et al., 2001; Moon and Balabramanian, 2001). They found that respondents with a higher perceived risk to human health or the environment from GM foods were more likely to oppose them, while those who perceived benefit were more likely to have a favourable attitude. The controversies over modern technologies usually centre on their risks, the potential and as yet unknown hazards that they may pose to human health, the environment and society. Higher levels of perceived risk decreased the likelihood of purchasing GM foods (Harrison et al., 2004). The beliefs that there is potential for negative environmental impact associated with production processes or agricultural practices and perception that there is uncertainty associated with unintended human or animal health effects may contribute to public concerns. Recent research indicated that, although the public is concerned with the outcomes of technical risk assessments, they are also concerned by the uncertainty related to these outcomes, suspecting that risk assessments are based on an insufficient level of scientific knowledge (Lassen *et al.*, 2002; Yeung and Morris, 2001). Uncertain outcomes that are attractive will be perceived as less risky, while unattractive outcomes will appear as more risky.

Most recently, the role of affective processes in risk perception was emphasised (Slovic *et al.*, 2002). Public scepticism has been framed as a risk issue, and public opposition to GM foods is attributed to the public's mis-perception of risk. Gaskell *et al.* (2004) identified four different groups of people based on a two-by-two classification of risk and benefit perception: no benefit and risk ('sceptical'); benefit and risk ('trade off'); benefit and no risk ('relaxed'); no benefit and no risk ('uninterested'). The characteristics of these four groups would suggest that the three groups might make judgements about GM foods in different ways, implementing different decision-taking strategies.

Risk perception seems to be linked to social factors and individual differences have been identified in risk perceptions related to environmental and food-related hazards. Several authors have attempted to analyse demo-graphic and other factors that influence consumers' perceptions of riskiness (Jussaume and Judson, 1992; Polacheck and Polacheck, 1989). The presence of children in a household may have an effect on consumer concern about the risks associated with food consumption. Some studies founded that gender was a significant, but weak, predictor of risk perceived (Siegrist, 2000). Women seem to express higher levels of concern about potential environmental and techno-logical risks than men (Davidson and Freudenburg, 1996). Risk perception also appears to differ geographically. A study on attitudes towards nutrition showed that women displayed greater anxiety about diet than men did, along with more concern about health issues (Rozin *et al.*, 1999). People in northern European countries tended to be risk averse, while those in southern European countries tended to make decisions based on food quality characteristics (Bredahl, 2000). Since consumers cannot directly measure food safety risks for themselves, food safety issues are also a matter of trust (Kennedy, 1988). A lack of trust has been cited in numerous studies as a critical factor in the gap between scientists' risk assessment and the public's perception of risk (Slovic, 1999). Trust has also been found to be strongly linked to risk perception, indicating positive relation-ships between distrust in regulatory agencies and risk perceptions (Slovic, 1993).

5.3 Ethical concerns associated with foods and agricultural technologies

Ethical (and moral) values are deeply held beliefs about what is right and wrong. Ethical issues arise when these values conflict with one another over an action, policy or technology. Ethical issues associated with agricultural biotechnology, however, are social: people have beliefs about biotechnology that conflict with those of others.

Empirical research suggested that measures of moral issues are very relevant to the social psychology of eating (Rozin, 1990) and findings from some studies revealed the important role of moral considerations in the domain of food choice (Raats et al., 1995). A measure of moral obligation for family's health was highly significant for predicting the consumption of milks with different levels of fats (Raats et al., 1995) and the mothers' attitudes towards giving their children foods containing certain types of chemicals (Shepherd and Raats, 1996).

More recently, ethical issues relating to the food chain have been of increasing concern including biotechnology, environmental degradation, food safety and animal welfare (Mepham, 1996). Ethical concerns associated with food and agricultural technologies play an important role in consumers' decisions and actions. The application of genetic engineering to food production, for example, has raised a number of moral and ethical concerns and consumers have particular ethical objections to GM foods (Eurobarometer, 2003; Miles and Frewer, 2001; Millar et al., 2001). Hence, moral and ethical acceptability is a better predictor of public acceptance of specific application of gene technology than risks to human health or the environment (Gaskell et al., 2001).

Attitudes towards the use of gene technology in food production were found to be predicted not only from a consideration of advantages and disadvantages of the technology but also from a consideration of level of perceived moral obligation and ethical consideration (Frewer et al., 1997; Sparks et al., 1995). Consumers are becoming more concerned with the processes by which food is produced in addition to its content, yet they are often not able to distinguish between products according to their production characteristics. For instance, animal-friendly eggs or biotech foods cannot be distinguished from conventional ones either before or after consumption, but consumers may nonetheless wish to choose between them because of how they were produced. The inability to identify certain product characteristics, which stops the consumer from making rational and optimal decisions, as well as reducing his or her autonomy, raises ethical issues. Providing the consumer with the necessary product information may resolve this ethical dilemma. Labelling is often recommended for food products (Kalaitzandonakes, 2004) and it is requested by the European consumers with increasing interest (Eurobarometer, 2003). Studies suggested that the major reasons why consumers choose eco-labelled food products (environment-friendly products) or organic foods are ethical consideration for the environment and/or for their own health (Chinnici et al., 2002; Grankvist et al., 2004; Wandel and Bugge, 1997).

In general, organic foods are generally regarded as natural products (Makatouni, 2002; Zanoli and Naspetti, 2002), as opposed to industrially processed or 'high-tech' foods, which tend to be considered, if not unnatural, then at least less natural (Bredahl, 1999; Holm and Kildevang, 1996), and this seems to be one of the reasons for the positive attitudes and ethical consideration towards organic foods generally found in consumer surveys (Hill and Lynchehaun, 2002; Magnusson *et al.*, 2001). Furthermore, subjects who attached high importance to the purchase criterion 'environmental consequences' were more affected by the labels than participants who valued environmental consequences less. Torjusen *et al.* (2001) found that traditional food quality aspects such as freshness and taste (called 'traditional traits') were common to all consumers, while those who purchased organic foods were more concerned about ethical, environmental and health concerns (called 'reflection traits'). The 'reflective' consumer was concerned about many parts of the food system: how food is produced, processed and handled, and how these circumstances affect people, animals and nature.

There is plenty of indirect evidence indicating that people's moral and ethical reasoning does not stop when they intend to buy a food product. In recent decades the environmental consequences of our packaging consumption, particularly the amount of household waste it leads to, have become the focus of political and public attention (Bech-Larsen, 1996). Consumers are more aware of the detrimental consequences of their packaging choices for others and of actions they can take in order to avoid these consequences. Awareness of these issues is the precondition for the development of moral norms for environmentally sensitive actions (Heberlein, 1972). Consumer boycotts of firms or countries that are caught in unethical practices, the business success of companies selling documented environmentally and/or socially improved products, and consumers' expressed preferences for such products in surveys are all signs that consumers let their moral views influence their buying decisions (Peattie, 1995). Hence, a necessary condition for the spread of moral environmental reasoning to buying decisions is that the individual feels a high degree of concern for an environmental issue that is associated with the particular buying decision, or feels a high degree of environmental concern in general (Thøgersen and Andersen, 1996).

Empirical studies have found that moral and ethical perceptions about multinational corporations also had measurable impacts on consumer acceptance of GM foods (Bredahl, 1999; Gaskell *et al.*, 1998; Moon and Balabramanian, 2001). Several works revealed that purchase motives are attributed to environmental/ethical concern, as well as to quality/health consciousness and exploratory food buying behaviour because of specific product attributes (Browne *et al.*, 2000; Davis *et al.*, 1995). Some studies also reveal a variety of other purchase motives that seem to reflect national interests, such as 'support to organic farmers' for the German consumers (Worner and Meier-Ploeger, 1999) or 'animal welfare' for the British (Meier-Ploeger and Woodward, 1999).

Even though studies of the influence of moral and ethical concerns on consumer buying decisions with environmental implications are rare, there is plenty of indirect evidence indicating that moral and ethical concerns influence environmentally important consumer buying decisions. Several empirical studies support the claim that environmental attitudes regarding everyday consumer behaviour are morally based (Black *et al.*, 1985; Guagnano *et al.*, 1995).

Environmental impacts on ecosystem processes may have ethical significance for consumers because they challenge the equilibrium of an ecological zone. Transgenic fish, for example, might eventually displace indigenous species from their habitat, and entire agricultural landscapes could be 'contaminated' by GM crops. Several studies indicate that consumers associate genetic modification applied to agriculture with 'tampering with nature' or that 'unintended effects are unpredictable and thus unknown to science' (Miles and Frewer, 2001). Genetic modification in agriculture is also considered unnatural, implying high risk (the underlying dimension being the threat to nature) (Alvesleben, 2001). Naturalness is, however, a vague concept that is notoriously difficult to define.

Qualitative and quantitative research highlighted moral concerns related to GM foods, regarding issues such as animal welfare, the power balance between producers and consumers, democracy, and disparity between the industrialised world and the third world (Bredahl, 1999; Grunert et al., 2001; Lassen et al., 2002; Miles and Frewer, 2001). Safety becomes a matter of moral and ethical concern when further questions about responsibility and justifiability are raised. Ethical concerns are associated, for example, with effects on human health from environmental exposure, such as air or water borne pathogens. The issue of farm animal welfare is increasingly being seen as important throughout the developed world (Thompson, 1997). A positive link between moral issues and people's stated willingness to pay for policy to address farm animal welfare was found (Bennett et al., 2002), even though the fact that the consumers tend to buy the cheapest meat does not automatically mean that they are not interested in animal welfare (Velde et al., 2002). Ethical principles questioning the killing or even the raising of animals were commonly mentioned by vegetarians and avoiders of meat as reasons for avoiding meat (Santos and Booth, 1996).

5.4   Future trends

Human food choice behaviour takes account of factors that are not objective and irrational. While consumers are faced with numerous different food hazards, their perceptions of the risks associated with them do not always correspond to scientific risk estimates, although the levels of risk associated with many of these hazards are scientifically uncertain. Past risk communication strategies assumed that these differences were essentially a consequence of consumer ignorance of 'the facts' and that presentation of 'the truth' would cure misjudgements. However, although cognitive limitations occasionally do hamper consumer judgements of risks, at other times, consumers' judgements are often understandable and consider many other qualitative factors than are accounted for in a standard risk assessment. It is therefore important for scientists, communicators and policy makers to consider the basis of these qualitative consumer beliefs, both to pre-empt future food-related health and policy crises and to communicate more effectively before, during and after such events. Trust in institutions is likely also to be an important determinant of whether increases or reductions in risk perceptions occur. Research should be conducted in order to investigate public responses to uncertainty in risk management processes and how to communicate these risks with the public. Admitting that uncertainty is inherent in risk assessment processes is likely to result in increased public trust.

It is important to understand how ethics impact on food choice decisions. Mepham (2000) argues that society should consider the ethical implication of a novel food along three dimensions: well-being, autonomy and justice. In the case of the genetically manipulated animal, well-being refers to animal welfare issues, autonomy to behavioural freedom, and justice to the extent to which the integrity of a particular animal is determined by its original animal characteristics. In the case of consumers, well-being refers to safety and acceptability, autonomy to consumers' ability to choose foods, and justice to the universal affordability of foods across different population groups, as well as a basic right to be informed about what you eat. For producers (farmers), well-being equates to adequate income and working conditions, autonomy to the right to decide whether or not to adopt a particular working practice, and justice to fair working practices and fair competitions.

Concern over food safety may have an effect on future consumption levels. Changes in lifestyles have led to an increase in consumption of food outside the home and, consequently, the consumer's awareness of hygiene. Finally, the growing of transformed food products, such as ready-to-eat and prepared meals is placing a greater emphasis on the retailer to adhere to hygienic standards to meet consumers' expectations.

5.5   Implications for new product development

In market-oriented product development the consumer is both the starting point and the final goal of the process. The interdependency between consumers' needs and wants on the one hand and technologies and research on the other has to be implemented systematically and should receive more attention in the modelling of food production innovation. Consumers' social and ethical considerations must be taken into account in decision-making for new product development. As many of the qualities of new products, such as the health effect and methods of production, are invisible to the consumers, much emphasis should be laid on the mechanism of consumer communication. Consumers are becoming more concerned with the processes by which food is produced, and unbiased, clear public information on new technologies is increasingly needed and demanded. However, it is important that public information is not equated with product promotion. Recent food scandals, and lack of transparency on food policy decisions, bred suspicion and led to loss of consumer confidence in food safety and in those who have responsibility for food systems. It is essential that the benefits to the consumer and any areas of risk are both understood and accepted before products enter the market. The development of a precautionary approach to the way that food risks are analysed is essential for restoring consumer confidence and the acceptance of new products.

Greater openness and transparencies are important elements to increase their confidence in the food system. Information is important in enabling members of the public to make their own informed decisions and to compare across alternatives. For example, the possibility of consumers distinguishing products less harmful to the environment from other products could allow them to gradually push less environmentally friendly products out of the market. Such a competitive advantage could give companies an incentive to develop more environmentally positive products that are acceptable from an ethical point of view. Finally, the system should be flexible enough to accommodate the changes in consumer perception over time, while providing for an appropriate level of protection through the involvement and interaction of all stakeholders in the decision-making process.

5.6   Sources of further information and advice

KORTHALS M (2006) 'The ethics of food production and consumption', in L Frewer and H van Trijp (eds), *Understanding Consumers of Food Products*. Cambridge: Woodhead Publishing Limited.

FREWER L, & MILES S (2001) 'Risk perception, communication and trust. How might consumer confidence in the food supply be maintained?', in L Frewer, E Risvik and H Schifferstein (eds), *Food, People and Society. A European Perspective of Consumers' Food Choices*. Springer-Verlag.

ZWART H (2000) 'A short history of food ethics', *Journal of Agricultural and Environmental Ethics*, **12**(2): 113–126.

THOMPSON P B (2000) 'Agricultural biotechnology, ethics, risk and individual consent', in *Encyclopedia of Ethical, Legal, and Policy Issues in Biotechnology*, New York: Wiley.

5.7   References

ADAM B (1999), 'Industrial food for thought: timescapes of risks', *Environ Values*, **8**(2), 219–238.

ALVESLEBEN R. (2001), 'Beliefs associated with food production methods', in Frewer L and MacFie H J H, *Food Choice, Acceptance and Consumption*. London: Blackie Academic and Professional, 346–364.

BECH-LARSEN T (1996), 'Danish consumers attitudes to the functional and environmental characteristics of food packaging', *J Cons Pol*, **19**, 339–363.

BENNETT R, ANDERSON J, & BLANEY R J P (2002), 'Moral intensity and willingness to pay concerning farm animal welfare issues and the implications for agricultural policy', *J Agr Env Ethics*, **15**, 187–202.

BLACK J S, STERN P C, & ELWORTH J T (1985), 'Personal and contextual influences on household energy adaptations', *J Appl Psychol*, **70**, 3–21.

BREDAHL L (1999), 'Consumers cognitions with regard to genetically modified foods. Results of a qualitative study in four countries', *Appetite*, **33**, 343–360.

BREDAHL L (2000), *Determinants of Consumer Attitudes and Purchase Intentions with Regard to Genetically Modified Foods – Results of a Cross-national Survey*, Working Paper no. 69, Aarus, School of Business, Mapp.

BROWNE A W, HARRIS P J C, HOFNY-COLLINS A H, PASIECZNIC N, & WALLACE R R (2000), 'Organic production and ethical trade: definition, practice and links', *Food Pol*, **25**, 69–89.

CHINNICI G, DAMICO M, & PECORINO B (2002), 'A multivariate statistical analysis on the consumers of organic products', *Br Food J*, **104**, 187–199.

COX D N, ANDERSON A S, REYNOLDS J, McKELLAR S, LEAN M E J, & MELA D J (1998), 'Take Five, a nutrition education intervention to increase fruit and vegetable intakes: impact on consumer choice and nutrient intakes', *Br J Nutr*, **80**, 123–131.

DAVIDSON D J, & FREUDENBURG W R (1996), 'Gender and environmental risk concerns: a review and analysis of available research', *Environ Behav*, **28**, 302–339.

DAVIS A, TITTERINGTON A J, & COCHRANE C (1995), 'Who buys organic food? A profile of the purchasers of organic food in N. Ireland, *Br Food J*, **97**(10), 17–23.

EUROBAROMETER (2003), *The Europeans and Biotechnology*, Report no. 58.0, Brussels, European Commission.

FREWER LJ, SHEPHERD R, & SPARKS P (1994), 'The interrelationship between perceived knowledge, control and risk associated with a range of food-related hazards targeted at the individual, other people and society', *J Food Safety*, **14**, 19–40.

FREWER L J, HOWARDS C, & SHEPHERD R (1997), 'Public concerns in the United Kingdom about general and specific applications of genetic engineering: risks, benefits and ethics', *Sci Technol Human Values*, **2**, 98–124.

FREWER L J, LASSEN J, KETTLITZ B, SCHOLDERER J, BEEKMAN V, & BERDAL K G (2004), 'Societal aspect of genetically modified foods', *Food Chem Toxicol*, **42**, 1181–1193.

GAIVORONSKAIA G, & HVINDEN B (2005), 'Food risk perception and responsibility in the eyes of consumers with allergic reactions to food', SCARR (Social Contexts and Responses to Risk) Launch Conference, *Learning about Risk*, Canterbury, University of Kent.

GASKELL G, BAUER M, & DURANT J (1998), 'Public perception of biotechnology in 1996', Eurobarometer 46.1 in Durant J, Gaskell G, and Bauer M, *Biotechnology in the Public Sphere. A European Sourcebook*. London: Science Museum, 189–217.

GASKELL G, ALLUM N, WAGNER W, NIELSEN T H, JELSØE E, KOHORING M, & BAUER M (2001), 'In the public eye: representations of biotechnology in Europe', in Gaskell G and Bauer M. *Biotechnology 1996–2000 – The Years of the Controversy*, London: Science Museum, 53–79.

GASKELL G, ALLUM N, WAGNER W, KRONERGER N, TORGERSEN H, HAMPEL J, & BARDES J (2004), 'GM foods and misperception of risk perception', *Risk Anal*, **24**(1), 185–194.

GATENBY S (1996), 'Healthy eating: consumer attitudes, beliefs and behaviour', *J Human Nutr Diet*, **9**, 384–385.

GLANZ K, BASIL M, MAIBACH E, GOLDENBERG J, & SNYDER D (1998), 'Why Americans eat what they do: taste, nutrition, cost, convenience, and weight control concerns as influences on food consumption', *J Am Diet Assoc*, **98**(10), 1118–1126.

GRANKVIST G, DAHLSTRAND U & ANDERS B (2004), 'The impact of environmental labelling on consumer preference negative vs. positive labels', *J Cons Pol*, **27**, 213–230.

GRUNERT K G, LAHTEENMÄKI L, NIELSEN N A, POULSEN J B, UELAND Ø, & ASTROM A (2001), 'Consumer perceptions of food products involving genetic modification – results from a qualitative study in four Nordic countries', *Food Qual Pref*, **12**, 527–542.

GUAGNANO G A, STERN P C, & DIETZ T (1995), 'Influences on attitude–behavior relationships. A natural experiment with curbside recycling', *Environ Behav*, **27**, 699–718.

HARRIS P (1996) 'Sufficient grounds for optimism: the relationship between perceived controllability and optimistic bias', *J Soc Clin Psychol*, **15**, 9–52.

HARRIS P, & MIDDLETON W (1994), 'The illusion of control and optimism about health: on being less at risk but no more in control than others', *Br J Soc Psychol*, **33**, 369–386.

HARRISON R W, BOCCALETTI S, & HOUSE L (2004), 'Risk perceptions of urban Italian and United States consumers for genetically modified foods', *AgBioForum*, **7**(4), 195–201 (available online: http://www.agbioforum.org)

HEBERLEIN T A (1972), 'The land ethic realized: some social psychological explanations for changing environmental attitudes', *J Soc Issues*, **28**, 479–487.

HILL H, & LYNCHEHAUN F (2002), 'Organic milk: Attitudes and consumption patterns', *Br Food J*, **104**, 526–542.

HOLM L, & KILDEVANG H (1996), 'Consumers views on food quality. A qualitative interview study', *Appetite*, **27**, 1–14.

JUSSAUME R A, & JUDSON D H (1992), 'Public perception about food safety in United States', *Rural Soc*, **57**(2), 235–249.

KALAITZANDONAKES N (2004), 'Another look at biotech regulation', *Regulation*, **27**(1), 44–50.

KENNEDY D (1988), 'Humans in the chemical decision chain', Carter H O & Nuckton C F, *Chemicals in the Human Food Chain: Sources, Options and Public Policies*, Agricultural Issues Center, University of California, Davis, 9–19.

LASSEN J, MADSEN K H, & SANDØE P (2002), 'Ethics and genetic engineering – lessons to be learned from genetically modified foods', *Bioprocess Biosystems Eng*, **24**, 263–271.

LEK Y, & BISHOP G D (1995), 'Perceived vulnerability to illness threats: the role of disease type, risk factor perception and attributions', *Psychol Health*, **10**, 205–217.

MAGNUSSON M K, ARVOLA A, HURSTI U-K K, ÅBERG L, & SJÖDEN P-O. (2001), 'Attitudes towards organic foods among Swedish consumers', *Br Food J*, **103**, 209–226.

MAKATOUNI A (2002), 'What motivates consumers to buy organic food in the UK?', *Br Food J*, **104**, 345–352.

MEIER-PLOEGER A, & WOODWARD L (1999), 'Trends between countries', *Ecol Farm*, **20**, 15–16.

MEPHAM B (1996), *Food Ethics*. London, New York: Routledge.

MEPHAM B (2000), 'A framework for the ethical analysis of the novel foods. The ethical matrix', *J Agr Environ Ethics*, **12**(2), 165–176.

MILES S, & FREWER L J (2001), 'Investigating specific concerns about different food hazards', *Food Qual Prefer*, **12**, 47–61.

MILLAR K M, TOMKINS S M, & MEPHAM, T B (2001), 'Food biotechnologies and retail ethics: a survey of UK retailers' views on the use of the two dairy technologies', *Int J Food Sci Technol*, **36**, 845–854.

MOON W, & BALABRAMANIAN S K (2001), 'Public perceptions and willingness-to-pay a premium for non-GM foods in the US and UK', *AgBioForum*, **4**(3&4), 221–231.

PEATTIE K (1995), *Environmental Marketing Management: Meeting the Green Challenge*. London: Pitman.

POLACHECK D E, & POLACHECK S W (1989) 'An indirect test of children's influence of effectiveness in parental consumer behaviour', *J Cons Affairs*, **23**(1), 91–110.

PRATKANIS A R, & ARONSON E (1992), *The Age of Propaganda: The Everyday Use and Abuse of Persuasion*. New York: WH Freeman.

RAATS M M, SHEPHERD R, & SPARKS P (1995), 'Including moral dimension of choice within the structure of the Theory of Planned Behaviour', *J Appl Soc Psychol*, **25**, 484–494.

ROZIN P (1990), 'Social and moral aspects of food and eating', in Rock I, *The Legacy of Solomon Asch: Essays in Cognition and Social Psychology*, Hillsdale, NJ: Lawrence Erlbaum, 97–110.

ROZIN P, FISCHLER C, IMADA S, SARUBIN A, & WRZESNIEWSKI A (1999), 'Attitudes to food and the role of food in life in the USA, Japan, Flemish Belgium and France: possible implications for the diet-health debate', *Appetite*, **33**, 163–180.

SANTOS M L S, & BOOTH D A (1996), 'Influences on meat avoidance among British students', *Appetite*, **27**, 197–205.

SENAUER B (1992), 'Consumer food safety concerns', *Cereal Foods World*, **37**(4), 298–303.

SERAGELDIN W H (1999), 'Biotechnology and food security in the 21st century', *Science*, **285**, 387–389.

SHEPHERD R, & RAATS M M (1996), 'Attitudes and beliefs in food habits', in Meiselman H L, Risvik E, & Schifferstein H, *Food, People and Society*, Berlin: Springer, 381–399.

SIEGRIST M (2000), 'The influence of trust and perceptions of risks and benefits on the acceptance of gene technology', *Risk Anal*, **20**, 195–203.

SIEGRIST M (2003), 'Perception of gene technology, and food risks: results of a survey in Switzerland', *J Risk Res*, **6**(1), 45–61.

SLOVIC P (1987), 'Perception of risk', *Science*, **236**, 280–285.

SLOVIC P (1993), 'Perceived risk, trust and democracy', *Risk Anal*, **13**(6), 675–682.

SLOVIC P (1999) 'Trust, emotion, sex, politics and science: surveying the risk-assessment battlefield', *Risk Analysis*, **19**(4), 689–701.

SLOVIC P, FISCHHOFF B, & LICHTENSTEIN S (1980), 'Facts and fears: understanding perceived risk', in Schwing R C & Alberts Jr W A, *Societal Risk Assessment: How Safe is Safe Enough?*, New York: Plenum Press.

SLOVIC P, FINUCANE M L, PETERS E, & MACGREGOR D (2002), 'Risk as analysis and risk as feelings: some thoughts about affect reason, risk and rationality', *Risk Anal*, **24**(2), 311–322.

SPARKS P, & SHEPHERD R (1994), 'Public perceptions of the potential hazards associated with food production and food consumption: an empirical study', *Risk Anal*, **14**, 799–806.

SPARKS P, SHEPHERD R, & FREWER L J (1995), 'Assessing and structuring attitudes toward the use of gene technology in food production. The role of perceived ethical obligation', *Basic Appl Soc Psychol*, **16**, 267–285.

THØGERSEN J, & ANDERSEN A K (1996), 'Environmentally friendly consumer behavior: the interplay of moral attitudes, private costs, and facilitating conditions', in Hill R P and Taylor C R, *Marketing and Public Policy Conference Proceedings, Vol. 6*, Chicago: American Marketing Association, 80–96.

THOMPSON P B (1997), 'Science policy and moral purity: the case of animal biotechnology', *Agr Human Values*, **14**, 11–27.

THOMPSON P B (2001), 'Risk, consent and public debate: some preliminary considerations for ethics of food safety', *Int J Food Sci Technol*, **77**(2), 372–377.

TORJUSEN H, LIEBLEIN G, WANDEL M, & FRANCIS C A (2001), 'Food system orientation and quality perception among consumers and producers of organic food in Hedmark county, Norway', *Food Qual Prefer*, **12**, 207–216.

VAN DER PLIGT J (1994), 'Healthy thoughts about unhealthy behaviour', *Psychol Health*, **9**, 187–190.

VELDE H T, AARTS N, & WOERKUM C V (2002), 'Dealing with ambivalence: farmers' and consumers' perceptions of animal welfare in livestock breeding', *J Agr Environ Ethics*, **15**, 203–219.

WANDEL M, & BUGGE A (1997), 'Environmental concern in consumer evaluation of food quality', *Food Qual Prefer*, **8**, 19–26.

WARREN V A, HILLERS V N, & JENNINGS G E (1990), 'Beliefs about food supply safety: a study of co-operative extension clientele', *J Am Diet Ass*, **90**, 713–714.

WEINSTEIN N D (1980), 'Unrealistic optimism about future life events', *J Person Soc Psychol*, **39**, 806–820.

WEINSTEIN N D (1987), 'Unrealistic optimism about susceptibility to health problems; conclusions from a community-wide sample', *J Behav Med*, **10**, 481–499.

WEINSTEIN N D (1989), 'Optimistic biases about personal risks', *Science*, **246**, 1232–1233.

WEINSTEIN N D, & KLEIN WM (1995), 'Resistance of personal risk perceptions to debiasing interventions', *Health Psychol*, **14**, 132–140.

WELKENHUYSEN M, EVERS-KIEBOOMS G, DECRUYENAERE M, & VAN DEN BERGHE H (1996), 'Unrealistic optimism and genetic risk', *Psychol Health*, **11**, 479–492.

WOODBURN M J, & RAAB C A (1997), 'Household food preparers' food-safety knowledge and practices following widely publicized outbreaks of foodborne illness', *J Food Protect*, **60**(9), 1105–1109.

WORNER F, & MEIER-PLOEGER A (1999), 'What the consumer says', *Ecol Farming*, 14–15.

WORSFOLD D, & GRIFFITH C J (1997), 'Food safety behaviour in the home', *Br Food J*, **99**, 97–104.

YEUNG R M W, & MORRIS J (2001), 'Food safety risk: consumer perception and purchase behaviour', *Br Food J*, **103**(3), 170–186.

ZANOLI R, & NASPETTI S (2002), 'Consumer motivations in the purchase of organic food', *Br Food J*, **104**, 643–653.

6

Consumer attitudes to food innovation and technology

M. Siegrist, ETH Zurich, Switzerland

6.1 Introduction

Development of new products is expensive and risky for the food industry. Most new food products are failures in terms of consumer acceptance and disappear from the market shortly after their introduction. The failure rate for new food products is between 60 and 80% (Grunert and Valli, 2001). It is important, therefore, to know what factors facilitate and what factors hinder consumer acceptance of new food products.

The sensory qualities of foods are important to their success in the marketplace, and the price must also be right, of course. One must not ignore the possibility, however, that other factors, such as the technologies used in processing novel foods, may affect their acceptance. The introduction of GM (genetically modified) food, for example, has not been successful in Europe. In most European countries, consumers are opposed to GM foods (Gaskell *et al.*, 2000). Nanotechnology will be the next innovation that will be important in the food sector. This technology can be used to alter foods or to create innovative packaging materials. It seems possible, for example, to change the texture of certain foods utilising nanotechnology. Or the technology can be used to produce new packaging materials that have, for example, anti-bacterial coatings (Kaiser and Tang, 2004). Survey results suggest that Europeans are more sceptical regarding this new technology than people in the United States (Gaskell *et al.*, 2004). How the public will react when nano-food is introduced in the marketplace remains to be seen.

Attitudes toward food technology and food innovations may play an important role in the acceptance of novel foods. The present chapter reviews the research that has examined the influence of attitudes on the acceptance of food innovations and technology. Consideration of this knowledge at an early stage of product development may help reduce the failure rate of new food products.

6.2   Methods and models for analysing consumer attitudes to food innovation and technology

Psychological research on influence has employed the concept of attitudes to explain public reactions toward new technologies (Frewer *et al.*, 2004). Attitudes are evaluations of objects in our environment. Attitudes present a summary evaluation of an object (Ajzen, 2001). These evaluations can vary from positive to negative, and they are experienced as affect. Typical evaluative dimensions are good–bad, pleasant–unpleasant and likeable–dislikeable (Ajzen, 2001). Positive attitudes are associated with approaching behaviour and negative attitudes are associated with avoidance behaviour.

The expectancy–values model is the most popular conceptualisation of attitude (Fishbein and Ajzen, 1975). According to this model, readily accessible beliefs or attributes associated with an object determine the attitude toward the object. The subjective value of the attribute is multiplied by the strength of the association between the object and the attribute. The products for all accessible attributes are summed. This summative index is directly proportional to a person's attitude (Ajzen, 1991). The theory of planned behaviour, or variations of it, have been widely used to explain people's intentions to buy new foods (Cook *et al.*, 2002; Saba and Vassallo, 2002).

In recent social psychological and cognitive models, two distinct processing modes have been identified (Smith and DeCoster, 2000). Based on these dual-mode models, Slovic and colleagues (Finucane *et al.*, 2000; Slovic *et al.*, 2002, 2004) draw a distinction between the experiential system and the analytic system. The analytic system uses probabilities or formal logic in making decisions. The experiential system, on the other hand, has a strong affective basis. It is an intuitive, fast, mostly automatic system. These intuitive feelings are our primary means of evaluating risks (Slovic *et al.*, 2004). The experiential system helps us to quickly decide whether something is good or bad. Slovic and colleagues assume that the affective reactions evoked by stimuli serve as cues for judgements. According to this view, perceived benefits and perceived risks are shaped by the affect associated with a technology. This phenomenon is known as the 'affect heuristic'. Slovic and colleagues use affect as it is employed in the concept of attitude (e.g. Ajzen, 2001), to mean overall degree of positivity or negativity toward the attitude object.

Slovic's (1987) psychometric paradigm has been widely used to study why people perceive various hazards differently. Results of this research suggest that feelings of dread are the major factor affecting public perception and acceptance of risk for a broad range of hazards (Slovic *et al.*, 2004). Food-related hazards, such as BSE or pesticide residuals, are perceived as dreaded risks, while food colouring or saturated fats are perceived as non-dreaded risks (Fife-Schaw and Rowe, 2000; Kirk *et al.*, 2002). Slovic and colleagues (2004) suggested that the importance of the dreadfulness of a hazard for perceived risks can be viewed as evidence of 'risk as feelings'. Affect or attitudes seem to determine risk perception.

Attitudes help us to make sense of and give meaning to our experiences. It has been shown that existing attitudes can affect the evaluation of new information (Prislin *et al.*, 1998). The influence of existing beliefs on the meaning of new information was demonstrated in a study by Eiser and colleagues (1995). Participants were asked about global warming and about the cause of an oil tanker collision in the English Channel. Although these two topics were unconnected, answers to questions about them were closely linked. Thinking about one of the issues primed people to think about the second issue in ways that differed from non-primed conditions.

Information conveyed by risk communication is, therefore, mediated by the attitudes people hold. Scholderer and Frewer (2003) examined the effects of various information strategies on consumer attitude change. Results indicated that the information strategies used by the researchers decreased consumers' acceptance of GM foods compared with the control group. The authors concluded that the information material was more likely to activate pre-existing attitudes than the no-additional-information condition in the control group. The activation of the pre-existing attitudes resulted in an increased consistency of the beliefs and choices expressed by the participants. People's attitudes toward GM foods seem to be so strong that new information is overridden. Informing the public about new technologies may often fail to increase acceptance unless other factors (such as personal or societal benefits, and the values placed on these) are also addressed.

Methods for measuring implicit attitudes have recently been developed (Greenwald *et al.*, 1998). In almost all studies examining attitudes toward foods, however, explicit measures have been used. Various instruments to measure attitudes toward foods have been proposed. Roininen *et al.* (1999) describe a scale that measures the importance of health and taste characteristics. Other scales measure attitudes toward new foods and food technology (Huotilainen and Tuorila, 2005). Most of these scales are not pure attitudinal measurements. They include mixtures of attitudinal items, behavioural intentions and beliefs.

People's attitudes toward food are related to their other attitudes and beliefs. The dichotomy between nature and technology, for example, is important for a better understanding of the acceptance of food innovations. People tend to have confidence in natural food and the way it is produced, but they are suspicious toward new foods and new food technologies (Huotilainen and Tuorila, 2005). Assessments of the naturalness of foods seem to be correlated with sensory appeal (Steptoe *et al.*, 1995). Natural food is associated with better looks and better taste compared with foods containing additives or artificial ingredients.

Attitudes toward GM technology are influenced by more general environmental attitudes (Siegrist, 1998; Sparks *et al.*, 1995). The attitude of favouring the protection of nature because of its intrinsic value had a negative impact on acceptance of GM technology. Valuing nature because of its usefulness and benefits to humans, however, had a positive influence on acceptance of GM technology. In a similar study, general attitudes or world views had an important influence on the perception of GM technology (Siegrist, 1999).

The concept of attitudes is a psychological approach toward a better understanding of the acceptance or non-acceptance of novel food. However, a psychological view may be too narrow. Attitudes toward a new food technology will be influenced not only by the innovation itself but also by the surrounding social, economic and political environments (Henson, 1995). Various dynamic social processes may generate public concern about hazards that are judged as low risks by experts, to the neglect of hazards that they judge as high risks (Kasperson et al., 2003). Such a process of the social amplification of risk perceptions can be observed in the domain of GM foods in Europe.

6.3   Outline of consumer attitudes to food innovation and technology

Novel foods and new food technologies may be more acceptable to the public if there are tangible benefits to the consumer (Frewer et al., 2003). A Swiss study (Siegrist, 2000) examined laypeople's perceptions of GM applications in the domains of food and medicine. Results suggested that acceptance of GM products was largely determined by perceived risk and perceived benefit. Standardised path weights show that perceived benefits are much more important for the acceptance of GM products than perceived risks. A Swedish study reported similar findings (Magnusson and Hursti, 2002). Tangible benefits – products that are better for the environment, for example, or products that are healthier – increased people's stated willingness to purchase GM products. The importance of perceived benefits for the acceptance of GM food was also demonstrated in experimental studies. In one study, participants received information about genetically engineered soybeans (Brown and Ping, 2003). Two groups received information that differed in the presence or absence of a consumer benefit. Results showed that participants who were informed about a GM application with a consumer benefit perceived lower personal risks than those who were informed about an application without a consumer benefit.

Recent studies suggest, however, that benefit alone does not guarantee acceptance. Cox et al. (2004) observed a low intention to consume GM food, even though they communicated clear benefits to the consumer. It should also be emphasised that consumers are not a homogeneous group. In other words, consumers differ in what they perceive as benefits. Organic food, for example, may constitute a benefit for one segment of consumers but not for others. In sum, results of these studies suggest that perceived benefits may have an impact on how GM food is assessed. However, the acceptance of novel food cannot be reduced to perceived risks and perceived benefits.

6.3.1   Attitudes toward specific ingredients

Consumers may hesitate to purchase food products because they contain certain food additives or food colourings. Little research has examined attitudes toward food additives and food colourings, however (for an exception see Kajanne and Pirttilä-Backman, 1996). There are a few studies in which the psychometric paradigm was utilised to examine how people perceive various food hazards. In these studies, participants used a variety of rating scales to evaluate a set of hazards (Fischhoff *et al.*, 1978). They assessed, for example, how well the hazard is known to science and the degree of dread associated with it. In the studies focusing on food hazards, a very heterogeneous set of hazards was presented, ranging from GM food to *Salmonella*.

Based on studies utilising the psychometric paradigm, it can be concluded that food colourings and food additives are perceived as unknown risks and as hazards with low dreadfulness (Fife-Schaw and Rowe, 2000; Sparks and Shepherd, 1994). Results of another study suggest that growth hormones are perceived in a similar way (Kirk *et al.*, 2002). Low severity suggests that food additives and food colourings are not perceived as a source of concern or a problem for further generations. These ingredients, therefore, may not pose a serious problem for the acceptance of new foods.

6.3.2   Attitudes toward new processes

New processes enable innovations in the food sector. Processes such as food irradiation or high-pressure processing are methods for food preservation. Recombinant DNA technology is used to create new varieties – of plants, for example – such as golden rice, a variety with improved nutritional value (Ye *et al.*, 2000). Some of these new processes are not well accepted by consumers. Attitudes towards these new processes may help to explain why this is the case.

Food irradiation offers a number of benefits for consumers (Henson, 1995). This method kills microorganisms in food, and it is a method of food preservation. A number of countries have approved the use of irradiation of specific doses on certain foods (Henson, 1995). However, there are still countries in which food irradiation is not approved.

The benefits associated with food irradiation are not tangible to the consumer; they must be explicitly communicated. This may not be an easy task since radiation is strongly associated with nuclear power, a technology that tends to evoke negative associations and images (Slovic *et al.*, 1991). These negative attitudes may shape attitudes toward food radiation, helping to explain why a number of consumers perceive food irradiation as a risky technology (Bord and O'Connor, 1990). Furthermore, consumers must trust the industry that food irradiation is properly used. The public may fear that this technology lowers the quality of foods (e.g. contaminated foods are irradiated and resold).

In an experimental study, utilising a student sample from Brazil, the effects of a video about food irradiation on attitudes toward this technology was examined (Oliveira and Sabato, 2004). Results suggest that people hold more positive attitudes toward food irradiation after receiving information about it. Utilising correlational data, Bord and O'Connor (1990) found that knowledge is positively correlated with acceptance of irradiated food. However, greater fear of radiation resulted in less acceptance. In a similar vein, Bruhn (1998) concluded that, when provided with science-based information, a high percentage of consumers favour irradiated food.

High-pressure processing is a new method for increasing food safety with minimum quality loss (Ozen and Floros, 2001). This processing technology was developed to meet consumer demands for fresh products with reduced microbiological contamination. It is a non-conventional and new technology since high-pressure processing does not use heat to preserve food (Deliza et al., 2005). Results of focus group interviews showed that use of this new technology had a positive impact on the perception of the product (Deliza et al., 2003). Information about this new processing technology, emphasizing its benefits, had a positive influence on purchase intention (Deliza et al., 2005). Future studies must show whether these results from Brazil can be generalised to developed countries.

In sum, knowledge seems to have an impact on the acceptance of food irradiation and high-pressure processing. However, these technologies may not evoke strong feelings. Risk communication may have an effect when people do not hold strong convictions related to the technology (Earle and Siegrist, 2006). As a consequence of GM technology being likely to evoke more affective responses by consumers, risk communication and knowledge may not positively affect its acceptance. Results of surveys examining public perception of biotechnology suggest that more knowledgeable persons tend to have more extreme attitudes than less knowledgeable persons (Durant et al., 1998). Those attitudes, however, may be positive or negative.

A good deal of research regarding new processes related to food has focused on GM foods. Frewer et al. (1997a), for example, used conjoint analysis to examine attitudes toward various processing technologies. Conjoint analysis is a statistical method that is based on multi-attribute decision theory. The results showed that genetic modification was the least acceptable production method; the traditional method was most acceptable. Additional results showed that the benefits of the product could compensate for the fact that it was produced by a less preferable method. Consumers may accept a food-processing technology, even though they have negative attitudes toward it, when the product is associated with tangible benefits. The study focused on genetic manipulation of microorganisms used for the production of cheese. Manipulation of microorganisms is perceived as less problematic than other applications, such as GM animals (Siegrist and Bühlmann, 1999). Therefore, tangible benefits may not result in higher acceptance for all food products; instead, it is contingent on consumer acceptability of specific applications.

Attitudes toward new technologies are shaped by the perceived benefits associated with them. In addition, consumers are susceptible to framing effects. The use of new technologies is accepted for some products, but it is not accepted for other products. Such effects have been demonstrated in various studies in the domain of GM foods. Perception and acceptance varies according to the type of application (Frewer *et al.*, 1997b; Gaskell *et al.*, 1999). Results have clearly indicated that people in Europe and in the United States have more positive attitudes toward medical applications than toward agricultural applications (Gaskell *et al.*, 1999). Furthermore, differing applications in the food domain are perceived completely differently (Siegrist and Bühlmann, 1999). In this study, several scenarios described various applications of gene technology drawn from the domains of agriculture, food, drugs and medicine. Participants rated the similarity of the different applications. Results of multidimensional scaling showed that two dimensions were relevant for the perception of gene technology. The first was related to the nature of the application (food related/medical application). Medical applications were viewed more positively than food applications. The second dimension was related to the organisms involved (animals, plants/microorganisms). The golden rice application (a rice variety with an enhanced level of vitamin A) was located between the medical applications and the agricultural applications. These results suggest that framing applications in a certain way may alter attitudes in a more positive or a more negative direction.

Various factors seem to affect attitudes toward gene technology. People who trusted institutions involved in using or regulating gene technology judged the benefits to be greater and the risks lower for this technology (Siegrist, 2000). Since most people possess only limited knowledge of gene technology (Durant *et al.*, 1998), the importance of trust should be of no surprise. One way people cope with a lack of knowledge is to rely on trust to reduce the complexity of risk management decisions (Earle and Cvetkovich, 1995). A causal model that has been proposed to explain acceptance of gene technology and other technologies is shown in Fig. 6.1. This model has been successfully tested (Siegrist, 1999, 2000; Siegrist *et al.*, 2000). Based on the results, one can conclude that trust in institutions, or in persons doing genetic modification research or using modified products, is an important factor influencing the perception of gene technology. Trust has an impact on perceived risk as well as on perceived benefit. Acceptance of, or willingness to buy, GM foods is directly determined by the perceived risk and the perceived benefit. In other words, trust has an indirect impact on the acceptance of GM foods. Perceived value similarity seems to be an important antecedent of trust (Siegrist *et al.*, 2000). People tend to trust persons who share their values, and distrust those who do not share their values. If the value similarity approach of trust is correct, trust in gene technology can be increased if a technology is framed to reflect the public's salient values (e.g. medical application and not food application).

An important segment of consumers are those who are willing to buy more expensive organic foods. There are at least two motives that can be identified to explain why some people show a preference for organic foods. Self-reported purchase of organic foods was related to perceived benefit for human health and to environmental concern (Magnusson *et al.*, 2003). The results of a recent study suggest, however, that even when the healthfulness of natural and artificial foods is specified to be equivalent, most of the people with a preference for natural

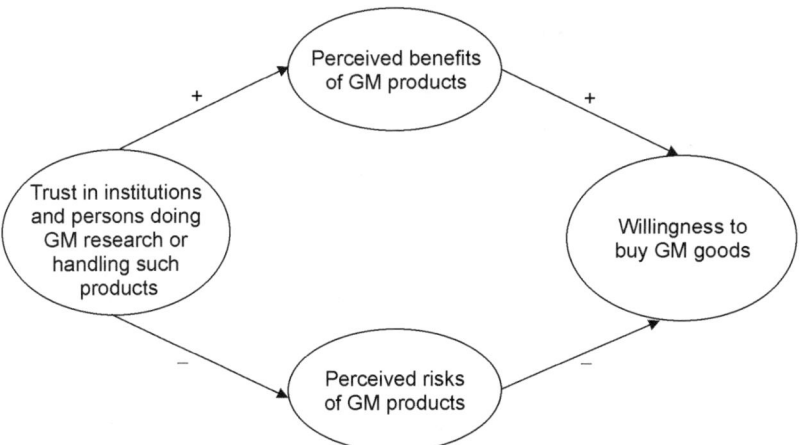

Fig. 6.1   Model explaining acceptance of GM foods.

food continue to prefer it (Rozin *et al.*, 2004). Perceived naturalness or lack of naturalness seems to be a factor that influences attitudes toward genetic engineered foods. Results of a study by Tenbült and colleagues (2005) suggest that the more a product is seen as natural, the less acceptable will be a genetic engineered version of that product. Similarly, a study by Siegrist (2003) showed that consumers considered it more important to have baby food and unprocessed food free of gene technology than to have processed foods, such as chocolate, frozen foods or convenience food, free of gene technology.

6.3.3   Attitudes toward new foods

New varieties of foods have been introduced in the market recently. The most important new categories are probably functional food and convenience food. Functional foods are products that promise consumers improvements in targeted physiological functions (Diplock *et al.*, 1999). Convenience food saves or reduces some kind of effort (Scholderer and Grunert, 2005). Time, physical energy or mental energy can be saved, and the saving can occur at different stages of home food preparation.

Consumers cannot directly experience the benefits of functional food. Producers must explicitly communicate the benefits. This makes trust crucial for the acceptance of functional food (Siegrist and Cvetkovich, 2000) because consumers must trust the health claims provided by the producers (Verbeke, 2006). Participants with stronger health benefit beliefs in functional food products showed more acceptance for functional food than participants with weaker health benefit beliefs (Verbeke, 2005).

Urala and Lähteenmäki (2004) measured attitudes toward functional food. The authors analysed responses toward 42 functional food-related statements. Results of a factor analysis showed that the following seven factors account for consumer's attitudes toward functional foods:

1. Reward from using functional food.
2. Confidence in functional food.
3. Necessity for functional food.
4. Functional food as medicine.
5. Functional food as part of a healthy diet.
6. Absence of nutritional risk in functional food.
7. Taste of functional food.

Perceived reward from using functional food was the best predictor of consumers' stated willingness to use functional food. Perceived risks had no effect, indicating that perceived benefits are more important than perceived risks. However, different results were observed for various types of functional food (e.g. probiotic juice, energy drinks). Consumers do not perceive functional foods as a homogeneous food category (Urala and Lähteenmäki, 2003, 2004). Further, results from Urala and Lähteenmäki (2004) suggest that attitudes toward functional food are different from general health interests.

Attitudes influence acceptance of functional food. However, other factors such as price or taste are important as well. Several studies suggest that consumers are not willing to compromise taste for possible health benefits (Tuorila and Cardello, 2002; Verbeke, 2006). Consumers no longer believe that good taste and healthiness are mutually exclusive.

Natural foods are valued more than functional foods. A majority of US consumers prefers to eat more fruits and vegetables, as opposed to functional foods, in order to obtain phytochemical health benefits (Childs and Poryzees, 1997). Not all food carriers are comparably compelling to consumers. Carriers with a good health image (e.g. yoghurt) are more attractive than carriers lacking such an image (e.g. chewing gum) (van Kleef et al., 2005). This study further suggests that consumers have more positive attitudes towards physiology-related health benefits (e.g. osteoporosis) than toward benefits that are psychology-related (e.g. stress).

The markets for functional foods and convenience foods do not overlap (Shiu et al., 2004). An individual is not likely to be a heavy user of both convenience and functional foods. Further, the consumption of convenience food is influenced by situational factors (Verlegh and Candel, 1999) as well as by socio-demographic factors (Shiu et al., 2004). Little is known, however, about the influence of attitudes on the consumption of convenience food.

6.4   Understanding consumer choice

Most studies measured the intention or the willingness to buy food products. It is not the intention to buy a certain product that is of interest, however; it is the actual purchasing behaviour. Unfortunately, purchasing behaviour cannot be investigated before a product is introduced in the marketplace. In the developmental phase of food innovations one therefore has to rely on data about willingness to purchase such new products. There are other reasons why, in the domain of GM foods, for example, researchers are forced to measure intentions to buy instead of actual purchasing behaviour. In many countries, GM products are not labelled. As a consequence, even consumers with no intention to buy GM foods may end up purchasing them. In other countries, GM foods are not available to consumers because grocery stores do not stock such items. Stated intentions, unfortunately, are not a very good proxy for actual behaviour. It can be expected that attitudes toward food innovations are much better predictors for reported 'willingness to buy' new products compared with actual purchase behaviours. The reason for this is that not only consumer attitudes, but also situational factors, determine consumer purchasing behaviour.

6.4.1   Impact of attitudes to food innovation and technology on purchasing behaviour

Frewer *et al.* (1996) measured the likelihood of purchase for different product categories. A container of yoghurt, a tomato and a chicken drumstick were presented as real products. Products were shown in pairs of photographs, one was labelled as a GM product; the other was labelled as conventional product. Results showed that participants were significantly less likely to purchase GM products compared with conventional equivalents.

Social and situational factors must not be neglected in studying food choice, preferences and purchasing behaviour. It has been shown, for example, that social situations are a major influence on the intent to consume TV dinners (Verlegh and Candel, 1999). In some situations (e.g. dinner alone), consuming convenience food is perceived as appropriate; few people, however, serve TV dinners when entertaining friends. Attitudes toward new foods may be moderated by social or situational factors.

6.4.2   Impact of attitudes to food innovation and technology on willingness to try new foods

Much research on acceptance of new food products has focused on abstract situations (e.g. description of new products). Only a few studies have examined people's reactions toward realistic products and their willingness to taste such products. Townsend and Campbell (2004) conducted an experiment in which participants were asked to taste three apples and to decide which of them was traditionally grown, organically grown or GM. In fact, all apples had been traditionally grown. About half the participants indicated that they would buy GM food, the other half of the respondents stated that they would not buy GM food. Purchasers and non-purchasers differed significantly in their attitudes toward GM food. However, 86% of the non-purchasers were happy to taste an apple that was labelled as a GM product. Owing to this ceiling effect, attitudes were of little value in determining the willingness to taste new foods.

In a study with participants from several Scandinavian countries, participants were offered five pieces of cheese as a reward for taking part in the study

(Lähteenmäki *et al.*, 2002). They could choose among cheese labelled as a GM product and cheese labelled as a traditional product. Two-thirds of the respondents took at least one piece of GM cheese. Attitudes toward the use of gene technology were the best predictor of participants' choice behaviour.

The rated willingness to try different new foods does not seem to be a one-dimensional construct (Backstrom *et al.*, 2004). The willingness to try modified dairy products (e.g. functional yogurts, fat-free yogurts) loaded on a different factor than the stated willingness to try GM products. Adherence to technology was a strong predictor of willingness to try GM foods. It was a weak predictor, however, for willingness to try functional dairy products.

Attitudes toward new foods may not only have an impact on willingness to try new foods, but they may also influence the actual liking of a product. Caporale and Monteleone (2004) examined the effect of specific product information on the perceived quality of the product. In an experiment, partici-pants received information about selected aspects of the manufacturing process (i.e. use of GM yeast, organic methods, traditional technology). Results sug-gested that information about the manufacturing process can have an impact on the perception of a product. GM organisms are seen as unnecessary in food manufacturing, and these negative attitudes toward GM organisms decreased the actual liking of the beers compared with identical beers that were labelled as being produced by traditional technology.

Consumers in general are more positive about familiar than about unfamiliar foods (Raudenbush and Frank, 1999). Personality variables may also shape consumers' attitudes toward new foods. It has been shown that some people have a stronger tendency to avoid new foods than other people, a phenomenon that has been labelled food neophobia. Results of several studies suggest that food neophobia negatively influences consumers' willingness to try new foods (Raudenbush and Frank, 1999; Tuorila *et al.*, 1998, 2001).

In sum, a number of studies examined the influence of attitudes toward new food technologies on the willingness to try new foods. Attitudes toward food innovations not only influenced the likelihood of trying new foods, but they also influenced how much consumers liked the taste of products.

6.5   Understanding consumer attitudes to innovation and technology for food product development

A number of factors must be taken into account for successful product develop-ment. Sensory quality and price are important factors, of course. However, sociodemographic variables such as income, household size, education, age, gender and cultural-background influence food-related behaviour (Axelson, 1986), and may, therefore, affect the success of new food products. In other words, attitudes toward food technology are one factor among many that may have an impact on the success of novel products.

Results of the research examining consumer attitudes toward new food technologies may contribute to a better understanding of which factors negatively influence acceptance of new food products. New processes, such as food irradiation or genetic modification, seem to be the least acceptable technologies. In Europe, especially, it seems very difficult to market such food products. Specific ingredients and new foods such as functional food, however, are easier to market. Nevertheless, attitudes may help to explain why some consumers buy new food products, and why, at the same time, other consumers are hesitant to buy the very same food products.

Results of surveys about GM foods conducted in European countries and in the United States suggest that considerable differences in public opinion on this subject exist (Gaskell *et al.*, 1999). Cultural differences must, therefore, be taken into account when consumer attitudes toward new food technologies are examined. According to Gaskell *et al.* (1999) two explanations may account for the lower consumer acceptance of GM foods in Europe compared with the United States. First, people in the United States have more trust in regulatory authorities than in Europe. As a consequence, US consumers are less concerned about new and unfamiliar technologies than European consumers. Second, there seems to be a greater prevalence of menacing food images (e.g. eating GM foods may result in genetic infection) in Europe than in the United States. This may be the result of recent food safety scares in Europe (e.g. BSE or 'mad cow disease').

It is common sense that GM foods with tangible benefits for the consumer will be easier to market than GM foods without obvious consumer benefits. For the industry it might be tempting to assume that attitudes toward GM foods will be more positive if a GM product with a desirable benefit is on the market. However, novel foods that have clear health benefits may not be appealing to all consumers. Thus, introducing such novel foods is unlikely to result, generally, in more positive attitudes toward GM food (Frewer *et al.*, 2004). It is more likely that, for some products, GM food is accepted, but not for other products.

6.6   Future trends

In most of the reviewed studies, beliefs about, and not attitudes toward, different foods were measured. Few researchers clearly distinguished between attitudes and beliefs. In one such study, beliefs about genetic engineered food influenced attitudes toward genetic engineered food (Dreezens *et al.*, 2005). However, based on this study, not much can be concluded about the causal relationship between attitudes and beliefs. Beliefs could be post-hoc rationalisations of decisions and attitudes. The affect dimension, good–bad, is probably the primary factor that influences purchasing behaviour in the food domain. In many cases, food purchasing can be viewed as an automatic process. Once people have established an attitude toward certain foods, it will be difficult to change these attitudes. Childhood experiences may be important, because it is very likely that in this phase of life attitudes toward foods are formed that are difficult to change later on. Future studies should focus more on the emotional aspects of attitudes and less on the beliefs. These studies should also examine factors that influence attitudes toward food innovation and technology.

In most studies, explicit measures have been used. However, explicit attitudes toward foods may be distorted by social desirability or self-presentation biases. In recent years, indirect measures have been used as an alternative to direct measures. A very popular indirect measure is the implicit association test (IAT), which has been used successfully in numerous studies (Greenwald *et al.*, 1998). The IAT was used to predict brand preferences, product usage and brand recognition in a blind taste test (Maison *et al.*, 2004). Other methods, such as affective priming, were used to indirectly measure food attitudes (Lamote *et al.*, 2004). It is premature to conclude whether these indirect measures will be valuable for a better understanding of attitudes toward food innovation and technology. The results so far are promising, however, and future research should address the question of whether indirect measures can better predict food choice behaviour than direct measures of food attitudes.

6.7   Sources of further information and advice

The journals *Appetite* and *Food Quality and Preference* regularly publish articles related to attitudes to food innovation and technology (http://www.sciencedirect.com/).

The Institute of Food Technologists provides information about new developments and acceptance of new technologies. The institute publishes the journal *Food Technology* (http://www.ift.org/cms/).

Articles about consumer choice, preferences, concerns and related topics can be found in the *British Food Journal* (http://www.emeraldinsight.com).

Articles related to consumer attitudes to food innovation and technology are occasionally published in journals focusing on risk perception and risk assessments (e.g. *Risk Analysis*, *Journal of Risk Research*).

The European Commission is monitoring public opinion in the Member States. Some of the surveys are related to food issues (e.g. attitudes toward gene technology) (http://europa.eu.int/comm/public_opinion/index_en.htm).

6.8   References

AJZEN, I. (1991). The theory of planned behavior. *Organizational Behavior and Human Decision Processes*, **50**, 179–211.

AJZEN, I. (2001). Nature and operation of attitudes. *Annual Review of Psychology*, **52**, 27–58.

AXELSON, M. L. (1986). The impact of culture on food-related behavior. *Annual Review of Nutrition*, **6**, 345–363.

BACKSTROM, A., PIRTTILA-BACKMAN, A. M. and TUORILA, H. (2004). Willingness to try new foods as predicted by social representations and attitude and trait scales. *Appetite*, **43**(1), 75–83.

BORD, R. J. and O'CONNOR, R. E. (1990). Risk communication, knowledge, and attitudes: explaining reactions to a technology perceived as risky. *Risk Analysis*, **10**, 499–506.

BROWN, J. L. and PING, Y. C. (2003). Consumer perception of risk associated with eating genetically engineered soybeans is less in the presence of a perceived consumer benefit. *Journal of the American Dietetic Association*, **103**(2), 208–214.

BRUHN, C. M. (1998). Consumer acceptance of irradiated food: theory and reality. *Radiation Physics and Chemistry*, **52**(1–6), 129–133.

CAPORALE, G. and MONTELEONE, E. (2004). Influence of information about manufacturing process on beer acceptability. *Food Quality and Preference*, **15**(3), 271–278.

CHILDS, N. M. and PORYZEES, G. H. (1997). Foods that help prevent disease: consumer attitudes and public policy implications. *Journal of Consumer Marketing*, **14**, 433–447.

COOK, A. J., KERR, G. N. and MOORE, K. (2002). Attitudes and intentions towards purchasing GM food. *Journal of Economic Psychology*, **23**(5), 557–572.

COX, D. N., KOSTER, A. and RUSSELL, C. G. (2004). Predicting intentions to consume functional foods and supplements to offset memory loss using an adaptation of protection motivation theory. *Appetite*, **43**, 55–64.

DELIZA, R., ROSENTHAL, A. and SILVA, A. L. S. (2003). Consumer attitude towards information on non conventional technology. *Trends in Food Science and Technology*, **14**(1), 43–49.

DELIZA, R., ROSENTHAL, A., ABADIO, F. B. D., SILVA, C. H. O. and CASTILLO, C. (2005). Application of high pressure technology in the fruit juice processing: benefits perceived by consumers. *Journal of Food Engineering*, **67**(1–2), 241–246.

DIPLOCK, A. T., AGGETT, P. J., ASHWELL, M., BORNET, F., FERN, E. B. and ROBERFROID, M. B. (1999). Scientific concepts of functional foods in Europe: consensus document. *British Journal of Nutrition*, **81**(4), S1–S27.

DREEZENS, E., MARTIJN, C., TENBULT, P., KOK, G. and DE VRIES, N. K. (2005). Food and values: an examination of values underlying attitudes toward genetically modified- and organically grown food products. *Appetite*, **44**(1), 115–122.

DURANT, J., BAUER, M. W. and GASKELL, G. (EDS.). (1998). *Biotechnology in the Public Sphere*. London: Science Museum.

EARLE, T. C. and CVETKOVICH, G. T. (1995). *Social Trust: Toward a Cosmopolitan Society*. Westport, CT: Praeger.

EARLE, T. C. and SIEGRIST, M. (2006). Morality information, performance information, and the distinction between trust and confidence. *Journal of Applied Social Psychology*, **36**, 383–416.

EISER, J. R., REICHER, S. D. and PODPADEC, T. J. (1995). Global changes and local accidents: consistency in attributions for environmental effects. *Journal of Applied Social Psychology*, **25**, 1518–1529.

FIFE-SCHAW, C. and ROWE, G. (2000). Extending the application of the psychometric approach for assessing public perceptions of food risks: some methodological considerations. *Journal of Risk Research*, **3**, 167–179.

FINUCANE, M. L., ALHAKAMI, A., SLOVIC, P. and JOHNSON, S. M. (2000). The affect heuristic in judgments of risks and benefits. *Journal of Behavioral Decision Making*, **13**, 1–17.

FISCHHOFF, B., SLOVIC, P., LICHTENSTEIN, S., READ, S. and COMBS, B. (1978). How safe is safe enough? A psychometric study of attitudes towards technological risks and benefits. *Policy Sciences*, **9**, 127–152.

FISHBEIN, M. and AJZEN, I. (1975). *Belief, Attitude, Intention, and Behavior*. Reading, MA: Addison-Wesley.

FREWER, L. J., HOWARD, C. and SHEPHERD, R. (1996). The influence of realistic product exposure on attitudes towards genetic engineering of food. *Food Quality and Preference*, **7**(1), 61–67.

FREWER, L. J., HOWARD, C., HEDDERLEY, D. and SHEPHERD, R. (1997a). Consumer attitudes towards different food-processing technologies used in cheese production – the influence of consumer benefit. *Food Quality and Preference*, **8**(4), 271–280.

FREWER, L. J., HOWARD, C. and SHEPHERD, R. (1997b). Public concerns in the United Kingdom about general and specific applications of genetic engineering: risk, benefit, and ethics. *Science, Technology, & Human Values*, **22**, 98–124.

FREWER, L., SCHOLDERER, J. and LAMBERT, N. (2003). Consumer acceptance of functional foods: issues for the future. *British Food Journal*, **105**, 714–731.

FREWER, L., LASSEN, J., KETTLITZ, B., SCHOLDERER, J., BEEKMAN, V. and BERDAL, K. G. (2004). Societal aspects of genetically modified foods. *Food and Chemical Toxicology*, **42**(7), 1181–1193.

GASKELL, G., BAUER, M. W., DURANT, J. and ALLUM, N. C. (1999). Worlds apart? The reception of genetically modified foods in Europe and the US. *Science*, **285**, 384–387.

GASKELL, G., ALLUM, N., BAUER, M., DURANT, J., ALLANSDOTTIR, A., BONFADELLI, H., *et al.* (2000). Biotechnology and the European public. *Nature Biotechnology*, **18**, 935–938.

GASKELL, G., TEN EYCK, T., JACKSON, J. and VELTRI, G. (2004). Public attitudes to nanotechnology in Europe and the United States. *Nature Materials*, **3**, 496.

GREENWALD, A. G., MCGHEE, D. E. and SCHWARTZ, J. L. K. (1998). Measuring individual differences in implicit cognition: the implicit association test. *Journal of Personality and Social Psychology*, **74**, 1464–1480.

GRUNERT, K. G. and VALLI, C. (2001). Designer-made meat and dairy products: consumer-led product development. *Livestock Production Science*, **72**, 83–98.

HENSON, S. (1995). Demand-side constraints on the introduction of new food technologies – the case of food irradiation. *Food Policy*, **20**(2), 111–127.

HUOTILAINEN, A. and TUORILA, H. (2005). Social representation of new foods has a stable structure based on suspicion and trust. *Food Quality and Preference*, 16, 565–572.

KAISER, H. and TANG, X. (2004). *Nanotechnology 2015 and Converging Markets.* Tübingen: HKC22.com.

KAJANNE, A. and PIRTTILÄ-BACKMAN, A. M. (1996). Toward an understanding of Laypeople's notions about additives in food: clear-cut viewpoints about additives decrease with education. *Appetite*, **27**(3), 207–222.

KASPERSON, J. X., KASPERSON, R. E., PIDGEON, N. and SLOVIC, P. (2003). The social amplification of risk: Assessing fifteen years of research and theory. In N. Pidgeon, R. E. Kasperson and P. Slovic (Eds.), *The Social Amplification of Risk* (pp. 13–46). Cambridge: Cambridge University Press.

KIRK, S. F. L., GREENWOOD, D., CADE, J. E. and PEARMAN, A. D. (2002). Public perception of a range of potential food risks in the United Kingdom. *Appetite*, **38**, 189–197.

LÄHTEENMÄKI, L., GRUNERT, K., UELAND, O., ASTRÖM, A., ARVOLA, A. and BECH-LARSEN, T. (2002). Acceptability of genetically modified cheese presented as real product alternative. *Food Quality and Preference*, **13**, 523–533.

LAMOTE, S., HERMANS, D., BAEYENS, F. and EELEN, P. (2004). An exploration of affective priming as an indirect measure of food attitudes. *Appetite*, **42**, 279–286.

MAGNUSSON, M. K. and HURSTI, U. K. K. (2002). Consumer attitudes towards genetically modified foods. *Appetite*, **39**(1), 9–24.

MAGNUSSON, M. K., ARVOLA, A., HURSTI, U. K. K., ABERG, L. and SJÖDEN, P. O. (2003). Choice of organic foods is related to perceived consequences for human health and to environmentally friendly behaviour. *Appetite*, **40**(2), 109–117.

MAISON, D., GREENWALD, A. G. and BRUIN, R. H. (2004). Predictive validity of the implicit association test in studies of brands, consumer attitudes, and behavior. *Journal of Consumer Psychology*, **14**, 405–415.

OLIVEIRA, I. B. and SABATO, S. F. (2004). Dissemination of the food irradiation process on different opportunities in Brazil. *Radiation Physics and Chemistry*, **71**(1–2), 495–499.

OZEN, B. F. and FLOROS, J. D. (2001). Effects of emerging food processing techniques on the packaging materials. *Trends in Food Science* and *Technology*, **12**, 60–67.

PRISLIN, R., WOOD, W. and POOL, G. J. (1998). Structural consistency and the deduction of novel from existing attitudes. *Journal of Experimental Social Psychology*, **34**, 66–89.

RAUDENBUSH, B. and FRANK, R. A. (1999). Assessing food neophobia: the role of stimulus familiarity. *Appetite*, **32**, 261–271.

ROININEN, K., LÄHTEENMÄKI, L. and TUORILA, H. (1999). Quantification of consumer attitudes to health and hedonic characteristics of foods. *Appetite*, **33**(1), 71–88.

ROZIN, P., SPRANCA, M., KRIEGER, Z., NEUHAUS, R., SURILLO, D., SWERDLIN, A., *et al.* (2004). Preference for natural: instrumental and ideational/moral motivations, and the contrast between foods and medicines. *Appetite*, **43**, 147–154.

SABA, A. and VASSALLO, M. (2002). Consumer attitudes toward the use of gene technology in tomato production. *Food Quality and Preference*, **13**(1), 13–21.

SCHOLDERER, J. and FREWER, L. J. (2003). The biotechnology communication paradox: experimental evidence and the need for a new strategy. *Journal of Consumer Policy*, **26**, 125–157.

SCHOLDERER, J. and GRUNERT, K. G. (2005). Consumers, food and convenience: the long way from resource constraints to actual consumption patterns. *Journal of Economic Psychology*, **26**(1), 105–128.

SHIU, E. C. C., DAWSON, J. A. and MARSHALL, D. W. (2004). Segmenting the convenience and health trends in the British food market. *British Food Journal*, **106**, 106–127.

SIEGRIST, M. (1998). Belief in gene technology: the influence of environmental attitudes and gender. *Personality and Individual Differences*, **24**, 861–866.

SIEGRIST, M. (1999). A causal model explaining the perception and acceptance of gene technology. *Journal of Applied Social Psychology*, **29**, 2093–2106.

SIEGRIST, M. (2000). The influence of trust and perceptions of risks and benefits on the acceptance of gene technology. *Risk Analysis*, **20**, 195–203.

SIEGRIST, M. (2003). Perception of gene technology, and food risks: result of a survey in Switzerland. *Journal of Risk Research*, **6**, 45–60.

SIEGRIST, M. and BÜHLMANN, R. (1999). Die Wahrnehmung verschiedener gentechnischer Anwendungen: Ergebnisse einer MDS-Analyse. *Zeitschrift für Sozialpsychologie*, **30**, 32–39.

SIEGRIST, M. and CVETKOVICH, G. (2000). Perception of hazards: the role of social trust and knowledge. *Risk Analysis*, **20**, 713–719.

SIEGRIST, M., CVETKOVICH, G. and ROTH, C. (2000). Salient value similarity, social trust, and risk/benefit perception. *Risk Analysis*, **20**, 353–362.

SLOVIC, P. (1987). Perception of risk. *Science*, **236**, 280–285.

SLOVIC, P., FLYNN, J. H. and LAYMAN, M. (1991). Perceived risk, trust, and the politics of nuclear waste. *Science*, **254**, 1603–1607.

SLOVIC, P., FINUCANE, M., PETERS, E. and MACGREGOR, D. G. (2002). The affect heuristic. In

T. Gilovich, D. Griffin and D. Kahneman (Eds.), *Heuristics and Biases: The Psychology of Intuitive Judgment* (pp. 397–420). Cambridge: Cambridge University Press.

SLOVIC, P., FINUCANE, M. L., PETERS, E. and MacGREGOR, D. G. (2004). Risk as analysis and risk as feelings: some thoughts about affect, reason, risk, and rationality. *Risk Analysis*, **24**, 311–322.

SMITH, E. R. and DECOSTER, J. (2000). Dual-process models in social and cognitive psychology: conceptual integration and links to underlying memory systems. *Personality and Social Psychology Review*, **4**, 108–131.

SPARKS, P. and SHEPHERD, R. (1994). Public perceptions of the potential hazards associated with food production and food consumption: an empirical study. *Risk Analysis*, **14**, 799–806.

SPARKS, P., SHEPHERD, R. and FREWER, L. J. (1995). Assessing and structuring attitudes toward the use of gene technology in food production: the role of perceived ethical obligation. *Basic and Applied Social Psychology*, **16**, 267–285.

STEPTOE, A., POLLARD, T. M. and WARDLE, J. (1995). Development of a measure of the motives underlying the selection of food: the food choice questionnaire. *Appetite*, **25**, 267–284.

TENBÜLT, P., DE VRIES, N. K., DREEZENS, E. and MARTIJN, C. (2005). Perceived naturalness and acceptance of genetically modified food. *Appetite*, 45, 47–50.

TOWNSEND, E. and CAMPBELL, S. (2004). Psychological determinants of willingness to taste and purchase genetically modified food. *Risk Analysis*, **24**, 1385–1393.

TUORILA, H. and CARDELLO, A. V. (2002). Consumer responses to an off-flavor in juice in the presence of specific health claims. *Food Quality and Preference*, **13**, 561–569.

TUORILA, H., ANDERSSON, A., MARTIKAINEN, A. and SALOVAARA, H. (1998). Effect of product formula, information and consumer characteristics on the acceptance of a new snack food. *Food Quality and Preference*, **9**, 313–320.

TUORILA, H., LÄHTEENMÄKI, L., POHJALAINEN, L. and LOTTI, L. (2001). Food neophobia among the Finns and related responses to familiar and unfamiliar foods. *Food Quality and Preference*, **12**, 29–37.

URALA, N. and LÄHTEENMÄKI, L. (2003). Reasons behind consumers' functional food choices. *Nutrition* and *Food Science*, **33**, 148–158.

URALA, N. and LÄHTEENMÄKI, L. (2004). Attitudes behind consumers' willingness to use functional foods. *Food Quality and Preference*, **15**(7–8), 793–803.

VAN KLEEF, E., VAN TRIJP, H. C. M. and LUNING, P. (2005). Functional foods: health claim-food product compatibility and the impact of health claim framing on consumer evaluation. *Appetite*, 44, 299–308.

VERBEKE, W. (2005). Consumer acceptance of functional foods: socio-demographic, cognitive and attitudinal determinants. *Food Quality and Preference*, **16**(1), 45–57.

VERBEKE, W. (2006). Functional foods: consumer willingness to compromise on taste for health. *Food Quality and Preference*, 17, 126–131.

VERLEGH, P. W. J. and CANDEL, M. J. J. M. (1999). The consumption of convenience foods: reference groups and eating situations. *Food Quality and Preference*, **10**(6), 457–464.

YE, X., AL-BABILI, S., KLÖTI, A., ZHANG, J., LUCCA, P., BEYER, P., *et al.* (2000). Engineering the provitamin A (*β*-carotene) biosynthetic pathway into (carotenoid-free) rice endosperm. *Science*, **287**, 303–305.

Part II

Determining consumers' food-related attitudes for food product development

7

Methods to understand consumer attitudes and motivations in food product development

D. Buck, Product Perceptions Ltd, UK

7.1 Introduction

Products achieve market success when they meet a genuine need at an appropriate price; when their benefits are communicated well to the right consumers; when they are distributed sufficiently widely to be easily available for purchase; and when they deliver to the expectations generated by their positioning. Consumer research is important in getting all of these elements right and in maximising the chance of success. This chapter considers ways of conducting consumer research to understand consumer attitudes and motivations and the role these play in food product development.

Both qualitative and quantitative research methods can be used variously in these contexts. Qualitative methods tend to be useful in all exploratory phases and in achieving depth of understanding while quantitative methods can confirm hypotheses emerging out of qualitative research and define the extent and size of opportunities.

7.2 Qualitative methods

7.2.1 Overview

Qualitative research is an established discipline, drawing on diverse roots such as social psychology and anthropology. As an industry, it grew rapidly in the 1970s and has become ever more frequently used to elicit valuable insights into attitude, behaviour, emotion and motivation, based on extended interviewing of small numbers of respondents. This is conducted either singly with individuals, one at a time, or in groups. These are the essential methods of qualitative research: so-called depth interviews or group discussions. However, within these two principal approaches, researchers might use a variety of techniques to elicit specific insights. These include projective and enabling techniques; imagery associations; laddering; regression to the point of consumption; and reconvened memory.

There are, of course, other forms of qualitative research important in developing food products. Of these, observation and ethnography are of foremost importance and will be considered further within this chapter. In the context of product development research, qualitative research does not claim to be statistically representative but offers insights into the aforementioned attitudes and behaviours with respect to products and brands and consumers' interactions with them.

7.2.2   Depth interviews

Depth interviews are typically conducted one on one, between an experienced qualitative researcher and a single respondent. The interview can take anything from 30 minutes to a couple of hours. Usually, the interview is recorded on to tape or minidisk, so that the interviewer can maintain the flow of interaction and give full attention to the interview rather than to taking notes or completing some form of questionnaire. Although the interview will follow some structure, the interviewer is able to follow interesting threads and develop themes relevant to the subject.

The depth interview is of particular benefit where insight is required into individual behaviour or attitude without the influence of peers. The length of the interview and the rapport established between interviewer and subject allows investigation of complicated or lengthy processes such as more complex buying decisions in full detail. The individual approach also means that more personal, even intimate, information can be discussed.

Laddering
Depth interviewing lends itself particularly well to techniques such as laddering. Laddering is a practical means of investigating the means–end chain theory. This recognises that people are attracted to specific product or brand attributes because the consequences of these attributes satisfy basic values of importance to them as individuals. However, respondents may not be aware of the connection between them and are unlikely to offer the information unprompted. Laddering is a questioning framework to unpeel these levels of information and identify the basic values and needs.

Starting with the product attributes of interest and importance to the respondent, the interviewer moves to a more abstract level in getting the subject to voice the benefits that he or she gets from each attribute. Finally, the interviewer probes further to identify the emotional values to which the benefits contribute.

For example, the attribute 'inclusion of solely organic ingredients' in a food product may generate consequences of 'consuming fewer additives' and 'requiring farmers to use fewer chemicals', thereby satisfying emotional values of 'family welfare' and 'ecological concern'.

Means–end chains are personal to individuals, so it is possible that another respondent will reach the same value via a quite different chain. Hence, they may appreciate the attribute 'cheapest possible ingredients' with the consequence 'best use of weekly budget' to reach the same value: 'family welfare'.

Laddering across a number of depths builds up such structures and groups together respondents with similar chains. It allows us to attach a perspective to the patterns that arise, and to understand who will be attracted to products and why. It also provides a framework for re-positioning or re-developing a product to maximise consonance and appeal. Software such as Laddermap and Skim Analytical's SOAP (Structured Open Association Pattern) are available to identify and link the different levels of information.

Repertory grid

The repertory grid approach also works particularly well in depth interviews. Originally devised by Kelly (1955) in a clinical psychology application, it enables researchers to gain insight into how individuals see the world. In the context of product development, stimuli (say brands, but equally it could be packs or products) are presented in groups of three. Also known as the method of triads, respondents split the three stimuli into a pair and a single brand and describe the similarity between the paired brands and the differences to the single brand. The reasons for splitting the three stimuli are usually referred to as 'constructs'. This is repeated across different combinations of three brands until the respondent has identified all possible constructs. Often, as a final step, respondents score the brands across all the constructs.

Repertory grid analysis tells us how consumers see brands or products and on which dimensions they differentiate them. If new brands or products are introduced into the process it identifies the competitive set and tells us how well the new entry meets the identified needs.

Depth interviewing plays an important role in informing product development but it does suffer some disadvantages. It is costly and takes a considerable time both in the interviews themselves and in transcribing and analysing the recordings. For these reasons and in many cases because of the more interactive dynamic operating in a group discussion, many researchers prefer the latter approach.

7.2.3 Group discussions

Groups are valued within the development process because they *can* generate bigger ideas than any individual can create. With careful moderation, they can develop ideas in ways that never could have been predicted in advance. They can promote interaction between respondents to spark creativity and spontaneity.

In overall terms they are quicker to conduct than depth interviews and clients can directly view the discussion in specialist facilities with one-way mirror viewing facilities (indeed these days, clients can observe groups from the computer on their desk thanks to web-cam and Internet viewing capabilities).

Groups usually comprise six to nine selected respondents. Often recruitment for groups specifies a purposive approach to ensure that respondents meet specific criteria. Hence, the eight respondents for a group discussing frozen fish might include two single person householders, two mothers of children at primary school, two mothers of teenage children and two empty nesters. Of the two respondents of each type, one might be a current buyer of the client's products; the other might be a lapsed buyer, who nowadays buys other brands.

For best results, the moderator convening the group will be a trained qualitative researcher, skilled in steering the discussion to address all the areas of importance. He or she will ensure that individuals in the group all have opportunity to participate equally and, therefore, that no one individual dominates the group or is subjugated completely by other members. The moderator will observe and note both verbal and non-verbal responses between respondents in addition to those directly between the moderator and respondents.

Moderators work to a discussion guide: a document agreed between the client or sponsor of the research and the moderator. It sets out the principal areas of interest and a rough time plan, or at least a sequence for when to introduce new concepts or activities to the group. This is not an entirely prescriptive document. It is quite possible for respondents to follow a theme not envisaged prior to the group. Indeed part of the moderator's job is to recognise important areas and steer the group to explore these in full detail. However, some aspects of the guide may be fixed, such as, for example, the agreed order of introduction of the concept and specific product formulations.

Groups provide a less intimidating environment for respondents than one-on-one interviews. The moderator will go to some lengths to ensure respondents are at their ease and that their opinions are valued. Respondents tend to be more open because at least someone else in the group usually shares their views. Even when this is not the case, respondents are encouraged in their views by the non-judgemental atmosphere in the group.

Moderators use a variety of approaches to gain understanding within the group. Those of most relevance to product development are described below.

Product placement
If prototype products are available, respondents often taste them during the course of the group. As an alternative, they may be placed with respondents for evaluation before or after the group. Groups pre-placed with products will bring information on palatability and overall product acceptability while those post-placed allows investigation of synergy with the concept. Post-placement usually takes place in conjunction with reconvening the group for a second discussion session.

The value of placing the products with respondents outside the group is that respondents taste the products at an appropriate time and in an appropriate place. Respondents assess the products individually with no peer influence. It can be a considered, full evaluation rather than just a taste or a sip of the product. Sharing the results of their evaluations with other respondents in the convened group provides a rich opportunity for discussion about product acceptability, context(s) of use, deficiencies and benefits, and so on.

Product/brand mapping
Here we are referring to a physical mapping process during the group discussion using props representing brands or maybe using actual products themselves. Here, individual respondents or subgroups or the entire group work together to place physically, brands or products they perceive as similar, close together and far away from brands they perceive are very different. After the exercise, respondents responsible for the map describe the dimensions of difference they have used to combine/separate brands. They may be asked to remap the brands using different criteria or a different context of use.

Respondents might map brands or products in a category, or they might use cross-category brands or products that compete for a particular type of consumption occasion. For example in investigating a peckish-snack consumption occasion, respondents may map bagged potato snacks, sweet biscuits, chocolate countlines, fresh fruit and so on. In this instance, the first map produced by respondents would be most likely to be an obvious separation on inter-category criteria. Here it is the re-mapping that is likely to tell us more about the emotional differences between the products and their specific benefits in use for specific occasions.

Such exercises provide an instant overview of respondents' views of brands or products in a category. It can challenge the way developers and marketers perceive the market and thereby identify gaps in the market. If new products are included in the exercise it indicates where they fit and how directly existing brands compete with the new entrant.

Regression to the initial point of consumption
Moderators often use this device in product development contexts. They ascertain the first ever point of consumption of the product category. Alternatively, they may ascertain the first occasion of consumption when respondents started to really appreciate a product rather than eat or drink it solely to impress peers or at least to accede to peers' behavioural norms. So, for example in understanding appeal of a new coffee drink it is revelatory to establish the circumstances of initial trial in the category. For some it will represent a rite of passage or class-boundary change. For others it may be linked with introduction via an early girlfriend or boyfriend. Hence, for some people the emotional backdrop to appreciation may be 'respect' or 'achievement' while for others may be 'sensual' or 'naughty'. This understanding helps shape the communication of the new product or even may influence the formulation of the product itself.

Reconvened groups

Researchers may use this technique in conjunction with product placement, discussed earlier. Reconvening the same group of respondents is useful to expose respondents to ideas, concepts or products; then give them an opportunity to consider issues or products outside the confines of the group. Finally the group is reconvened to share and discuss their new experiences with the moderator and co-respondents.

Another objective in reconvening a group is to undertake exercises such as sequential-recycling or rapid-iteration development. In this approach a group is reconvened on a number of occasions with the reactions from the group used to modify the stimuli to counter the group's criticisms in the meantime. These techniques are at their best for creative development of concepts but are also used with products to speed up the development process. In either context, researchers advise that it would be far better to convene matched, fresh groups at each occasion rather than reconvening the same respondents. In this way, the danger is minimised of developing a product that optimally meets the requirements of just that particular group of respondents.

Disadvantages of group discussions

Group discussions do have some disadvantages. Peer pressure, despite the best efforts of moderators, may inhibit respondents' open exposition of views or encourage them to exaggerate their experiences. In addition, the intense nature of the group may influence respondents to register greater interest in a new product or concept than they would in normal circumstances.

Below we consider two uses of qualitative techniques in food product development where quantitative techniques are inappropriate or would fail to deliver to the objectives. Finally, their benefits and limitations are summarised.

7.2.4   Qualitative case study 1: Screening potential development routes for a known opportunity where the execution is unclear

Known opportunity

People want to eat more fruit but do not like the inconvenience of fresh fruit (e.g. potential for sticky juice to spray, dribble onto clothes, need to peel, need to dispose of peel/core).

Potential routes for development

Depending on the company developing the product, perhaps it could be some sort of fruit bar; perhaps an extra thick juice or purée in a bottle with a 'sports-cap'; perhaps it could be a suckable lozenge with new gel technology to release flavour and texture at different rates. Alternatively, it could be something else entirely.

Research approach

At this stage in the product's development, it makes no sense to use quantitative methods. There are too many variables and not enough understanding to set up a useful structured quantitative task. This will come later when the major elements of product format are fixed.

Qualitative research here, will firstly investigate:

- the virtues of fruit as perceived by consumers;
- which fruits best exhibit which virtues;
- how well do current, marketed, fruit products capture these virtues;
- how we can narrow the focus? Perhaps to deliver 'freshness of taste without cloying sweetness' together with ability to 'eat on the move'.

R&D may have developed some prototype products at this stage, e.g. for a fruit bar execution there may be versions that combine fruit jelly or fruit pieces with marshmallow or with cereal. There may be different consistencies or fruit varieties in combination. There may be different sweetness levels and so on.

For some groups, respondents could have these prototype products placed for trial prior to the group convening; others could taste prototypes subsequent to the understandings about fruit virtues being investigated and after the product-concept has been introduced.

From qualitative research we would expect to understand which virtues are most valued by consumers and focus the product format to best deliver those virtues. The product will be unfinished at this stage, but within the much-narrowed range of variables, it will now be possible to use quantitative approaches to fine tune the formulation and evaluate overall appeal to the target population.

7.2.5   Qualitative case study 2: understanding consumers' perceptions of 'smoothness' to guide developers in the formulation of a cream liqueur

Research issue

Smoothness in this context is a very desirable perception. But what kind of smoothness is required by consumers? Do they mean a taste or a texture or both? Does it relate purely to viscosity or coating characteristics? How can diverse stimuli, all potentially offering some perception of smoothness, help the developer to achieve his or her goal?

In this instance, a quantitative approach of presenting respondents with the stimuli and evaluating each against a standard questionnaire would suffer badly in meeting the objective. Undoubtedly, halo effects would mislead the true smoothness measures. The approach also would lack the ability to empower respondents to describe accurately their reactions.

Potential approaches

In investigating this issue via group discussions, moderators might utilise product-mapping exercises: asking respondents to group products and then explain their grouping strategy.

Moderators might also conduct enabling projective techniques. These unlock parts of respondents' brains that otherwise might remain unaccessed. For example, it may be effective to ask respondents to assign or associate images torn out from magazines or supplied items of different textures with the various products or product groupings. In this way it could be that some products or product groups are associated with a smooth pebble while others are associated with velvet. Clearly it requires skill to employ these techniques effectively and lots of interpretation to draw useful insights from their outcome but dimensions of difference can be identified that never would have emerged from straight questioning.

In this instance, it would be helpful to integrate some form of sensory evaluation into the process (see Chapter 13). Investigating consumers' perceptions of smoothness in the groups while having a sensory profile of products is very powerful. It allows a link between objective information on how the products are perceived by the senses together and the emotional descriptive response. This informs the product developer as well as the marketing team.

The value of this parallel approach of conducting sensory evaluation alongside qualitative research is immense. Some companies take this approach further: in the UK: Product Perceptions' Sensory Fusion and mmr's SensoEmotional Optimisation seek to align sensory properties of products with their emotional value to consumers, to drive purchase towards receptive consumers. In so doing, they subsequently build adoption of products into those consumers' repertoires.

7.2.6 Observation and ethnography

Observation is a very powerful research technique. The opportunity to observe and record real (as opposed to reported) behaviour provides strong insights for product developers, into, for example, the use (and 'normal abuse') of products, the nature of the competitive set and unmet consumer needs. When observed behaviour is compared with reported behaviour from the same individuals, interesting insights emerge into subjects' perceptions (and sometimes self-delusions) of their world.

Observation can be passive or active. Passive observation involves merely recording actions and behaviours as they occur in a given situation. In this, there is no active observer effect, although the very presence of the observer may influence the respondent's behaviour. In some situations an observer's presence may be completely incidental to behaviour. In retail store observation, where the observer is not 'shadowing' a single subject but instead passively watching the selection and purchase of products by a number of subjects, the presence of the observer is unlikely to have any major effect.

Where a single subject is the sole focus of the observer, some level of observer effect must exist but there are strategies available to minimise it, such as extended observation. Here, the subject becomes so used to having an observer present that they revert to natural behaviour. Similarly, where the approach is applicable, subjects rapidly become conditioned to the presence of video-cameras. Obviously the use of a fixed video camera either running continuously, or triggered by some event (e.g. water flow, movement, switching a kettle on) has limitations in terms of the breadth of information that can be gathered and specialist agencies such as everydaylives (www.edlglobal.net) have emerged with creative video-recording solutions for valid and useful ethnographic interpretation.

Extended observation of a single subject at a time in a retail context often takes place in the form of accompanied shopping, where the observer shadows a respondent throughout a shopping trip in which the product or products of interest are likely to be considered, if not actually purchased. Again, observation can be passive: just recording events and their sequence, or more usually subjects can be questioned, later, about their strategies and decisions. Accompanied shops are of obvious interest in a marketing context: in understanding product/brand appeal, shelf-standout, identification of the competitive set and so on, and as such are relatively widely used.

In food product development, however, observation as an approach tends to be underused, despite offering a uniquely useful source of information right through the product development cycle, from identifying unmet consumer needs at the very start, to evaluating how a product fits into people's lives, much later on in the development process.

In personal as opposed to video observation, an ethnographer shadows a respondent in their own home or other familiar environment (their place of work perhaps) to observe an event or sequence of events such as preparation of a meal or snack, or perhaps a beverage, in order to identify and record real behaviour associated with the process or product. In order to overcome the 'observer-effect' it may be necessary for the observer almost to take up residence and observe over a number of days. However, ethnographers tend to have an ability to become assimilated quickly into a home and minimise the effect of their presence. Such specialists often also use a gentle questioning style that collects relevant background information close to behavioural events' occurrence but does not lead the subject to modify their current or subsequent behaviour. From this approach, it is possible to get much closer to real behaviour, and the motives and needs behind it, than could ever be achieved simply by asking respondents about their behaviour in a group discussion or a depth interview, or in a quantitative usage and attitude (U & A) study.

For example, the background information for a study on convenience foods showed that a frequent response among higher social grade mothers of school-age children reported use of ready meals 'only in emergencies'. Observation conducted among some of these subjects tells a contrasting story. First of all, despite their reported low use, subjects' freezers were very well stocked with ready meals. Observation across a number of evenings provided the answer. Subjects were not lying: they *did* use ready meals only in emergencies. However, it emerged that *many* evenings *were classed* as emergencies. For example, on one night: 'I don't get home from Pilates until 4:30 and the kids are starving', and the next night: 'well, Johnny has football practice and Trish goes to Girl Guides', all turned out to be emergencies, as did many such evenings filled with all the normal toing and froing of a young, busy family!

Similarly, respondents may be reluctant in a straight interviewing situation to report problems (i.e. unmet needs, to a product developer) because they have devised strategies, sometimes subconscious, for dealing with them. Hence, for example a respondent may not comment in a standard market research interview on the extent to which salt blocks up their salt-cellar when their kitchen-diner is somewhat steamy from cooking. However, observation records them frequently resorting to opening the filling-aperture of the salt cellar, tipping some into a hand, sprinkling some onto their meal and then disposing of the rest in a waste bin. It is only when the problem is pointed out to them, usually when a new product (in this case: 'guaranteed free-running salt') is launched with a flourish that they recognise that a minor but real irritation can be avoided.

Observation, because of its focus and intensity also allows the recording of non-verbal signals that can tell us more than recorded answers: perhaps a transient, delighted, facial expression as a cooking aroma emerges when removing a saucepan lid, or a disapproving scowl as a subject wrestles with unyielding packaging. These spontaneous reactions can be much more useful than considered responses to an interviewer's questions.

Observation is a strong technique but it can be costly to conduct properly, which partially explains its under-use in product development. However, it can provide unique insights into products and related consumer behaviour that no other form of research can deliver.

7.3   Quantitative methods

7.3.1   Overview

The quantitative methods available are much more specific to the issue or context to be investigated. Hence U & A studies are used to identify needs or opportunities, or attitudes to the category of interest. Conjoint or trade-off studies quantify the value consumers put on various aspects of the product or concept (see Chapter 15). Specific approaches are available also to investigate the product itself, its pricing and its packaging. Wider-ranging studies may test many elements of the offer as an integrated package such as a concept-product test; a market-mix test or even a market potential study: designed to estimate product sales-volumes by measuring all the contributing elements.

7.3.2   Usage and attitude studies

U & A studies are used to identify needs or opportunities, by investigating behaviours and attitudes concerned with the category of interest. Hence, in the context of product development, U & A studies address a number of business issues such as: is there an unmet need for a new product? Is there an opportunity to refine my product to increase its use through greater convenience or perception of product quality?

Typical U & A studies are conducted among a large number of eligible respondents, often 500+, or even thousands, depending on objectives.

Interviews taking 25–45 minutes are common, with the questionnaire administered by a trained interviewer throughout. Respondents first provide demographic data before being questioned on a range of behavioural and attitudinal areas of investigation. These might include category use and purchase; brand image; frequency and weight of purchase; occasions of use; how product is prepared, cooked, served: to whom, when, with what accompaniments; and so on. The two main frameworks to collect this information: the traditional and the event-centred approaches are described below.

Traditional approach
The traditional approach to designing a U & A study involves building the questionnaire in a number of subsections, each collecting information across relevant areas using a 'general recall' approach. For example, a U & A study on orange juice drinking may have sections on purchase and consumption of orange juice in the home and out of the home. Specifically for the latter there may be subsections for consumption in licensed premises, in other establishments, and 'on the go'. Each respondent will contribute data to all relevant sections and subsections.

For drinking in licensed premises, the questionnaire may ask how often the respondent drinks orange juice in licensed premises, how many they drink per visit, whether they ask for a specific brand and so on. Now, for an individual respondent, there may be many and varied consumption occasions of this type. Some may be with friends, some alone, some with a partner. Some may be before going to a restaurant to eat or before going clubbing. Some may be occasions when the respondent is driving later, some not. Some may be very recent and others distant in memory. Without having a very specific and an extremely tedious questionnaire to investigate all of those possible situations separately, we ask a more general question. In effect, we rely on the respondent mentally weighing all the different types of relevant occasion to come up with an 'averaged' response. Hence, the answers will reflect respondents' recall and perceptions of 'average' behaviour or attitudes distilled across all such drinking occasions.

The benefit of this approach is that it is very efficient in collecting a lot of information. Every respondent contributes to every relevant section of the questionnaire. The responses from respondents are 'averages' but this process is an implicit, intuitive process, personal to each respondent and using their individual view of their own world. However, the downside is that respondents may distil their responses to a question in many different ways. Some might be reflecting the most recent experience. Others may respond based on whichever type of occasion is predominant to them. Some responses may suffer from classic recall biases such as telescoping or overclaim. Some responses will be biased towards social norms.

While it does not attempt to remove all such problems, a potentially superior approach for product development purposes is to use an 'event-centred' approach.

Event-centred approach

As the title suggests, this approach moves away from general questioning to focus on very recent purchase or consumption occasions. The rationale is that by so doing, respondents will be drawing from very recent memory. In addition, recall of behaviour and attitudes will be specific to individual occasions. Results therefore will be more likely to be valid and specific.

Typically, in this approach respondents are asked to think about the last two or three consumption occasions of a specific product or products within a category. Depending on objectives, these may be any occasions or they may be constrained by context: e.g. 'the last three times you drank orange juice in a bar, club or restaurant'.

For each occasion in turn, and starting with the most recent the respondent answers a battery of questions. These might include physiological state (hungry, thirsty, jaded, etc.); mood (happy, bored, content, excited, etc.); which products were considered; which was chosen; was it available (and if not what alternative was chosen); did the chosen product meet expectations; would the same product be chosen for similar occasions in the future; and so on.

The approach builds up quantitative information on frequency of use; the ratio of being chosen to being in the choice set; need states by occasion type and ability of products to meet specific need states. It can be very enlightening in revealing choice-sets for certain need states. For instance, a manufacturer may want to develop an ice cream to meet an indulgent/reward need. It may be a surprise to find that their ice cream is competing not only with other ice creams but with a much wider repertoire of competitors. Certainly these may include other obvious confectionery indulgence products such as chocolate count-lines. However, in emotional terms competitors may include products completely outside the category (such as a glass of wine), or may not even involve products at all (e.g. a luxurious bath).

The method lends itself well to modelling techniques so that potential changes to the product, its positioning, or distribution and salience can be used to predict increased share of choice.

The downside to this approach is that by limiting respondents to just a few recent events, less common usage occasions generate little information. Therefore, sample sizes need to be high to ensure sufficiently robust data for analysis. Other factors need consideration too. For instance, if a product is used differently depending on season, then an event-based study in just one season could mislead the interpretation.

Segmentation

Another major benefit in using U & A studies food product development is the ability to segment respondents (and hence infer segmentation in the target population) on the basis of behaviours, attitudes, needs and so on. The power of a large quantitative sample means that consumer segments can be identified validly and robustly and can be profiled in terms of demographics or behavioural or attitudinal variables other than those explicitly used to create the segments.

The value of segmentation is enormous. By its use we can determine the proportion of the population with specific needs or behaviours or attitudes and we can establish their characteristics. This knowledge can direct product development routes, give an early indication of viability, and define the positioning necessary to ensure that those consumers most likely to respond positively to the product try it in the first place.

For example, say a manufacturer is developing a new pasta sauce to include only very high-quality ingredients and be sold at a considerable premium, the research to establish reactions to it is only relevant if the sample comes from the true target market sample. Some large food companies have constructed general segmentation schemes derived from very large base studies. These identify fundamental food behaviour or attitudinal segments, e.g. experimentalists, food-faddists, ingredients-scrutinisers, premium-only snobs, cost-conscious buyers, etc. Discriminant analysis is then used to identify a much smaller number of questions to ask, to assign efficiently the segment a 'new' individual belongs to. Hence in subsequent, smaller-scale surveys conducted for specific product developments, it becomes possible to ensure that reactions to a new product are tested among the relevant groups of interest. Smaller food companies have to rely on similar schemes offered by agencies who specialise in food and drink research.

Gap analysis
Often, U & A studies, regardless of approach will include a brand-image matrix. Respondents identify, or rate, the extent to which brands are associated with which image descriptors. The data are mapped perceptually, most commonly using correspondence analysis. Inspecting and interpreting this map gives a broad view of which brands over- and under-trade in key image attributes. Areas of the map in which no current brand resides – 'gaps' – provide clues to potential product development.

Figure 7.1 shows a brand image perceptual map for individual cartons of fruit juice drinks conducted some years ago. The map has been simplified for ease of interpretation, to show just six brands and 11 image characteristics. For the reduced data, the market structure in perceptual terms is clear from the map.

Capri Sun and Um Bongo are clearly seen as fun and directed towards children. Capri Sun is rather more stylish but also less good value. Kia Ora and Quosh operate in the good value area. Ribena and Robinsons are seen as healthier, higher quality and containing a higher juice content. The attributes 'refreshing' and 'fruity', are equally associated with all the brands. Hence, they sit in the middle of the map. For this market, they are 'entry-level' attributes.

Aside from a means of understanding the general structure, the map can be used to generate hypotheses about potentially unfilled gaps in the market. Identifying the gaps is relatively easy from the map: finding the appropriate product and positioning to fill the gap, and establishing its viability is somewhat more difficult. In the figure, a potential gap exists in the top left-hand quadrant for a product somewhere between Ribena and Capri-Sun. To fill the gap, the

Fig. 7.1   Perceptual map of individual fruit-drinks.

product would need to be stylish, healthy, high quality with a high juice content but would also be fun and as much 'for me' as 'for kids'. This type of gap analysis identified the opportunity that led to the development of brands such as Oasis and Fruitopia.

7.4   Qualitative vs quantitative: are they really in competition?

Qualitative and quantitative approaches should not be regarded as straight alternatives. The most appropriate method to adopt is normally dictated by the research objectives. The strongest research is often a combination of qualitative and quantitative phases to answer appropriate aspects of the business issues.

For example, to choose the best of three or four potential formulations, a quantitative product test may be chosen. It provides evidence on which formulation wins among respondents who are representative of the target population. It also establishes whether the absolute levels of scores are sufficiently high compared with category norms, to believe the product will be well accepted in normal use. Although such a quantitative test normally will include a range of rating scales to provide diagnostic information on why any one product does well or badly, the addition of even quite limited qualitative research can enhance the understanding manifold. Simply conducting a small number of mini-depth interviews alongside the quantitative research will not only support the rated diagnostics but also provide the linkage between them, attach variable priority to them, even start to identify the emotional background and motivations linked with them.

A recent quantitative test of cereals aimed at very young children revealed that one formulation was much less liked than others and failed the action-standard for continued development. The quantitative test diagnostics revealed that the disliked formulation was too big but it was rated better for flavour. Mini-depth interviews got to the real truth. The cereal pieces were indeed too big but so much so that the kids could not easily get them into their mouths. The children felt a sense of failure and feared retribution because of the problem of eating the cereal without making a mess. Once they got the cereal into their mouths they were happy to concede it had a very good flavour indeed. Subsequently, the product was formed in a shape easier to put into the mouth and, in a retest, was more acceptable than any other formulation.

7.5   Future trends

7.5.1   Content analysis

Advances in analysis of qualitative data include content analysis and other means of quantifying some of the output. Software exists to count the usage of words or phrases in respondents' transcripts (e.g. Wordstat, T-Lab). Although this information may in some cases be insightful, many qualitative researchers are very wary of such analyses. They claim that use of language is very personal and so inter-group validity is likely to be poor. Groups are restricted in size and rarely recruited to be representative of a target population. Hence, while advocates of this quantification of qualitative output claim that analysis is based on the tens of thousands of words generated by a group, in the end it is still only eight to ten purposively recruited individuals who have uttered them.

7.5.2   Eye-tracking

Recently, increasing attention is being paid to techniques that record direct (as opposed to reported) human responses. Eye tracking is one such method (for a description of methods see Duchowski, 2003). A respondent sees a visual stimulus, presented usually via a monitor or large screen and specialised equipment tracks and records their eye-movements as they scan the stimulus. This is repeated with different stimuli and among other respondents. After the data have been collected, the eye-tracks are simplified, compared and examined for pattern similarity. While the effects are of interest in themselves, careful interpretation is important in trying to infer cause. Some agencies offering eye-tracking incorporate additional techniques to measure affect, such as skin-conductivity (tracking: sweating, temperature and heart-rate) or pupil-dilation to gauge emotional response. Developments are also afoot to integrate eye-tracking with functional magnetic resonance imaging (fMRI; see below).

Eye-tracking has strong applications in print and electronic media research: particularly in website design. In food-choice contexts, it also has obvious relevance for pack design and in shelf display investigations. For pack design research, knowing which elements or areas of a pack are of primary importance in consumers' visual information gathering can be fundamental to ensuring better communication of positioning or messages. Similarly, knowing which elements or areas are secondary, or those that are scanned hardly at all, provides vital information for improving the pack. In shelf-display research, objective information on how many brands are considered, for how long, and the order of consideration can provide vital clues on stand-out and context of choice.

7.5.3   Neuromarketing

Following research by Gerald Zaltman at Harvard and, later, a seminal study by researchers at the Neuroscience Department of Baylor College of Medicine (McClure *et al.*, 2004), the latest hot topic in the objective measurement of consumer response is the application of neuroscience to subjects' reactions to stimuli. With less intrusive means of registering brain activity it is possible to identify which areas of the brain are stimulated by ideas, concepts and products and what this means in terms of cognition, memory and emotion. In the context of brand and product development, this is encompassed by the term neuromarketing. Clearly, it is unlikely that qualitative researchers are going to equip themselves with electrodes or fMRI scanners to use regularly on their respondents; however, there are huge benefits to researchers from advances in neuroscience. It offers the opportunity to validate consumer market research approaches by correlating their results with analysis of brain activity. It will also help researchers to understand people more effectively in some fundamental areas. These include investigating theories such as 'hardwiring' where repeated cognitive processing, perhaps of a strong marketing message for instance, leads to cells becoming literally fused together. In turn, this leads to very strong memories and entrenched views (Whitehill, 2005). This knowledge helps to inform decisions on when and how to re-brand products in the marketplace. (For further information see: Zaltman, 2003.)

7.5.4   Mood and emotion measurement

Another significant advance is in quantitative collection of mood and emotional responses. This is clearly relevant in evaluating products designed or marketed to appeal to, or generate, specific emotions (in the past couple of years for example, there have been launches of herbal and green teas, and ice creams on a mood 'platform'). However, this is becoming more important generally, with regard to all products. Researchers have acknowledged that often it is possible to differentiate two equally liked products on the basis of specific emotional responses. The next stage, of measuring the implications of this response in terms of purchase or re-purchase rates has not been widely reported but it is a clear area of interest for the future. Recently a computer-based personal interviewing approach: Premo, has been developed by Pieter Desmet at Delft University to investigate emotional response to products (see, for example, Desmet *et al.*, 2001). This uses facial expressions and sounds from an animated character to depict emotions. It has been used very successfully in evaluating small appliances such as mobile phones, larger products such as cars, and could be extended to food and drink products, and fragrances.

7.6   References/sources of further information

DESMET, P.M.A., OVERBEEKE, C.J., TAX, S.J.E.T. (2001). Designing products with added emotional value: development and application of an approach for research through design. *The Design Journal*, **4**(1), 32–47.

DUCHOWSKI, A.T. (2003). *Eye Tracking Methodology: Theory and Practice*, London, Springer.

GORDON, W., LANGMAID, R. (1988). *Qualitative Market Research – a practitioner's and buyer's guide*, Aldershot, Gower.

KELLY, G.A. (1955). *Psychology of Personal Constructs, Volumes I and II*, New York, Norton.

McCLURE, S.M., LI, J., TOMLIN, D., CYPERT, K.S., MONTAGUE, L.M., READ MONTAGUE, P. (2004). 'Neural correlates of behavioural preference for culturally familiar drinks', *Neuron*, **44**(2), 379–387.

VAN KLEEF, E., VAN TRIJP, H. C. M., LUNING, P. (2005). 'Consumer research in the early stages of product development: a critical review of methods & techniques', *Food Quality and Preference*, **16**(3), 181–201.

WHITEHILL, C. (2005). 'You don't have to be a brain scientist', *In Depth*, Association for Qualitative Research, www.aqr.org.uk/indepth/summer2005

ZALTMAN, G. (2003). *How Customers Think: Essential Insights into the Mind of the Market*, Harvard, MA, Harvard Business School Press.

8

Using means–end chains to understand consumers' knowledge structures

A. Krystallis, National Agricultural Research Foundation, Greece

8.1 Introduction

Means–end chains (MECs) are hierarchical cognitive structures that model the basis for personal relevance by relating consumers' product knowledge to their self-knowledge. The lower levels of a means–end hierarchy contain relatively concrete knowledge about *product attributes* and their perceived linkages to the *functional consequences* of product use. These functional consequences may be associated with more abstract knowledge about the *psychological and social consequences* of product use. Finally, some means–end chains may connect these psychosocial consequences to abstract self-knowledge about the consumer's life goals and *values*.

Each of the key constructs in a MEC structure (attributes–consequences–values) can be further dichotomised to allow a more detailed analysis of the consumer knowledge structures: *concrete attributes* represent tangible, physical characteristics of the product; *abstract attributes* represent intangible, subjective characteristics; *functional consequences* are directly experienced tangible outcomes of product use; *psychological consequences* are more personal and less tangible outcomes; *instrumental values* are the cognitive representations of preferred modes of behaviour; *and terminal values* represent preferred end-states. Walker and Olson (1991) have suggested that in this six-level MEC, the three lower levels constitute the product-knowledge of consumers, while the three higher levels constitute consumers' self-knowledge (see Fig. 8.1).

Van Kleef (2006) includes MEC analysis in the consumer research methods to be followed in early stages of new product development (NPD), especially in the *opportunity identification* stage. In the numerous studies of NPD over the

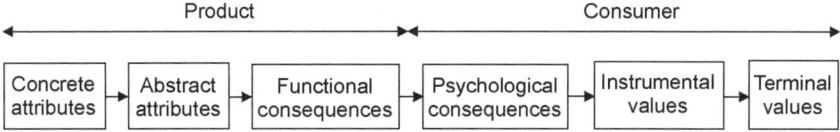

Fig. 8.1   Six-level means–end chains (from Walker and Olson, 1991).

years, agreement has developed that understanding consumer needs is of greatest strategic value, especially in the early stages of the NPD process. During these early stages, the aim is to search for novel product ideas. Successful NPD strongly depends on the quality and quantity of these ideas. Presumably, consumer research should improve this quality. Consumer research for opportunity identification reflects a more creative, proactive side of product development (van Kleef, 2006).

Effective consumer research for NPD must, among other characteristics, provide a detailed insight into the relation between product characteristics and consumers' need fulfilment. Consumer research for NPD is often thought of as existing of historical purchase information or product evaluations. However, understanding consumer behaviour encompasses much more than just getting insight into how consumers evaluate and purchase products and services (Jacoby, 1979, in van Kleef, 2006). Consumers tend to see products as more self-relevant or involving to the extent that their product knowledge about product attributes and their functional consequences of use is connected to their desirable psychosocial consequences (or benefits), goals and values (Gutman, 1982; Walker and Olson, 1991). To develop a superior new product, consumer research needs to identify consumers' product attribute perceptions, use benefits and personal values that provide the underlying basis for choosing products (van Kleef, 2006). As such, MEC analysis makes clear which crucial factors affect consumer perceptions, preferences and choices, and what trade-offs need to be made.

The purpose of this chapter is to introduce MEC analysis theory and the laddering data collection technique to the reader as a consumer research technique for NPD. The main part of the chapter is dedicated to the description of the MEC methodology. The method's broader conceptual model is introduced (Section 8.2), followed by a detailed description of the laddering technique (Section 8.3): interview environment, probing techniques, analysis of stages and criteria to assess the validity of the results. The chapter proceeds by emphasising the value of MEC and laddering for early stages of NPD (Section 8.4). Then follows a detailed presentation of the MEC-related literature of the last 25 years that points out past and current trends in the MEC research in various products and services, together with a short commentary on likely future trends (Section 8.5) and a short descriptive section of sources of further information (Section 8.6). The chapter closes with a case study taken from the food industry (Section 8.7), as an illustrative example of the value of the MEC analysis methodology in the marketing and consumer behaviour research area.

8.2   Conceptual model of MEC analysis theory

Several early attempts were made to provide a theoretical and conceptual structure connecting consumers' values to their behaviour (e.g. Young and Feigin, 1975; Vinson *et al.*, 1977). Part of these attempts can be subsumed under the title 'means–end chain'. 'Means' are objects (products) or activities in which people engage. 'Ends' are valued states of being, such as happiness, security or accomplishment. A 'means–end chain' is a conceptual construct that seeks to explain how a product or service selection facilitates the achievement of desired end-states. Such constructs form a model that consists of elements that represent the major consumer processes that link values to behaviour.

According to Gutman (1982), the MEC model is based on two fundamental assumptions about consumer behaviour: (a) values, defined here as desirable end-states of existence, play a dominant role in guiding choice patterns; and (b) people cope with the tremendous diversity of products that are potential satisfiers of their values by grouping them into sets or classes so as to reduce the complexity of choice. This suggests that, in addition to the product-class type of product categories, consumers are capable of creating categories based on product functions. In addition to these two assumptions about consumers' behaviour that are essential to the model, there are two other assumptions of a more general nature: (c) all consumer actions have consequences (although all consumers would not agree that the same actions in the same situation produce the same consequences); and (d) consumers learn to associate particular consequences with particular situations (Gutman, 1982).

'Consequences' may be defined as any result (physiological or psychological) accruing directly or indirectly to the consumer from his/her behaviour. Consequences can be desirable or undesirable. There is a literature in marketing dealing with desirable consequences called benefits (e.g. Haley, 1968; Myers, 1976) or consumer motivations, which are the advantages consumers enjoy from the consumption of products. Benefits differ from attributes in that people receive benefits whereas products have attributes. Consequences may be *physiological* in nature (satisfying hunger, thirst or other need), *psychological* (self-esteem, improved outlook for the future) or *sociological* (enhanced status, group membership). *Direct* consequences come directly from the thing consumed or from the act of consumption. *Indirect* consequences can occur when other people react favourably or unfavourably to us because of our consumption behaviour.

The central aspect of the model is that consumers choose actions that produce desired consequences (benefits or motives) and minimise undesired ones (Gutman, 1982). It is suggested that values provide consequences with positive or negative valences. Therefore, the *values–consequences linkage* is one of the critical linkages in the model. To the extent that values are ordered in importance, they also give respective importance to consequences. An act of purchasing or consumption must occur in order for the desired consequences to be realised. Thus, a choice among alternative products has to be made. In order to make this choice, the consumer has to learn which products have attributes that will produce these desired consequences. Therefore, the second important *linkage* in the model is that between *consequences and product attributes.*

It is essential for consumers to reduce the complexity inherent in the multi-tude of alternatives with which they are faced. Although grouping is determined by the object's properties, the choice of properties to be focused on is influenced by values. Consumers group products in different categories depending on which features they emphasise and which they ignore. This means that values are translated from their context at the more abstract or inclusive levels of the chain to the less abstract, where products are categorised into classes. Consumers are more likely to be in agreement on what physical characteristics products possess than at high levels of abstraction or inclusiveness, where categories are based on the function of value-producing consequences. This categorisation process takes place at each level of the means-end chains as categories of greater *inclusiveness* are formed at higher and higher levels of the chain (according to Gutman, 1982, 'inclusiveness' refers to the degree of similarity among objects in a category).

To summarise, the MEC model may be conceptualised as shown in Fig. 8.2. The method of studying how consumers organise their thinking about specific product alternatives has been named the 'categorisation process'. It is hypothe-sised that consumers create arrays of products (categories) that will be instru-mental in helping them achieve their desired consequences (motives), which in turn move consumers towards valued end-states. If this connection can be made, marketing will be in a better position to understand how personal values influence everyday consumer choices (Gutman, 1982). The marketplace is full of many more objects than individuals have values. Therefore, 'ends' are few, but 'means' are many (Vinson *et al.*, 1977).

According to Grunert and Grunert (1995), the MEC analysis concept can be regarded from two different points of view: the *motivational* and the *cognitive structure* view: the motivational view is that MEC is concerned with obtaining insight into consumer's buying motives and attitudes towards purchasing or consuming, e.g. in the way basic motives are linked to shopping behaviour. The process of data collection for MEC models (to be described later) can give valuable insights by prompting consumers to reflect on their buying motives in a way not typical for daily shopping behaviour. Such insight is necessarily qualitative in its character. The main criterion for evaluating the usefulness of the approach would be to what extent users of the results feel that they have achieved a better understanding of consumers that gives inspiration to managers and helps them make better business decisions.

The cognitive structure view is that MEC analysis is a model of consumers' consumption-relevant cognitive structure, e.g. of the way consumption-relevant knowledge is stored and organised in human memory. A basic hierarchical model is assumed, in which cognitive categories of different levels of abstraction are interlinked in chains and networks. It is assumed that behavioural motivation is derived by linking cognitive categories corresponding to concrete products with cognitive categories at a high degree of abstraction, like values. It

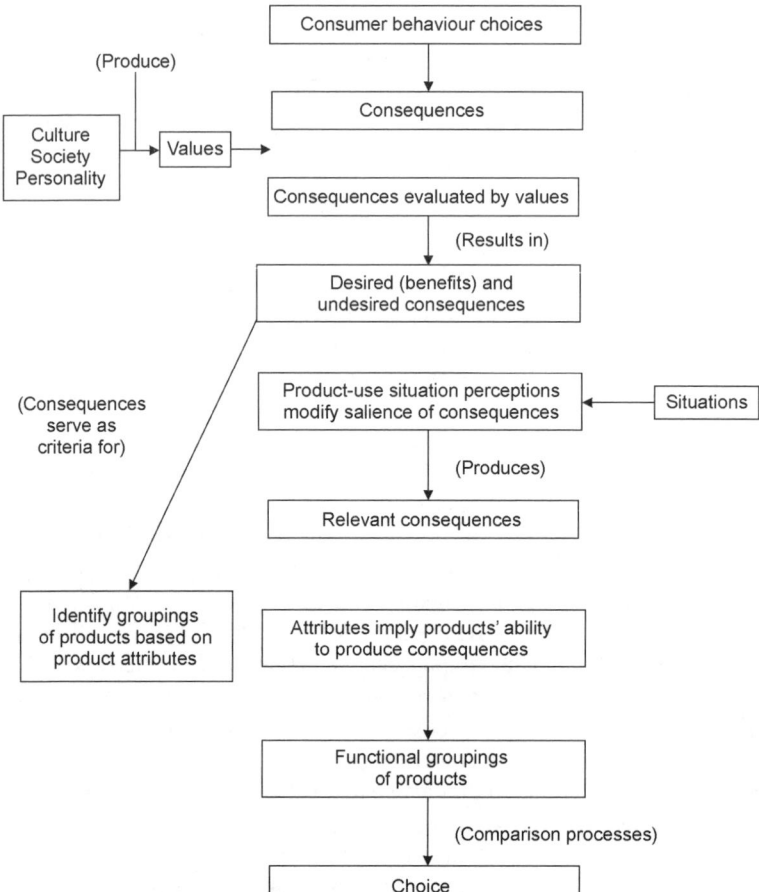

Fig. 8.2   Conceptual model for means–end chains (from Gutman, 1982).

should, then, be possible to explain and/or predict actual behaviour with regard to these concrete objects by specifying how, in a given situation, parts of the cognitive structure are retrieved and used to guide behaviour.

As a model of personal relevance, the MEC approach implies that marketing strategies should create and/or reinforce connections between product attributes (the 'means'), self-relevant consequences (or consumers' attitudes and motivations) and personal values (the 'ends'). To develop effective marketing strategies that build personal relevance, marketing managers need to understand the factors that underlie consumers' perceptions of importance and self-relevance (Olson, 1995). A basic task for marketing managers is to develop an overall strategy in the form of a competitive positioning for their product. The MEC approach treats this strategic position as a distinctive MEC that establishes the basis for the perceived personal relevance of the product. Once the positioning strategy is determined, all marketing decisions about price, product features, distribution channels, sales promotions and advertising should contribute towards creating and reinforcing this means–end strategy.

8.3   The 'laddering' interviewing technique

Laddering refers to an in-depth, one-on-one interviewing technique used to develop an understanding of how consumers translate the attributes of products into meaningful associations with respect to self, following MEC analysis theory. Laddering involves a tailored interviewing format using primarily a series of directed probes, typified by the '*why is this important to you?*' question, with the express goal of determining sets of linkages between the key conceptual elements across the range of attributes (A), consequences (C) and values (V). Distinctions at the different levels of abstraction, represented by the A-C-Vs, provide more personally relevant ways in which products are grouped and categorised.

The analysis of laddering data across respondents first involves summarising the key elements by standard content-analysis procedures, while bearing in mind the levels of abstraction (A-C-V). Then, a *summary table* can be constructed representing the number of connections between the elements ('integration' phase). From this summary table, dominant connections can be graphically represented in a tree diagram, a *hierarchical value map* (HVM). This type of cognitive map, unlike those output from traditional factor analysis or multi-dimensional scaling methods, is structural in nature and represents the linkages across levels of abstraction without reference to specific products or brands (Reynolds and Gutman, 1988).

Interpretation of this type of qualitative, in-depth information permits an understanding of consumers' underlying personal motivations with respect to a given product class in that the underlying reasons why an attribute or a consequence is important can be uncovered (Reynolds and Gutman, 1988). Each unique pathway from an attribute to a value represents a possible perceptual orientation with respect to viewing the product category. Herein lays the opportunity to differentiate a specific product, not by focusing on a product attribute but rather by communicating how it delivers higher level consequences (Olson and Reynolds, 1983).

8.3.1   Laddering method process description[1]
Initially the attributes of the product used by the consumer to judge, evaluate and compare are elicited via several available techniques (e.g. exploratory qualitative research, repertory grid, triadic sorting). Then, these attributes serve as a

---

1. Sections 8.3.1 to 8.3.3 are based on the following references: Olson and Reynolds (1983); Reynolds and Jamieson (1985); Reynolds and Gutman (1988); Reynolds and Craddock (1988); Zeithaml (1988); Gutman (1991); Valette-Florence and Rapacchi (1991); Baker and Knox (1994); Grunert (1995); Grunert and Grunert (1995); Pieters *et al.* (1995); Reynolds *et al.* (1995); Aurifeille and Valette-Florence (1995); Gengler *et al.* (1995); Audenaert and Steenkamp (1997); Nielsen *et al.* (1998).

starting point for the depth interview in which the consumer is continuously probed with some form of the question *'why is that important to you?'* This type of questioning forces the consumer up on the ladder of abstractness, until the value level is reached or the consumer indicates that he/she is unable to provide any reason for his/her answer at the previous level.

The initial task of the analysis is to content-analyse all of the elements from the ladders. The goal of content analysis is to reduce the bulk of the raw data by finding common patterns of meaning. The first step is to record the entire set of ladders across respondents on a separate coding form ('categorisation' and 'comparison' phases). Having inspected them for completeness and having developed an overall sense of the types of elements elicited, the next step is to develop a set of summary codes ('abstraction' phase) that reflects everything that was mentioned and summarises the responses and words that 'go together'. If the coding is too broad, too much meaning is lost. The goal at this level of the analysis is to focus on meanings central to the purpose of the study, remembering that it is the relationships between the elements that are the focus of interest, not the elements themselves.

Once the master codes are finalised, numbers are assigned to each code. These numbers are then used to score each element in each ladder producing a matrix ('integration' phase), with rows representing an individual respondent's ladder (one respondent can have multiple ladders and thus multiple rows), with the sequential elements within the ladder corresponding to the consecutive column designations. Thus, the number of the columns corresponds to the number of elements in the longest ladder.

The next step is the straightforward one of constructing a matrix called the *implication matrix*, which displays the number of times each element in a given row precedes other elements in the same row. Such a matrix reflects the number of elements one is trying to map, and includes all attributes, benefits and values elicited through the interviews. Two types of relations may be represented in this matrix: direct and indirect relations. Direct relations refer to implicative relations between adjacent elements. The designations of (A) through (E) for the elements refer simply to the sequential order within the ladder. That is:

Attribute (A) → Consequence (B) → Consequence (C) → Value (D) → Value (E)

The A–B relation is a direct one, as is B–C, C–D and D–E. However, within any given ladder there are many more indirect relations, A–C, A–D, A–E, B–D, and so on. It is useful to examine both types of relations in determining what paths are dominant in an aggregate map of relationship among elements.

Another option in constructing the overall matrix of relations among elements is whether to count each mention of a relationship among elements that an individual respondent makes or to count a relation only once for each respondent, no matter how many times each respondent mentions it. Often, of all the cells having any relations, only half will be mentioned by as many as three respondents. The numbers in the implication matrix are expressed in fractional form, with direct relations to the left of the decimal and indirect or total to the right. It is this 'crossing over' from the qualitative nature of the interviews to the quantitative way of dealing with the information obtained that is one of the unique aspects of laddering and clearly the one that sets it apart from other qualitative methods. This summary score matrix, then, serves as the basis for determining the dominant pathways or connections between the key elements as well as providing the ability to summarise by subgroups (e.g. different segments of respondents with varying dominant pathways or segments resulting from cluster analysis with totally different summary matrices).

In constructing the HVM, 'chains' have to be reconstructed from the aggregate data. To avoid confusion, the term 'ladders' refers to the elicitation from individual respondents, while the term 'chains' is used in reference to the sequence of elements that emerge from the aggregate implication matrix. To construct a HVM from the matrix of aggregate relations, one begins by considering adjacent relations, that is, if A→B and B→C and C→D, then a chain A–B–C–D is formed. There does not necessarily have to be an individual with an A–B–C–D ladder for an A–B–C–D chain to emerge from the analysis. An HVM is gradually built up by connecting all the chains that are formed by considering the linkages in the matrix of relations among elements.

The most typical approach is to try to map all relations above several different cut-off levels (usually between three and five relations, given a sample of 50 to 60 individuals). The use of *multiple cut-offs* permits the researcher to evaluate several solutions, choosing the one that appears to be the most informative and most stable set of relations. It is typical that a cut-off of four relations with 50 respondents and 125 ladders will account for as many as two-thirds of all relations among elements. Indeed, the number of relations mapped in relation to the number of relations in the square implication matrix above cut-off level can be used as an index of the ability of the map to express the aggregate relationships. In establishing a cut-off level, one may count only the direct linkages in any cell or one may count the total number of linkages, direct and indirect.

Researchers should strive to design HVMs that are both simple and meaningful. By representing means–end data in a graphical format, the chaotic data set is organised and transformed into meaningful information. A unique advantage of graphics is that a large amount of quantitative data can be both displayed and absorbed. The reader should be able to identify the attributes, consequences and values in an HVM at a glance. Organising information by location on the diagram and using shading or colour is particularly effective for reducing search time (see Fig. 8.3). Recognition can be enhanced using shapes and patterns. In an HVM it would be desirable to immediately identify which concepts were mentioned most and least often.

Finally, a good design will facilitate inferences. The graphic representation of means–end data in an HVM may allow the researcher to 'see' things that may not be evident in the raw data or summary statistics, like evidence of potential market segments. These segments 'emerge' from the graphic through pattern recognition. Some researchers (e.g. Kumar and Rust, 1989) advocate the use of a

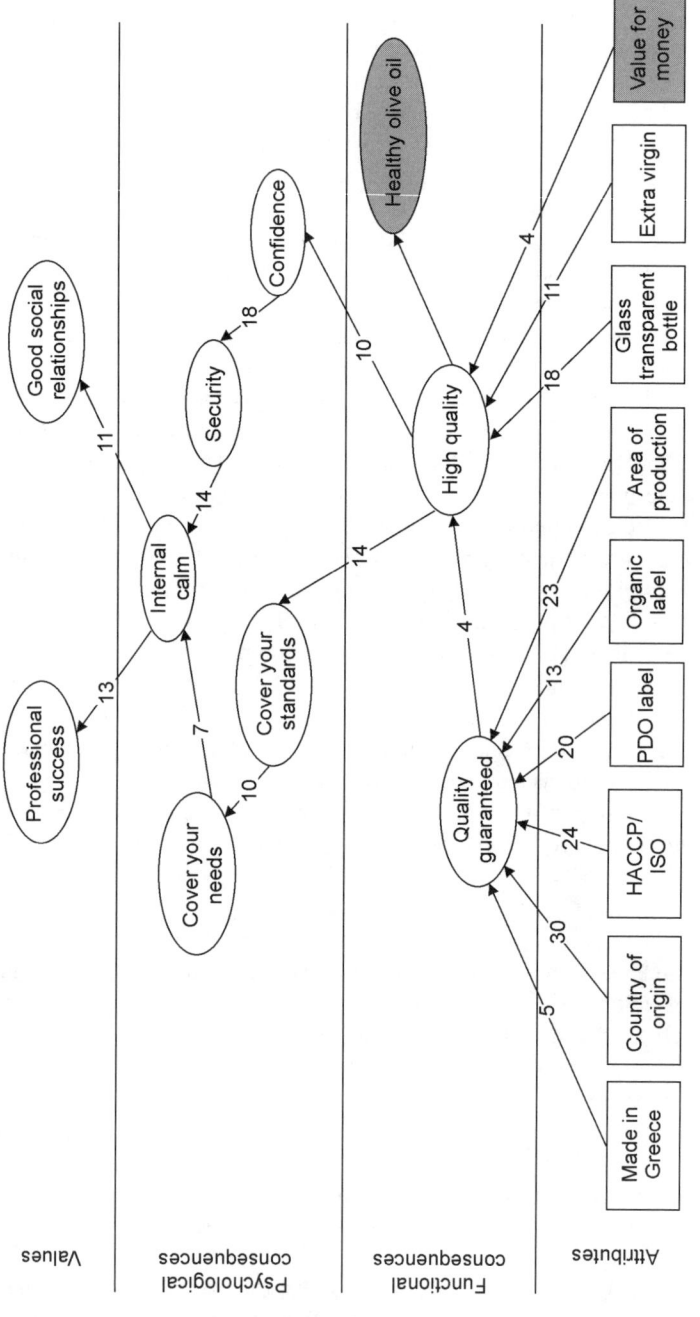

Fig. 8.3 The 'quality orientation' part of a hierarchical value map for olive oil, $n = 40$, cut-off level $= 3$ (HACCP, hazard analysis critical control point; ISO, International Organization for Standardization; PDO, protected denomination of origin) (from Krystallis and Ness, 2004).

'visual' approach to market segmentation. By taking advantage of the inherent human ability to recognise patterns, graphics may facilitate the discovery of the underlying structure of the data. In the hands of a knowledgeable academic researcher, a marketing analyst, or a brand manager, the HVM can be an effective tool to identify patterns or clusters. Coupled with experience, the researcher may recognise strategic opportunities or develop ideas for building theory ('integration' phase). The HVM may be used to communicate the results to others and may also serve as a means of supporting a conclusion.

8.3.2   Laddering interviewing technique description

An interviewing environment must be created such that the respondents are not 'threatened' and are thus willing to be introspective and look inside themselves for the underlying attitudes and motivations behind their perceptions of a given product class. This process can be enhanced by suggesting in the introductory comments that there are no right or wrong answers, thus relaxing the respondent, and further reinforcing the notion that the entire purpose of the interview is simply to understand the ways in which the respondent sees this particular product. Put simply, the respondent is positioned as the expert. The goal of the questioning is to understand the way in which the respondent sees the world, where the world is the product domain comprising relevant actors, behaviours and contexts.

Importantly, interviewers must position themselves as merely trained facilitators of this discovery process. In addition, owing to the rather personal nature of the probing process, Reynolds and Gutman (1988) advise creating a slight sense of vulnerability on the part of the interviewer. This can be accomplished by initially stating that many of the questions may seem somewhat obvious and possibly even stupid, associating this predicament with the interviewing process, which requires the interviewer to follow certain specific guidelines.

Obviously, as with all qualitative research, the interviewer must maintain control of the interview, which is somewhat more difficult in this context owing to the more abstract concepts that are the focus of the discussion. This can be best accomplished by minimising the response options, in essence being as direct as possible with the questioning, while still following what appears to be an 'unstructured' format. By continually asking the 'why is this important to you?' question, the interviewer reinforces the perception of being genuinely interested and thus tends to command the respect and control of the dialogue. By creating a sense of involvement and caring in the interview, the interviewer is able to get below the respondent's surface reasons and rationalisations to discover the fundamentals underlying the perceptions and behaviour. Understanding the respondent involves putting aside all internal references and biases while putting oneself in the respondent's place. Reynolds and Gutman (1988) claim that rapport established before the actual in-depth probing, as well as maintained during the course of the interview, is critical. The interviewer must instil confidence in the respondent so the opinions expressed are perceived as simply being recorded rather than judged.

Table 8.1 Basic problems of laddering (from Reynolds and Gutman, 1988)

Problem	Solution recommended
The respondent really does not 'know' the answer	(a) Ask what would happen if the attribute or consequence was not delivered (negative laddering): the 'non-conscious' reason is then discovered in relating the absence of a construct to what must be delivered if that negative is to be avoided. (b) Rephrase the question in a situational context: an answer is in this way typically 'discovered' owing to the ability to deal with specific circumstances.
Issues that become too sensitive	(a) Move the conversation into a third person format: create a role-playing exercise. (b) The interviewer to reveal a relevant personal fact (typically fabricated) about him/herself: it makes the respondent feel less inhibited by comparison. (c) Make a note of the problem area and come back to the issue when other relevant information is uncovered later in the interview.

The two most common problems of laddering and their recommended solution can be found in Table 8.1. The most commonly used probing techniques are illustrated in Table 8.2. The central idea is to keep the focus of the discussion on the person rather than on the product or service. This is not an easy task because typically at some point the respondent realises that the product seems to have disappeared from the conversation. Typically, two or three ladders can be obtained from roughly three-quarters of the respondents interviewed. Approximately one-quarter of the respondents, depending on the level of involvement in the product class, cannot go beyond one ladder. The time required from attribute distinctions to final ladders varies substantially, of course, but 60–75 minutes represents a typical standard (Reynolds and Gutman, 1988).

8.3.3 Criteria for assessing the predictive validity of the laddering data
There are four criteria that can be identified to measure consumers' cognitive structures as emerged through a laddering interview. The more an instrument fulfils these four criteria, the higher will be the predictive validity of the laddering data.

Criterion 1: The raw data should be a result more of the respondent's cognitive structures than of the researcher's cognitive structures and processes
Any data collection method that does not allow the respondent to use some kind of natural speech is proven to violate this criterion. Measuring cognitive structures by Fishbein-type rating scales (see Fishbein and Ajzen, 1975) violates it, because the cognitive categories are specified in advance, and not formulated by the respondent. The same goes for the card-sorting tasks used in some studies

Table 8.2   Most commonly used laddering techniques (from Reynolds and Gutman, 1988)

Technique	Application
1. Evoking the situational context	Respondents are providing associations while thinking of a realistic occasion in which they would use the product.
2. Postulating the absence of an object or a state of being	Respondents are encouraged to consider what it would be like to lack an object or to not feel a certain way.
3. Negative laddering	Enquiring into the reasons why respondents do not do certain things or do not want to feel certain ways (using the device of making the opposite assumption).
4. Age-regression contrast probe	Moving respondents backward in time encourages respondents to think critically about and be able to verbalise their feelings and behaviour.
5. Third person probe	Respondents are asked how others they know feel in similar circumstances.
6. Redirecting techniques: silence, communication check	Silence on the part of the interviewer can be used to make the respondent keep trying to look for more appropriate of definitive answer. A communication check simply refers to repeating back what the respondent has said and asking for clarification and more precise expression of the concept.

(e.g. Valette-Florence and Rapacchi, 1991). Only when there is very strong past evidence of high similarity of the cognitive categories used by respondent and researcher will that type of closed questioning not violate this first criterion. Hence, the first criterion calls for 'open' methods, in which each respondent can relate his/her own cognitive categories. We can immediately acknowledge that the laddering interview as described above in its typical format allows the respondents to answer in their own words, freely generating cognitive categories, thus meeting criterion 1.

Criterion 2: The data collection should not involve strategic processes not typical of the target situation (respondent's answers should be the result of a retrieval process and not of a 'problem-solving' one)
One can make sure that the strategic processes in the data collection situation (laddering) are very similar to those expected from the target situation (consumption) or one tries to find a data collection situation with minimal strategic processing (problem-solving). An assumption about the cognitive processes occurring during laddering can be as follows: at each step in the laddering process the interviewer's repeated question 'why is it important to you?' activates the cognitive category that the respondent has named last. Activation spreads from this category throughout the network, causing retrieval of additional categories, if the associations between the categories are strong enough.

Presumably, the respondent would then answer with the category that has received the highest activation ('spreading activation' theory, Anderson, 1983, in Grunert and Grunert, 1995). At some point no strong associations exist between the category last named and any category not yet named. Then, the respondent will be unable to answer, and the laddering would stop. If the respondent's retrieval process in the data collection works in this way, data collection would be mainly determined by automatic processes, and criterion 2 would be met.

However, it seems plausible that deviations from this 'ideal' laddering process can occur, especially when the cognitive structure with regard to the product in question is either especially weak or especially sophisticated. When the cognitive structure is weak, the more the task changes from retrieval to a problem-solving one, the more the criterion 2 is violated. In this case, the predictive validity of the results is likely to be lower, because it becomes increasingly unlikely that the strategic processes in the data collection situation will resemble those in a consumption situation.

In a post-laddering interview, Grunert and Grunert (1995) report that several respondents mentioned that the laddering task led them to a considerable amount of strategic processing, especially when they got to the more abstract levels. They said that the laddering task made them think of connections they had never thought about before, and that they gained new insights during the process. This seems to indicate that the laddering task actually *changed* their cognitive structure. Respondents also said that they felt some pressure to come up as high as possible in the ladder, and that they felt silly when they stopped the ladder at the concrete level. Some actually said that they thought the interviewer should press them to give more answers as long as possible, because it gives a more interesting experience. Other respondents said that, when they had reached the value level, they had a feeling of having come 'up too far'.

Strategic processes (thinking) on behalf of the respondent are time-consuming and will, therefore, be noticeable to the attentive interviewer as pauses, breaks, unfinished sentences, etc. To minimise unwanted strategic processing, the interviewer should end this line of questioning, thus avoiding construction of new associations. Obviously, the interviewer has considerable influence on the amount of strategic processing occurring in this way. For instance, the interviewer can try to guide the respondent into answering from a specific situational context. This suggests that the respondent uses a strategic process of thinking to provide an answer. By providing a more or less relaxed interview atmosphere, automatic retrieval may be facilitated. When the cognitive structure in respondents' minds is especially sophisticated in relation to a specific product, one retrieval process may result in the retrieval of several cognitive categories at the same level of abstraction. How this affects the interview will again depend on how the interviewer reacts.

The interview where the natural flow of speech of the respondent is restricted as little as possible has been designated as 'soft' laddering. In contrast, 'hard' laddering refers to an interview and data collection technique where the respondent is forced to give answers in such a way that the sequence of the answers reflects increasing levels of abstraction. Data collection techniques that do not involve personal interviews at all, such as self-administered question-naires and computerised data collection devices, are all equivalent to the 'hard' laddering, which make it impossible to detect strategic processing of the respondent. 'Softer' methods are being proposed whenever the researcher can expect the cognitive structure of the respondent to be weak, owing to low involvement and/or little experience with the product, or to be very elaborate, owing to high involvement and much experience, such as when the respondent is an expert in the respective area.

Criterion 3: Coding should preferably be based on cognitive categories widely shared among consumers, researchers, and users of research results, and not on the researcher's idiosyncratic cognitive categories
Coding means grouping different answers together, a fact that, by necessity, is based on the researcher's estimate of the semantic distance between the various answers, and hence on his/her cognitive structures. This gives the researcher considerable influence on the results. Concerning laddering and this third criterion, two problem areas are being discussed by the literature: first, that the distinction between attributes, consequences and values should be based on a conceptual definition of these terms. The laddering literature is surprisingly void of such definitions. In practice, many borderline cases turn up. Making such categorisations in a uniform way is heavily dependent on the availability of context information,[2] which will again depend on the way the laddering is executed, with the 'hard' form usually providing very little context information. Only when the interviews are taped and transcribed is the full context available in coding. The less one knows in advance about how respondents think about the topic to be researched, the more difficult it is to relate the respondent to his/her background and thus to understand his/her definitions of attributes, consequences and values. Thus, it is more important to devise a data collection method that helps the researcher understand the meanings in the respondents' answers, such as methods employing natural speech (focus groups, in-depth interviews).

Second, to find the 'right' level of abstraction is the other problem associated with coding. The difference between two answers is rarely purely lexical. To define them as synonyms and group them into the same category, the category has to be at a more abstract level than the answers themselves. For example 'excellent taste' and 'pretty good taste' may both be sorted into 'tasty'. It is at the consequence and value levels where the real difficulties start. Can 'joy' and 'not being depressed' both be coded into 'well-being'? Such rather broad categories usually have to be created, if a technically manageable implication matrix is to result.

---

2. It reflects the more general qualitative research problems of *indexicality*, according to which it is possible to understand or make sense of a respondent's answer only by relating it to the respondent's individual background.

Having parallel coders is of course the most common remedy used in research practice. But it may be helpful to draw on some of the experiences and tools developed within the realm of computer-assisted content analysis (Catterall and Maclaran, 1996). The basic idea developed is that of *iterative coding*. This means that a first coding is performed, and the implications of this coding are made transparent by aids such as key-words-in-context lists, leftover lists, and insertion of codes into the text database. Based on these aids, the coding is revised, and the implications are analysed in the same way. This procedure continues until the coding appears satisfactory. Of course, the decision about what can be regarded as satisfactory still rests to a large extent on face validity considerations and, therefore, on the judgement of the individual researcher. However, such procedures provide documentation for how the coding has proceeded, thus increasing the intersubjectivity of the process.

Criterion 4: Data reduction should be based on theory about cognitive structure and processes

Data reduction in analysing laddering encompasses two main steps: *aggregation* and *condensation* (Grunert and Grunert, 1995). *Aggregation* involves the step from the individual to the collective. The HVM, the main output from a laddering analysis, is a characterisation of a group of respondents. An HVM is not only a device that allows us to see the major results from a laddering study without having to go through all the individuals' ladders, but also – according to a more ambitious view – it is an estimate of cognitive structure for that group of respondents. At the individual level laddering data are not rich enough to estimate a respondent's cognitive structure because the cognitive structure itself is not a collection of single chains, but an interrelated net of associations. However, when we obtain ladders from a group of *homogeneous* respondents, then the set of ladders obtained taken together will yield an estimate of this group's cognitive structure and have predictive validity.

Condensation refers to reducing the HVM to a small subset of the associations between cognitive categories that have shown up in the data. In principle, based on the implication matrix one could draw a map that shows all the cognitive categories that resulted from the coding process, and in which two cognitive categories are linked whenever the corresponding cell in the implication matrix has a non-zero entry. In practice, this is seldom possible or desirable, since one tries to find an HVM that includes the most important links. This is achieved by two means: first, by specifying the condition that the network has to be *non-redundant* and, second, by specifying, as we have seen, a *cut-off level*. *Non-redundancy* means that, if category 1 at abstraction level A is linked to category 2 at abstraction level B, which again is linked to category 3 at abstraction level C, in the HVM there should be no direct link between categories 1 and 3, because such a link would be redundant.

In trying to develop a better understanding of the laddering data, the first step should be an explicit stand on what a HVM is supposed to do. If it is meant as an estimate of cognitive structure, then the next step would be to spell out clearly the assumptions made about non-redundancy. The technical problem then will be to aggregate only respondents whose cognitive structures can be regarded as homogeneous with regard to the product in question. This can be achieved by applying clustering methods to the laddering data based on a specific clustering criterion before aggregation (e.g. applying cluster analysis through dummy variables that reflect the type of benefits that emerged can lead to the identification of respondent clusters at the benefit level), or using a prespecified homogeneous sample.

8.4   MEC consumer research and new product development

NPD can originate from new technology or new market opportunities. But irrespectively of where opportunities originate from, when it comes to successful new products it is the consumer who is the ultimate judge (Cooper and Kleinschmidt, 1987; Brown and Eisenhardt, 1995; both in van Kleef et al., 2005). In order to develop successful new products, companies should gain a deep understanding of the consumer. Consumer research can be carried out during each of the basic stages of the NPD process, namely opportunity identification, development, testing and launch (Urban and Hauser, 1993, in van Kleef, 2006). Despite the importance of the later stages, it is increasingly recognised that successful NPD strongly depends on the quality of the *opportunity identification* stage (Cooper, 1998, in van Kleef, 2006).

Large parts of the conducted research in NPD consist of focus groups, surveys and the study of demographic data. This is considered to be one of the reasons for the relatively low new product success rates (Wind and Mahajan, 1997). The failure of methods to reach their full potential in NPD is perhaps the result of the limited and confused way in which they have been evaluated and made clear to potential users. In contrast to the significant attention paid to techniques such as, for example, posterior product testing, analysis of strengths and weaknesses of consumer research methods for the *opportunity identification* stage of NPD has received only little attention. Van Kleef et al. (2005) have critically reviewed, among other common consumer research methods and techniques, MEC analysis and laddering for application in the *opportunity identification* stage of the NPD process, based on the following criteria.

Type of information source for need evaluation (product- vs need-driven methods; consumer familiarity with stimuli provided for data generation)
According to van Kleef et al. (2005), in the laddering interviewing technique the stimulus used as information source to elicit consumer needs is *product-driven*: laddering interviewing typically begins with distinctions made by the individual participant concerning perceived differences among attributes of different brands or products. Product-driven methods elicit consumer needs within an existing framework of what is already available in the market. Consequently, reactions to existing products are relatively predictable and results can easily be

translated into corresponding product requirements. A disadvantage, however, in starting too early in the NPD process with concrete products is that it may destroy creativity and thinking 'out of the box' (van Kleef *et al.*, 2005).

Moreover, the stimuli presented in laddering are generally *familiar* to participants. In the case where a consumer has minimal experience with the product, it is difficult to retrieve the relevant attributes to evaluate the product. Owing to limited cognitive capacities of the human mind, people often make heuristic decisions when encountering complex stimuli. As a consequence, decisions are made by a rule of thumb and consumers' opinion about new products may not have a high predictive ability (van Kleef *et al.*, 2005).

Task format of method/technique (possibility to evaluate multiple products vs a single product; type of consumer response; type of data collection tool structure)
The task format of laddering can be characterised by the evaluation of *multiple products*, after which the interviewer obtains the needs by *directly* 'why'-probing the participant. According to van Kleef *et al.* (2005), methods that include a set of competing products available in the market have the advantage that they represent the task that consumers typically perform in real market conditions. However, when consumers compare very different kind of products, they do so at higher levels of comparison. For example, a CD recorder and two tickets to the ballet are two dissimilar alternatives that can be compared through MEC at the more abstract level of benefits (e.g. potential for fun and enjoyment).

In terms of the type of consumer response, the actual laddering interview starts after both *similarity* and *preference judgements*, depending on the type of technique used to elicit distinctions between products. Van Kleef *et al.* (2005) argue that similarity questions that identify perceptual differences between products resulting from participants' comparison processes are useful information for technical product development. In contrast, asking a consumer about his/her preferences evokes a different thinking process, resulting in other aspects of the product considered important. Before giving a preference judgement, consumers will imagine the benefits the product will deliver for them. This information is very important for NPD, as consumer needs arising from preference judgements have a higher predictive validity for purchase than consumer needs arising from perceptual judgements (van Kleef *et al.*, 2005).

Furthermore, the data collection in laddering is *unstructured* because the participant's subsequent answers determine the direction of the interview. In an unstructured interview, the questions to be asked are not necessarily presented in exactly the same wording to every participant, who is free to respond in his/her own words. The advantage is (van Kleef *et al.*, 2005) that detailed responses can be queried for, which may provide the researcher with new insights and ideas for NPD. A shortcoming of this type of research is that the idiosyncratic information obtained does not lend itself to direct use in subsequent analysis. The categorisation and condensation steps of MEC analysis that preclude subjective interpretations of the data affect the way the latter are interpreted and a research bias can occur as a result from selective observation and recording of information.

Actionability of the output (actionability for marketing; actionability for product technical development)
Finally, MEC approaches, by linking the type of concrete product charac-teristics, use benefits and values within consumers' cognitive structures for a product class, provide information for *marketing purposes* (e.g. creative phase, advertising) for incremental new products. Marketing-oriented tasks involve the creative phase of finding new product ideas. The more abstract consumer needs are, the more freedom in creativity is allowed. Information through MEC about which benefits consumers are seeking in a particular product enlarges the solution space and prevents thinking within the 'box' of current product delivery. In this way, it can serve as a source of inspiration and create a shared understanding in the NPD group (van Kleef *et al.*, 2005).

8.5   Past and future trends in MEC-related research

Since the introduction of the laddering methodology into the international consumer research domain 25 years ago, numerous applications have been published. Figure 8.4 illustrates the yearly frequency of appearance of published MEC and laddering involving papers from 1981 to 2005 in the marketing and consumer behaviour literature. These publications can be roughly divided into two types:[3] (a) concept-related and (b) applications in various products and services.

(a) **MEC concept-related**. In this category one can identify two subcategories:
    (a1) Conceptual papers that *describe the principles of the method and set their theoretical foundations,* as well as *provide detailed suggestions on laddering implementation, data collection, coding and analysis.* Here belong most of the early published works (Gutman, 1982, 1991; Jolly *et al.*, 1988; Walker and Olson, 1991; Aurifeille and Valette-Florence, 1992, 1995; Pieters *et al.*, 1995) that analyse the principles of a 'mainstream' version of the method, as described previously.
    (a2) Papers that *set the agenda of MEC's drawbacks and suggest remedies* (Valette-Florence and Rapacchi, 1991; Mulvey *et al.*, 1994; Pieters *et al.*, 1994; Grunert and Grunert, 1995; Grunert *et al.*, 1995a). The majority of the works in this sub-category published after 1995 *refine*

---

3. The classification is only indicative of the range of MEC-related publications that can be found in the literature and not based on solid scientific criteria, apart from their main objective. As a result, a number of papers especially in the subcategories b1 and b2 are classified in more than one category and are therefore marked with an asterisk; however, these papers are counted only once in the summary Figs 8.4 and 8.5.

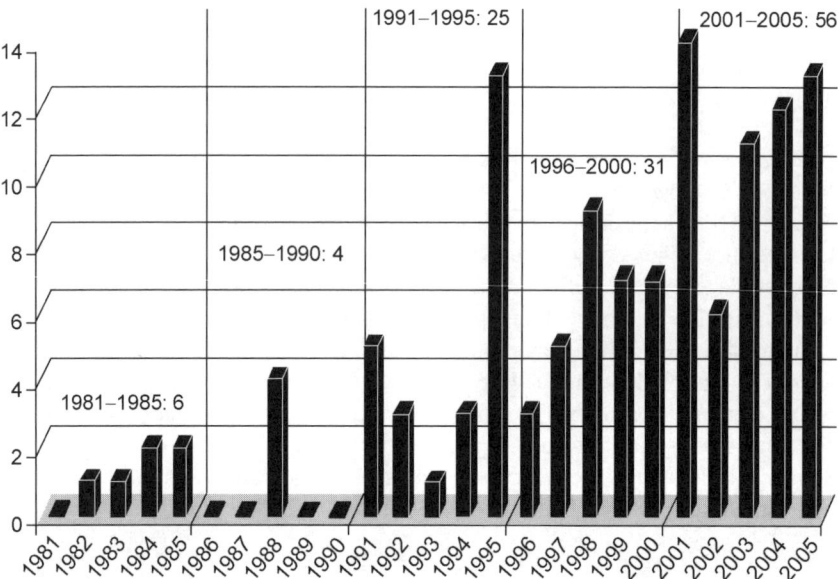

Fig. 8.4 Frequency of appearance of published MEC and laddering applications in the marketing and consumer behaviour literature, 1981–2005 (total: 120) (papers 'in press' are classified in 2005).

the theoretical basis of the method or even criticise its conceptual assumptions and methodological underpinnings (Gengler *et al.*, 1995; Sorensen *et al.*, 1996; Bech-Larsen *et al.*, 1997; Gutman, 1997; ter Hofstede *et al.*, 1998; Valette-Florence, 1998; Valette-Florence *et al.*, 1998; Bech-Larsen and Grunert, 1998; Bagozzi and Dabholkar, 2000; Cohen and Warlop, 2001; Grunert *et al.*, 2001a; Manyiwa and Crawford, 2001; Huber *et al.*, 2004; Lepard *et al.*, 2004; Russel *et al.*, 2004a,b; Scholderer and Grunert, 2004; Miles and Rowe, 2004; Mort and Rose, 2004; Lin and Fu, 2005).

(b) **MEC applications**. The primary objective of these surveys is to develop a cognitive value map indicating the interrelation of attributes, consequences and personal values for a given product or service. In this type, one can identify three additional subcategories:

(b1) *Applications of the mainstream MEC theory in food-related research*:
- Various food production methods and food types, such as: genetically modified foods (Bredahl, 1999; Grunert *et al.*, 2001b; Grantham, 2005), organic foods (Bech-Larsen and Grunert, 1998*; Makatouni, 2002; Zanoli and Naspetti, 2002; Fotopoulos *et al.*, 2003; Naspetti and Zanoli, 2003; Sirieix and Schaer, 2005), functional foods (Jonas and Bechmann, 1998; Urala and Lähteenmäki, 2003; Morris *et al.*, 2004), fresh foods (Vannoppen *et al.*, 2001); convenience foods (de Boer and McCarthy, 2003); fair trade foods (de Ferran and Grunert, 2007); and baby foods (Gengler *et al.*, 1999).

- Consumer perceptions about various food-related issues, such as food hazard perceptions (Miles and Frewer, 2001) and food-related health and hedonic perceptions (Roininen *et al.*, 2000).
- Various food products, such as beverages (Gutman, 1984), yoghurt (Grunert and Sorensen, 1996), fruits (Jaeger and MacFie, 2001; Vannoppen *et al.*, 2002), vegetable oils (Bech-Larsen *et al.*, 1996; Valli, 1997; Nielsen *et al.*, 1998; Krystallis and Ness, 2004), meat (Bech-Larsen and Grunert, 1998*; Le Page *et al.*, 2005), fish (Nielsen *et al.*, 1995; Sorensen *et al.*, 1996*; Valette-Florence *et al.*, 2000), and wine (Judica and Perkins, 1992; Hall and Winchester, 1999; Fotopoulos *et al.*, 2003*; Overby *et al.*, 2004).

(b2) *Papers that extend the applicability of the method to other marketing and consumer behaviour research areas*:

- Advertising research (Olson and Reynolds, 1983; Reynolds and Gutman, 1984, 1988; Reynolds and Jamieson, 1985; Reynolds and Craddock, 1988; Reynolds and Rochon, 1991; Gengler and Reynolds, 1995; Reynolds *et al.*, 1995; Reynolds and Whitlark, 1995; Bech-Larsen, 2000; Jaeger and MacFie, 2001*; Reynolds and Olson, 2001; Li, 2003).
- Market segmentation (Baker and Knox, 1994; Valli, 1997*; Valette-Florence, 1998*; Valette-Florence *et al.*, 1998*; Botschen *et al.*, 1999; ter Hofstede *et al.*, 1999; Lin and Yeh, 2000) and consumers' lifestyle research (Scholderer *et al.*, 2002; Brunso *et al.*, 2004).
- Retailing (Gutman and Alden, 1985*; Vannoppen *et al.*, 2001* and 2002*; Devlin *et al.*, 2003; Skytte and Bove, 2004; Lin and Fu, 2005*; Vincent-Wayne and Harris, 2005).
- Various topics in business-related consumer research, such as consumer ethics (Pitts *et al.*, 1991; Bagozzi and Dabholkar, 1994; Smeesters *et al.*, 2003) and smoking motivation (Kaciak and Cullen, 2005); and management, economics and business research (Audenaert and Steenkamp, 1997; Botschen and Hemetsberger, 1998; Grunert and Bechmann, 1999; Langerak *et al.*, 1999; Overby *et al.*, 2004*; Grantham, 2005).
- Consumer-perceived quality (Gutman and Alden, 1985; Zeithaml, 1988; Grunert, 1995) and consumer choice (Claeys *et al.*, 1995; Grunert *et al.*, 1995b; Dibley and Baker, 2001; Grunert and Bech-Larsen, 2004).

(b3) *Applications of the mainstream MEC theory in non-food products and services*:

- Information technology, such as Internet site evaluation (Aschmoneit and Heitman, 2001, 2002, 2003), software evaluation (Wong and Ross, 2001; Wong, 2002), e-commerce and e-banking (Subramony, 2002; Brown, 2005; Hiltunen *et al.*, 2005).
- High-tech industry, such as mobile telephony (Heirman *et al.*, 2004) and automobiles (Valette-Florence *et al.*, 2003).

- Garment industry (Hines and O'Neil, 1995; Huang *et al.*, 2003).
- Service industry, such as healthcare (Doucette and Wiederholt, 1992; Jensen, 2005), real-estate (Coolen and van Montfort, 2001), leisure and tourism (Klenosky *et al.*, 1993, 1998; Frauman *et al.*, 1998; Goldenberg *et al.*, 2000; Zins, 2000; Frauman and Cunningham, 2001; Haras *et al.*, 2005; King, 2005) and energy (Boutin *et al.*, 1997).

It can be easily concluded from the above rough classification that since the mid-1990s, and especially the year 1995 onwards, the MEC bibliography flourished. As can be observed in Fig. 8.5, conceptual papers that described the principles of the mainstream version of the method and provided detailed suggestions on laddering implementation, data collection, coding and analysis (seven papers) appeared in the marketing and consumer behaviour literature until 1995; they constitute the start-up phase of the MEC method life cycle. A few years before (1991 onwards to now), papers that refine the theoretical basis of the method or criticise its assumptions and methodological underpinnings start to appear (25 papers), constituting the next – maturity – phase of the method. Almost simultaneously, a number of applications of the mainstream MEC theory in food (30 papers) and non-food products and services-related research (24 papers) appeared in the literature. Finally, almost all along the 25-year history of MEC publications one can find numerous papers that extend the applicability of the method to other consumer-related research areas, and espe-

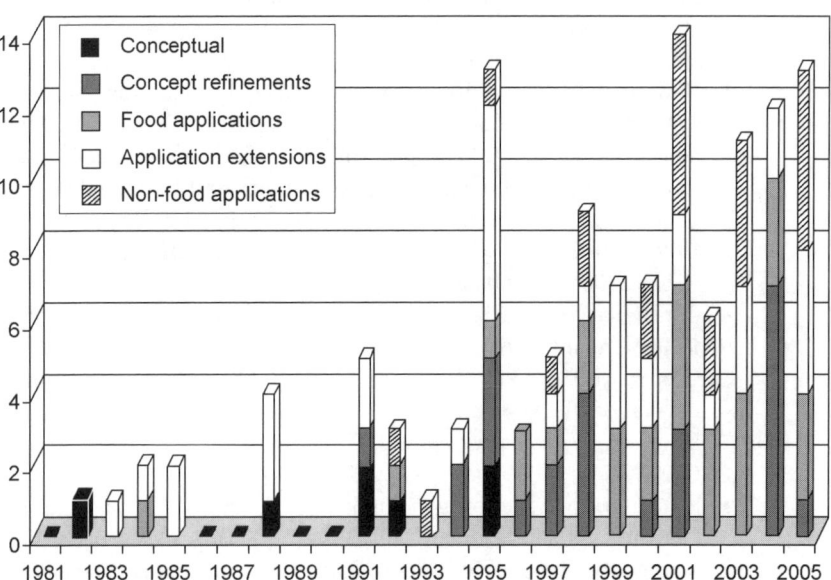

Fig. 8.5 Frequency of appearance of published MEC and laddering applications in the marketing and consumer behaviour literature per publication type, 1981–2005 (total: 120) (papers 'in press' are classified in 2005).

cially advertising development and evaluation, market segmentation and retailing management (36 papers). The three latter categories of MEC applications and extensions constitute the largest part of the MEC-related work found in the literature, representing three-quarters of the 122 papers mentioned here, while the publications of all types appearing after the year 2000 represent almost half of all MEC publications of the last 25 years.

Miles and Rowe (2004) detail various recent uses of the laddering techniques and the specific method variations employed in the identified empirical papers (Table 8.3). As it can be seen from column two of Table 8.3, the studies detailed have used laddering as an exploratory tool, often to consider the nature of cross-cultural differences, usually regarding beliefs about certain products. Two types of attribute elicitation procedures (free/direct elicitation and triadic sorting) have been more commonly used (column three), while the free/direct elicitation method is usually employed when the paper and pencil method is used. Regarding the use of laddering *per se*, there have been more uses of the interview than the paper and pencil approach. In the interview studies the use of validation (e.g. through inferential statistics) is rare, only being described in one paper. According to Miles and Rowe (2004), validation is not generally necessary for paper and pencil laddering, as a larger, potentially more-representative, sample of the population of interest may be obtained. Finally, it is clear that results from laddering interviews have tended to be described graphically, using the HVM, and then described from observation, rather than using inferential statistics.

Despite its popularity, MEC theory suffers from problems of unconfirmed validity (Scholderer and Grunert, 2004). In particular, the nomological status of its central construct, the MEC, is unknown. Previous reviews list a multitude of reasons that may be responsible for this (Grunert and Grunert, 1995; Grunert *et al.*, 1995a, 2001a). The foremost one is a lack of theoretical rigor (Scholderer and Grunert, 2004): while some researchers interpret MECs as memory structures (Gutman, 1982; Reynolds and Gutman, 1988), others see them as models of motivation (Pieters *et al.*, 1995; Gutman, 1997; Cohen and Warlop, 2001).

Moreover, the term laddering in the marketing community very quickly became a somewhat generic term, representing merely a qualitative, in-depth interviewing process (Morgan, 1984), without reference to either its theoretical underpinnings or the rather critical distinction between the interview process and analytical methods used to derive meaning from the resulting data (Durgee, 1985). The methodological problems of laddering have also being identified in the literature (Valette-Florence and Rapacchi, 1991; Mulvey *et al.*, 1994; Pieters *et al.*, 1994; Grunert and Grunert, 1995). Given the value of this type of in-depth understanding of the consumer, and in particular its potential with respect to the specification of more appropriate marketing segmentation and positioning strategies, a comprehensive documentation of this research approach is needed.

Despite the fact that both conceptual and methodological problems of MEC and laddering emerged early, a common conceptual and methodological framework that would help the development of coherent theory is lacking. Future research on the method should persist on this direction.

Table 8.3 Methodological details of ten recent food-related applications of MEC and the laddering technique (1998–2003) (adapted from Miles and Rowe, 2004)

Reference	Task	Elicitation	Laddering details	Representation & analysis	Comments
1. Fotopoulos et al. (2003)	Exploratory (perceptions of wine in organic and non-organic buyers)	Attribute list (importance on 1–3 scale when purchasing wine; important attributes used)	Interviews, no validation, two judges coded content	HVMs Used LadderMap software	Authors stated they used a 'soft laddering' approach. Difference in cognitive structures between organic and non-organic wine buyers.
2. Grunert et al. (2006)	Exploratory (cross-cultural perceptions of GM and non-GM foods)	Ranking (ranking product descriptions according to buying intentions)	Interviews, no validation, individually coded in each country, then one researcher synchronised codes from different countries	HVMs Used LadderMap software, multiple correspondence analysis	Found that 'non-GM' is a value in itself; differences dependent on the extent of modification.
3. Jaeger and MacFie (2001)	Exploratory (perceptions of apples)	Free/direct elicitation (attributes considered when purchasing apples)	Interviews, no validation, two judges coded content	HVMs Used LadderMap software	Study included an appendix to provide data on MEC for use on study on expectations of advertising.
4. Miles and Frewer (2001)	Exploratory (perceptions of five food hazards)	Free/direct elicitation (what comes to mind about food hazard)	Interviews, validation by PCA of questionnaire data from larger sample, two judges coded content	HVMs (refer to as 'results diagrams')	Hazards characteristics and concerns elicited instead of A-C-V chains. PCA loadings matched pattern of associated characteristics for three of the five hazards.
5. Roininen et al. (2000)	Exploratory (consumer perceptions of health/hedonic aspects of food)	Free sorting (sorting food into four groups re health/hedonic characteristics)	Interviews, no validation	HVMs, multiple correspondence analysis	Unclear number of content coders. Laddering not to value level. After free sorting, attributes are generated for one 'representative' food in each category.

Study	Purpose	Elicitation method	Data/validation	Analysis	Findings
6. Bredahl (1999)	Exploratory (cross-cultural perceptions of GM in foods)	Ranking (attributes of GM foods)	Interviews, no validation, one researcher synchronised codes from different countries	HVMs, Multiple correspondence analysis	More complex cognitive structures for some nationalities than others as seen in differences in HVMs.
7. Grunert and Beckman (1999)	Exploratory (cross-cultural perceptions of food products)	Triadic sorting (attributes of food products)	Interviews, no validation	HVMs Used LadderMap software	Differences in HVM stricture between the groups, suggesting one had lower habitualisation in shopping behaviour than others.
8. Langerak et al. (1999)	Exploratory (experts establishing sequence for NPD approaches)	Attribute list (selected which nine NPD acceleration approaches they would use)	Interviews, no validation	Abstractness, centrality and prestige indices calculated	Does not follow classic attempt to establish A-C-V chains. Links between objectives generated rather than typical A-C-V chains.
9. ter Hofstede et al. (1999)	Exploratory (cross-cultural perceptions of yoghurt)	Not clear: states 'the relevant attributes … were elicited' (p. 7)	Interviews, no validation per se, but data used to develop questionnaire for larger sample	A 'segmentation model' applied to questionnaire data revealing four segments charted in 'probabilistic MEC maps'	Laddering interviews used to help define A-C and C-V matrices for inclusion in a questionnaire in which respondents indicated links between concepts 'association pattern technique'). Aim of paper assesses the segmentation model.
10. Nielsen et al. (1998)	Exploratory (cross-cultural perceptions of vegetable oils)	Free sorting into three groups, then ranking within the groups (attributes of vegetable oils)	Interviews, no validation	HVMs, correspondence analysis, used LadderMap software	Differences in structures between the national groups in terms of knowledge and preferences.

8.6   Sources of further information

The book by Reynolds and Olson (2001), *Understanding Consumer Decision Making: The Means–End Approach to Marketing and Advertising Strategy* (Mahwah, NJ, Lawrence Erlbaum Associates, Inc.) is of special value. The goals of the authors of this book are to help business managers and academic researchers understand the MEC perspective and the methods by which it is operationalised and to demonstrate how to use the MEC approach to develop better marketing and advertising strategy. The book, which is a collection of chapters by different authors, seeks to address a number of problems with the MEC approach; most of these problems are also mentioned here. The authors of the various chapters discuss methodological issues regarding interviewing and coding, present applications of the MEC approach to marketing and advertising problems, and describe its conceptual foundations. The book contains a mix of original and previously published articles in roughly a 65:35 ratio and it is divided into five parts: (I) Introduction; (II) Using laddering methods to identify MECs; (III) Developing and assessing advertising strategy; (IV) The MEC approach to developing marketing strategy; and (V) Theoretical perspectives for MEC research. The target audience for the book includes academic researchers in marketing and related fields, graduate students in business, marketing research professionals, and business managers.

Of special – though mostly academic – interest is also the special issue of the *International Journal of Research in Marketing* (Olson, 1995, 'Introduction to the special issue of means-end chains', vol. 12, pp. 189–191), yet most of the papers included in that issue are mentioned in the present chapter. Finally, on the web one can find various interesting sites of academic institutions with mainly teaching material on MEC and laddering (e.g. 'What is a value chain?' at https://marketing.byu.edu/htmlpages/courses/657/laddering.htm or 'Attribute-value chain' at http://diegm.uniud.it/create/Handbook/techniques/List/Attribute.php).

Finally, special emphasis merits the software tool which facilitates the analysis of laddering data and especially the generation of the implication matrix and the HVM. Perhaps the newest and most user-friendly software tool is MECanalyst (www.skymax-dg.com/mecanalyst/meceng.html). MECanalyst was developed in 2002 by Prof. Raffaele Zanoli and Dr Simona Naspetti of the University of Ancona (IT) in cooperation with Leonardo Cigolini Gulesu and Antonio Ruccia of SKYMAX DG (IT). This software performs a significant amount of operations, making the greatest part of the data processing and map construction work nimbler and more detailed at the same time. It allows managing and storing an unlimited amount of data (i.e. number of respondents, number and/or length of the ladders/chunks expressed by each respondent). It has an option which facilitates content analysis: its in-built 'pattern recognition' tool allows reducing and simplifying the coding phase. MECanalyst is based on Microsoft Windows^{TM} applications (such as Powerpoint^{TM} and Excel^{TM}) for output generation, a fact that makes it easier to manage the processing of results with reference to both the graphic module, i.e. when presenting the final maps, and the implication matrix.

8.7   Case study: The Greek wine industry and its re-direction to quality organic wine production – what does MEC analysis tell us?

The wine market setting …

A number of emerging trends in consumer wine preferences can be observed internationally. In the traditional consumer countries, wine has lost its function as a daily component of diet; wine consumption is now linked with pleasure, conviviality, psychological satisfaction, refinement and cultural interest. This phenomenon started in late 1980s with a strong interest of the major import markets in internationally known wine varieties (Chardonnay, Cabernet, Merlot, etc.). The world wine market is nowadays a 'two-way' market, as a consequence of an increasing supply of 'new world' wines (South Africa, California, Chile, Australia, etc.) and the evolution of EU consumer preferences towards differentiated consumption patterns.

Traditionally, the international image of Greek wines was poor, despite the country's long experience of winemaking. The wine sector in Greece features the worst structural situation of all the traditional wine-producing countries: vine growing is distributed throughout all regions with increased cost implications and the emphasis is traditionally on bulk self-consumption or bottled low-quality wine distributed to strictly local markets. Commercial wines represent only 40% of consumption, with an astonishing 60% assigned to bulk wine consumption.

However, Greek climatic conditions and the existence of a broad viticulture potential have recently generated ample room for differentiation. The number of 'Vins Délimite de Qualité Supérieure' is 215, in addition to 74 regional wines. In 2002, 12% of Greek wine production was made up of quality VDQS wines. Innovative wine producers demonstrate excellent performance, proof of which lies in the increasing interest of the foreign markets. As a result, average export price per litre, although still below the EU average, increased by 26% from 1993's level of €1.23. Over the last decade, total domestic consumption increased by 4%, mainly because of an increase in the consumption of quality red wines. Hence, there is strong evidence that the Greek wine sector can have a market potential domestically and internationally, if oriented towards the satisfaction of newly emerged, quality-based, consumer preferences.

… what is known about organic wine consumers …

Confusion exists as to what defines an 'organic' wine. Experts seem to agree that organic wine must start from organically grown grapes. The use of sulphur dioxide (SO_2) is the most controversial issue to be tackled. Greek as well as European legislation has not yet provided any directive for organic wine standards; therefore the relevant term cannot be put on EU wine bottles. In Greece, the area devoted to the organic cultivation of vineyards was approximately 1500 ha in 2000, 80% of which comprised organic cultivation of raisin-producing varieties of grape. The wine-producing organic vineyards produced about 7000 hl of more than 75 'organic' wine labels, some of which are VDQS. In 1998, sales of 'organic' wine represented 10% of Greek organic market.

Previous research demonstrated that Greek 'organic' wine consumers did not seem to be very demanding, owing to the low level of domestic competition, non-existence of foreign competition as yet, and the fact that, for most consumers, it is a new and unknown product. However, levels of overall satisfaction indicated that there is significant potential for further improvement of the product. Greek consumers' satisfaction with organic wine was found to be achieved mainly through product and image criteria, such as quality (colour, aroma, taste and finesse), reliability and retail price, as opposed to promotion and place. Customers seemed to be indifferent towards the issue of price, which was the least important satisfaction criterion of all.

... what was also learned through consumer values research ...

A qualitative consumer survey was undertaken in 2000, to contribute to the redirection of the Greek wine sector towards high-quality, competitive products, such as 'organic' wines. The survey aimed to understand Greek consumers' attribute preferences, motivations and cognitive structures with regards to conventional wine as opposed to wine produced from organic grapes. The MEC methodology in a sample of 49 wine consumers divided in two subgroups, 28 organic food buyers and 21 organic food non-buyers, was adopted. In Figs 8.6 and 8.7 the HVMs of the relationships between wine attributes, consumer consequences and values of the two subgroups can be seen.

Focusing on the consequence level of organic buyers, the HVM's most important area is constituted by concepts related to what can be termed as *the pursuit of quality* (see Figs 8.6 and 8.7). In particular, this is related to 'high quality' and 'value of tradition'. Organic buyers prefer a large variety of different wine attributes (such as its region of production, 'Appellation d'Origine Contrôlée' – AOC – sign, country-of-origin, number of bottles per year, etc.) because they perceive that these attributes satisfy their search for high quality. In turn, this search is strongly associated with the value of 'searching for pleasure in life'. The second most important consequence is that termed *healthiness–long life*. This is constituted by 'healthiness–long life', 'purity–chemical-free' (from the attribute of 'organic label') and 'ethical production'. This is a cognitive structure strongly related to the organic concept and which is mostly responsible for organic food's purchase intention of the buyers. The remaining areas that constitute the organic buyers' HVM are: 'acquisition of information' 'aesthetic attractiveness–relaxation', 'good taste', 'environmental consciousness', 'control–attention to the production process' and 'ethnocentrism'.

The same concepts form the areas of the non-buyers' HVM; however, their order of importance is different. In descending order these are: acquisition of information, healthiness–long life, aesthetic attractiveness, high quality, good taste, control–attention to the production process, environmental consciousness, ethnocentrism and distinctiveness–innovative character. In other words, the

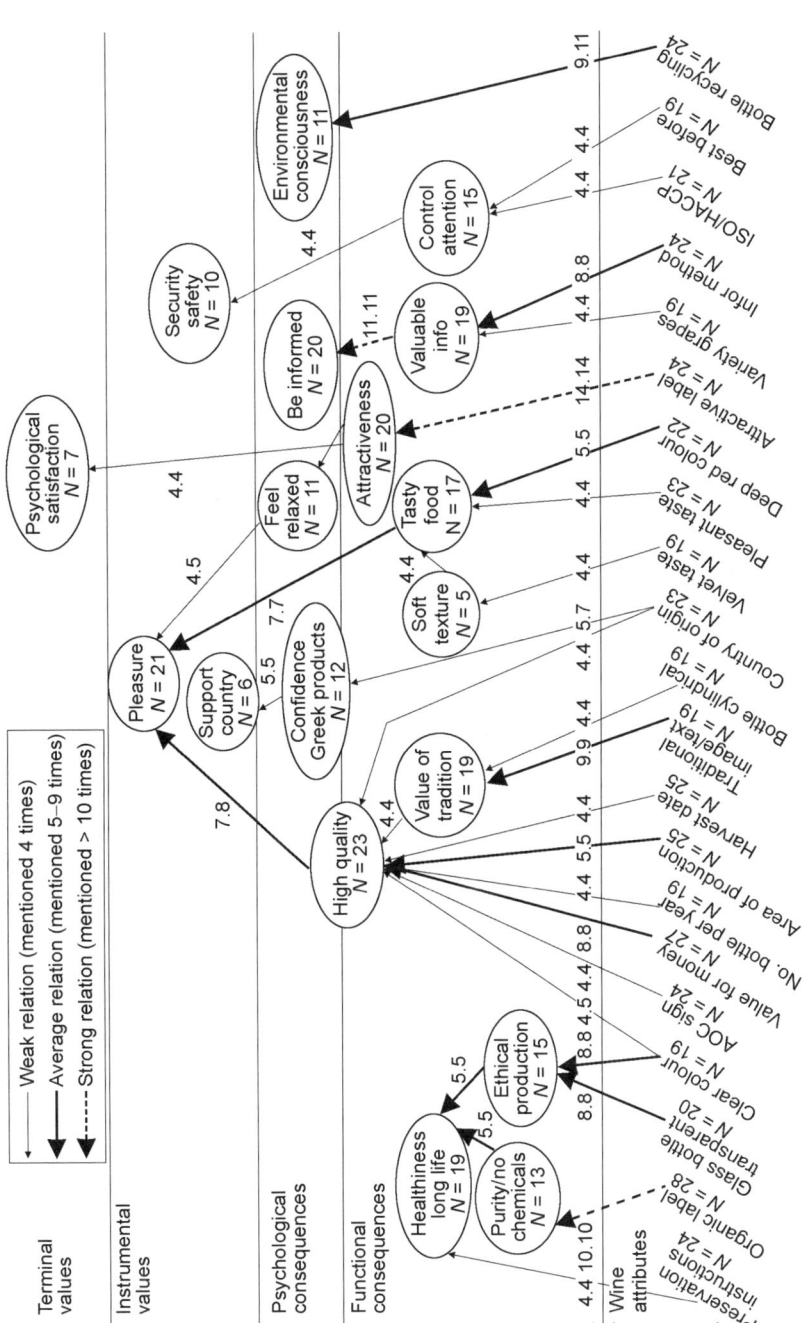

Fig. 8.6 The hierarchical value map, organic buyers ($n = 28$, cut-off level = 3).

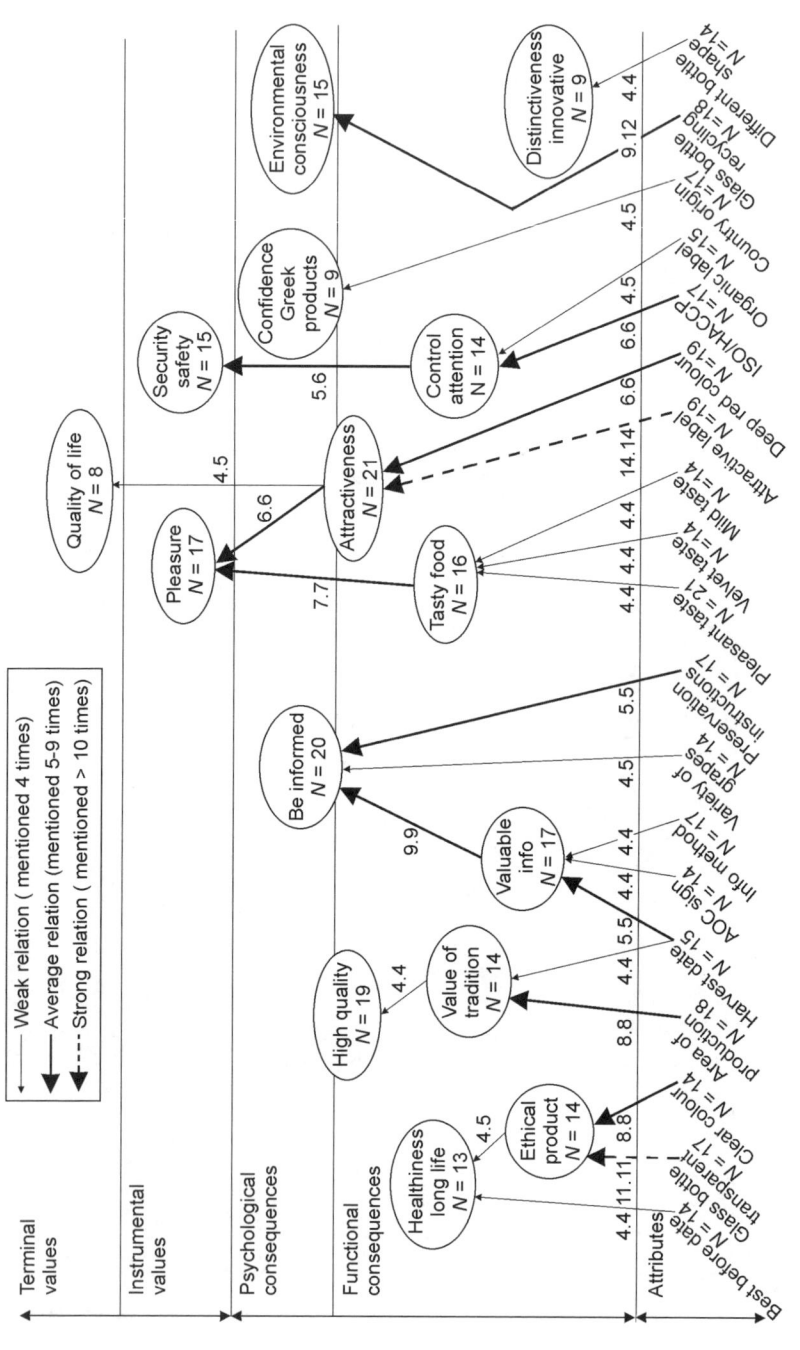

Fig. 8.7 The hierarchical value map, organic non-buyers ($n = 21$, cut-off level = 3).

motives related to information about wine and its aesthetic features constitute more important leverages of purchase, while that of wine's perceived quality constitutes less important one for the non-buyers, compared with the buyers.

Additionally, the composition of the common areas in the two maps is different. While the buyers find the organic label a health-related aspect of wine, the non-buyers mostly relate it to the control–attention paid during the production process, a much weaker motive, possibly not enough to translate the favourite attitude towards the organic label to organic food purchase intention. On the other hand, the AOC sign in wine (wines of superior quality) is a powerful purchasing leverage, since it belongs to the strongest motivational area in each group.

At the elicited value-level, the main concept related to the consequences for both subgroups is that of 'searching for pleasure in life', which imposes the motives of tastiness and attractiveness to the two subgroups and that of high quality to the buyers. This proves that pleasure is the rationale behind wine consumption, irrespective of the organic production method.

... and the implication for wine marketing

Healthiness, quality, information, attractiveness and *good taste* are the five main motives of wine purchase, while *pleasure* is the value-leverage of wine consumption. Additionally, the distinction between organic buyers and non-buyers derives from the differences in the evaluation of these motives in consumers' cognitive structures and the different motives with which wine's organic character is associated.

Each of the HVMs' perceptual orientations could be seen as a potential organic wine positioning strategy. This could be accomplished by benchmarking the strengths and weaknesses of the product. For instance, in the organic buyer HVM's quality orientation the linkages between a number of wine attributes (such as the 'AOC sign' and 'country-of-origin') and the 'high-quality' motive are weak. Therefore, one positioning option would be to build a strong association there, in the context of 'an AOC wine means high-quality wine'. That context would then need to be defined in terms of another, higher-order meaning, such as 'pleasure'. The result would be a strategic positioning that communicates to the consumer (e.g. via advertising) that the higher-order 'need of pleasure in life' can be fulfilled by wine's discriminating characteristic of 'AOC' through the satisfaction of high-quality expectation.

'Organic' wine products seem able to satisfy the expectations of customers who search for wines of novel origin and style. This wine consumption trend is consistent and matches a more general trend that embraces all the food sectors and is characteristic of many consumer segments which are not purely need-driven. As in the entire food sector, the wine sector has been influenced by the philosophy of organisation and by the new paradigms of industrial quality management such as hazard analysis and critical control point (HACCP) or the International Organization of Standardization (ISO).

In addition, a significant number of innovations have been introduced in grape production (for example, organic production), winery equipment and winemaking technologies, production organisation and marketing. Greek innovative wineries need to understand that it is no longer enough to offer wine as a traditional product. Rather, they should improve their understanding of the market needs and subsequently the achievement of competitive advantage. Linked to this market orientation is an innovation in production organisation that may be defined as a project-oriented and creative approach. Since the market accepts and even prefers organically produced wines, it appears increasingly necessary to organise the production process starting from the idea about a product as such, transforming the idea into 'product specifics', and planning the process, including all technological factors consistent with the product and the market where the product is to be sold.

The aim of consumer research methods in early stages of the NPD process is to make the voice of the consumer heard up front to facilitate the design of consumer-relevant new products. Research on success and failure factors in NPD (e.g. Cooper, 1988, in van Kleef, 2006) have identified consumer research as a key success factor, yet often overlooked or underdeveloped. The various consumer research methods available in NPD's early stages as reviewed by van Kleef *et al.* (2005) primarily differ in their degree of actionability for marketing vs. R&D and their ability to develop 'out of the box' ideas. The important implication is that the methods are not direct substitutes. Rather, their appropriateness depends on the purpose for which they are implemented (support marketing vs support R&D) and the innovation strategy, which is pursued (winning in existing well-defined markets vs building a new market through radically new products).

MEC and laddering, as justified by van Kleef *et al.* (2005) and explained in Section 8.4, are particularly appropriate for the *opportunity identification* phase of the NPD process for incremental new products, meaning products that are repositionings or updated versions of existing products, as the organic wine case study described above. The optimisation of products is a continually needed activity to keep up with competitors and stay cost-efficient. MEC and laddering are product-driven and consumer needs are primarily elicited with familiar stimuli. Consequently, they provide insights that are limited by the particular product(s) included in the study. Consumers can give reliable judgements about new products that are relatively similar to familiar products. Hence, the advantage of laddering lies in its capacity to capture current needs and desires and optimise existing products accordingly. Overall, laddering is more appropriate for marketing than R&D purposes as it reveals more abstract product use benefits and consumer values. These more abstract latent elements of a person's cognitive structure are closer to what drives consumer choice behaviour. However, they are too abstract and allow too many degrees of freedom for unambiguous translation into tangible product design.

8.8   References

ASCHMONEIT P and HEITMANN M (2001), 'Introduction to the means-end chain framework for product design and communications strategy for internet applications', Working paper, February 11, http://netacademy.org

ASCHMONEIT P and HEITMANN M (2002), 'Customer centred community application design', *Int J Media Manag*, **4**, 13–20.

ASCHMONEIT P and HEITMANN M (2003), 'Consumers cognition towards communities: customer-centred community design using the means-end chain perspective', Proceedings of the 37th International Conference on System Sciences (ICSS), 6–9 January, Hawaii.

AUDENAERT A and STEENKAMP J-B E M (1997), 'Means-end chain theory in agricultural market research', in Wierenga B van Tilburg A Grunert K Steenkamp J-B E M and Webel M, *Agricultural Marketing and Consumer Behaviour in a Changing World*, London, Kluwer Academic Publishers, 217–230.

AURIFEILLE J and VALETTE-FLORENCE P (1992), 'A 'chain-constrained' clustering approach in means–end analysis: an empirical illustration', in Grunert K, *Marketing for Europe, Marketing for the Future*, Proceedings of the 21st European Marketing Academy (EMAC) conference, Aarhus, 49–64.

AURIFEILLE J and VALETTE-FLORENCE P (1995), 'Determination of the dominant means–end chains: a constrained clustering approach', *Int J Res Mark*, **12**, 267–278.

BAGOZZI R P and DABHOLKAR P A (1994), 'Consumer recycling goals and their effects on decisions to recycle: a means–end chain analysis', *Psych Mark*, **1**, 313–340.

BAGOZZI R P and DABHOLKAR P A (2000), 'Discursive psychology: an alternative conceptual foundation to means–end chain theory', *Psych Mark*, **17**, 535–586.

BAKER S and KNOX S (1994), 'Getting the measure of brand-derived values cross pan-European segments: a cross-cultural application of means–end analysis', *Marketing: Unity in Diversity*, Proceedings of the 1994 Marketing Education Group annual conference, Vol. I, University of Ulster, UK.

BECH-LARSEN T (2000), 'Model-based development and testing of advertising messages – a comparative study of two campaign proposals based on the MECCAS model and a conventional approach', MAPP Working Paper 74, The Aarchus School of Business, July.

BECH-LARSEN T and GRUNERT, K.G (1998), 'Integrating the theory of planned behaviour with means–end chain theory', in Anderson P, *Marketing Research and Practice*, Proceedings of the 27th European Marketing Academy (EMAC) conference, 20–23 May, Stockholm, 305–313.

BECH-LARSEN T, NIELSEN N A, GRUNERT K G and SORENSEN, E (1996), 'Means-end chains for low involvement food products – a study of Danish consumers' cognitions regarding different applications of vegetable oil', MAPP Working Paper 41, The Aarchus School of Business, August.

BECH-LARSEN T, NIELSEN N A, GRUNERT K G and SORENSEN, E (1997), 'Attributes of low involvement products – a comparison of five elicitation techniques and a test of their nomological validity', MAPP Working Paper 43, The Aarhus School of Business, February.

BOTSCHEN G and HEMETSBERGER A (1998), 'Diagnosing means–end structures to determine the degree of potential marketing program standardisation', *J Bus Res*, **42**, 151–159.

BOTSCHEN G, THELEN E M and PIETERS R (1999), 'Using means–end structures for benefit segmentation: an application to services', *Eur J Mark*, **33**, 38–58.

BOUTIN E, FERRANDI J-M and VALETTE-FLORENCE P (1997), ' "Means-end chain model" and automatic network drawing as commercial awareness tools', *Int J Info Sc Dec Mak*, April, 19–35.

BREDAHL L (1999), 'Consumers' cognitions with regard to genetically modified foods. Results from a qualitative study in four countries', *Appetite*, **33**, 343–360.

BROWN S A (2005), 'Creating value in online collaboration in e-commerce', *Iss Info Syst*, **VI**, 87–93.

BRUNSO K, SCHOLDERER J and GRUNERT K G (2004), 'Closing the gap between values and behaviour: a means–end theory of lifestyle', *J Bus Res*, **57**(6), 665–670.

CATTERALL M and MACLARAN P (1996), 'Using computer programs to code qualitative data', *Mark Intel Plan*, **14**, 29–33.

CLAEYS C, SWINNEN A and ABEELE P V (1995), 'Consumers' means–end chains for "think" and "feel" products', *Int J Res Mark*, **12**, 193–208.

COHEN, J B and WARLOP L (2001), 'A motivational perspective on means–end chains', in *Understanding Consumer Decision Making: the Means–End Approach to Marketing and Advertising Strategy*, Mahwah, NJ, Lawrence Erlbaum.

COOLEN H and VAN MONTFORT X (2001), 'Meaning-based representations of preferences for housing attributes' Paper presented at the Housing Studies Association Autumn 2001 Conference *Housing Imaginations: New Concepts, New Theories, New Research*, 3–5 September, Cardiff University, UK,

DE BOER M and MCCARTHY M (2003), 'Means-end chain theory applied to Irish convenience food consumers', Paper presented at the 83d European Association of Agricultural Economists (EAAE) Seminar, 4–6 September, MAICh, Chania, Greece

DE FERRAN F and GRUNERT K G (2007), 'French fair trade product buyers' purchasing motives: an exploratory study using means–end chains analysis', *Food Qual Pref*, **18**(2), 218–229.

DEVLIN D, BIRTWISTLE G and MACEO N (2003), 'Food retail positioning strategy: a means–end chain analysis', *Brit Food J*, **105**, 653–670.

DIBLEY A and BAKER S (2001), 'Uncovering the links between brand choice and personal values among young British and Spanish girls', *J Cons Beh*, **1**, 77–93.

DOUCETTE W R and WIEDERHOLT J B (1992), 'Measuring product for prescribed medication using a means–end chain model', *J Health Care Mark*, **12**, 48–54.

DURGEE J F (1985), 'Depth-interview techniques for creative advertising', *J Adver Res*, **25**, 29–37.

FISHBEIN M and AJZEN I (1975), *Belief, Attitude, Intention and Behaviour: An Introduction to Theory and Research*, Boston, MA, Addison-Wesley Publishing Co.

FOTOPOULOS CH, KRYSTALLIS A and NESS M (2003), 'Wine produced by organic grapes in Greece: using means–end chains analysis to reveal organic buyers' purchasing motives in comparison to the non-buyers', *Food Qual Pref*, **14**, 549–566.

FRAUMAN E and CUNNINGHAM P H (2001), 'Using a means–end chain approach to understand the factors that influence greenway use', *J Park Recr Admin*, **19**, 93–113.

FRAUMAN E, NORMAN W C and KLEONSKY D B (1998), 'Using means–end theory to understand visitors within a nature-based interpretive setting: a comparison of two methods', *Tour Anal*, **2**, 161–174.

GENGLER CH E and REYNOLDS T J (1995), 'Consumer understanding and advertising strategy: analysis and strategic translation of laddering data', *J Advert Res*, **35**, 19–33.

GENGLER CH E, KLENOSKY D B and MULVEY M S (1995), 'Improving the graphic representation of means–end results', *Int J Res Mark*, **12**, 245–256.

GENGLER CH E, MULVEY M S and OGLETHORPE J E (1999), 'A means–end analysis of mothers' infant feeding choices', *J Pub Pol Mark*, **18**, 172–188.

GOLDENBERG M A, KLENOSKY D B, O'LEARY J T and TEMPLIN T J (2000), 'A means–end investigation of ropes course experiences', *J Leis Res*, **32**, 208–224.

GRANTHAM S (2005), 'Using funnel-based interview methods in focus groups to develop means–end chain ladders that explore perceptions of unfamiliar technical innovations', Paper presented at the Society of Risk Analysis (SRA) annual meeting, 4–7 December, Orlando, FL.

GRUNERT K G (1995), 'Food quality: a means–end perspective', *Food Qual Pref*, **6**, 171–176.

GRUNERT K G and BECH-LARSEN T (2004), 'Explaining choice option attractiveness by beliefs elicited by the laddering method', *J Econ Psych*, **25**(1), 97–123.

GRUNERT K G and BECHMANN S C (1999), 'A comparative analysis of the influence of economic culture on East and West Germany consumers' subjective product meanings', *Appl Psych Int Rev*, **48**, 367–490.

GRUNERT K G and GRUNERT S C (1995), 'Measuring subjective meaning structures by the laddering method: theoretical considerations and methodological problems', *Int J Res Mark*, **12**, 209–225.

GRUNERT K G and SORENSEN E (1996), 'Perceived and actual key success factors: a study of the yogurt market in Denmark, Germany and the United Kingdom', MAPP Working Paper 40, The Aarhus School of Business, January.

GRUNERT K G, GRUNERT S C and SORENSEN E (1995a), 'Means-end chains and laddering: an inventory of problems and an agenda for research', MAPP Working Paper 34, The Aarhus School of Business, November.

GRUNERT K G, SORENSEN E, JOHANSEN L B and NIELSEN N A (1995b), 'Analysing food choice from a means–end perspective', *Eur Advanc Cons Res*, **2**, 366–371.

GRUNERT K G, BECKMANN S C and SORENSEN E (2001a), 'Means–end chains and laddering: an inventory of problems and an agenda for research', in *Understanding Consumer Decision Making: The Means–End Approach to Marketing and Advertising Strategy*, Mahwah, NJ, Lawrence Erlbaum.

GRUNERT K G, LÄHTEENMÄKI L, NIELSEN N A, POULSEN J B, UELAND O and ASTROM A (2001b), 'Consumer perceptions of food products involving genetic modification – results from a qualitative study in four Nordic countries', *Food Qual Pref*, **12**, 527–542.

GUTMAN J (1982), 'A means–end model based on consumer categorisation processes', *J Mark*, **46**, 60–72.

GUTMAN J (1984), 'Analysing consumer orientations towards beverages through means–end chain analysis', *Psych Mark*, **1**, 23–43.

GUTMAN J (1991), 'Exploring the nature of linkages between consequences and values', *J Bus Res*, **22**, 143–148.

GUTMAN J (1997), 'Means–end chains as goal hierarchies', *Psych Mark*, **14**, 545–560.

GUTMAN J and ALDEN D (1985), 'Adolescents' cognitive structure of retail stores and fashion consumption: a means-end chain analysis of quality', in Jacoby J and Olson J, *Perceived Quality*, Boston, MA, Lexington Books.

HALEY R (1968), 'Benefit segmentation: a decision-oriented research tool', *J Mark*, **32**, 30–35.

HALL J and WINCHESTER M K (1999), 'An empirical confirmation of segments in the Australian wine market', *Int J Wine Mark*, **11**, 19–35.

HARAS K, BUNTING C and WITT P A (2005), 'Different strokes for different folks: a case study of ropes course programs using the means–end analysis', in Delamere T, Randall C and Robinson D, Proceedings of the 11th Canadian Congress on Leisure Research, 17–20 May, Nanaimo, Canada.

HEITMANN M, PRYKOP C and ASCHMONEIT P (2004), 'Using means–end chains to build mobile brand communities', Proceedings of the 37th International Conference on System Sciences (ICSS), 6–9 January, Hawaii.

HILTUNEN T, LAUKKANEN T and HILTUNEN M (2005), 'The influence of trial in consumer resistance to switching electronic banking channel from ATM to internet', in Retail Distribution Channels and Supply Chain Management, Australian and New Zealand Marketing Academy (ANZMA) 2005 Conference, 5–7 December, The University of Western Australia.

HINES J D and O'NEAL G S (1995), 'Underlying determinants of clothing quality: the consumers' perspective', Cloth Text Res J, 13, 227–223.

HUANG Y Y, CHOU T J and CHOU Y H (2003), 'A qualitative investigation of customer values: the means–end chains of garments buying behaviours', Manag Res, 3, 39–70.

HUBER F, BECHMANN S C and HERRMANN A (2004), 'Means-end analysis: does the affective state influence information processing style?', Psych Mark, 21, 715–737.

JAEGER S R and MACFIE H J H (2001), 'The effect of advertising format and means–end information on consumer expectations for apples', Food Qual Pref, 12, 189–205.

JENSEN T B (2005), 'Nurses' perception of an EPR implementation process based on a means–end chain approach', Department of Marketing Informatics and Statistics, Working Paper 05-1, The Aarhus School of Business.

JOLLY J P, REYNOLDS T J and SLOCUM JR JW (1988), 'Application of the means–end theoretic for understanding the cognitive bases of performance appraisal', Org Behav Hum Dec Proc, 41, 153–179.

JONAS M S and BECHMANN S C (1998), 'Functional foods: consumer perceptions in Denmark and England', MAPP Working Paper 55, The Aarhus School of Business, October.

JUDICA F and PERKINS W S (1992), 'A means–end approach to the market for sparkling wines', Int J Wine Mark, 4, 10–20.

KACIAK E and CULLEN C W (2005), 'Consumer purchase motives and product perceptions: a "hard" laddering study of smoking habits of Poles', Int Bus Econ Res J, 4, 69–85.

KING C (2005), 'Means–end chains analysis of recreational scuba diver values', Paper presented at the 2005 American Association of Active Lifestyle and Fitness (AAALF) Congress, 14–17 April, Chicago, IL.

KLENOSKY D B, GENGLER CH E and MULVEY M S (1993), 'Understanding the factors influencing ski destination choice: a means-end analytic approach', J Leis Res, 25, 362–379.

KLENOSKY D B, FRAUMAN E, NORMAN W C and GENGLER CH E (1998), 'Nature-based tourists' use of interpretive services: a means-end investigation', J Tour Stud, 9, 26–36.

KRYSTALLIS A and NESS M (2004), 'Motivational and cognitive structures of consumers in the purchase of quality food products: the case of Greek olive oil', J Int Cons Mark, 16, 7–36.

KUMAR V and RUST R T (1989), 'Marketing segmentation by visual inspection', J Adver Res, 29, 23–29.

LANGERAK F, PEELEN E and NIJSSEN E (1999), 'A laddering approach to the use of methods and techniques to reduce the cycle time of new-to-the-firm products', *J Prod Innov Manag*, **16**, 173–182.

LE PAGE A, COX D N, RUSSELL C G and LEPPARD P I (2005), 'Assessing the predictive value of means–end chain theory: an application to meat product choice by Australian middle-aged women', *Appetite*, **44**, 151–162.

LEPARD P, RUSSEL C G and COX D N (2004), 'Improving means-end chain studies by using a ranking method to construct hierarchical value maps', *Food Qual Pref*, **15**, 489–497.

LI C-F (2003), 'A conceptual model of web ad message design', Proceedings of the 2003 Association of Collegiate Marketing Educators (ACME) conference, 4–8 March, Houston, TX, 116–122.

LIN C F and YEH M Y (2000), 'Means–end chains and cluster analysis: an integrated approach to improving marketing strategy', *J Targ Meas Anal Mark*, **9**, 20–35.

LIN CH and FU CH-S (2005), 'A conceptual framework of intangible product design: applying means–end measurement', *Asia Pacif J Mark Logist*, **17**, 15–29.

MAKATOUNI A (2002), 'What motivates consumers to buy organic food in the UK? Results from a qualitative survey', *Brit Food J*, **104**, 345–352.

MANYIWA S and CRAWFORD I (2001), 'Determining linkages between consumer choices in a social context and the consumers' values: a means–end approach', *J Cons Behav*, **2**, 54–70.

MILES S and FREWER L (2001), 'Investigating specific concerns about different food hazards', *Food Qual Pref*, **12**, 47–61.

MILES S and ROWE G (2004), 'The laddering technique', in Breakwell G M, *Doing Social Psychology Research*, Oxford, BPS Blackwell, 305–343.

MORGAN A I (1984), 'Point of view: Magic Town revisited (a personal perspective)', *J Adver Res*, **24**, 49–51.

MORRIS D, McCARTHY M and O'REILLY S (2004), 'Customer perceptions of calcium enriched orange juice', University College Cork, Agribusiness Discussion Paper No. 42, July

MORT G S and ROSE T (2004), 'The effect of product type on value linkages in the means–end chain: implications for theory and method', *J Cons Behav*, **3**, 221–234.

MULVEY M S, OLSON J C, CELSI R L and WALKER B A (1994), 'Exploring the relationships between means–end knowledge and involvement', *Adv Cons Res*, **21**, 51–57.

MYERS J (1976), 'Benefit structure analysis: a new tool for product planning', *J Mark*, **40**, 23–32.

NASPETTI S and ZANOLI R (2003), 'Do consumers care about where they buy organic products? A means–end study with evidence from Italian data', Paper presented at the 83rd European Association of Agricultural Economists (EAAE) Seminar, 4–6 September, MAICh, Chania, Greece.

NIELSEN N A, SORENSEN E and GRUNERT K G (1995), 'Consumer motives for buying fresh or frozen plaice – a means–end chain approach', in Luten J B, Borresen T and Oehlencshlager J, *Seafood from Producer to Consumer, Integrated Approach to Quality*, Proceedings of the 25th West European Fish Technology Association (WEFTA) anniversary conference, Noordwijkerhout, Netherlands, Elsevier Science Ltd, 31–46.

NIELSEN N A BECH-LARSEN T and GRUNERT K G (1998), 'Consumer purchase motives and product perceptions: a laddering study on vegetable oil in three countries', *Food Qual Pref*, **9**, 455–466.

OLSON J C (1995), 'Introduction to the special issue of means–end chains', *Int J Res Mark,* **12**, 189–191.

OLSON J C and REYNOLDS T J (1983), 'Understanding consumers' cognitive structures: implications for advertising strategy', in Percy L and Woodside A, *Advertising and Consumer Psychology* Boston, MA, Lexington Books, 77–90.

OVERBY J W, GARDIAL S F and WOODRUFF R B (2004), 'French versus American consumers' attachment of value to a product in a common consumption context: a cross-national comparison', *J Acad Mark Sci*, **32**, 437–460.

PIETERS R, BAUMGARTNER H and STAD H (1994), 'Diagnosing means–end structures: the perception of wordprocessing software and the adaptive-innovative personality of managers', Proceedings of the 23rd European Marketing Academy (EMAC) conference, Maastricht, Netherlands, 749–762.

PIETERS R, BAUMGARTNER H and ALLEN D (1995), 'A means–end chain approach to consumer goal structures', *Int J Res Mark*, **12**, 227–244.

PITTS R E, WONG J K and WHALEN D J (1991), 'Consumers' evaluative structures in two ethical situations: a means–end approach', *J Bus Res,* **22**, 119–130.

REYNOLDS T J and CRADDOCK, P (1988), 'The application of the MECCAS model to the development and assessment of advertising strategy', *J Advert Res*, **28**, 43–54.

REYNOLDS T J and GUTMAN J (1984), 'Advertising is image management', *J Advert Res*, **24**, 27–37.

REYNOLDS T J and GUTMAN J (1988), 'Laddering theory, method, analysis, and interpretation', *J Advert Res*, **28**, 11–31.

REYNOLDS T J and JAMIESON L F (1985), 'Image representations: an analytic framework', in Jacoby J and Olson J, *Perceived Quality*, Boston, MA, Lexington Books.

REYNOLDS T J and OLSON J C (2001), *Understanding Consumer Decision Making: The Means–End Approach to Marketing and Advertising Strategy*, Mahwah, NJ, Lawrence Erlbaum Associates, Inc.

REYNOLDS T J and ROCHON J P (1991), 'Means–end based advertising research: copy testing is not strategy assessment', *J Bus Res*, **22**, 131–142.

REYNOLDS T J and WHITLARK D B (1995), 'Applying laddering data to communications strategy and advertising practice', *J Advert Res*, **35**, 9–17.

REYNOLDS T J, GENGLER C E and HOWARD D J (1995), 'A means–end analysis of brand persuasion through advertising', *Int J Res Mark*, **12**, 257–266.

ROININEN, T T, LÄHTEENMÄKI L and TUORILA H (2000), 'An application of means-end chain approach to consumers' orientation to health and hedonic characteristics of foods', *Ecol Food Nutr*, **39**, 61–81.

RUSSELL C G, FLIGHT I, LEPARD P J, VAN LAWICK J, A SYRETTE J A and COX D N (2004a), 'A comparison of paper-and-pencil and computerised methods of "hard" laddering', *Food Qual Pref*, **15**, 279–291.

RUSSELL C G, BUSSON A, FLIGHT I, BRYAN J, VAN LAWICK J A and COX D N (2004b), 'A comparison of three laddering techniques applied to an example of a complex food choice', *Food Qual Pref*, **15**, 569–583.

SCHOLDERER J and GRUNERT K G (2004), 'Do means–end chains exist? Experimental tests of their hierarchicity, automatic spreading activation, directionality, and self relevance', Paper presented at the Association for Consumer Research (ACR) 2004 North American Conference, 7–10 October, Portland, OR.

SCHOLDERER J, BRUNSØ K and GRUNERT K G (2002), 'Means–end theory of lifestyle: a replication in the UK', *Advanc Consu Res*, **29**, 551–557.

SIRIEIX L and SCHAER B (2005), 'Buying organic food in France: shopping habits and trust', Paper presented at the 15th International Farm Management Association (IFMA) Congress, 14–19 August, São Paolo, Brazil.

SKYTTE H and BOVE K (2004), 'The concept of retailer value: a means–end chain analysis', *Agribus*, **20**, 323–345.

SMEESTERS D, WARLOP L, CORNELISSEN G and ABEELE P V (2003), 'Consumer motivation to recycle when recycling is mandatory: two exploratory studies', *Tijdschrift voor Economie en Management*, **XLVIII**, 451–468.

SORENSEN E, GRUNERT K G and NIELSEN N A (1996), 'The impact of product experience, product involvement and verbal processing style on consumers' cognitive structures with regard to fresh fish', The Aarhus School of Business, MAPP Working Paper 42, October.

SUBRAMONY D P (2002), 'Why users choose a particular web sites over others: introducing a "means–end" approach to human–computer interaction', *J Electr Comm Res*, **3**, 144–161.

TER HOFSTEDE F, AUDENAERT A, STEENKAMP J-B E M and WEDEL M (1998), 'An investigation into the association pattern technique as a quantitative approach to measuring means–end chains', *Int J Res Mark*, **15**, 37–50.

TER HOFSTEDE F, STEENKAMP J-B E M and WEDEL M (1999), 'International market segmentation based on consumer-product relations', *J Mark Res*, **35**, 1–17.

URALA N and LÄHTEENMÄKI L (2003), 'Reasons behind consumers' functional food choice', *Nutr Food Sci*, **33**, 148–158.

VALETTE-FLORENCE P (1998), 'A causal analysis of means-end hierarchies in a cross-cultural context: methodological refinements', *J Bus Res,* **42**, 161–166.

VALETTE-FLORENCE P and RAPACCHI B (1991), 'Improvements in means–end chain analysis: using graph theory and correspondence analysis', *J Advert Res*, **31**, 30–45.

VALETTE-FLORENCE P, SIRIEIX L, GRUNERT K and NIELSEN N (1998), 'Comparing means-end hierarchies: a multi-dimensional perspective applied to a cross-cultural context', in Anderson P, *Marketing Research and Practice*, Proceedings of the 27th European Marketing Academy (EMAC) conference, 20–23 May, Stockholm.

VALETTE-FLORENCE P, SIRIEIX L, GRUNERT K and NIELSEN N (2000), 'Means–end chains analyses of fish consumption in Denmark and France: a multidimensional perspective', *J Euromark*, **8**, 15–27.

VALETTE-FLORENCE P, FERRANDI J-M, MERUNKA D and BACHELET D (2003), 'Introducing a new customer segmentation in the automotive market: a means–end perspective', Paper presented at the 2003 ESCP-EAP University marketing conference, www.escp-eap.net/conferences/marketing/pdf_2003/

VALLI C (1997), 'International food market segmentation: a conceptual framework for the operationalization of segmentation results in a pan-European context', in Loader, R J, Henson, S J and Trail W B, *Globalisation of the Food Industry: Policy Implications*, Conference proceedings, The University of Reading, 141–160.

VAN KLEEF E (2006), *Consumer research in the early stages of new product development. Issues and applications in the food domain,* PhD thesis, Wageningen University Publication.

VAN KLEEF E, VAN TRIJP HCM and LUNING P (2005), 'Consumer research in the early stages of new product development: a critical review of methods and techniques', *Food Qual Pref,* **16**, 181–201.

VANNOPPEN J, VERBEKE W, VAN HUYLENBROECK G and VIAENE J (2001), 'Consumer valuation of short market channels for fresh food through laddering', *J Int Food Agrib Mark*, **12**, 41–70.

VANNOPPEN J, VERBEKE W and VAN HUYLENBROECK G (2002), 'Consumer value structures towards supermarket versus farm shop purchase of apples from integrated production in Belgium', *Brit Food J*, **104**, 828–844.

VINCENT-WAYNE M and HARRIS G (2005), 'The importance of consumers' perceived risk in retail strategy', *Eur J Mark*, **39**, 821–837.

VINSON D E, SCOTT J E and LAMONT L M (1977), 'The role of personal values in marketing and consumer behaviour', *J Mark*, **41**, 44–50.

WALKER B A and OLSON J C (1991), 'Means-end chains: connecting products with self', *J Bus Res*, **22**, 111–118.

WIND J and MAHAJAN V (1997), 'Issue and opportunities in new product development: an introduction to a special issue', *J Mark Res*, **34**, 1–12.

WONG B (2002), 'The appropriateness of Gutman's means–end chain model in software evaluation', Proceedings of the 2002 International Symposium on Empirical Software Engineering (ISESE), Nara, Japan, 3–4 October.

WONG B and ROSS J D (2001), 'Cognitive structures of software evaluation: a means-end chain analysis of quality', in Bomarius F and Komi-Sirvio S, *Product Focused Software Process Improvement*, Proceedings of the 3rd international conference on Product Focused Software Process Improvement (PROFES) 2001, September, Kaiserslautern, Springer.

YOUNG S and FEIGIN B (1975), 'Using the benefit chain for improved strategy formulation', *J Mark*, **39**, 72–74.

ZANOLI R and NASPETTI R (2002), 'Consumer motivations in the purchase of organic food: a means–end approach', *Brit Food J*, **104**, 643–652.

ZEITHAML V A (1988), 'Consumer perceptions of price, quality and value: a means–end model and synthesis of evidence', *J Mark*, **52**, 2–22.

ZINS A H (2000), 'Two means to the same end: hierarchical value maps in tourism – comparing the association pattern technique with direct importance ratings', *Tour Anal*, **5**, 119–123.

9

Consumer attitude measures and food product development

K. Brunsø and K. G. Grunert, University of Aarhus, Denmark

9.1   Introduction

The success of a new product in the marketplace depends largely on consumers' interest and willingness to buy the specific product in question. Thus, it is crucial for a company to know what consumers like or need, e.g. to understand the preferences and attitudes guiding food choices, as well as how to translate this information into new product concepts.

While there have been numerous approaches to analysing preferences and attitudes to food products, a major distinction can be made between approaches covering preferences and attitudes for specific food products and approaches detecting more abstract and general preferences and attitudes (van Raaij and Verhallen, 1994; Brunsø *et al.*, 2004b). What finally determines a specific food choice will depend on a combination of general lifestyle attitudes and product-related attitudes as well as situational factors in the marketplace (Steenkamp, 1990).

In this chapter we introduce two highly recognised approaches to the understanding of consumer perception of food products and food-related attitudes, namely means–end chain analysis and food-related lifestyle segmentation. Furthermore, the two approaches represent the above-mentioned distinction between approaches covering preferences and attitudes for specific food products and approaches detecting more abstract and general preferences and attitudes as mentioned above. The first approach, the means–end chain approach, has a product-specific perspective, while the food-related lifestyle approach is based on a more abstract perspective on understanding consumers, their food consumption and preferences. Finally, the two methodological approaches for measuring consumers' food-related attitudes will be discussed in relation to consumer-led food product development, and recommendations will be given for future product development activities.

9.2   The means–end chain approach

The means–end chain approach aims at measuring consumer preferences and attitudes towards specific products at various levels of abstraction (Gutman, 1982, 1991; Olson and Reynolds, 1983; Valette-Florence, 1997). The model is based on the assumption that consumers' subjective product perception is established by associations between *product attributes* and more abstract, more central cognitive categories such as *values,* which can motivate behaviour and create interest in product attributes. Therefore, a *product attribute* is not relevant in and by itself, but only to the extent that the consumer expects the attribute to lead to one or more desirable or undesirable *consequences*. In turn, the relevance and desirability of these consequences are determined by the consumer's own *personal values,* and the consumer is motivated to choose a product if it leads to desirable consequences, thereby contributing to the attainment of personal values (Grunert, 1995).

Means–end chains are the *links* that a consumer establishes between product perceptions and abstract motives or values. They show how a product characteristic/attribute (e.g. 'light') is linked to consequences of consumption (e.g. 'being slim'), which may lead to the attainment of important life values (e.g. 'higher self-esteem'). Thus, when a consumer inspects the colour of a piece of meat (a product characteristic) it may be because s/he believes it to be related to the taste of the piece of meat when prepared (expected quality), and the taste will lead to enjoyment while eating (abstract purchase motive).

In order to measure means–end chains the laddering methodology was developed which uncovers the hierarchical relationship between product-related attributes, more abstract attitudes and general life values, motives and goals (Reynolds and Gutman, 1988; Grunert *et al.*, 2001). A laddering interview consists of two steps. First, relevant concrete product attributes must be identified. This can be done by means of direct questioning of consumers, triadic sorting or other methods used to identify the attributes that consumers find most important (Bech-Larsen and Nielsen, 1999). Next, after identifying the most salient attributes, respondents are asked which attributes they prefer. They are then asked 'why do you prefer?' and the answer is used as the basis for the next 'why do you prefer' question and so forth until the respondent is unable to answer. This technique attempts to trace lines of causal reasoning from the concrete to the abstract, pushing the respondent up a 'ladder of abstraction'. After the interview, data are recorded as individual means–end chains, which are content analysed and condensed into so-called *hierarchical value maps*, showing the strongest links between attributes, consequences and values for a group of respondents.

Besides being a valuable tool for gaining information about consumer attitudes, results from means–end chain analysis have recently shown to be a very helpful input for new product development. The information serves well as the basis for the product development process and also for stating and keeping project development goals, especially in the early stages of new product development (Costa *et al.*, 2004; Søndergaard, 2005).

9.2.1   Applications of the means–end chain approach

In the following we present results from a number of means–end chain studies that shed light on how consumers mentally link concrete food product attributes to desired and perceived consequences, and how this in turn is linked to more fundamental purchase motives and life values. This can be investigated using the laddering method, as explained in the methodological overview above, as it reveals consumers' cognitive structures and associations that are useful for product development purposes.

Motives for eating fish

A study of consumers' perceptions of fish and motives for buying seafood was carried out in Denmark, involving 90 laddering interviews (Nielsen *et al.*, 1997; Valette-Florence *et al.*, 2000). Consumers with varying levels of experience with fish products were interviewed, and Fig. 9.1 shows the hierarchical value map for consumers with the most experience in buying, preparing and eating fresh fish. The figure illustrates consumers' reasons for and against choosing a fresh, gutted plaice as the main ingredient for an evening meal to be prepared at home. The attributes mentioned by consumers can be seen at the bottom of the map, the self-relevant consequences linked to these attributes in the middle, and the life values at the top.

According to the study, *healthiness and physical well-being* is one of the most important reasons why these consumers buy fresh fish, along with *enjoyment*. According to these consumers, the fact that fresh fish is an *unprocessed product* (a natural product), contains *vitamins and minerals*, and is *low in fat* are all attributes which contribute to *wholesomeness and physical well-being*. And, as the map shows, healthiness is important because being healthy fulfils basic life values, the most important of which are *good health/a long life* and *the family's welfare*. But also *freshness* is important, leading to *good taste*, which has the desired consequence of *enjoying eating it* and this leads to *happiness and well-being* – a life value of high importance to consumers.

This hierarchical value map reveal consumers' most central and important attitudes when it comes to consumption of fish, and shows that health, naturalness and taste lead to attainment of desired life values.

Apples – leading to success

In another study on apples involving 50 laddering interviews of young Danes (18 to 35) (Bech-Larsen, 2001), respondents were asked to mention differences

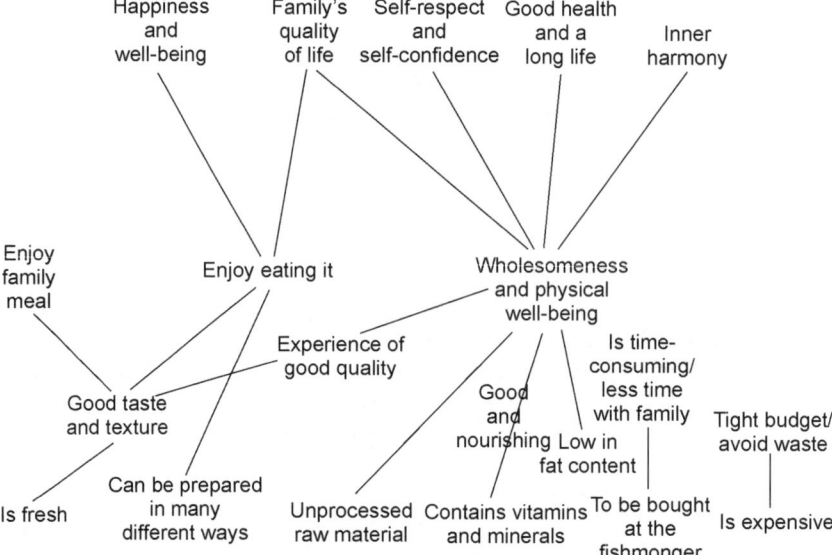

Fig. 9.1   Hierarchical value map for fresh-gutted plaice, the most experienced fish eaters
($n = 45$, cut-off $= 4$) (from Nielsen, N.A., Sørensen, E. and Grunert, K.G. (1997)
'Consumer motives for buying fresh or frozen plaice: a means end chain approach', in
Seafood from Producer to Consumer: Integrated Approach to Quality, J. B. Luten, T.
Børresen and J. Oehlenschläger (eds), Amsterdam, Elsevier: 31–43).

between apples and alternative foods. For each difference, the respondent was
asked whether it had any importance, and if yes this formed the point of
departure for the next laddering question. The resulting hierarchical value map is
shown in Fig. 9.2.

Perceptions relating to the wholesomeness of apples appear on the left-hand
side of the hierarchical value map. *Organic/not sprayed*, *wholesome* and
vitamins all contribute to the feeling of being *healthy* and *not ill*, leading to a
long, healthy life, a *good feeling* and a *high life quality*. But it is also worth
noting that by no means does the health aspect dominate: enjoyment and
convenience also play a major role. Apples are perceived as *easy to eat*, leading
to the convenience aspect *a quick meal* that eventually may lead to life values of
success and career. And *fresh taste* and *juicy* lead to good *taste* and *enjoyment*
and *pleasure* – also important consequences for consumers.

This hierarchical value map can be used as input to new ways of producing
and promoting apples with a higher emphasis on convenience, since this is
perceived as a possible mediator for success.

Healthy oil?

Since fresh fish and apples are probably more likely to be regarded as basic and
unprocessed foods, we also include an example of a product category which is
not regarded as completely unprocessed, namely vegetable oil (Nielsen *et al.*,

Fig. 9.2 Hierarchical value map for apples ($n = 50$, cut-off $= 6$) (from Bech-Larsen, T. (2001) 'Model-based development and testing of advertising messages: a comparative study of two campaign proposals based on the MECCAS model and a conventional approach', *International Journal of Advertising: The Quarterly Review of Marketing Communications* 20(4): 499–519).

1998). As part of a large study, interviews were made with 90 Danish women between 20 and 50, who had at least one child living at home, and who shopped for and used edible oil on a regular basis. Seven types of oil were presented to the respondents, who were asked to rank them into three groups (most preferred, least preferred and in-between). The respondents were then asked to give reasons for the grouping. Laddering interviews were carried out based on the important attributes, and the resulting hierarchical value map can be seen in Fig. 9.3.

As can be seen from the hierarchical value map, health is quite an important aspect for Danish consumers when buying vegetable oil, as illustrated by the large number of links to and from *healthy body, physical well-being*. Health is mainly related to the content of *unsaturated* fat and *cholesterol*, but also to the oil being a *pure* and *natural* product. Moreover, the health-related concepts are closely associated with several personal values, which indicates a high involvement in health.

But also *taste* is central to consumers – influencing *cooking results* which again lead to *enjoyment* and *happiness for family and self*. Most central to the perception of the quality of vegetable oil is both *price* and the perception of the attributes *pure and natural product*.

Even though vegetable oil must be regarded as a processed product, we see that the aspect of naturalness still plays an important role to consumers in their perception of desired qualities.

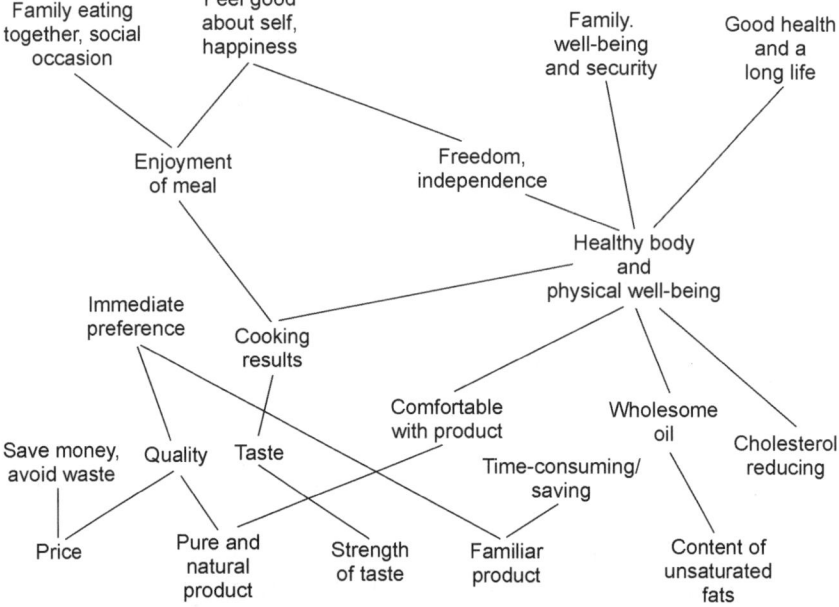

Fig. 9.3 Hierarchical value map for vegetable oil (cut-off = 15) (from Nielsen, N.A., Bech-Larsen, T. and Grunert, K.G. (1998) 'Consumer purchase motives and product perceptions: A laddering study on vegetable oil in three countries', *Food Quality & Preference* 9(6): 455–466).

Pork – organic or not

The means–end chain approach can also be used for cross-cultural comparisons, giving valuable input regarding consumers' perceived attributes, consequences and life values in different countries. In a pork study, consumers in Great Britain and Denmark were asked to imagine that they had to choose between ordinary and organic pork, and then asked to explain both the difference between the two types of meat and why the product attributes mentioned were important to them (Bech-Larsen and Grunert, 1998). The resulting hierarchical value maps are shown in Figs 9.4 and 9.5, cut-off levels express the minimum number of consumers that actually mentioned the relationships drawn.

As can be seen from the results, Danes regard *good, natural, non-animal feeding of pigs, leaner meat* and *less, no use of antibiotics, chemicals* as important attributes, which are expected to lead to *physical well-being, avoid illness,* which in turn are expected to lead to *good health and a long life.* As in the other examples of fish, apples and vegetable oil, we find that the health aspect is central to consumers, and that naturalness is closely related to health. For both the British and the Danish respondents, there seems to be at least four different reasons for choosing or not choosing organic pork: animal welfare, budgetary restraints, health and enjoyment. Concerns for animal welfare seem to be more important to British than to Danish consumers, while budgetary

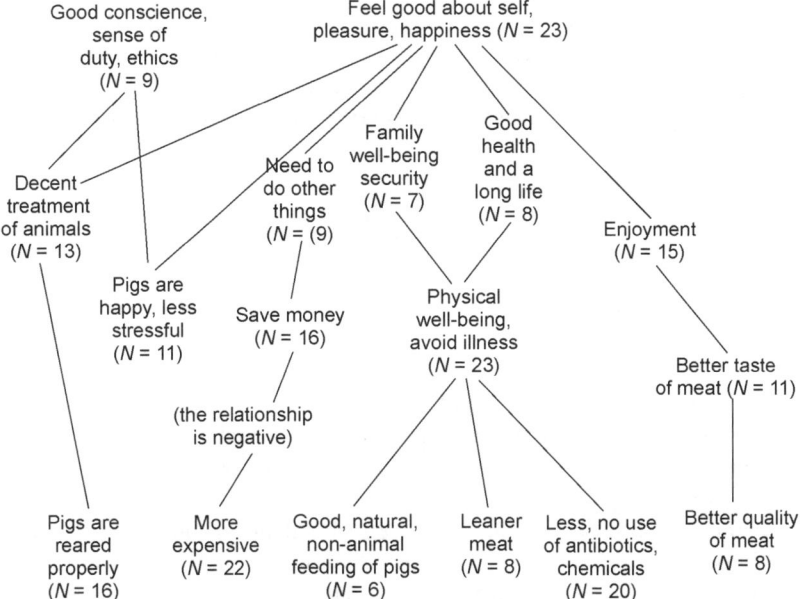

Fig. 9.4   Hierarchical value map for Danish consumers' perception of organic/conventional pork (*n* = 30, cut-off = 4) (from Bech-Larsen, T. and Grunert, K.G. (1998) *Integrating the Theory of Planned Behaviour with Means-End Chain Theory: A study of possible improvements in predictive ability*, 27th EMAC Conference, Stockholm).

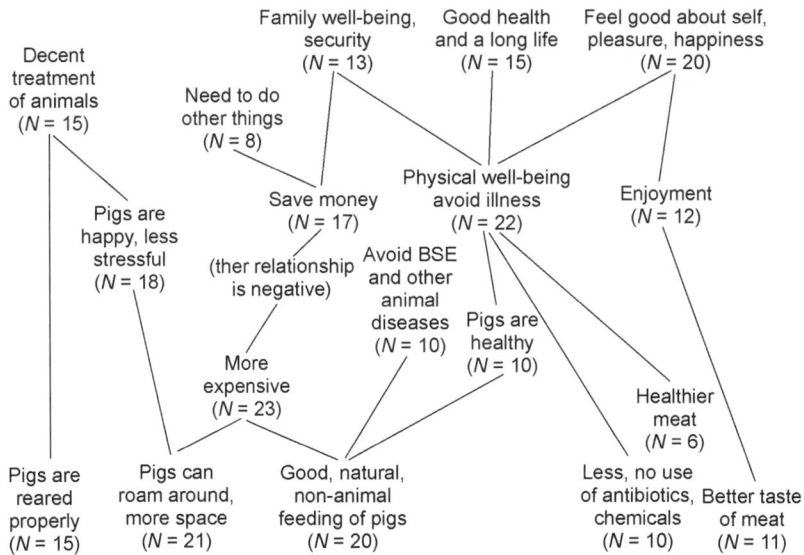

Fig. 9.5   Hierarchical value map for UK consumers' perception of organic/conventional pork (*n* = 30, cut-off = 5) (from Bech-Larsen, T. and Grunert, K.G. (1998) *Integrating the Theory of Planned Behaviour with Means-End Chain Theory: A study of possible improvements in predictive ability*, 27th EMAC Conference, Stockholm).

restraints seem to be important both in Great Britain and in Denmark. This is by far the most important motive for *not* eating organic pork, according to the laddering studies.

The means–end study shows two important things. Firstly, consumers make a whole range of positive inferences from the label 'organic', and these refer not only to the concern for the environment and health, but also to animal welfare and better taste. Secondly, positive inferences do not necessarily lead to a purchase if consumers do not think that the trade-off between give and get components is sufficiently favourable, e.g. if the 'give' component *price* is too high compared with the 'get' components (here *health, taste and animal welfare*).

9.2.2   Discussion of means–end chain results

Together, the examples presented above give interesting insight into how consumers evaluate central aspects of food products. Also, the results indicate that there are some food-related attitudes that are very central in relation to food in general – food-related attitudes that are found in all the examples across consumers, food products and countries: hedonic-related attitudes, health-related attitudes, naturalness-related attitudes and convenience-related attitudes.

For most people, food is, and has always been, a matter of pleasure. The hedonic characteristics of food – primarily *taste*, but also appearance and smell – thus constitute a central aspect for consumers and their food choices. This has always been the case, of course, but in recent decades consumers have shown an increasing interest in the other three aspects as well. The means–end chain analyses show that health-related attitudes have become a very central and important aspect to many consumers, and that consumers form preferences based on health-related attitudes motivated by expectations of both a longer life and a higher quality of life, as have also been found in other studies of food-related attitudes (Roininen *et al.*, 2001). Futhermore, the issue of naturalness is quite visible in the hierarchical value maps, indicating that consumers attach increasing importance to the way food is produced, i.e. the *production process* has become a dimension of quality, even when it has no immediate bearing on the taste or healthiness of the product. This dimension covers organic production, production which takes animal welfare into consideration, and issues of genetically modified food products. But as shown in the means–end chain examples, still much of consumer interest in the production process focuses on the perceived 'naturalness' of how the product was produced. Finally, the issue of convenience seems to be of increasing importance to consumers. From a consumer point of view, convenience is more than just easy shopping or quick consumption. Convenience means the saving of time, physical or mental energy at one or more stages of the overall meal process: planning and shopping, storage and preparation of products, consumption, and the cleaning up and disposal of leftovers (Gofton, 1995). And the means–end chain results indicate that the time factor, as well as other ease aspects leading to convenience, are becoming more and more important.

The four food-related attitudes presented should not be regarded as independent – there are clearly both overlaps and interrelationships. These interrelationships are not unambiguous, and vary across products, e.g. consumers sometimes perceive good taste and healthiness to be positively correlated, and negatively correlated at other times. Taste is sometimes perceived to be related to the naturalness and at other times is not. Such inferences are typical for consumer attitudes and product perceptions and must be taken into account in the food product development process in order to avoid potential conflicts and to use this information to develop and promote the right food quality.

9.3   The food-related lifestyle approach

The lifestyle construct and the resulting measurement have been widely used in marketing as a valuable source of information for advertising strategy, segmentation, and product development. In the food-related lifestyle approach, lifestyle is regarded as a mental construct that explains, but is not identical with, actual behaviour. The food-related lifestyle approach has been developed following an extensive review of existing instruments such as VALS, Rokeach's List of Values (LOV), and Schwartz's Motivational Types of Values (see Brunsø and Grunert, 1995; Grunert *et al.*, 1997). The approach proposed for analysing food-related lifestyle is based on general assumptions of cognitive psychology (see, e.g., Anderson, 1983; Peter and Olson, 1990), and food-related lifestyle is defined as a set of mental constructs or cognitive categories or scripts, and their associations, which relate a set of food products to a set of values. According to Grunert *et al.* (1997) this proposed definition:

- makes lifestyle distinct from values;
- makes lifestyles transcend individual brands or products, but possibly specific to a product class so that it makes sense to talk about a food-related lifestyle;
- places lifestyles clearly in a hierarchy of constructs at different levels of abstraction, where lifestyles have an intermediate place between values and product/brand perceptions or attitudes;
- covers both factual and procedural knowledge, i.e. both subjective perceptions, based on information and experience, of which products contribute to the attainment of life values, and acquired procedures concerning how to obtain, use, or dispose of products; and
- refers to enduring dispositions to behave, not to single behaviour acts.

Thus, food-related lifestyle represents a 'means–end chain approach' to lifestyle – focusing on how consumers link generic food-related attitudes to the achievement of desired consequences (Brunsø *et al.*, 2004a).

Based on the theoretical approach presented in Brunsø and Grunert (1995) and in Grunert *et al.* (1997), a cross-culturally valid instrument for measuring food-related lifestyles was developed. The resulting instrument is a 69-item questionnaire covering 23 dimensions measured by three items each rated on 7-point Likert-type scales.

The 69 items are shown in the Appendix, and the 23 dimensions cover the following areas:

- *Ways of shopping*: importance of product information, attitude towards advertising, joy of shopping, speciality shops, price criterion, shopping list.
- *Cooking methods*: involvement in cooking, looking for new ways, convenience, family involvement, spontaneity, woman's task.
- *Quality aspects*: health, price–quality relationship, novelty, organic products, tastiness, freshness.
- *Consumption situations*: snacks versus meals, social event.
- *Purchasing motives*: self-fulfilment, security, social relationships.

9.3.1 Applications of the food-related lifestyle approach

So far the instrument has been applied in a number of European countries with the purpose of generating data that allow a better understanding of food consumers in these countries with regard to how they use food products to attain life values. Throughout these studies the food-related lifestyle instrument has been extensively tested for cross-cultural validity, i.e. for its ability to obtain results that can be compared even though respondents come from different countries, cultures and language regions. These tests have confirmed that food-related lifestyle is cross-culturally valid in Western Europe (Scholderer *et al.*, 2004). Also food-related lifestyle has been applied in Singapore drawing on a limited sample (Askegaard and Brunsø, 1999), revealing that cross-cultural validity could only be established for a reduced number of items. Last but not least the food-related lifestyle instrument has been applied in Adelaide, Australia, drawing on a convenience sample, and results proved interesting similarities and differences when compared with the European and Singaporean studies (Reid *et al.*, 2001).

Extensive research on consumers' food-related lifestyle in a number of European countries (Grunert *et al.*, 2001b), and also some countries outside Europe (Askegaard and Brunsø, 1999; Reid *et al.*, 2001), has established a number of basic food consumer segments, which are described below together with general demographic characteristics.

- *The uninvolved food consumer.* Food is not a central element in the lives of these consumers. Consequently, their purchase motives for food are weak, and their interest in food quality is limited mostly to the convenience aspect. They are also uninterested in most aspects of shopping, do not use speciality shops, and do not read product information, limiting their exposure to and processing of food quality cues. Even their interest in price is limited. They have little interest in cooking, tend not to plan their meals, and snack a great deal. Compared with the average consumer, these consumers are single, young, have part or full-time jobs, average to low incomes, and tend to live in big cities.

- *The careless food consumer.* In many ways, these consumers are quite similar to the uninvolved food consumers in the sense that food is not very important to them, and, with the exception of convenience, their interest in food quality is correspondingly low. The main difference is that these consumers are interested in novelty: they like new products and tend to buy them spontaneously, so long as they do not require any great effort in the kitchen or new cooking skills. In general the careless food consumer is, as the uninvolved food consumer, young and often lives in big cities. But in contrast to the uninvolved consumers, these consumers are more educated and they lie in the upper income brackets.

- *The conservative food consumer.* For these consumers the security and stability achieved by following traditional meal patterns is a major purchase motive. They are very interested in the taste and health aspects of food products, but are not particularly interested in convenience, since meals are prepared in the traditional way and regarded as part of the woman's tasks. The conservative food consumers have the highest average age and they are the least educated. Households are on average small, and their household income is in general lower than that of the other segments. These consumers tend to live in rural areas.

- *The rational food consumer.* These consumers process a lot of information when shopping: they look at product information and prices, and they use shopping lists to plan their purchases. They are interested in all aspects of food quality. Self-fulfilment, recognition and security are major purchase motives for these consumers, and their meals tend to be planned. Compared with the average food consumer, this segment has a higher proportion of women with families. The level of education and income in this segment differ from country to country, but they tend to live in medium-sized towns and a relatively large proportion of these consumers are not wage earners.

- *The adventurous food consumer.* While these consumers have a somewhat above-average interest in most quality aspects, this segment is mainly characterised by the effort they put into meal preparation. They are very interested in cooking, look for new recipes and new ways of cooking, involve the whole family in the cooking process, are not interested in convenience and reject the notion that cooking is the woman's task. They want quality, and demand tasty food products. Self-fulfilment in food is an important purchase motive. Food and food products are important elements in these consumers' lives. Cooking is a creative and social process for the whole family. The adventurous food consumer is in general from the younger part of the population, and household size is above average. The adventurous food consumers have the highest educational level and have high incomes. They tend to live in big cities.

Table 9.1 shows the distribution of segments in a number of European countries – all studies were conducted within the last decade based on representative samples of 1000 to 1200 respondents in each country. Also, it should

Table 9.1   Food-related lifestyle segments across Europe

	Germany (%)	France (%)	GB (%)	Denmark (%)	Spain (%)
The uninvolved	21	15	7	11	16
The careless	11		17	23	
The rational	26	17	15	11	26
The conservative	18	17	14	11	26
The adventurous	24	11	22	24	20
The moderate/pragmatic		21	25		
The hedonistic		21			
The eco-moderate				20	
The enthusiastic					12

be noted that the segments described above are based on the cross-cultural segments we found in most countries. In addition to these, a particular country may have idiosyncratic segments which differ a little from the basic types described above. Thus, the figure shows a *pragmatic* segment in France, covering consumers with a lot of interest in both health and organic food, as well as in convenience and snacking, a *hedonistic* segment in France, which resembles the adventurous segment, but with a stronger emphasis on the pleasurable aspects of food, a *moderate* segment in Great Britain, expressing an average attitude to all aspects of food-related lifestyle, and an *eco-moderate* segment in Denmark, where organic production is *the* aspect of food these consumers are really interested in.

These cross-cultural segments have also been found to be stable over time, and may thus be used for monitoring trends over time (Brunsø *et al.*, 1996). In addition, analysis has shown that the different segments have different preferences, different perception of food quality and are interested in different types of product information revealing a need for targeting product development activities and marketing communication towards the specific consumer segments (Grunert *et al.*, 2001b). For instance, conservative food consumers are very reluctant to try out new food products while adventurous food consumers experiment a lot in the kitchen and have an above average interest in organic food products. At the conceptual level, linking food products to values as proposed in the definition (testing of the 'means–end chain approach' to lifestyle) has been explored and established (Brunsø *et al.*, 2004a), and has also resulted in interesting relations between the value and the food-related lifestyle level (Brunsø *et al.*, 2004b).

Below we show how the different food-related lifestyle segments from the European countries evaluate the importance of the central food-related attitudes described earlier in Section 9.2.2 (hedonic-related attitudes, health-related attitudes, naturalness-related attitudes and convenience-related attitudes).

Figure 9.6 shows that in general the central food-related attitudes are quite important to most consumers, but also that the emphasis placed on taste, health,

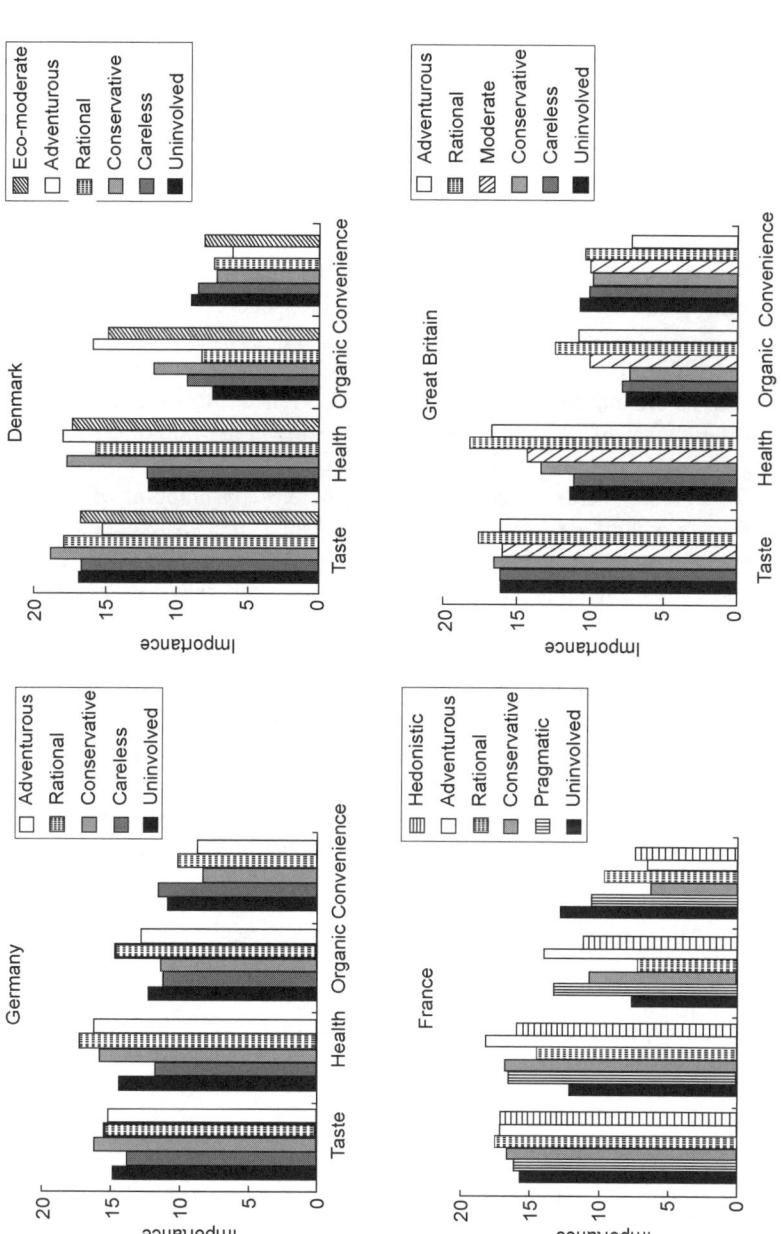

Fig. 9.6 Quality dimensions and consumer segments in four countries. The columns show the means for the importance of the few quality dimensions across countries and segments (min. 3, max. 21).

naturalness and convenience varies across both segments and countries. As can be seen in the figure, taste and health are very important attitudes, both across countries and across segments. Besides, process and convenience-related attitudes are also quite important to many consumers in the four countries, but with more variation among the six segments.

9.3.2   Is the behaviour of the segments distinctive? Danish segments and organic consumption

While it is interesting to have profiles of both cross-cultural as well as national segments, the real value of the lifestyle segments will depend on whether they are indeed distinctive, e.g. that they respond differently towards new products with specific qualities depending on the specific value's significance to the segment in question. We can expect interest in and consumption of organic products to differ between segments. Table 9.2 illustrates variations in the consumption of organic food products across food-related lifestyle segments in Denmark (Grunert et al., 2001b). The table shows the proportion of the specific segments that often buy the various organic products.

As can be seen, there are big differences in the consumption of various organic food products across segments. As expected, the eco-moderate consumers have the highest consumption rates for some products (e.g. milk, bread, coffee), and together with the adventurous food consumers they have by far the highest consumption levels for the rest of the products. But we also know from the segment profiles that their purchase motives differ: the eco-moderates are first and foremost interested in their own (and their family's) health, while environmental concerns play a minor role. The opposite is true for the adventurous food consumers – they are very concerned about the environment, whereas their own health is of secondary importance. As can also be seen, for most products, the uninvolved and careless food consumers have the lowest consumption levels, with the conservative and rational food consumers lying in

Table 9.2   Food-related lifestyle segments and organic consumption in Denmark. From Grunert et al. (2001b)

Organic products	The uninvolved (%)	The careless (%)	The conserva-tive (%)	The rational (%)	The eco-moderate (%)	The adventurous (%)
Buy often:						
Potatoes	5.8	6.7	19.6	19.7	41.9	45.1
Eggs	23.3	24.9	39.9	32.0	66.2	66.3
Milk	12.6	25.8	24.6	31.4	70.0	65.7
Bread	8.0	18.3	36.7	23.2	49.1	44.0
Ice cream	1.2	2.1	3.7	8.2	8.2	11.2
Coffee	2.5	4.1	6.3	6.5	20.1	16.9

the middle. In general we can conclude that the interest in and consumption of organic products differ considerably across consumers, and segmenting consumers according to their food-related lifestyle is a useful way of understanding these individual differences.

9.4   General discussion – the issue of consumer-led food product development

This chapter, which presents two approaches to analysing consumer attitudes and perception of food as well as a number of examples and results, also shows that food attitudes and preferences are complex. Consumers have various and sometimes contradictory wishes with regard to food, which differ from one consumer to the next. Manufacturers have the difficult task of understanding these wishes, trying to determine which of them can be fulfilled in a profitable way, translating them into production processes, and then communicating back to consumers about the qualities being offered.

In order to do so, there are a few issues that are important to take into account. Firstly, from the consumer point of view, attitudes and food-related preferences are subjective. The attitudes and preferences of a food product are in the mind of the consumer – some aspects of the product are perceived as good and others as bad. Some, e.g. good taste, are good because they lead to immediate hedonic reinforcement of the food purchase. Others are good because the consumer believes them to have positive consequences, e.g. for a better health, for the environment, or for society at large. Objectively such convictions may be wrong, but they, nevertheless, constitute reality for the consumer who perceives quality and makes choices. Therefore it is a challenge for manufacturers working professionally with food to understand and accept the mind of the consumer. Also, there is no doubt that price is an important parameter in consumer choice, and that the trade-off between price and product quality is an important aspect in consumer food choice. But unwillingness to pay for a certain quality does not necessarily imply a lack of interest in quality. If a consumer is unwilling to pay for a specific quality, it may be due to many other reasons than lack of interest in quality:

- the product does not, in an objective sense, have the specific quality the manufacturer claims;
- the consumer does not desire the specific quality (enough), i.e. the consumer does not consider the quality improvement to be worth the price differential;
- the consumer does not realise that the product actually has the specific attributes or quality.

Secondly, consumers differ. Some are not even interested in food. Others are interested, but in very different ways. Such differences lead to different patterns of quality perception, attitude formation and food choice. These can be analysed, and grouping consumers into segments is a major improvement over undifferentiated approaches to 'the consumer'. Segmentation, when done properly, is a powerful tool to a better understanding of consumers, which, unfortunately, has been routinely ignored in the analysis of consumer attitudes and food-related preferences.

So in spite of the complexity of consumers' attitudes, motives and lifestyles, this does not mean that it is impossible to understand and analyse consumer food quality perception, attitudes and choice, and the examples given throughout the chapter are provided in order to show how this challenge can be overcome. The next challenge is how to use results on food-related attitudes from consumer studies for food product development; this is discussed below.

9.4.1   Two steps to consumer-led product development

The means–end chain approach and the food-related lifestyle segments can be used as valuable input to food product development. We propose a two-step procedure where the first step, the selection of target segment(s), should be taken before even starting the new product development process. Ideally the identification of a target consumer segment should be based on the understanding of the existing consumer base (if any), followed by the specification of a desired future consumer segment: who will be the attractive consumers of tomorrow – and how can they be characterised?

Thus, for the specification of a future market segment, knowledge of food-related lifestyle segments, their motives and priorities may serve to understand and select a future market segment. Having selected a future potential target segment, a means–end chain analysis should be made next to produce input for the early stages of the product development process – especially for the idea generation phase. Below we discuss the two-step procedure by describing product development perspectives for the cross-cultural lifestyle segments followed by an introduction to the means–end chain approach as input for product development.

9.4.2   Food-related lifestyle – who to target?

Studies have shown that the *uninvolved* food consumer exists in all countries. From a food-product development perspective, this segment may not seem to be very interesting, but the people in it still buy food and eat every day – and try to do so with as little effort as possible. This segment is a group of consumers that find life challenges in other areas than food, and they can be characterised by a low degree of stability and loyalty towards specific products or brands, and they perceive few differences between food products. This makes them more sensitive to price differences, and therefore price differentiation seems to be the easiest way to appeal to these consumers.

In general, *careless* food consumers are not very interested in food – with the exception of novelty – and they prefer easy ways of cooking. This segment can be characterised as 'variety seeking' consumers who like to try out new food products, and often they are willing to pay extra for novelty. But new products should not require too much meal preparation effort, as their interest in new products does not include interest in new ways of cooking – it doesn't! In general, food is not particularly important to these consumers, but new and easy meal solutions will appeal to them.

The *rational* food consumer segment puts a lot of emphasis on evaluating food quality, especially on the functional characteristics, i.e. healthiness, freshness, organic qualities and naturalness overall. These aspects are very important to this segment, and consumers are willing to put extra effort into information search. Therefore, new products emphasising these qualities will appeal immensely to this segment – if the communication of product qualities is put forward in a trustworthy and understandable way. Also, new products aimed at this segment must not be too novel in relation to meal preparation – cooking traditions and traditions in relation to food in general are of high importance to these consumers. All in all, this segment is interested in good quality in relation to the above-mentioned areas, and is also willing to pay more if the quality meets expectations.

The *conservative* food consumer uses food especially as a means to achieving security and stability – life values that are very important to this segment. As long as this goal is fulfilled these consumers are quite interested and involved in food, and put effort into both food shopping and cooking – desiring predictability and stability. They are quite loyal towards brands, product types and shops, and can develop strong preferences if their demands are met. Product development targeting this segment must therefore try to transfer well-known aspects to the new product in order to appeal to these consumers. But once they are convinced, they tend to be very loyal consumers.

Last but not least, *adventurous* food consumers are very interested in most aspects of food shopping and cooking, and they find food-related activities stimulating and see them as part of their self-fulfilment in life, for instance as adventures, challenges and as emphasising social relationships. These consumers want to be stimulated in relation to creativity and novelty, and are not afraid of trying out new products, food types or recipes. They are interested in aspects such as healthiness, freshness and organic products, but the best way to approach this segment is to encourage self-expression, creativity and social togetherness.

Having selected a target segment, the product development process can start, i.e. ideas may be generated with the particular segment in mind, followed by idea screening and testing of concepts and prototypes, ideally using respondents from the chosen food-related lifestyle segment. A recent research project about using means–end chain analysis for new product development found that information about consumers' high-priority means–end chains for a product category can give companies the necessary knowledge to develop products that offer consumers the desired consequences and which help them attain central life values (Søndergaard, 2005). Since a means–end chain analysis gives information about consumer attitudes on three levels of abstraction: attitudes, consequences and values, this is an ideal point of departure for new product idea generation. The product development team gets information about the motivations underlying consumer preferences helping them to create ideas for products leading to superior value for consumers. The use of means–end chain analysis thus provides more relevant information for, especially, the idea generation process and provides a very good basis for discussions and for agreeing on and maintaining goals during the process (Costa *et al.*, 2004; Søndergaard, 2005). The information provided by a means–end chain analysis reveals both the negative consequences perceived by consumers, which should be avoided, and the positive consequences and values that should be emphasised in a new food product. Finally the results showed that when taking a means–end approach for product development in the food industry, it may be an advantage to let respondents taste real products using their reactions as input for laddering interviews (Søndergaard, 2005).

Future research should focus on the use of several approaches for studying consumer attitudes as input for product development. All too often, consumers are involved only in the last phases of the product development process, e.g. when it comes to concept testing or testing of real products – while, as emphasised here, input from consumer studies with different perspectives covering both more general and more product specific attitudes can and should be included and applied *much* earlier in the product development process.

9.5   Sources of further information and advice

As already stated above, consumer-led product development should apply consumer studies from the outset of the product development process in order to select and target a specific consumer market. The first requirement is segmentation, and here we have introduced and described the food-related lifestyle approach which has been developed and targeted towards food markets; examples of applications of the approach are given. The items included in the food-related lifestyle instrument are shown in the Appendix, and recent discussions of the applicability of the instrument can be found in the following references:

BRUNSØ, K., SCHOLDERER, J. & GRUNERT, K. G. (2004). 'Testing relationships between values and food-related lifestyle: results from two European countries'. *Appetite* **43**(2): 195–205.

GRUNERT, K. G., BRUNSØ, K., BREDAHL, L. & BECH, A. (2001). Food-related lifestyle: a segmentation approach to European food consumers. *Food, People and Society: A European Perspective of Consumers' Food Choices*. L. J. Frewer, E. Risvik, H. N. J. Schifferstein and R. von Alvensleben. London, Springer Verlag: 211–230.

Basic theoretical assumptions and testing can be found in:

BRUNSØ, K., SCHOLDERER, J. & GRUNERT, K. G. (2004). 'Closing the gap between values and behaviour: A means–end theory of lifestyle'. *Journal of Business Research* **57**: 665–670.

Also the means–end chain approach has been shown to be a very useful approach to consumer-driven product-development processes, and further information and discussion of the application of the means–end chain approach in general and in a product development process in particular can be found in the following references:

SØNDERGAARD, H. A. (2005). 'Market-oriented new product development: how can a means–end chain approach affect the process?' *European Journal of Innovation Management* **8**(1): 79–90.

COSTA, A. I. A., DEKKER, M. & JONGEN, W. M. F. (2004). 'An overview of means–end theory: Potential application in consumer-oriented food product design'. *Trends in Food Science &Technology* **17**(7–8): 403–415.

9.6   References

ANDERSON, J. R. (1983). *The Architecture of Cognition*. Cambridge, MA, Harvard University Press.

ASKEGAARD, S. & BRUNSØ, K. (1999). 'Food-related life styles in Singapore: Preliminary testing of a Western European research instrument in Southeast Asia'. *Journal of Euromarketing* **7**(4): 65–86.

BECH-LARSEN, T. (2001). 'Model-based development and testing of advertising messages: a comparative study of two campaign proposals based on the MECCAS model and a conventional approach'. *International Journal of Advertising: The Quarterly Review of Marketing Communications* **20**(4): 499–519.

BECH-LARSEN, T. & GRUNERT, K. G. (1998). *Integrating the Theory of Planned Behaviour with Means–end Chain Theory: A study of possible improvements in predictive ability*. 27th EMAC Conference, Stockholm.

BECH-LARSEN, T. & NIELSEN, N. A. (1999). 'A comparison of five elicitation techniques for elicitation of attributes of low involvement products'. *Journal of Economic Psychology* **20**(3): 315–341.

BRUNSØ, K. & GRUNERT, K. G. (1995). 'Development and testing of a cross-culturally valid instrument: food-related life style'. *Advances in Consumer Research*. F. Kardes and M. Sujan. UT, Association for Consumer Research. **22**: 475–480.

BRUNSØ, K., BREDAHL, L. & GRUNERT, K. G. (1996). *Food-related Lifestyle Trends in Germany. A Comparison 1993–1996*. 25th EMAC Conference, Budapest, Budapest University of Economic Sciences.

BRUNSØ, K., SCHOLDERER, J. & GRUNERT, K. G. (2004a). 'Closing the gap between values and behaviour: a means–end theory of lifestyle'. *Journal of Business Research* **57**: 665–670.

BRUNSØ, K., SCHOLDERER, J. & GRUNERT, K. G. (2004b). 'Testing relationships between values and food-related lifestyle: results from two European countries'. *Appetite* **43**(2): 195–205.

COSTA, A. I. A., DEKKER, M. & JONGEN, W. M. F. (2004). 'An overview of means–end theory: potential application in consumer-oriented food product design'. *Trends in Food Science & Technology* **15**(7–8): 403–415.

GOFTON, L. (1995). 'Dollar rich and time poor? Some problems in interpreting changing food habits'. *British Food Journal* **97**(10): 11–16.

GRUNERT, K. G. (1995). 'Food quality: a means–end perspective'. *Food Quality and Preference* **6**(3): 171–176.

GRUNERT, K. G., BRUNSØ, K. & BISP, S. (1997). Food-related lifestyle: development of a cross-culturally valid instrument for market surveillance. *Values, Lifestyles, and Psychographics*. L. R. Kahle and L. Chiagouris. Mahwah, NJ, Lawrence Erlbaum: 337–354.

GRUNERT, K. G., BECKMANN, S. C. & SØRENSEN, E. (2001a). Means–end chains and laddering: An inventory of problems and an agenda for research. *Understanding Consumer Decision-making: The Means–End Approach to Marketing and Advertising Strategy*. T. C. Reynolds and J. C. Olson. Mahwah, NJ, Lawrence Erlbaum: 63–90.

GRUNERT, K. G., BRUNSØ, K., BREDAHL, L. & BECH, A. (2001b). Food-related lifestyle: a segmentation approach to European food consumers. *Food, People and Society: A European Perspective of Consumers' Food Choices*. L. J. Frewer, E. Risvik, H. N. J. Schifferstein and R. von Alvensleben. London, Springer Verlag: 211–230.

GUTMAN, J. (1982). 'A means–end chain model based on consumer categorization processes'. *Journal of Marketing* **46**(2): 60–72.

GUTMAN, J. (1991). 'Exploring the nature of linkages between consequences and values'. *Journal of Business Research* **22**: 143–149.

NIELSEN, N. A., SØRENSEN, E. & GRUNERT, K. G. (1997). Consumer motives for buying fresh or frozen plaice: a means end chain approach. *Seafood from Producer to Consumer: Integrated Approach to Quality*. J. B. Luten, T. Børresen and J. Oehlenschläger. Amsterdam, Elsevier: 31–43.

NIELSEN, N. A., BECH-LARSEN, T. & GRUNERT, K. G. (1998). 'Consumer purchase motives and product perceptions: a laddering study on vegetable oil in three countries'. *Food Quality & Preference* **9**(6): 455–466.

OLSON, J. C. & REYNOLDS, T. J. (1983). Understanding consumers' cognitive structures: Implications for advertising strategy. *Advertising and Consumer Psychology*. L. Percy and A. G. Woodside. Lexington, MA, Lexington Books: 77–90.

PETER, J. P. & OLSON, J. C. (1990). *Consumer Behaviour and Marketing Strategy*. Homewood, IL, Irwin.

REID, M., LI, E., BRUWER, J. & GRUNERT, K. G. (2001). 'Food-related lifestyles in a cross-cultural context: comparing Australia with Singapore, Britain, France and Denmark'. *Journal of Food Products Marketing* **7**(4): 57–75.

REYNOLDS, T. J. & GUTMAN, J. (1988). 'Laddering theory, methods, analysis, and interpretation'. *Journal of Advertising Research* **18**(1): 11–31.

ROININEN, K., TUORILA, H., ZANDSTRA, E. H., DE GRAAF, C. & VEHKALAHTI, K. (2001). 'Differences in health and taste attitudes and reported behaviour among Finnish, Dutch, and British consumers: a cross-national validation of the Health and Taste Attitude Scales (HTAS)'. *Appetite* **37**(1): 33–45.

SCHOLDERER, J., BRUNSØ, K., BREDAHL, L. & GRUNERT, K. G. (2004). 'Cross-cultural validity of the food-related lifestyles instrument (FRL) within Western Europe'. *Appetite* **42**: 197–211.

SØNDERGAARD, H. A. (2005). 'Market-oriented new product development: how can a means–end chain approach affect the process?' *European Journal of Innovation Management* **8**(1): 79–90.

STEENKAMP, J.-B. E. M. (1990). 'Conceptual model of the quality perception process'. *Journal of Business Research* **21**(4): 309–333.

VALETTE-FLORENCE, P. (1997). A causal analysis of means–end hierarchies: implications in advertising strategies. *Values, Lifestyles, and Psychographics*. L. Kahle and L.

Chiagouris. Mahway, NJ, Lawrence Erlbaum: 199–216.

VALETTE-FLORENCE, SIRIEIX, P., L., GRUNERT, K. G. & SØRENSEN, N. A. (2000). 'Means–end chain analyses of fish consumption in Denmark and France: a multidimensional perspective'. *Journal of Euromarketing* **8**(1/2): 15–27.

VAN RAAIJ, W. F. & VERHALLEN, T. M. M. (1994). 'Domain-specific market segmentation'. *European Journal of Marketing* **28**(10): 49–66.

Appendix

The food-related lifestyle instrument

1. To me product information is of high importance. I need to know what the product contains.

completely disagree   1   2   3   4   5   6   7   completely agree

2. **(Do not answer this question if you live alone)** The kids or other members of the family always help in the kitchen; for example they peel the potatoes and cut the vegetables.

completely disagree   1   2   3   4   5   6   7   completely agree

3. I only buy and eat foods which are familiar to me.

completely disagree   1   2   3   4   5   6   7   completely agree

4. Shopping for food does not interest me at all.

completely disagree   1   2   3   4   5   6   7   completely agree

5. I find taste in food products important.

completely disagree   1   2   3   4   5   6   7   completely agree

6. Usually I do not decide what to buy until I am in the shop.

completely disagree   1   2   3   4   5   6   7   completely agree

7. It is important for me to know that I get quality for all my money.

completely disagree   1   2   3   4   5   6   7   completely agree

8. Well-known recipes are indeed the best.

completely disagree   1   2   3   4   5   6   7   completely agree

9. I make a point of using natural or ecological food products.

completely disagree   1   2   3   4   5   6   7   completely agree

10. I eat before I get hungry, which means that I am never hungry at meal times.

completely disagree   1   2   3   4   5   6   7   completely agree

11. I compare product information labels to decide which brand to buy.

completely disagree   1   2   3   4   5   6   7   completely agree

12. I like buying food products in speciality stores where I can get expert advice.

completely disagree   1   2   3   4   5   6   7   completely agree

13. I compare prices between product variants in order to get the best value for money.

completely disagree   1   2   3   4   5   6   7   completely agree

14. We use a lot of ready-to-eat foods in our household.

completely disagree   1   2   3   4   5   6   7   completely agree

15. I notice when products I buy regularly change in price.

completely disagree   1   2   3   4   5   6   7   completely agree

16. I always buy organically grown food products if I have the opportunity.

completely disagree   1   2   3   4   5   6   7   completely agree

17. I find that dining with friends is an important part of my social life.

completely disagree   1   2   3   4   5   6   7   completely agree

18. I don't like spending too much time on cooking.

completely disagree   1   2   3   4   5   6   7   completely agree

19. I dislike anything that might change my eating habits.

completely disagree   1   2   3   4   5   6   7   completely agree

20. I have more confidence in food products that I have seen advertised than in unadvertised products.

completely disagree   1   2   3   4   5   6   7   completely agree

21. When cooking I first and foremost consider taste.

completely disagree   1   2   3   4   5   6   7   completely agree

22. I prefer fresh products to canned or frozen products.

completely disagree   1   2   3   4   5   6   7   completely agree

23. In our house, nibbling has taken over and replaced set eating hours.

completely disagree   1   2   3   4   5   6   7   completely agree

24. I look for ways to prepare unusual meals.

completely disagree   1   2   3   4   5   6   7   completely agree

25. I do not see any reason to shop in speciality food stores.

completely disagree   1   2   3   4   5   6   7   completely agree

26. It is the woman's responsibility to keep the family healthy by serving a nutritious diet.

completely disagree   1   2   3   4   5   6   7   completely agree

27. Going out for dinner is a regular part of our eating habits.

completely disagree   1   2   3   4   5   6   7   completely agree

28. I look for ads in the newspaper for store specials and plan to take advantage of them when I go shopping.

completely disagree   1   2   3   4   5   6   7   completely agree

29. I compare labels to select the most nutritious food.

completely disagree   1   2   3   4   5   6   7   completely agree

30. I don't mind paying a premium for ecological products.

completely disagree   1   2   3   4   5   6   7   completely agree

31. I always plan what we are going to eat a couple of days in advance.

completely disagree   1   2   3   4   5   6   7   completely agree

32. Nowadays the responsibility for shopping and cooking ought to lie just as much with the husband as with the wife.

completely disagree   1   2   3   4   5   6   7   completely agree

33. A familiar dish gives me a sense of security.

completely disagree   1   2   3   4   5   6   7   completely agree

34. **(Do not answer this question if you live alone)** My family helps with other mealtime chores, such as setting the table and doing the dishes.

completely disagree   1   2   3   4   5   6   7   completely agree

35. To me the naturalness of the food that I buy is an important quality.

completely disagree   1   2   3   4   5   6   7   completely agree

36. I like to know what I am buying, so I often ask questions in stores where I shop for food.

completely disagree   1   2   3   4   5   6   7   completely agree

37. Recipes and articles on food from other culinary traditions make me experiment in the kitchen.

completely disagree   1   2   3   4   5   6   7   completely agree

38. Over a meal one may have a lovely chat.

completely disagree   1   2   3   4   5   6   7   completely agree

39. It is important to me that food products are fresh.

completely disagree   1   2   3   4   5   6   7   completely agree

40. I love to try recipes from foreign countries.

completely disagree   1   2   3   4   5   6   7   completely agree

41. I always check prices, even on small items.

completely disagree   1   2   3   4   5   6   7   completely agree

42. I enjoy going to restaurants with my family and friends.

completely disagree   1   2   3   4   5   6   7   completely agree

43. I like to have ample time in the kitchen.

completely disagree   1   2   3   4   5   6   7   completely agree

44. I am influenced by what people say about a food product.

completely disagree   1   2   3   4   5   6   7   completely agree

45. We often get together with friends to enjoy an easy-to-cook, casual dinner.

completely disagree   1   2   3   4   5   6   7   completely agree

46. Shopping for food is like a game to me.

completely disagree   1   2   3   4   5   6   7   completely agree

47. Before I go shopping for food, I make a list of everything I need.

completely disagree   1   2   3   4   5   6   7   completely agree

48. I prefer to buy meat and vegetables fresh rather than pre-packed.

completely disagree   1   2   3   4   5   6   7   completely agree

49. I try to avoid food products with additives.

completely disagree   1   2   3   4   5   6   7   completely agree

50. It is more important to choose food products for their nutritional value rather than for their taste.

completely disagree   1   2   3   4   5   6   7   completely agree

51. Being praised for my cooking adds a lot to my self-esteem.

completely disagree   1   2   3   4   5   6   7   completely agree

52. Frozen foods account for a large part of the food products I use in our household.

completely disagree   1   2   3   4   5   6   7   completely agree

53. I just love shopping for food.

completely disagree   1   2   3   4   5   6   7   completely agree

54. I am an excellent cook.

completely disagree   1   2   3   4   5   6   7   completely agree

55. When I serve a dinner to friends, the most important thing is that we are together.

completely disagree   1   2   3   4   5   6   7   completely agree

56. I prefer to buy natural products, i.e. products without preservatives.

completely disagree   1   2   3   4   5   6   7   completely agree

57. I consider the kitchen to be the woman's domain.

completely disagree   1   2   3   4   5   6   7   completely agree

58. Information from advertising helps me to make better buying decisions.

completely disagree   1   2   3   4   5   6   7   completely agree

59. I use a lot of mixes, for instance baking mixes and powder soups.

completely disagree   1   2   3   4   5   6   7   completely agree

60. I make a shopping list to guide my food purchases.

completely disagree   1   2   3   4   5   6   7   completely agree

61. What we are going to have for supper is very often a last-minute decision.

completely disagree   1   2   3   4   5   6   7   completely agree

62. Cooking is a task that is best over and done with.

completely disagree   1   2   3   4   5   6   7   completely agree

63. Eating is to me a matter of touching, smelling, tasting and seeing, all the senses are involved. It is a very exciting sensation.

completely disagree   1   2   3   4   5   6   7   completely agree

64. I always try to get the best quality for the best price.

completely disagree   1   2   3   4   5   6   7   completely agree

65. I eat whenever I feel the slightest bit hungry.

completely disagree   1   2   3   4   5   6   7   completely agree

66. Cooking needs to be planned in advance.

completely disagree   1   2   3   4   5   6   7   completely agree

67. I like to try new foods that I have never tasted before.

completely disagree   1   2   3   4   5   6   7   completely agree

68. **(Do not answer this question if you live alone)** When I do not really feel like cooking, I can get one of the other members of my family to do it.

completely disagree   1   2   3   4   5   6   7   completely agree

69. I like to try out new recipes.

completely disagree   1   2   3   4   5   6   7   completely agree

10

Measuring consumer expectations to improve food product development

A. V. Cardello, US Army Natick Soldier, R, D & E Center, USA

10.1 Introduction

10.1.1 The role of expectations in consumer-driven product development

Consumer-driven product development requires the developer to have a detailed understanding of what the consumer desires in a product. Market researchers and others who support consumer-driven product development use a variety of qualitative and quantitative techniques to obtain useful information for this purpose. In a recent review of consumer research methods for application in early stages of new product development, van Kleef *et al.* (2005) categorized the available techniques for uncovering the 'voice of the consumer.' One critical factor used by van Kleef *et al.* (2005) to categorize these methods was the actionability of the data obtained, i.e. how easily the data can be used by the product developer to create a product. The ability to engineer into the product the exact characteristics, attributes, and benefits that the consumer expects to find in the product is the key to achieving new product success.

Although early stage product development research can be used to identify what the consumer expects from a product, this is only part of what is necessary to ensure product success. The other part is an assessment of (1) how packaging, labeling, product information, and other extrinsic product characteristics influence consumer expectations, (2) how well the product actually meets expectations, and (3) how failing to meet consumer expectations affects acceptance, use, and repeat choice of the product. These latter elements of effective product development research are often ignored. Sometimes this results from management decisions to rush the product to market prematurely, with the intent of letting 'the market' determine the success or failure of the product. However, a more common reason is lack of knowledge of the scientific measurement tools and theoretical models by which to compare actual product characteristics with what the consumer expects.

10.1.2 Chapter goals

Ultimately, product success depends on the consumer's liking or disliking of the product. The antecedent factors that contribute to product liking are numerous and complex. Many derive from the *intrinsic* properties of the food, i.e. its physicochemical and associated sensory properties, while others derive from its *extrinsic* factors, e.g. product name, packaging, labeling, pricing, and promotional information. The former operate primarily through human *sensory and perceptual* systems, while the latter operate through *cognitive and psychological* mechanisms. This chapter will focus on an important theoretical construct that links both intrinsic and extrinsic product factors to consumer needs and desires, in order to better predict product liking. This construct is consumer 'expectations.'

The chapter will begin by describing the concept of 'expectations,' how it has been used in the psychological literature, and its theoretical bases. This will be followed by a discussion of the tools and methods that can be used to assess consumer expectations of product characteristics and some of the more recent literature showing the importance of consumer expectations to food choice and acceptance. Lastly, the chapter will address important theoretical and methodological issues, including emerging research in neuropsychology that offers important insights into the biological mechanisms that underlie expectations. The chapter will not attempt to review the large literature on simple information effects on product acceptance that does not address the mediating role of expectations. Rather, the chapter will focus on the critical theories, methods, and issues that the product developer needs to know in order to understand expectations research and how it applies to consumer-driven product development.

10.2 Expectations in psychology and consumer behavior

10.2.1 Expectations as a psychological construct

What are expectations? Dictionaries offer several definitions. These include (1) 'to look for as likely to occur or appear,' (2) 'to look for as due, proper or necessary,' and (3) 'to suppose, presume or guess.' All of these definitions imply a *psychological anticipation* that something will occur or be experienced. In general terms, we can define an expectation as a belief that an *object* possesses a particular *attribute* or that a *behavior* will result in a particular *consequence*. Operationally, we might define it in terms of the *perceived probability* or the *anticipated magnitude* of these attributes or consequences.

In modern psychology, the construct of expectations was introduced by Tolman (1938, 1951), who used the concept of 'expected consequences of behavior' as an explanatory variable to account for animal learning behavior, e.g. rats in a maze 'anticipate' that certain behaviors will result in a reward (or no reward) and behave in accordance with those expected consequences. Tolman's notion was one of the first formal representations in psychology to posit that behavior occurs in response to *expected pleasure*. The concept of expectations in psychology was further advanced by Meehl and MacCorquodale (1951) and MacCorquodale and Meehl (1953, 1954) who expanded the concept to human learning and added the notion that expectations are a multiplicative product of the strength of the expectation and its valence, giving early form to expectancy $\times$ value models in psychology (see Feather, 1982, for a review). This early theorizing about the role of expectations in human behavior was followed by a series of theoretical formulations that utilized expectations as a construct to account for social learning (Rotter, 1955), motivation (Atkinson, 1954, 1957, 1958), and attitude formation (Fishbein and Ajzen, 1975).

At the same time that the concept of expectations was evolving as an explanatory variable in areas of learning and motivation, other theoretical trends in psychology were converging to produce practical models of the effect of expectations on both perceptions and the affective responses to them. In the area of human perception, it long had been known that perceptions were not one-to-one translations of sensory stimuli into perceptual images. However, most accounts of the effects of context on perception treated these effects as resulting from physiological adaptation or psychological biases. However, Helson (1948, 1964) put forward a theory of perception that proposed that every stimulus was evaluated in a comparative way to an internal, momentary state of the organism. His 'adaptation-level theory' viewed all perceptual events as resulting from a comparison of the external stimulus to an internal *adaptation level* that served as a neutral point on the perceptual dimension. The adaptation level was considered to be a direct consequence of the individual's past experiences with stimuli along that dimension and, thus, was considered more cognitive than physiological in nature. Helson (1955) referred to this adaptation level as '*psychological* homeostasis' as opposed to earlier formulations that focused on 'physiological homeostasis.'

The notion that stimuli and other information in the environment are interpreted through comparison with internal cognitive states was also a central concept in Festinger's theory of cognitive dissonance. Working in the area of cognition and emotion, Festinger (1957) proposed that cognitions (beliefs, attitudes, values, perceptions) and behaviors must be psychologically consistent with one another in order to avoid psychological dissonance (a state of psychological discomfort). He proposed that when cognitions or behaviors were in conflict, cognitive dissonance arose, which, by its nature, produced a *negative drive state* that motivated the individual to change one or more of the cognitions or behaviors to make them more consistent with one another. In Festinger's words,

> the simplest definition of dissonance can, perhaps, be given in terms of *expectations*. In the course of our lives we have all accumulated a large number of expectations about what things go together and what things

do not. When such an expectation is not fulfilled, dissonance occurs (Festinger, 1962, p. 94).

In response to this dissonance, he hypothesized that 'a person can change his opinion; he can change his behavior, ... he can even distort his perception and his information about the world around him' (Festinger, 1962, p. 93).

The above lines of theorizing led to the conclusion that pre-existing experiences, beliefs, and expectations about stimulus objects are strong determinants of perceptions, attitudes, and behaviors toward them, and that these influences operate through a comparative mechanism that compares and contrasts the stimulus object to pre-existing internal cognitive states. In cases where there is a *mismatch* between incoming sensory information and pre-existing expectations, the perception of the stimulus may change, the cognitions may change, or both.

As the important role of expectations in human behavior and perception gained attention in psychology, powerful demonstrations of its effects were reported in areas as diverse as education, where teachers' expectancies about students' academic potential were found to influence their perceptions of student achievement (Rosenthal and Jacobson, 1968), in human performance, where confidence in the expectation of performing well was shown to influence both physical and cognitive performance (Aronson and Carlsmith, 1962; Wiggins and Brustad, 1996), and in medicine, where the importance of patient beliefs and expectations about potential drug benefits produced powerful 'placebo' effects on physiological functioning (Honigfeld, 1964; Lowinger and Dobie, 1969; Beck, 1977; Shapiro and Morris, 1978; Ross and Olson, 1981). Thus, it was not unexpected that cognitive expectations about everyday objects and events would soon come to play an important explanatory role in consumer behavior.

10.2.2   Expectations in consumer product behavior

Early research in consumer behavior
Beginning in the 1970s, researchers in the area of marketing and consumer behavior utilized cognitive expectations as an explanatory variable in the study of perceptions of service quality and consumer *satisfaction*. Business management researchers quantified the discrepancies between consumer expectations of business service and the actual service provided by businesses in order to predict the effects of these discrepancies on perceived service quality and customer satisfaction. One product of these efforts was the development of 'gap' models that focused on gaps between consumer expectations and actual service quality, e.g. the SERVQUAL model of quality service (Parasuraman *et al.*, 1985). Still other researchers were motivated to understand the effects of persuasive communications on perceived satisfaction with products. The paradigm for the latter research was to manipulate expectations of product quality using persuasive information (either accurate or false) about brands, price, functionality, etc. and then to present the product for trial and evaluation in terms of perceived satisfaction. Using predictive models derived from earlier research in learning and motivational psychology, these researchers uncovered important effects on

satisfaction that resulted when product quality did not meet expectations (for seminal research and reviews in this area, see Insko, 1967; Oliver, 1977a,b, 1980, 1993; Latour and Peat, 1979; Swan and Trawick, 1981; Oliver and DeSarbo, 1988; Anderson and Sullivan, 1993).

Although the dependent measure commonly used in this research was 'consumer satisfaction,' a review of how the construct of satisfaction has been used in the literature found that the most common definition is of a 'summary *affective* response of varying intensity' (Giese and Cote, 2000). Thus, as an affective dimension, the construct of satisfaction is quite similar to other affective and/or evaluative responses to food and consumer products. This similarity led the present author and others to propose that liking/disliking and, perhaps, other consumer evaluative dimensions of food can be analyzed using the same models that have been applied to product satisfaction (Cardello, *et al.*, 1985; Cardello and Sawyer, 1992; Cardello, 1994, 1995; Deliza and MacFie, 1996).

Disconfirmed expectations and food behavior
The importance of meeting consumer expectations for trial, use, or repeat purchase has become especially clear in the past decade as researchers have begun to demonstrate the powerful effects of disconfirmed expectations on food acceptance and choice behaviors. Figure 10.1 is taken from Deliza and MacFie (1996) and depicts the chain of events hypothesized to occur during consumer product choice. At the top, pre-existing information and experiences lead to certain expectations about a product. Upon encountering the product in a retail setting, the extrinsic (non-sensory) attributes of the product can alter these pre-existing expectations. If the resultant expectancy is low, the product will not be selected or used. However, if the expectation is high, the product will be chosen and used/consumed, at which time the consumer will experience its varied attributes. The perception of the product's attributes will then confirm or disconfirm the held expectations. If the product attributes fail to meet expectations (negative disconfirmation), negative affect (and rejection) will result and future expectations for the product will decline. However, if product characteristics meet or exceed expectations, satisfaction will result, leading to repeat usage and elevating future expectations.

The above descriptive model provides an outline of the important theoretical elements involved in consumer choice behavior. However, in order to understand the empirical methods by which a product developer can improve consumer expectations for his/her product, to assess whether the product meets these expectations, and to quantify the market impact of failing to meet them, an explanation of the basic research methods and theoretical models in the area is needed.

Empirical research methods
For product development applications in the food and beverage industry, it is instructive to examine two studies that used very different experimental paradigms to assess the role of cognitive expectations on the perception of taste

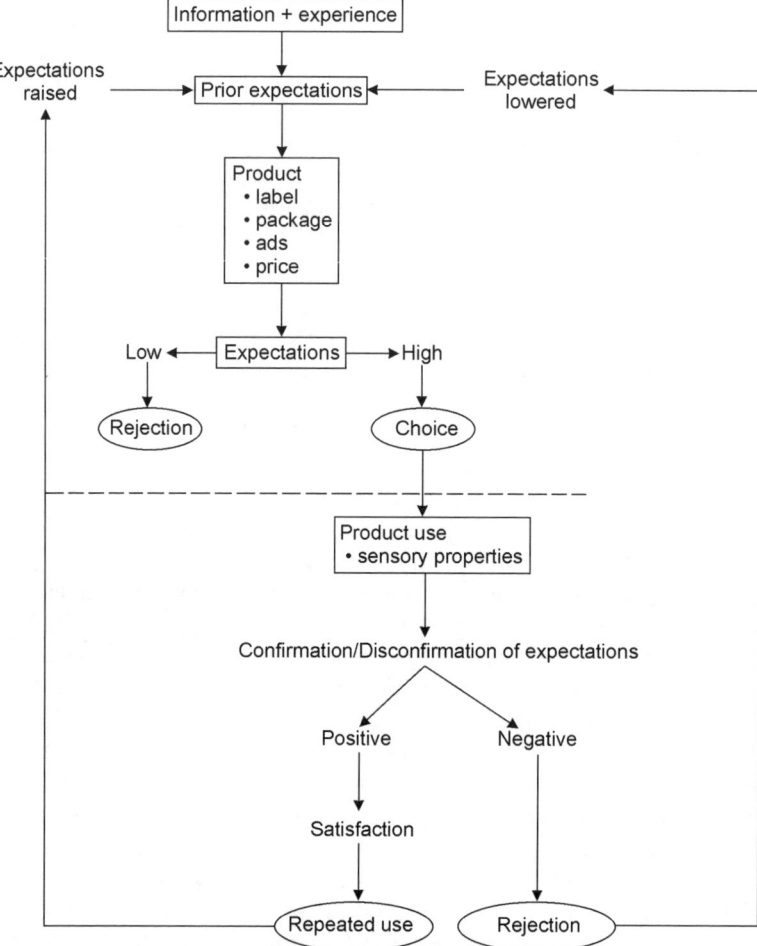

Fig. 10.1   Schematic model of the effects of expectations on product selection and evaluation (reprinted from *Journal of Sensory Studies*, vol. 11, R. Deliza and H.J.H. MacFie, 'The generation of sensory expectations by external cues and its effect on sensory perception', 103–128, copyright 1996, with permission from Blackwell).

stimuli – one an early study focusing on sensory and hedonic responses to model tastants, the other a more recent study examining the hedonic responses to a dessert product.

   In the first study (Carlsmith and Aronson, 1963), the experimenters administered a series of iso-intense solutions of sucrose and quinine sulfate to subjects who rated them for their perceived intensity of sweetness or bitterness using a magnitude estimation procedure. Prior to presentation of each stimulus, the investigators created an expectation for the taste quality of the stimulus by providing subjects with a manual cue that the solution would be either sweet or bitter. In some cases the stimulus was *consistent* with the cue, i.e. sucrose was cued/sucrose was presented, but in other cases it was *inconsistent* with the cue, i.e. sucrose was cued/quinine was presented. Analysis of the data showed that when the subject had a strong expectancy for the solution, the effect on perceived taste intensity was different depending upon the taste quality of the stimulus. S*ucrose* solutions that *disconfirmed* an expectancy were rated less sweet than sucrose solutions that *confirmed* an expectancy, whereas *quinine* solutions that *disconfirmed* an expectancy were rated more bitter than quinine solutions that *confirmed* an expectancy. The conclusion reached by the authors was that the data were consistent with cognitive dissonance theory, i.e. disconfirmed expectations about both solutions resulted in *negative affect* toward them, lowering their perceived sweetness and/or increasing their perceived bitterness (both taste shifts reflecting reduced pleasantness).

The second study to consider was conducted by Cardello (2003). In this study, hedonic ratings to tasted samples of a chocolate pudding were obtained first in a blind taste test. Subjects returned some weeks later and were informed that they would be tasting chocolate pudding that had been processed by one of several different novel processing techniques, e.g. pulsed electric fields, high-pressure processing, irradiation, etc. Subjects were then asked to rate their expected liking/disliking for the pudding both before and after viewing it. Finally, subjects tasted the pudding and rated their liking/disliking of it. In reality, all the puddings were identical to those served in the blind test. An analysis of the data showed that the information about how the pudding was processed greatly influenced both their expectations and actual liking of it. However, unlike in the Carlsmith and Aronson (1963) study, disconfirmed expectations reduced liking *only* when the liking expectations were *lower* than the subject's baseline liking. When liking expectations were *higher* than the baseline liking, product liking increased.

Explanatory models
The findings of Carlsmith and Aronson (1963), that disconfirmed expectations produce negative affect, is consistent with a critical prediction of cognitive dissonance theory, i.e. that dissonant cognitions produce a negative affective state. The prediction that disconfirmed expectations invariably produce negative affect is referred to as the *generalized negativity* model of disconfirmed expectations. Although the notion that dissonance results in negative affect has garnered considerable empirical support in psychology (Elkin and Lieppe, 1986), many studies have shown that the negative affect arises *only* when the individual attributes his/her arousal to the attitude–behavior discrepancy (Keisler and Pallak, 1976; Cooper *et al.*, 1978; Cooper and Fazio, 1984). Thus, the discrepancy must be sufficiently large as to be salient to the consumer. For the product developer and marketing team, the implications of the negativity model are that *any product that fails to meet consumer expectations will suffer in the marketplace*, regardless of whether the product is better or worse than expected. It is the *discrepancy* itself that produces *negative affect* and that is *transferred* to the product experience.

In contrast, the results of the Cardello (2003) study support the notion that liking will change in the direction of the expectation, i.e. if expectations are higher than product performance, liking will increase; if expectations are lower than product performance, liking will decrease. The model that describes these data is the *assimilation* model, deriving its name from the fact that the affective response to the product assimilates (absorbs and incorporates) the level of the expectation (Hovland *et al.*, 1957; Sherif and Hovland, 1961; Olshavsky and Miller, 1972; Olson and Dover, 1976, 1979). The implications of this model for the product developer and marketing team are more optimistic, since the *impact on liking is driven by the expectation*. If the expectation is poor, the product will suffer. However, if the expectation is high, it will cause the product to be perceived as better than its actual attributes would suggest. Thus, the opportunity exists to *improve the acceptance of the product and its market share through creative marketing* that establishes a positive image and expectation for the product. Here lies the heart of all advertising strategies aimed at improving product 'image.'

Of course, these two outcomes are not exhaustive for any set of data. For example, one might imagine that a very high or very low expectation relative to actual product performance could produce a 'boomerang' effect on liking, much like the disappointment that results upon viewing a mediocre movie after reading rave reviews about it. Such a counter-polarizing influence on product acceptance is described by the *contrast* model of disconfirmed expectations (Hovland *et al.*, 1957; Sherif and Hovland, 1961; Dawes *et al.*, 1972). Fortunately for product developers and marketers, contrast effects have been reported only occasionally in the consumer literature, e.g. Cardozo (1965), Anderson (1973) and Cardello and Sawyer (1992). However, such effects are frequently observed in perceptual studies in which a test stimulus is presented within a context of other stimuli of either high or low stimulus intensity or affect (Riskey *et al.*, 1979; Lawless, 1983; McBride, 1985, 1986; Conner *et al.*, 1987; Rankin and Marks, 1991; Schifferstein and Frijters, 1992; Schifferstein, 1995; Diamond and Lawless, 2001), suggesting that these effects can occur more frequently in *product* contexts, as opposed to *information* contexts (see Section 10.4.4).

A fourth and final possibility for the impact of disconfirmed expectations on product acceptance is that, if the level of disconfirmation is low, i.e. the difference between product attributes and expectations is within a small *zone of indifference* around the confirmation point (Woodruff *et al.*, 1983), assimilation of the ratings occurs. However, if the disconfirmation is large (the product is quite different from expectations), contrast results. This hybrid model is termed the *assimilation–contrast* model (Hovland *et al.*, 1957) and its predictions have been compared with those of the curiosity hypothesis of McClelland *et al.* (1953) and Berlyne (1960) by Schifferstein (2001; Schifferstein *et al.*, 1999). Like the contrast model, the precautionary prediction of this model for product developers/marketers is that they must *avoid large discrepancies between what the consumer expects and what the product can deliver*, because consumer awareness of the discrepancy will create negative affect toward the product (Keisler and Pallak, 1976; Zanna and Cooper, 1974; Cooper *et al.*, 1978; Cooper and Fazio, 1984). Deliza and MacFie (1996) have discussed the issue of consumer awareness of the expectation–product discrepancy in sensory terms, suggesting that if a consumer is confident in his or her sensory abilities, he or she may be much less susceptible to persuasive information that creates expectations in conflicts with their sensory experiences.

Predicted outcomes and empirical findings

Figure 10.2, taken from Schifferstein (1997), schematically depicts the predictions of the four models. In this figure, the abscissa represents the level of expectation for the product and the ordinate represents product performance. For a product with an *actual* or inherent level of performance represented by the horizontal line, expectation and product performance 'match' (or are not perceptibly different) at the intersection of the horizontal and diagonal lines. Here, all four models predict that *perceived* product performance will approximate actual product performance. If actual performance is held constant but the level of expectation is varied, positive or negative disconfirmation will occur. If expectations are low, product performance will be *better* than expected and *positive* disconfirmation will occur (all points to the left of the intersect in Fig. 10.2). If expectations are high, the product will be *worse* than expected and *negative* disconfirmation will occur (points to the right of the intersect). Regardless of whether the disconfirmation is positive or negative, the degree of disconfirmation will be reflected in the vertical distance between the horizontal line and its point of intercept with the diagonal line.

Looking at the line for the 'assimilation' model in Fig. 10.2, we see that regardless of whether positive or negative disconfirmation occurs, perceived product performance assimilates (becomes similar to) the level of expectation (diagonal line). Alternatively, looking at the prediction line for the 'contrast' model, we see that perceived product performance becomes dissimilar to the expectation. The curvilinear line for the assimilation–contrast model reflects the fact that perceived product performance will assimilate the expectations under conditions of low positive or negative disconfirmation (within the latitude of acceptance defined by the left and right vertical lines in Fig. 10.2), but produce contrast under conditions of high disconfirmation. Lastly, the generalized negativity predictions are reflected in the two solid lines forming an upward arrow and joining at the center of Fig. 10.2.

In numerous studies conducted over many years in which food and beverages have been used as test products, the most commonly observed effects have been assimilation (Cardello and Sawyer, 1992; Deliza, 1996; Tuorila *et al.*, 1994a, Cardello *et al.*, 1996a; Lange *et al.*, 1999, 2000; Schifferstein *et al.*, 1999, Siret and Issanchou, 2000; Caporale and Monteleone, 2001, 2004; Hurling and Shepherd, 2003; DiMonaco *et al.*, 2004; Stefani *et al.*, 2006; Caporale *et al.*, 2006; Iaccarino *et al.*, 2006). In these studies, the magnitude of the assimilation has ranged from partial to complete (the product rating is elevated to match the

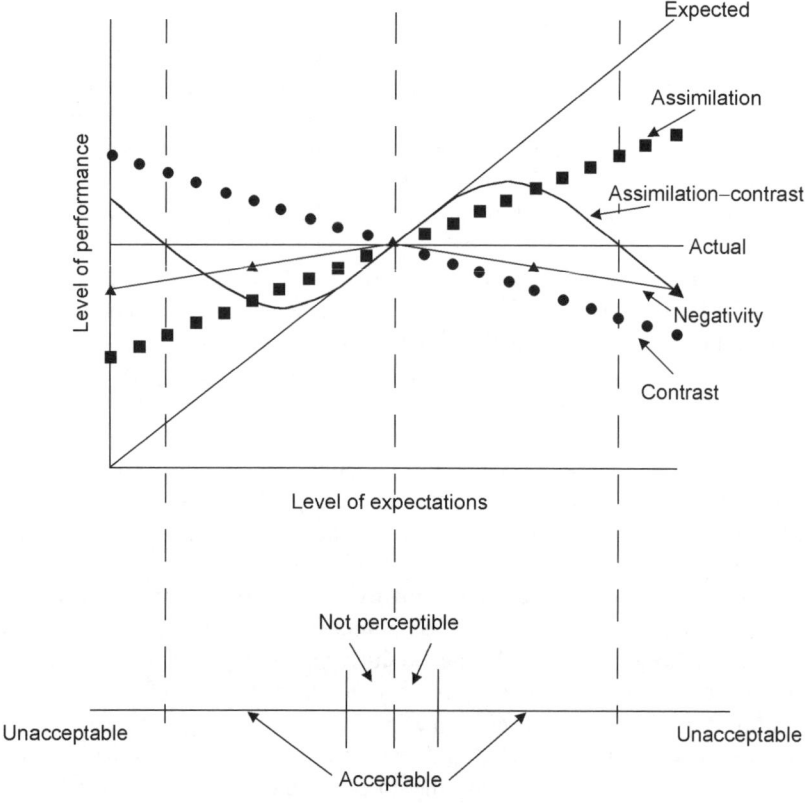

Fig. 10.2 Schematic representation of the predictions of the effects of disconfirmed expectations according to the assimilation, contrast, assimilation-contrast and negativity models (reprinted from the *Proceedings of the 26th EMAC Conference*, H. Schifferstein, The role of expecting disconfirmation in food acceptability, 2019–2025, copyright 1997, with permission from the European Marketing Academy).

magnitude of the expectation). As we shall see, such assimilation effects have been observed in other affective domains, and evidence is beginning to emerge concerning the likely physiological mechanisms that underlie both expectations and assimilation (Section 10.4.7).

10.3 Basic elements of conducting research on consumer expectations

For the product developer interested in applying expectations research to guide consumer-driven product development, there are a number of important methodological issues that must be considered. In the following section, these important elements of research are discussed.

10.3.1   Types of expectations to be measured

There are several types of expectations that can be measured in product development work. The first are *sensory*-based expectations, i.e. beliefs that the product will possess certain sensory attributes at specific intensities. For example, in the study by Carlsmith and Aronson (1963) sensory-based expectations were manipulated by cueing subjects to anticipate one taste quality and these expectations were then either confirmed or disconfirmed by presenting either the expected taste or a different one. Sensory expectations are critical to product development and marketing because the sensory attributes of a product are the primary drivers of its acceptance and, therefore, are more important than extrinsic variables in the formation of product expectations (Regan and Fazio, 1977; Fazio and Zanna, 1978; Smith and Swinyard, 1983; Steenkamp, 1989). The ill-fated attempt to introduce New Coke into the beverage market in the 1990s is often attributed to its sweeter taste, which, while found to be acceptable in taste tests, still violated the well-established sensory expectations for Coke, negatively influencing consumers' emotional attachment to the product.

Sensory expectations are ubiquitous, arising from all cognitive elements of the product. The product name, category, style, brand, nutrient composition, labeling, and advertising will all create specific sensory expectations for the product through prior associations and experience. Especially influential in setting sensory expectations are pictorial and photographic representations on the labels of products (Jaeger and MacFie, 2001) because they provide direct information about the product's expected color, shape, size, and even its likely textural and moistness properties.

While sensory expectations are essential for the product developer to meet during formulation of a product, relatively few studies have focused on the identification of these important drivers of acceptance. Perhaps this is because the initial interest in consumer expectations arose out of research by consumer and market researchers, not sensory scientists or product developers. As a result, the focus of this area of research has been on affect-based expectations that are more directly related to purchase behavior and that are amenable to manipulation using the type of extrinsic variables over which marketing and consumer researchers have control. Recent studies that have addressed sensory-based expectations include those by Cardello and Sawyer (1992), Tuorila *et al.* (1994a, 1998a), Kahkonen and Tuorila (1998, 1999), Kahkonen *et al.* (1997, 1999), Scharf and Volkmer (2000), Deliza *et al.* (2003) and Caporale *et al.* (2006). As will be discussed, the primary conclusion from these studies is that sensory expectations often exhibit assimilation effects, but under certain circumstances, contrast effects may be observed.

The second type of expectations are those that are *hedonic-* or *affect-*based, i.e. a belief that a product will be liked/disliked to a certain degree. These are the most commonly examined type of expectation, partly because they are easily manipulated by extrinsic product attributes, partly because they are relatively easy to measure, and partly because they form a direct link with other affect-based behavioral variables, e.g. choice, purchase, and consumption. Quality judgments, when made by consumers, are another type of expectation that falls into this category, e.g. Cardello *et al.* (1996a), Lange *et al.* (2000), and Acebron and Dopico (2000). Some interesting studies of hedonic expectations created by a variety of informational elements include those by Tuorila *et al.* (1994a), Cardello *et al.* (1996a), Lange *et al.* (1999), Kahkonen *et al.* (1997, 1999), Cardello (2003), Deliza *et al.* (2003), DiMonaco *et al.* (2004), Caporale *et al.* (2006), Stefani *et al.* (2006), and Iaccarino *et al.* (2006). As noted at the end of Section 10.2.2, the overwhelming majority of these studies have found assimilation of hedonic ratings to the expected liking of the product.

There are other types of expectation that can be examined as well. For example, there are expectations that are *ideational* in nature, i.e. beliefs that a product will elicit certain emotional or memory-based cognitions, e.g. a sense of nostalgia, of home-cooking, of health. Although these ideational expectations are important for many products, they often are *in*actionable for the product developer, because it is unclear how to reformulate the product to better meet these ideational expectations. Thus, few studies have been conducted on ideational expectations.

Recently, investigators have also begun to address *conative* expectations, e.g. expected likelihood to purchase, expected willingness to pay (see, for example, Lange *et al.*, 2000; Stefani *et al.*, 2006). Such expectations are a step removed from *product* expectations, but hold the promise of better predictive validity for market researchers. It is too early to tell if these conative expectations will behave in a similar manner to sensory and hedonic expectations, but the studies conducted to date using these measures have produced promising results.

Owing to the variety of expectations that can be held by the consumer, the product developer should take special care in determining whether sensory, hedonic, image-based, conative, or some combination of these expectations are the important drivers of his/her product's acceptance in the marketplace before undertaking any experimental work on the product.

10.3.2 Methods of quantifying expectations

In most cases, product developers do not aim to develop a product that is unique to itself but, rather, to develop a new product within some pre-existing category of products, e.g. an improved margarine. In such cases, the product developer is faced with a set of pre-existing or schema-based expectations for the product (Crocker, 1984; Tuorila *et al.*, 1998b) that are based on the consumers' past experiences with the same or similar products in the category (see Fig. 10.1). Such pre-existing expectations can be measured by having consumers rate the sensory, hedonic, ideational, or conative expectations for the product in response to the name of the food, e.g. 'margarine.' In fact, it has been argued that simple food attitude ratings (like/dislike) made in response to food names, may well constitute a form of generalized expected liking for the foods (Cardello, 2003). Such a logical equivalence between liking judgments in response to food names and expectations also was made in a study by Rousset and Jolivet (2002).

Of course, the most direct method for eliciting hedonic expectations for any product is to simply ask for a judgment of 'expected liking/disliking.' Expected liking/disliking can be operationally quantified using any acceptable psychophysical technique. Often a 9-point hedonic scale (Peryam and Pilgrim, 1957) or a linear graphic scale (Giovanni and Pangborn, 1983) is used, but a labeled affective magnitude (LAM) scale (Schutz and Cardello, 2001; Cardello and Schutz, 2004) or magnitude estimation procedure (Moskowitz, 1977) may also be used. Similarly, for sensory expectations, any acceptable stimulus intensity scale can be used (category scale, linear graphic scale, generalized Labeled Magnitude Scale (gLMS) (Green *et al.*, 1993, 1996), or magnitude estimation). If a complete profile of the expected sensory attributes of a product is desired, then an appropriate descriptive analytic technique that will enable the generation of an 'expected sensory profile' should be employed.

Interestingly, relatively few studies have examined either the reliability or validity of expectation measures. Reliability studies would do a great service to our understanding of how best to measure expectations. Validity measures are more difficult to obtain, owing to the private nature of conscious events. However, construct validity, established through behavioral and/or physiological measures, may well be possible in the near future (see Section 10.4.7).

In general, expectations will exist for almost all common foods. For certain items that are unique or, otherwise, novel to the consumer, there may be no pre-existing expectations, except for those that are established during the process of describing or presenting the product to the consumer. For these products, their sensory properties and schema-driven (Meyers-Levy and Tybout, 1989; Stayman *et al.*, 1992; Peracchio and Tybout, 1996) associations will be the primary drivers of their acceptance upon initial exposure (Pelchat and Pliner, 1995: Tuorila *et al.*, 1994b, 1998a,b).

Although rarely reported in the food expectation literature, it is advisable that a measure of belief strength or confidence in the expectation be obtained, e.g. Olson and Dover (1976), Smith and Swinyard (1988), or Marks and Kamins (1988). Consumer belief strength and confidence are important, because it is possible that the consumer may hold a very high expectation for a product, but the expectation may be weakly held, perhaps because of skepticism regarding the likelihood that the product will truly deliver what is promised. This skepticism may also result from distrust with the promotional material or with the source of the information. As noted previously, confidence in one's ability to accurately assess one's experiences will result in a greater reliance on the sensory components of a product experience and less on information that may run contrary to those perceptions (Goering, 1985; Deliza and MacFie, 1996). In addition, one must always consider that the demand characteristics of requesting an expectation judgment from consumers in response to promotional information may lead the consumer to report expectations that are in keeping with the information provided, but that, in reality, the consumer believes are unlikely to be met (see Section 10.4.2).

10.3.3   Modifying expectations using information

A fundamental question that has yet to be adequately addressed is how expectations are formed. However, we can begin to understand some of the factors important to expectation formation by examining how expectations are manipulated for research purposes. For example, in the study by Cardello (2003) cited earlier, the expectations of consumers were manipulated by informing them that the chocolate puddings to be presented were processed using particular food technologies. In the Carlsmith and Aronson (1963) study, expectations were manipulated by providing manual cues that alerted the subject that either a sweet or bitter stimulus was to follow. Although entirely different in nature, the common element of these studies is the use of 'information' to lead subjects to expect a particular sensory attribute or hedonic quality.

In almost all published studies on consumer expectations of foods and beverages, the expectation is manipulated using either verbal or non-verbal information. In product development and marketing *verbal* information may be as simple and unassuming as a product name, as powerfully influential as a brand label, as critical to health as a dietary label, as emotionally laden as a controversial processing technique, or as economically important as a full-blown advertising campaign. Each of these informational elements or arrays will influence the consumer's expectations for the product and his or her ultimate acceptance or rejection of it. Similarly, non-verbal information may be as subtle as the placement of the product on a store shelf, as functionally important as the shape of a bottle, as esthetically powerful as the color of a container, or as financially important as its price. The reader is referred to reviews and book chapters by Rozin and Tuorila (1993), Cardello (1994), Deliza and MacFie (1996), Jaeger and MacFie (2001), Hutchings (2003), Deliza *et al.* (2003), Wansink *et al.* (2005), and Jaeger (2006) for a sampling of the many forms of verbal and non-verbal (contextual) information that may influence consumer product expectations and the perception of food and foodservice settings.

In many cases, the content of the information is the primary interest of the study, because the goal is to identify product communications that can be used to elevate the expected quality or liking of the product, with the goal of improving choice, acceptance, and/or consumption. In recent years, numerous studies have been undertaken on such important informational elements as the nutritional labeling of foods (Tuorila *et al.*, 1994a, 1998b; Kahkonen and Tuorila, 1995; Kahkonen *et al.*, 1996, 1997, 1999), their geographic origins (Stefani *et al.*, 2006; Caporale *et al.*, 2006) or the methods of processing and preserving them (Siret and Issanchou, 2000; Cardello, 2003; Caporale and Monteleone, 2004; Iaccarino *et al.*, 2006). In some cases simple sensory information, e.g. visual (Hurling and Shepherd, 2003; Siret and Issanchou, 2000) or taste (Lange *et al.*, 2000) has been examined for its effect on liking expectations. In certain of these cases, label or ingredient information is compared with the sensory information (Siret and Issanchou, 2000; Lange *et al.*, 2000) to assess the differential effect of 'expected' versus 'experienced' attributes. In these studies the 'sensory' condition is often considered to be an information 'control' condition, because no verbal information is presented. However, the absence of verbal information does not preclude the consumer from holding strong expectancies about the product that are based on past experiences and other non-verbal information obtained about the product prior to or during the experimental test (Cardello and Sawyer, 1992). As such, these procedures merely establish a relative expectancy difference between the information and 'no information' conditions.

Jaeger and MacFie (2001) have addressed a number of important issues related to the content and format of advertising copy for food products. The focus of their analysis was the importance of pictorial information (versus textual) on product expectations and the potential importance of means–end chain (MEC) elements of information. Although their data did not support differential effects of MEC elements on product expectations, they observed differential responding to the pictorial information among individuals with high versus low need for cognition (see Section 10.4.5).

The findings of Jaeger and MacFie (2001), that certain elements of information may have differential effects on consumer expectations, is not unusual. In several studies that examined the effect of 'low-fat' labels on sensory and hedonic expectations, e.g. Tuorila *et al.* (1994a), Kahkonen *et al.* (1996), and Kahkonen and Tuorila (1998), the label manipulation altered sensory expectations for the product, lowered hedonic expectations, and had post-test effects on liking and sensory perception. However, in other studies using the same label with different foods (e.g. Kakhonen *et al.*, 1999), only sensory expectations were altered. Similarly, in the attempt to manipulate one type of sensory expectation, other sensory expectations may be altered (Cardello and Sawyer, 1992).

One critical factor in the effectiveness of any information element is the consumer's confidence in the source. If confidence in the source is lacking or if confidence in one's own beliefs is greater, the source may be discounted and little attitude change will occur (Horland and Wise, 1951). Also, the timing of information may affect its impact on the consumer. For example, Hurling and Shepherd (2003) have suggested that expectations formed closer in time to the eating occasion may be more influential than expectations formed earlier (see also Section 10.4.1). Similarly, Levin and Gaeth (1988) showed that information effects were greatest when no product was presented. When the information was presented *before* product exposure, the effects were of intermediate impact, and the information was least effective when presented *after* product exposure. These differences highlight the fact that the critical determinant of any manipulation is not the information alone, but the temporal and other interactions between the *information*, the *product*, and the *consumer*'s pre-existing knowledge and beliefs and/or other idiosyncratic factors. They also underscore the importance of *verifying* that the expectations that are presumed to be acting in any product situation are, in fact, operative for the consumers being investigated.

To operationally verify that an information manipulation had its intended effect on expectations, one of the measurement techniques discussed in Section 10.3.2 can be used. Although this step may seem obvious, it was often overlooked both in early discussions of 'expectation' effects on foods and beverages (Cardello *et al.*, 1985; Zellner *et al.*, 1988), and in more recent treatments (Hutchings, 2003; DiMonaco *et al.*, 2003). One still encounters studies in the literature where information is manipulated using popular versus less popular brands or high price versus low price labels. The investigator then simply *assumes* that the popular brand or high price label creates a more positive expectation and that this is what influences subsequent liking. Such speculations are inappropriate and utilize the expectation construct as a post-hoc explanatory variable, rather than as a predictive variable that can be used to guide product development.

When crafting the nature and content of informational elements to be used in setting consumer expectations for improved product acceptance, the product developer and marketing specialist should work together closely to ensure that the information conveys accurate, proper and timely content, uses a well-respected and relevant source, and is meaningful to the target audience. In all cases, the effect of the information on consumer expectations should be measured and not assumed.

10.3.4   Measuring inherent product characteristics

In order to know whether any product meets or fails to meet consumer expectations, it is necessary to measure the characteristics that are inherent in the product. This measurement process must occur independently of the information manipulation. Most often it is accomplished during a 'blind' evaluation of the product some period of time prior to the execution of the informational phase. Separation in terms of time (days, weeks, months) is necessary to ensure that the consumer does not recall the product attributes and his or her assessment of it when it is subsequently presented in the informational phase. Alternatively, the test sample(s) may be presented in a blind evaluation of multiple food products, e.g. Kahkonen *et al.* (1999). Such a procedure enables better disassociation of the baseline and post-informational ratings, but the potential for stimulus context effects must be considered. It is also best that the same scalar technique be used to obtain both the baseline product ratings and expectations, so that one can easily index the degree of discrepancy between the two.

10.3.5   Quantifying the disconfirmation experience

Affective disconfirmation

In order to assess the degree of *affective* (a) disconfirmation (positive or negative) created by a product, one can calculate a disconfirmation index (D). Depending upon the scalar technique used to obtain the expected (E) and baseline (B) liking ratings, one procedure is simply to subtract the value of the expectation (E) from the product's baseline rating on that dimension (B), i.e.

$$D_a = B_a - E_a$$

Using this subtractive measure, disconfirmation will have a positive value (positive disconfirmation) when expectations are lower than baseline product quality, but a negative value (negative disconfirmation) when expectations are greater than baseline product quality. In addition to this subtractive or 'inferred' pre-trial measure of disconfirmation, some investigators obtain direct, post-trial measures of 'perceived disconfirmation' by asking consumers to scale whether the product is better, worse or the same as expected (Cardello and Sawyer, 1992; Tuorila et al., 1998a; Schifferstein et al., 1999; Cardello et al., 2000). An advantage of this approach is that it avoids the co-linearity that results from the inferred disconfirmation index utilizing the baseline product rating, which is also used to measure the change in perceived product performance. Correlations between these two measures of disconfirmation have been reported to range from 0.54 to 0.8 (Oliver, 1977a; Trawick and Swan, 1980; Cardello and Sawyer, 1992). However, it must be kept in mind that perceived disconfirmation judgments are made after exposure to *both* the information and the product. Thus, they well may be confounded by the very expectancy effect that is being investigated.

Sensory disconfirmation
For individual *sensory* (s) attributes, disconfirmation can be calculated in the same manner as described above, i.e. $D_s = B_s - E_s$. However, when multiple sensory attributes are examined, an overall index of disconfirmation must be calculated by summing the individual differences between expected and baseline sensory attributes across all attributes, i.e.

$$D = \sum_{i=1}^{n} B_i - E_i$$

where i = attributes.

Without additional information about the relative importance or salience of the individual sensory differences to the overall disconfirmation experience, an unweighted average of the differences must be used. However, if data exist regarding the relative salience of the attributes, a weighted average can be calculated. For example, Schifferstein et al. (1999) used principal components analysis (PCA) factor scores obtained both in blind testing and after the expectation manipulation to derive weights for the total disconfirmation score. The factor scores were used in a regression model to predict overall quality ratings of the product and the regression weights for the factor scores were used as measures of relative importance in calculating the disconfirmation index.

Analogous measures of disconfirmation can be developed for both ideational and conative expectations.

10.3.6 The dependent variable(s)
The dependent variables in most studies of expectations are either food liking/disliking ratings or sensory product attribute ratings. These dependent variables are most commonly measured in the same manner as both the expectations and

the intrinsic product attributes, because it is only if all variables are measured on the same scale that disconfirmation can be easily quantified and compared with the *change* in product rating from its baseline level. Of course, if one is using conative expectations or simply wishes to examine how sensory or hedonic disconfirmation influences behavior or behavioral intent, other food-relevant dependent variables, e.g. buying intentions, willingness to pay, consumption, etc. can be measured (see Schifferstein *et al.*, 1999; Lange *et al.*, 2000; Stefani *et al.*, 2006). In fact, for certain types of information, e.g. health (Kahkonen *et al.*, 1996; Tuorila and Cardello, 2002; Bower *et al.*, 2003) and price (DiMonaco *et al.*, 2005), behavioral intentions may be more sensitive than affective variables. As previously noted, such behaviorally oriented variables are powerful and practical indices of the impact of expectations on consumer food behavior.

The product developer should select the dependent variables of the study in accordance with the specific product outcomes that are desired by management and/or the product management team.

10.3.7   Fitting models to the data

To determine whether changes in product acceptance (or sensory) ratings are better explained by the assimilation, contrast, generalized negativity, or assimilation–contrast models, one must compare the direction of change in the product's rating from pre-trial levels (B) to post-trial levels (P) with the level and direction (positive/negative) of disconfirmation (D). Since the assimilation and contrast models predict that liking and/or sensory attribute ratings will move toward (assimilation) or away (contrast) from the expectation (Fig. 10.2), it is possible to index the degree to which any set of data are fit by these models by regressing the change in product rating from baseline to post-test against the disconfirmation index. If these two variables are *linearly and negatively* correlated, the data conform to the assimilation model and the slope of the regression line will reflect the magnitude of assimilation. Figure 10.3 shows a plot of these two variables from a recent study of consumer expectations of foods processed by novel food technologies (Cardello, 2003). The high negative correlation between the two variables provides support for an assimilation model. The lack of any marked deviation from linearity at extremely high or low disconfirmation levels also eliminates the possibility that an assimilation–contrast model might better account for the data (see Fig. 10.2).

Since the *contrast* model predicts a shift of product ratings *away* from expectations, support will hold for this model if the disconfirmation and change in liking (or sensory attribute) ratings are *linearly and positively* correlated. If the linear correlation coefficient between disconfirmation and change in product rating is low, then it must be determined whether the data are better explained by the generalized negativity or assimilation–contrast models. To determine whether the generalized negativity model fits the data, the effects of positive and negative disconfirmation on product ratings must be assessed separately. If both positive and negative disconfirmation produce *lowered* product ratings, it

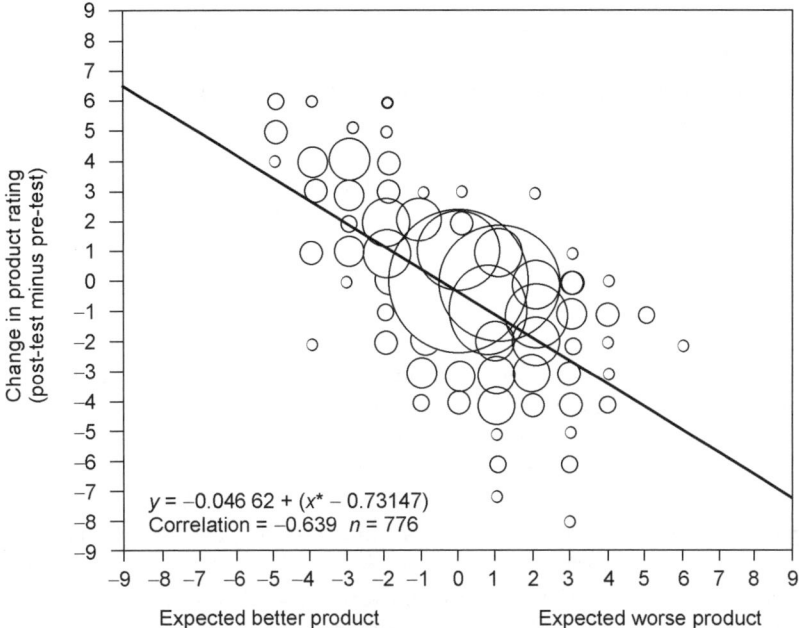

$$y = -0.046\,62 + (x^* - 0.73147)$$
$$\text{Correlation} = -0.639 \ \ n = 776$$

Fig. 10.3 Plot of the change in product liking ratings from baseline to post-test against the level of hedonic disconfirmation. Circles and their size represent the number of data points falling at the same location (reprinted from *Appetite*, vol. 40, A.V. Cardello, 'Consumer concerns and expectations about novel food processing technologies', 217–233, copyright 2003, with permission from Elsevier).

provides support for the generalized negativity model.

The assimilation–contrast model is more complicated to assess, because the predicted effects are dependent upon the absolute level of disconfirmation. As depicted in Fig. 10.2, marked deviations from linearity would be expected in the plot of disconfirmation against product rating change, such that at high disconfirmation levels, the slope of the data would shift lower and reverse direction.

10.4 Current issues and future trends

The preceding section outlined the methodological considerations required for conducting studies of consumer expectations for consumer-driven product development. In the following section, important theoretical issues that bear on the analysis and interpretation of expectations data are discussed, along with important areas for future research effort.

10.4.1 Temporal shifts in expectations

In almost all of the research cited in this chapter, expectations were manipulated in response to a discrete information event. However, one of the most important and least studied areas is *how expectations evolve over time*. One thing that is agreed upon by current investigators is that expectations arise from a combination of direct exposure to the product or product category and from information related to the product. Any product exposure that fails to meet expectations will cause the consumer to adjust his or her expectations in accordance with the product experience (Goering, 1985; Deliza and MacFie, 1996; Lange *et al.*, 1999). Goering (1985) discussed this theoretical ebb and flow in terms of 'learning' and related the systematic shifts in consumer product expectations to changes in market demand for the product and optimal pricing strategies.

In order to address the issue of changes in expectations with repeated exposure, a study was conducted in our laboratory to examine the effects of presentation of a random series of stimuli of varying intensity on sensory and hedonic expectations. Twenty consumers were presented six different concentrations of an orange drink (100%, 50%, 25%, 12%, 6% and 3%) in random order. After tasting the first sample, subjects were instructed to indicate their *expectations* of the flavor intensity *and* liking/disliking of subsequent samples by placing a hash-mark on analog line scales of flavor intensity (no flavor–strong flavor) and liking/disliking (dislike extremely–like extremely). Following the last stimulus in the series, an expectation judgment was made for a final sample of 100% orange drink.

In order to index how expectations changed over time, the *change* in expectation from its initial value was calculated and plotted as a function of the change in concentration from the initial stimulus concentration. Figure 10.4 shows these mean changes in expected flavor intensity and expected liking/disliking. As can be seen, changes in expected flavor intensity and expected liking/disliking tracked the changes in stimulus concentration from baseline. Whenever a higher concentration was presented, expected flavor intensity and liking increased. Whenever a lower concentration was presented, expectations decreased, with the change in expectation being monotonic with the directional shift in concentration from its initial value. It should also be noted that the mean value for the change in expected flavor intensity and expected liking/disliking at the *extrapolated* point at which there was no change in concentration was close to zero.

The data in Fig. 10.4 show that consumers' expectations closely track product experiences. The data support the notion that consumers in a market setting who are exposed to multiple products in a category over time experience a cascade of changing expectations about the likely sensory and/or hedonic properties of subsequently encountered products upon each new product exposure. As each product experience confirms or disconfirms an existing expectation, the expectation for the next product to be encountered changes accordingly. As suggested by Hurling and Shepherd (2003), this expectation cascade continues into the preparation and consumption stages of a food, with expectations about the final, eaten product changing repeatedly in response to visual cues from packaging and labeling, olfactory cues during preparation, and other contextual information gained prior to actually consuming the food.

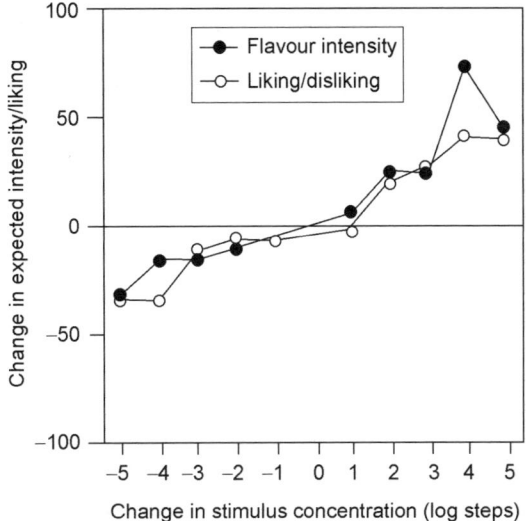

Fig. 10.4 Plot of changes in expectation ratings as a function of changes in stimulus concentration in a random series of stimulus exposures.

Since expectations can shift over time, it is incumbent upon the product developer and market research team to be alert to new sources of information available to consumers about their product, the product category, or new entries to the category, all of which may alter consumer expectations for the product. Constant vigilance in this regard is the price of continued product success.

10.4.2   Demand artifacts in expectations research

In their essence, expectations are intervening variables, i.e. they are *presumed* psychological entities that are measured using an overt response (verbal behavior). To the extent that the verbal behavior related to an expectation is elicited by a variable unrelated to the presumed intervening construct, the resultant data may be artifactual. A 'demand' artifact is an experimental effect produced by the subject's motivation to please the experimenter and to confirm what they believe is the experimental hypothesis (Orne, 1962). Sawyer (1975) summarized many of the demand artifacts that can occur in consumer research and suggested appropriate countermeasures to mitigate their effects. One demand characteristic that has received frequent attention in consumer research is the effect of asking behavioral intent questions on subsequent purchase behaviors (Morwitz *et al.*, 1993; Fitzsimons and Morwitz, 1996). In a similar manner, it can be questioned whether consumers who are asked to give an expectation judgment for a product are motivated to rate the tasted product in a manner that is consistent with what the consumer believes should be the effect of expectations on perceptual judgments. Might it be that evoking the overt expectation judgment creates an experimental demand for the consumer to confirm this expectation in their subsequent verbal reports about the product?

In a study designed to examine the effects of asking versus not asking subjects to provide overt judgments of their expectations, Cardello *et al.* (1996b) found contrast effects using a typical perceptual contrast paradigm, but no difference between subjects who provided overt expectation judgments and those who did not in the relative ratings of either the flavor intensity or liking/disliking of the target stimulus. If expectations are the powerful drivers of perceptions, as has been repeatedly shown in the literature, and if the act of eliciting the expectations is a critical factor in this process, then some difference in the ratings of perceived flavor intensity and liking would have been expected to be found. Although it could be argued that the primary drivers of perceived flavor intensity and liking in this experiment were *neither* overt nor covert expectations, but rather the stimulus context itself, i.e. an explanation in terms of range-frequency theory, this explanation begs the question as to why this contextual paradigm is immune to the effects of expectations (see Section 10.4.4). Future research should focus on controlled studies of potential demand characteristics in expectations research to ensure that the important effects observed to date are not dependent upon the manner in which expectations are elicited.

10.4.3   Asymmetries between positive and negative disconfirmation

There has been some important discussion in the literature regarding assimilation effects and whether or not positive versus negative disconfirmation is more likely to produce them. Schifferstein (1997; Schifferstein *et al.*, 1999) noted that prospect theory (Kahneman and Tversky, 1979) would predict that positive disconfirmation, which can be classified as a form of economic *gain*, is more likely to produce assimilation effects than negative disconfirmation, which is viewed as a loss by the consumer. Although he produced data supporting such asymmetry for hedonic expectations, inconsistent effects were found with sensory expectations (Schifferstein *et al.*, 1999). Other investigators have either failed to show any asymmetry between positive and negative disconfirmation (Lange *et al.*, 1999) or have found varying degrees of asymmetry in the opposite direction (Anderson and Sullivan, 1993; DeSarbo *et al.*, 1994; Siret and Issanchou, 2000; Caporale and Monteleone, 2004). The latter findings provide empirical support to the suggestion by Deliza (1996) that negative disconfirmation may be more likely to produce assimilation. Given the variance in findings, it appears that that asymmetry effects may well be product-, information- (Caporale and Monteleone, 2004), and/or subject-specific.

It is of some interest that asymmetries in assimilation effects have also been reported in other domains. For example, in studies of pain perception it has been shown that perceived pain assimilates the individual's expected pain level, but only in cases where the stimulus is greater in intensity than the expectation. When the stimulus is lower in intensity than expected, no effect is observed

(Koyama *et al.*, 2005). These results have led investigators to pursue the possibility that personality characteristics, such as pessimism versus optimism, may be a determining factor in this asymmetry. Such investigations may also prove fruitful in understanding asymmetries in food expectations, since the outlook of the consumer (on an optimism–pessimism dimension) may well determine his or her response to persuasive communication about a product and, in turn, the impact of positive versus negative disconfirmation on perception and behavior (see Section 10.4.5).

10.4.4   Where are the contrast effects?

Another important issue is why contrast effects are so rarely observed in expectations research. The assimilation–contrast model predicts that contrast effects will occur when the discrepancy between expectations and product performance is large. One might expect this to occur frequently, especially in research settings where expectations are grossly manipulated in order to maximize the likely effects on product ratings. Yet, even in cases where disconfirmation is arguably quite large, contrast effects are rare. Cardozo (1965) reported a contrast effect in the evaluation of ballpoint pens after subjects studied catalogs of high- or low-priced pens, but his data were compromised by the use of a context-dependent rating scale. Anderson (1973) also reported a contrast effect in the evaluation of ballpoint pens at the highest of five disconfirmation levels. However, examination of his data shows only a slight reversal at this level, and the perceived product performance was still higher than in the 'accurate' information condition (Table 2 in Anderson, 1973). Cardello and Sawyer (1992) also reported a contrast effect in sensory judgments of the bitterness of juices; however, the bitterness judgments were made in conjunction with liking judgments, the latter of which showed assimilation. The contrast effect may have been an artifact of the negative association between liking and bitterness. That is, since liking judgments increased in response to positive information, the bitterness judgments of those beverages may have co-varied in the opposite direction in keeping with the logical association between the two.

Bickart (1993) and Meyers-Levy and Tybout (1997) have suggested that contrast effects occur as a result of deliberate corrective action in response to the awareness of a discrepancy between persuasive information or expectations and actual product features. The latter authors have found contrast effects primarily among individuals who have a high need for cognition (NFC), suggesting that these individuals deliberate more on discrepancies brought to consciousness. The differential effect of NFC on expectations (Deliza and MacFie, 1996; Jaeger and MacFie, 2001) and on the likelihood of contrast effects (Meyers-Levy and Tybout, 1997) makes the study of personality characteristics in expectations research a potentially fruitful one for further investigation (see Section 10.4.5).

In contrast to this, Raghunathan and Irwin (2001), using a contextual product series that either increases or decreases in pleasantness, have suggested that when there is a 'domain match' between contextual and target stimuli (the contextual and target stimuli are perceived to be part of a single stimulus set) 'people will compare the target with the average of the items in the context, resulting in a contrast effect.' However, under conditions of 'domain mismatch' (the target stimulus is perceived to be qualitatively different from the contextual stimuli), 'mood-based assimilation occurs.' Thus, if consumers perceive a product to be in the same product class as other products, it will be *contrasted* with the 'market mix.' However, if the product is seen as distinct, then no perceptual comparison is made. However, assimilation effects may be observed in response to the hedonic or other cognitive-level characteristics of the contextual product set to which the consumer is exposed.

Contrast effects have also been repeatedly observed in perceptual studies in which a target stimulus is presented within a context of other stimuli of either higher or lower intensity or affect (Riskey *et al.*, 1979; Lawless, 1983; McBride, 1985, 1986; Conner *et al.*, 1987; Schifferstein, 1995). These results are commonly explained in terms of adaptation level (Helson, 1964) or range-frequency (Parducci, 1965, 1974) theories, which focus on the perceptual frame of reference and/or the human judgmental process that causes subjects to divide a given range of stimuli into equally spaced intervals on the stimulus dimension and then to evenly space the frequency of their judgments along this range. In a series of papers, Schifferstein has addressed this phenomenon in terms of the nature and characteristics of the contextual stimuli and their impact on the magnitude of these contrast effects (Schifferstein, 1994, 1995; Schifferstein and Frijter, 1992; Schifferstein and Oudejans, 1996). He and others have noted the relationship between these perceptual effects and the contrast effects observed in expectations research (Schifferstein and Oudejans, 1996; Cardello *et al.*, 1996b), suggesting that *expectations* established by the contextual stimuli may serve as a complementary explanation for the observed effects.

One prediction of assimilation–contrast theory is that contrast effects will occur under conditions of high disconfirmation. An important component of a high level of disconfirmation is a strongly held belief or awareness of the discrepancy between expectations and product attributes. One hypothesis to account for the contrast effects in perceptual studies is that the presentation of a series of real world stimuli of a given contextual nature creates a more robust expectation for subsequent stimuli than does mere information. This possibility derives support from the information integration response model of Smith and Swinyard (1982, 1983, 1988), which posits that product *information* establishes only a low degree of information acceptance, whereas product *exposure* establishes higher-order (more strongly held) beliefs (Marks and Kamins, 1988; Smith, 1993; Kopalle and Lehmann, 1995). If this is true, the repetitive nature of a contextual series of stimuli would further reinforce any expectations, as compared with the situation in information-based studies where no actual stimuli are presented prior to product trial.

If contextual stimuli create strong cognitive expectations about the sensory intensity of a target stimulus, then the discrepancy between this expected intensity and the (higher/lower) intensity of the target stimulus would be more salient and likely to rise to conscious awareness. Under these conditions, assimilation–contrast theory would predict a contrast effect. Thus, the perceptual contrast effects that have been reported in the literature may incorporate a component due to range-frequency effects and a component due to cognitive expectancy effects.

Due to their importance to consumer marketing, more research is needed into the factors that produce contrast effects. Unless the conditions that produce contrast effects are better understood, marketing attempts to create better expectations for a product will run the risk of creating an effect that is diametrically opposed to that which is intended. In keeping with the bulk of available evidence to date, the product developer and marketing specialist must ensure that product expectations are set realistically and that any positive expectations (relative to product characteristics) are not so large as to become obvious to the consumer. If the latter occurs, undesired contrast effects may result.

10.4.5   The role of individual differences

Consumers are not all alike. They each have their own beliefs, experiences, behavioral tendencies, and personalities. In a recent study of the sensory expectations for soppressata salami, Iaccarino et al. (2006) observed significant differences in the product expectations of consumers living in different geographic regions in which soppressata is produced. Clearly, these differences in product expectations are based on familiarity and experience with differing forms of the product. In reviews and studies of the factors that influence consumer expectations and disconfirmation effects, Deliza and MacFie (1996) and Deliza et al. (2003) identified several important personality characteristics that play a role. Among these personal variables were the self-confidence of the consumer in his or her sensory abilities and/or personal attributions (Lichtenstein and Burton, 1988; Deliza and MacFie, 1996), the consumer's 'need for closure' (Kruglanski et al., 1993), the degree to which the consumer 'self-monitors' their personal image (Snyder and DeBono, 1985), private body consciousness (Miller et al., 1981; Stevens, 1991), and 'need for cognition' (Petty and Cacioppo, 1981, 1986) (see below). For each of these, Deliza and MacFie (1996) offered predictions for the relative likelihood of observing different disconfirmation effects among individuals possessing these personal characteristics.

Another 'personal' variable of critical importance to expectations research is the 'involvement' of the individual in the product (Freedman, 1964). Brewer (1988) has argued that the level of involvement that an individual has with a product will determine whether 'category' or 'schema'-based processing versus 'piecemeal' processing of information will occur, and Lee (1995) has shown that involvement interacts with the schema congruity of the product to differentially produce assimilation or contrast effects. When product involvement is low, the consumer may not be motivated to think critically about information presented about the product. The effects of such product orientations are addressed in the elaboration likelihood model (ELM) of Petty and Cacioppo (1981, 1986), which predicts that individuals who are low in need for cognition will be more susceptible to simple cues in an informational array, whereas those who have a high need for cognition will respond through a more central route of information processing, resulting in greater effectiveness of the persuasive elements in an information message. In their study of packaging effects on sensory expectations, Deliza and MacFie demonstrated a significant influence of need for cognition on expectation generation. However, not all studies have found empirical support for the predictions of the ELM model (Jaeger and MacFie, 2001).

Lastly, for foods for which consumers may have safety concerns, recent evidence has shown that perceived concern about a product can have a negative impact on liking expectations (Cardello, 2003). To the extent that concern levels can be related to individual consumer characteristics, e.g. gender or risk-taking tendencies, these individual differences can influence expectations and subsequent product acceptance (Cardello, 2003).

Understanding individual differences in expectations is critical for segmenting the consumer population into specific target audiences. The product developer should strive to be aware of these individual differences and how they influence the perception and/or processing of product information, so that marketing communications about his/her product can be tailored to specific consumer segments, thereby maximizing expectations and their impact on purchase behavior.

10.4.6   Analyzing aggregate *versus* individual data

One consequence of idiosyncratic differences among consumers is that the evaluation of all expectation effects *must* be done at the level of the individual consumer. Examining and analyzing aggregate (mean) data will not provide useful insights. Owing to personality differences, idiosyncratic preferences and unique experiences with the foods and informational elements, every consumer will have a different expectation for any product. Thus, even if all consumers have the same perception of the intrinsic attributes of a product, the level of post-information disconfirmation will differ from one consumer to the next, perhaps being positive for some and negative for others. Add to this the fact that *not* all consumers will have the same initial perception of the product, and it becomes evident that the level and direction of disconfirmation may vary greatly from consumer to consumer. In such a situation, the difference in the mean expectation versus the mean baseline rating will reveal nothing about the levels and kinds of disconfirmation being experienced by the individuals. As a consequence, the same degree and direction of change of pre- to post-product ratings may reflect an assimilation effect for some consumers, but a contrast effect for others. In different data sets published by Lange *et al.* (1999), Cardello and Sawyer (1992), and Scharf and Volkmer (2000) this point was made clearly, because analyses of the individual data showed some consumers exhibiting assimilation effects and others exhibiting contrast effects, even at high levels of disconfirmation. If these data sets were to be aggregated, the idiosyncratic behavior of the consumers is lost, and mistaken conclusions could be drawn about the data.

10.4.7   The neuropsychology of expectations – a portal to the future?

The past few years have produced an explosion of discoveries in psychobiology and neuroscience that hold great promise for the future of consumer-driven product development. For example, 'neuromarketing' is a controversial new area of research that uses fMRI (functional magnetic resonance imaging) and other neurophysiological measures to uncover the brain mechanisms involved in consumer behavior. This new discipline has received great attention in the popular press, and its promise of finding product features that will stimulate 'buy' centers in the brain is appealing to product developers and marketers alike. Although the practical applications of this research are still far in the future, findings from the field of neuroscience are beginning to provide important insights into the likely brain mechanisms that underlie consumer expectations, product perceptions and decision-making. These findings may well revolutionize how we conduct consumer-driven product development in the future.

Much of the relevant research being conducted in the neurosciences uses fMRI technology, which measures the blood flow associated with neuronal activity in different parts of the brain. In a typical study using this technique, Ploghaus *et al.* (1999) found that the *expectation* of pain, created by associating colored light with a painful stimulus, activated sites within the medial frontal lobe, insular cortex, and cerebellum that are distinct from, but spatially adjacent to, brain areas mediating the *actual* pain experience itself. Similarly, Koyama *et al.* (2005) used fMRI to track brain locations associated with both the expectation *and* experience of pain. The latter investigators found that areas in the anterior insula and anterior cingulate cortex (rACC) were activated regardless of whether the actual stimulus was presented or just the expectation of it. In addition, as an interesting neural analogue to the asymmetry in assimilation effects, the investigators found that the perceived levels of pain assimilated the expected levels when the stimulus was more intense than expected, but not when the stimulus was less intense than expected. These and similar studies (e.g. Wager *et al.*, 2004) demonstrate that representations of both *expected* and *actual* sensory stimuli occur in adjoining and/or identical brain areas, creating the potential for expectation-related neuronal activity to interact with and modify neuronal activity in adjoining brain areas that encode the actual perception of the stimulus.

With regard to affective expectations, recent studies of neuronal responses in monkeys (Samejima *et al.*, 2005) have shown that striatal projection neurons in the putamen and claudate nucleus encode the expected reward associated with executing a behavior. These 'expected reward' neurons are distinct from the neurons that encode the behavioral actions themselves, yet they accurately *predict* those behaviors. Hsu *et al.* (2005) used fMRI to confirm these findings in humans, showing that striatal brain activity in humans is also associated with 'expected rewards'.

With regard to the neural substrates that underlie the affective response to foods, it has been demonstrated in humans that the 'reward value' of visual, olfactory, and taste stimuli is represented in the orbitofrontal cortex (OFC) (Rolls, 2000; O'Doherty *et al.*, 2001a,b; Hornack *et al.*, 2004). In studies by O'Doherty *et al.* (2002, 2003, 2006), the brain mechanisms involved in pre-dictive representation of food preferences have been examined. In one study, these investigators identified areas in the ventral midbrain and ventral striatum of humans which were associated with the anticipation of specific levels of liking/disliking. By pairing the presentation of visual shapes with juices of varying taste preferences, they demonstrated that mere presentation of the visual stimulus produced activation in the ventral midbrain and that this activity increased monotonically with the *expected preference* for the anticipated juice (O'Doherty *et al.*, 2006).

In a similar study of fMRI responses to cola beverages, it was found that areas of the ventromedial prefrontal cortex became activated in response to the presentation of Coca Cola and Pepsi in behavioral tests (McClure *et al.*, 2004). This same brain area has also been implicated in encoding stimulus reward values (Knutson *et al.*, 2001; O'Doherty *et al.*, 2003). However, when a visual stimulus of a Coca Cola can (versus a circle of light) was presented before the test stimulus, increased brain activity was observed in the dorsolateral prefrontal cortex (DLPC), hippocampus, and midbrain. The authors concluded that the latter brain areas are involved in the process of 'biasing perception based on prior affective behavior.'

In the most recent of this series of experiments, Sarinopolous *et al.* (2006) demonstrated that taste expectations can attenuate taste perception. Using fMRI, these investigators showed that the misleading information about a soon to be presented taste solution would be only mildly bitter resulted in neural activity in the anterior cingulate cortex (rACC), orbitofrontal cortex (OFC), and dorso-lateral prefrontal cortex (DLPC). In turn, activity in these areas predicted decreases in activity in the insula and amygdala in response to the presentation of a very bitter taste solution. The authors concluded that 'the (neural) network processing aversive taste may be suppressed by expectancy-related brain activity in specific regions of the rACC, OFC, and DLPC.'

On a more behavioral level, the notion that neuronal substrates can be activated in response to anticipated sensory or other experiences has also been documented in the study of what are called 'mirror neurons.' Mirror neurons are neurons activated by the simple anticipation of an experience or behavior and have been identified in humans and other primates (Gallese *et al.*, 1996; Rizzolatti *et al.*, 1996, 2001). For example, it has been demonstrated in monkeys that simply observing another monkey putting food in their mouth will activate the same neurons that respond to actual placement of food in the mouth (Rizzolatti *et al.*, 1996). Similarly, the anticipation of a sensory experience, such as being touched on the body, stimulates the same neuronal substrates as does the mere observation of someone else being touched (Keysers *et al.*, 2004) Using fMRI, it has also been shown that simply observing emotions in others, as reflected in their facial expressions, activates the same neural substrates as does the actual experience of those emotions (Wicker *et al.*, 2003). One of these brain areas, the anterior insula, is the same area that has been reported to respond to both actual and anticipated pain (Koyama *et al.*, 2005). Thus, it appears that mirror neurons may underlie both the *anticipation* of sensory, emotional, and behavioral events and their *actual* experience, again suggesting a direct neural link by which expectations can modulate experience.

In addition to the identification of neurons that can mediate expectations and assimilatory effects, there is also new research that points to a potential neural basis for the experience of disconfirmation and disappointment. In several recent studies in monkeys, neurons in the centromedial/parafascicular complex of the thalamus, which characteristically respond to multi-modal stimulation, have been shown to respond with greater activity when a stimulus is presented that is contrary to expectations (Matsumoto *et al.*, 2001; Minamimoto and Kimura, 2002). In the most recent of this series of studies (Minamimoto *et al.*, 2005) monkeys who were trained on a visual GO–NOGO task had individual neuronal activity monitored in the centromedial nucleus. When either the *expectation* or the *actual delivery* of a 'small' reward occurred during testing, these neurons increased their firing rate significantly, suggesting that these neurons are tuned to respond to both *expected* and *actual disappointment*.

As the field of neuroscience grows, much greater attention will be focused on the brain mechanisms that underlie the anticipation of sensory, behavioral, and emotional experiences and how these anticipatory experiences influence actual behavior. At some time in the not too distant future, it may be possible for product developers to formulate products with known influences on specific brain regions, neural mechanisms, and associated consumer behavior. Through research into the psychobiology of consumer expectations, future consumer-driven product research may well achieve levels of actionability that can only be imagined today.

10.5   Sources of further information and advice

The lengthy reference list found at the end of this chapter is a starting point for sources of information on the topic of product expectations. Additional sources of information include scientific journals in the areas of marketing, psychology, and consumer behavior and preferences. Among the marketing journals, a partial list of those that publish articles relevant to consumer expectations and product behavior includes the *Journal of Marketing, Journal of Marketing Research, Journal of Consumer Marketing, Journal of Marketing Communications,*

Journal of Marketing Theory and Practice, Marketing Science, and the *Journal of Food Products Marketing.* In psychology, such journals as the *Journal of Consumer Psychology, the Journal of Economic Psychology, Psychology and Marketing, Psychological Review, Journal of Personality and Social Psychology, Journal of Experimental Psychology: Learning, Memory and Cognition,* and the *Journal of Experimental Psychology: General* all publish relevant papers in this area. Lastly, in the areas of consumer behavior and product preferences, such journals as the *Journal of Consumer Behavior, Journal of Consumer Research, Journal of Consumer Satisfaction, Dissatisfaction and Complaining Behavior,* and *Advances in Consumer Research,* along with *Food Quality and Preference, Appetite,* and the *Journal of Sensory Studies* all publish relevant work on expectations and consumer product behavior.

Among the professional organizations that are most active in this area of research are the American Marketing Association (AMA), the Academy of Marketing Sciences (AMS), the American Psychological Association (APA), the American Psychological Society (APS), the Society for Consumer Psychology (SCP), and the Society for Judgment and Decision Making (SJDM). Many of these organizations sponsor research symposia at their annual meetings that address specific topics related to consumer expectations. In addition, the Pangborn Sensory Science Symposia series provides a continuing venue for research in this area. The AMA website (www.ama.org) also contains a new section designed to encourage PhD advisors to get students involved in marketing-related fields by submitting brief summaries of marketing theories that the student may be using. Among the relevant theories for which information can be found are cognitive dissonance theory, prospect theory, commitment-trust theory, and the elaboration likelihood model.

Finally, product developers should seek advice from knowledgable professionals in industry and academia. The reader is encouraged to contact the author of this chapter or any one of the many authors cited in the reference section to obtain advice, comments on research designs, or recommendations on how to conduct expectations research to support consumer-driven product development.

10.6   References

ACEBRON LB & DOPICO DC (2000), 'The importance of intrinsic and extrinsic cues to expected and experienced quality: an empirical application for beef', *Food Quality and Preference,* **11**, 229–238.

ANDERSON EW & SULLIVAN MW (1993), 'The antecedents and consequences of customer satisfaction for firms', *Marketing Science,* **12**, 125–143.

ANDERSON RE (1973), 'Consumer dissatisfaction: the effects of disconfirmed expectancy on perceived product performance', *Journal of Marketing Research,* **10**(2), 38–44.

ARONSON E & CARLSMITH M (1962), 'Performance expectancy as a determinant of actual performance', *Journal of Abnormal and Social Psychology,* **65**, 178–182.

ATKINSON JW (1954), 'Explorations using imaginative thought to assess the strength of human motives', in Jones MR, *Nebraska Symposium on Motivation* (Vol. 2), Lincoln, University of Nebraska Press, 56–112.

ATKINSON JW (1957), 'Motivational determinants of risk-taking behavior', *Psychology Review*, **64**, 359–372.

ATKINSON JW (1958), 'Toward experimental analysis of human motivation in terms of motives, expectancies, and incentives', in Atkinson JW, *Motives in Fantasy, Action and Society*, Princeton, Van Nostrand, 288–305.

BECK FM (1977), 'Placebos in dentistry: their profound potential effects', *Journal of the American Dental Association*, **95**, 1122–1126.

BERLYNE DE (1960), *Conflict, Arousal and Curiosity*, New York, McGraw-Hill.

BICKART, BA (1993), 'Carryover and backfire effects in marketing research', *Journal of Marketing Research*, **30**, 52–62.

BOWER JA, SAADAT MA & WHITTEN C (2003), 'Effect of liking, information and consumer characteristics on purchase intention and willingness to pay more for a fat spread with a proven health benefit', *Food Quality and Preference*, **14**, 65–70.

BREWER MB (1988), 'A dual-process model of impression formation', in Srull TK & Wyer Jr RS, *Advances in Social Cognition*, Hillsdale, NJ, Lawrence Erlbaum, 1–36.

CAPORALE G & MONTELEONE E (2001), 'Effects of expectations induced by information on origin and its guarantee on the acceptability of a traditional food: olive oil', *Sciences des Aliments*, **21**, 243–254.

CAPORALE G & MONTELEONE E (2004), 'Influence of information about manufacturing process on beer acceptability', *Food Quality and Preference*, **15**, 271–278.

CAPORALE G, POLICASTRO S, CARLUCCI A & MONTELEONE E (2006), 'Consumer expectations for sensory properties in virgin olive oils', *Food Quality and Preference*, **17**, 116–125.

CARDELLO, AV (1994), 'Consumer expectations and their role in food acceptance', in MacFie HJH & Thomas DMH (eds.), *Measurement of Food Preferences*, London, Blackie Academic, 253–297.

CARDELLO AV (1995), 'The role of ration image, stereotypes, and expectations on acceptance and consumption', in Marriott B (ed.), *Not Eating Enough: Strategies to Overcome Underconsumption of Field Rations*, Washington, DC, National Academy Press, 177–201.

CARDELLO AV (2003), 'Consumer concerns and expectations about novel food processing technologies: effects on product liking', *Appetite*, **40**, 217–233.

CARDELLO AV & SAWYER FM (1992), 'Effects of disconfirmed consumer expectations on food acceptability', *Journal of Sensory Studies*, **7**, 253–277.

CARDELLO AV & SCHUTZ HG (2004), 'Numerical scale-point locations for constructing the LAM (labeled affective magnitude) scale', *Journal of Sensory Studies*, **19**, 341–346.

CARDELLO AV, MALLER O, MASOR HB, DUBOSE C AND EDELMAN B (1985), 'Role of consumer expectancies in the acceptance of novel foods', *Journal of Food Science*, **50**, 1707–1714, 1718.

CARDELLO AV, BELL R & KRAMER FM (1996a), 'Attitudes of consumers toward military and other institutional foods', *Food Quality and Preference*, **7**, 7–20.

CARDELLO AV, MELNICK SM & ROWAN PA (1996b), 'Expectations as a mediating variable in context effects', in *Proceedings of the Food Preservation 2000 Conference*, Vol. 1, Hampton, VA, Science and Technology Corp., 259–309.

CARDELLO AV, SCHUTZ H, SNOW C & LESHER L (2000), 'Predictors of food acceptance, consumption and satisfaction in specific eating situations', *Food Quality and Preference*, **11**, 201–216.

CARDOZO RN (1965), 'An experimental study of customer effort, expectation, and satisfaction', *Journal of Marketing Research*, **2**, 244–249.

CARLSMITH, JM & ARONSON E (1963), 'Some hedonic consequences of the confirmation and disconfirmation of expectancies', *Journal of Abnormal and Social Psychology*, **66**, 151–156.

CONNER MT, LAND DG & BOOTH DA (1987), 'Effect of stimulus range on judgments of sweetness intensity in a lime drink', *British Journal of Psychology*, **78**, 357–364.

COOPER J & FAZIO RH (1984), 'A new look at dissonance theory', in Berkowitz L (ed.) *Advances in Experimental Social Psychology*, New York, Academic Press, 229–266.

COOPER J, ZANNA MP & TAVES PA (1978), 'Arousal as a necessary condition for attitude change following induced compliance', *Journal of Personality and Social Psychology*, **36**, 1101–1106.

CROCKER J (1984), 'A schematic approach to changing consumers' beliefs' *Advances in Consumer Research*, **11**, 472–477.

DAWES RM, SINGER, D AND LEMONS F (1972), 'An experimental analysis of the contrast effect and its implications for intergroup communications and the indirect assessment of attitude', *Journal of Personality and Social Psychology*, **21**(3), 281–295.

DELIZA R (1996), 'The effects of expectation on sensory perception and acceptance', Unpublished PhD thesis, University of Reading, UK.

DELIZA R & MACFIE, HJH (1996), 'The generation of sensory expectations by external cues and its effect on sensory perception', *Journal of Sensory Studies*, **11**, 103–128.

DELIZA R, MACFIE H & HEDDERLEY D (2003), 'Use of computer-generated images and conjoint analysis to investigate sensory expectations', *Journal of Sensory Studies*, **18**, 465–486.

DESARBO WS, HUFF L, ROLANDELLI MM & CHOI J (1994), 'On the measurement of perceived service quality', in Rust RT & Oliver RL (eds.), *Service Quality: New Directions in Theory and Practice*, London, Sage, 201–222.

DIAMOND J & LAWLESS HT (2001), 'Context effects and preference standards with magnitude estimation and the labeled magnitude scale', *Journal of Sensory Studies*, **16**, 1–10.

DIMONACO R, CAVELLA S, IACCARINO T, MINCIONE A & MASI P (2003), 'The role of the knowledge of color and brand name on the consumer's hedonic rating of tomato purees', *Journal of Sensory Studies*, **18**, 391–408.

DIMONACO R, CAVELLA S, DI MARZO S & MASI P (2004), 'The effect of expectations generated by brand name on the acceptability of dried semolina pasta', *Food Quality and Preference*, **15**, 429–438.

DIMONACO R, OLLILA S & TUORILA H (2005), 'Effect of price on pleasantness ratings and use intentions for a chocolate bar in the presence and absence of a health claim', *Journal of Sensory Studies*, **20**, 1–16.

ELKIN RA & LEIPPE MR (1986), 'Physiological arousal, dissonance, and attitude change: evidence for a dissonance–arousal link and a "Don't remind me" effect', *Journal of Personality and Social Psychology*, **51**, 55–65.

FAZIO RH & ZANNA MP (1978), 'Attitudinal qualities relating to the strength of the attitude–behavior relationship', *Journal of Experimental Social Psychology*, **14**, 398–408.

FEATHER NT (1982), *Expectations and Actions: Expectancy-Value Models in Psychology*, Hillsdale, NJ, Lawrence Erlbaum.

FESTINGER L (1957), *A Theory of Cognitive Dissonance*, Evanston, IL, Row and Peterson.

FESTINGER L (1962), 'Cognitive dissonance', *Scientific American*, **207**, 93–102.

FISHBEIN M & AJZEN I (1975), *Belief, Attitude, Intention and Behavior: An Introduction to Theory and Research*, Reading MA, Addison-Wesley.

FITZSIMONS GJ & MORWITZ VG (1996), 'The effect of measuring intent on brand-level purchase behavior', *Journal of Consumer Research*, **23**, 1–11.

FREEDMAN JL (1964), 'Involvement, discrepancy, and change', *Journal of Abnormal and Social Psychology*, **69**, 290–295.

GALLESE V, FADIGA L, FOGASSI L & RIZZOLATTI G (1996), 'Action recognition in the premotor cortex', *Brain*, **119**, 593–609.

GIESE, J L & COTE JA (2000), 'Defining consumer satisfaction', *Academy of Marketing Science Review*, #1. http://www.amsreview.org/articles/giese01-2000.pdf

GIOVANNI ME & PANGBORN RM (1983), 'Measurement of taste intensity and degrees of liking of beverages by graphic scales and magnitude estimation', *Journal of Food Science*, **48**, 1175–1182.

GOERING PA (1985), 'Effects of product trial on consumer expectations, demand, and prices', *Journal of Consumer Research*, **12**, 74–82.

GREEN BG, SHAFFER GS & GILMORE MM (1993), 'Derivation and evaluation of a semantic scale of oral sensation magnitude with apparent ratio properties', *Chemical Senses*, **18**, 683–702.

GREEN BG, DALTON P, COWART B, SHAFFER G, RANKIN K & HIGGINS J (1996), 'Evaluating the "labeled magnitude scale" for measuring sensations of taste and smell', *Chemical Senses*, **21**, 323–335.

HELSON H (1948), 'Adaptation-level as a basis for a quantitative theory of frames of reference', *Psychology Review*, **55**, 297–313.

HELSON H (1955), 'An experimental approach to personality', *Psychiatric Research Reports*, **2**, 89–99.

HELSON H (1964), *Adaptation Level Theory*, New York, Harper and Row.

HONIGFELD G (1964), 'Non-specific factors in treatment: II. Review of social-psychological factors', *Diseases of the Nervous System*, **25**, 225–239.

HORNAK J, O'DOHERTY J, BRAMHAM J, ROLLS ET, MORRIS RG, BULLOCK PR & POLKEY CE (2004), 'Reward-related reversal learning after surgical excisions in orbito-frontal or dorsolateral prefrontal cortex in humans', *Journal of Cognitive Neuroscience*, **16**(3), 463–478.

HOVLAND CI & WISE W (1951), 'The influence of source credibility on communication effectiveness', *Public Opinion Quarterly*, **15**, 635–650.

HOVLAND CI, HARVEY OH & SHERIF M (1957), 'Assimilation and contrast effects in reactions to communication and attitude change', *Journal of Abnormal and Social Psychology*, **55**, 244–252.

HSU M, BHATT M, ADOLPHS R, TRANEL D & CAMERER CF (2005), 'Neural systems responding to degrees of uncertainty in human decision-making', *Science*, **310**, 1680–1683.

HURLING R & SHEPHERD R (2003), 'Eating with your eyes: effect of appearance on expectations of liking', *Appetite*, **41**, 167–174.

HUTCHINGS JB (2003), *Expectations and the Food Industry: The Impact of Color and Appearance*, New York, Kluwer, Academic/Plenum Publishers.

IACCARINO T, DIMONACO R, MINCIONE A, CAVELLA S & MASI P (2006), 'Influence of information on origin and technology on the consumer response: the case of soppressata salami', *Food Quality and Preference*, **17**, 76–84.

INSKO CA (1967), *Theories of Attitude Change*, New York, Appleton-Century-Crofts.

JAEGER SR (2006), 'Non-sensory factors in sensory science research', *Food Quality and Preference*, **17**, 132–144.

JAEGER SR & MACFIE HJH (2001), 'The effect of advertising format and means–end information on consumer expectations of apples', *Food Quality and Preference*, **12**, 189–206.

KAHKONEN P & TUORILA H (1995), 'The role of expectations and information in sensory perception of low-fat and regular-fat sausages', *Appetite*, **24**, 298–299.

KAHKONEN P & TUORILA H (1998), 'Effect of reduced-fat information on expected and actual hedonic and sensory ratings of sausage', *Appetite*, **30**, 13–23.

KAHKONEN P & TUORILA H (1999), 'Consumer responses to reduced and regular fat content in different products: effects of gender, involvement and health concern', *Food Quality Preference*, **10**, 83–91.

KAHKONEN P, TUORILA H & RITA H (1996), 'How information enhances acceptability of a low-fat spread', *Food Quality and Preference*, **7**, 87–94.

KAHKONEN P, TUORILA H & LAWLESS H (1997), 'Lack of effect of taste and nutrition claims on sensory and hedonic responses to a fat-free yogurt', *Food Quality and Preference*, **8**, 125–130.

KAHKONEN P, HAKANPAA P & TOURILA H (1999), 'The effects of information related to fat content and taste on consumer responses to a reduced-fat frankfurter and a reduced-fat chocolate bar', *Journal of Sensory Studies*, **14**, 35–46.

KAHNEMAN D & TVERSKY A (1979), 'Prospect theory: an analysis of decisions under risk', *Econometrica*, **47**, 263–291.

KEISLER CA & PALLAK MS (1976), 'Arousal properties of dissonance manipulations', *Psychological Bulletin*, **83**, 1014–1025.

KEYSERS C, WICKERS B, GAZZOLA V, ANTON JL, FOGASSI L & GALLESE V (2004), 'A touching sight: SII/PV activation during the observation and experience of touch', *Neuron*, **42**, 1–20.

KNUTSON B, ADAMS CM, FONG GW & HOMMER D (2001), 'Anticipation of increasing monetary reward selectively recruits nucleus accumbens', *Journal of Neuroscience*, **21**, RC159.

KOPALLE PK & LEHMANN DR (1995), 'The effects of advertised and observed quality on expectations about new product quality', *Journal of Marketing Research*, **32**, 280–290.

KOYAMA T, MCHAFFIE JG, LAURIENTI PJ & COGHILL RC (2005), 'The subjective experience of pain: where expectations become reality', *Proceedings of the National Academy of Sciences of the United States of America*, **102**, 36, 12950–12955.

KRUGLANSKI AW, WEBSTER DM & KLEM A (1993), 'Motivated resistance and openness to persuasion in the presence or absence of prior information', *Journal of Personality and Social Psychology*, **65**, 861–876.

LANGE C, ROUSSEAU R & ISSANCHOU S (1999), 'Expectation, liking and purchase behaviour under economical constraint', *Food Quality and Preference*, **10**, 31–39.

LANGE C, ISSANCHOU S & COMBRIS P (2000), 'Expected versus experienced quality: trade-off with price', *Food Quality and Preference*, **11**, 289–297.

LATOUR SA & PEAT SA (1979), 'Conceptual and methodological issues in consumer satisfaction research', in Wilkie WL (ed.), *Advances in Consumer Research*, Association for Consumer Research, Vol. 6, 431–437.

LAWLESS H (1983), 'Contextual effects in category ratings', *Journal of Testing and Evaluation*, **11**, 346–349.

LEE M (1995), 'Effects of schema congruity and involvement on product evaluations',

Advances in Consumer Research, **22**, 210–215.

LEVIN IP & GAETH GJ (1988), 'How consumers are affected by the framing of attribute information before and after consuming the product', *Journal of Consumer Research*, **15**, 374–378.

LICHTENSTEIN DR & BURTON S (1988), 'The measurement and moderating role of confidence in attributions', *Advances in Consumer Research*, **15**, 468–475.

LOWINGER P & DOBIE S (1969), 'A study of placebo response rates', *Archives of General Psychiatry*, **20**, 84–88.

MACCORQUODALE K & MEEHL PE (1953), 'Preliminary suggestions as to a formalization of expectancy theory', *Psychology Review*, **60**, 55–63.

MACCORQUODALE K & MEEHL PE (1954), in Estes WK (eds), *Modern Learning Theory*, New York, Appleton-Century-Crofts.

MARKS LJ & KAMINS MA (1988), 'The use of product sampling and advertising: effects of sequence of exposure and degree of advertising claim exaggeration on consumers' belief strength, belief confidence, and attitudes', *Journal of Marketing Research*, **25**, 266–281.

MATSUMOTO N, MINAMIMOTO T, GRAYBIEL AM & KIMURA M (2001), 'Neurons in the thalamic CM–PF complex supply striatal neurons with information about behaviorally significant sensory events, *Journal of Neurophysiology*, **85**, 960–976.

McBRIDE RL (1985), 'Stimulus range influences intensity and hedonic ratings of flavour', *Appetite*, **6**, 125–131.

McBRIDE RL (1986), 'Hedonic rating of food: single or side-by-side sample presentation', *Journal of Food Technology*, **21**, 355–363.

McCLELLAND DC, ATKINSON JW, CLARK RA & LOWELL EL (1953), *The Achievement Motive*, New York, Appleton-Century-Crofts.

McCLURE SM, LI J, TOMLIN D, CYPERT KS, MONTAGUE LM & MONTAGUE PR (2004), 'Neural correlates of behavioral preference for culturally familiar drinks', *Neuron*, **44**, 379–387.

MEEHL PE & MACCORQUODALE K (1951), 'Some methodological comments concerning expectancy theory', *Psychology Review*, **58**, 230–233.

MERTON RK (1948), 'The self-fulfilling prophecy', *Antioch Review*, **8**, 193–210.

MEYERS-LEVY J & TYBOUT A (1989), 'Schema congruity as a basis for product evaluation', *Journal of Consumer Research*, **16**, 39–54.

MEYERS-LEVY J & TYBOUT A (1997), 'Context effects at encoding and judgment in consumption settings: the role of cognitive resource', *Journal of Consumer Research*, **24**, 1–14.

MILLER LC, MURPHEY R & BUSS AH (1981), 'Consciousness of body: private and public', *Journal of Personality and Social Psychology*, **41**, 397–406.

MINAMIMOTO T & KIMURA M (2002), 'Participation of the thalamic CM-PF complex in attention learning', *Journal of Neurophysiology*, **87**, 3090–3101.

MINAMIMOTO T, HORI Y & KIMURA M (2005), 'Complementary process to response bias in the centromedian nucleus of the thalamus', *Science*, **308**, 1798–1801.

MORWITZ VG, JOHNSON E & SCHMITTLEIN D (1993), 'Does measuring intent change behavior?', *Journal of Consumer Research*, **20**, 46–61.

MOSKOWITZ HR (1977), 'Magnitude estimation: notes on what, how, when and why to use it' *Journal of Food Quality*, **1**, 195–227.

O'DOHERTY J, KRINGELBACH ML, ROLLS ET, HORNAK J & ANDREWS C (2001a), 'Abstract reward and punishment representations in the human orbitofrontal cortex', *Nature Neuroscience*, **4**, 95–102.

O'DOHERTY J, ROLLS ET, FRANCIS S, MCGLONE F & BOWTELL R (2001b), 'The representation of pleasant and aversive taste in the human brain', *Journal of Neurophysiology*, **85**, 1315–1321.

O'DOHERTY JP, DEICHMANN R, CRITCHLEY HD & DOLAN RJ (2002), 'Neural responses during anticipation of a primary taste reward', *Neuron*, **33**, 815–826.

O'DOHERTY JP, CRITCHLEY H, DEICHMANN R & DOLAN RJ (2003), 'Dissociating valence of outcome from behavioral control in human orbital and ventral prefrontal cortices', *Journal of Neuroscience*, **23**, 7931–7939.

O'DOHERTY JP, BUCHANAN TW, SEYMOUR B & DOLAN RJ (2006), 'Predictive neural coding of reward preference involves dissociable response in human ventral midbrain and ventral striatum', *Neuron*, **49**, 157–166.

OLIVER RL (1977a), 'Effect of expectation and disconfirmation on postexposure product evaluations: an alternative interpretation', *Journal of Applied Psychology*, **62**(4), 480–486.

OLIVER RL (1977b), 'A theoretical reinterpretation of expectation and disconfirmation effects on postexposure product evaluations: experience in the field', in Day RL (ed.), *Consumer Satisfaction, Dissatisfaction and Complaining Behavior*, Bloomington, IN, Indiana University, School of Business, Division of Research, 2–9.

OLIVER RL (1980), 'A cognitive model of the antecedents and consequences of satisfaction decisions', *Journal of Marketing Research*, **17**, 460–469.

OLIVER RL (1993), 'Cognitive, affective, and attribute bases of the satisfaction response', *Journal of Consumer Research*, **20**, 418–430.

OLIVER RL & DESARBO WS (1988), 'Response determinants in satisfaction judgments', *Journal of Consumer Research*, **14**, 495–507.

OLSHAVSKY RW & MILLER JA (1972), 'Consumer expectation, product performance, and perceived product quality', *Journal of Marketing Research*, **9**, 19–21.

OLSON JC & DOVER P (1976), 'Effects of expectation creation and disconfirmation on belief elements of cognitive structure', in Anderson BB (ed.), *Advances in Consumer Research,* Association for Consumer Research, Vol. 3, 168–175.

OLSON JC & DOVER P (1979), 'Disconfirmation of consumer expectations through product trial', *Journal of Applied Psychology*, **64**, 179–189.

ORNE MT (1962), 'On the social psychology of the psychological experiment: with particular reference to demand characteristics and their implication', *American Psychologist*, **17**, 776–783.

PARASURAMAN A, ZEITHAML VA & BERRY LL (1985), 'A conceptual model of service quality and its implications for future research', *Journal of Marketing*, **49**, 41–50.

PARDUCCI A (1965), 'Category judgment: a range–frequency model', *Psychological Review*, **72**, 407–418.

PARDUCCI A (1974), 'Contextual effects: a range-frequency analysis', in Carterette EC & Friedman MP (eds.), *Handbook of Perception, Psychophysical Judgment and Measurement*, New York, Academic, 127–141.

PELCHAT ML & PLINER P (1995), ' "Try it. You'll like it" Effects of information on willingness to try novel foods', *Appetite*, **24**, 153–165.

PERACCHIO LA & TYBOUT AM (1996), 'The moderating role of prior knowledge in schema-based product evaluation', *Journal of Consumer Research*, **23**, 177–192.

PERYAM DR & PILGRIM FJ (1957), 'Hedonic scale method of measuring food preferences', *Food Technology*, **11**, 9–14.

PETTY RE & CACIOPPO JT (1981), *Attitudes and Persuasion: Classic and Contemporary Approaches*, Dubuque, IA, William C. Brown.

PETTY RE & CACIOPPO JT (1986), *Communication and Persuasion. Central and Peripheral Routes to Attitude Change*, New York, Springer-Verlag.

PLOGHAUS A, TRACEY I, GATI J, CLARE S, MENON RS, MATTHEWS PM & RAWLINS JNP (1999), 'Dissociating pain from its anticipation in the human brain', *Science*, **284**, 1979–1981.

RAGHUNATHAN R & IRWIN JR (2001), 'Walking the hedonic product treadmill: default contrast and mood-based assimilation in judgments of predicted happiness with a target product', *Journal of Consumer Research*, **28**, 355–368.

RANKIN KM & MARKS LE (1991), 'Differential context effects in taste perception', *Chemical Senses*, **16**, 617–629.

REGAN DT & FAZIO RH (1977), 'On the consistency between attitudes and behavior: look to the method of attitude formation', *Journal of Experimental Social Psychology*, **13**, 28–45.

RISKEY DR, PARDUCCI A & BEAUCHAMP GK (1979), 'Effects of context in judgments of sweetness and pleasantness', *Perception & Psychophysics*, **26**(3), 171–176.

RIZZOLATTI G, FADIGA L, GALLESE V & FOGASSI L (1996), Premotor cortex and the recognition of motor actions', *Cognitive Brain Research*, **3**, 131–141.

RIZZOLATTI G, FOGASSI L & GALLESE V (2001), 'Neurophysiological mechanisms underlying the understanding and imitation of action', *Nature Neuroscience Reviews*, **2**, 661–670.

ROLLS ET (2000), 'The orbitofrontal cortex and reward', *Cerebral Cortex*, **10**, 284–294.

ROSENTHAL R & JACOBSON L (1968), *Pygmalion in the Classroom: Teacher Expectation and Pupils' Intellectual Development*, New York, Rinehart and Winston.

ROSS M & OLSON JM (1981), 'An expectancy–attribution model of the effects of placebos', *Psychological Review*, **88**, 408–437.

ROTTER JB (1955), 'The role of the psychological situation in determining the direction of human behavior', in Jones MR (ed.), *Nebraska Symposium on Motivation, Lincoln*, University of Nebraska Press, 245–269.

ROUSSET S & JOLIVET P (2002), 'Discrepancy between the expected and actual acceptability of meat products, eggs and fish: the case of older consumers', *Journal of Sensory Studies*, **17**, 61–67.

ROZIN P & TUORILA H (1993), 'Simultaneous and temporal contextual influences on food acceptance', *Food Quality and Preference*, **4**, 11–20.

SAMEJIMA K, UEDA Y, DOYA K & KIMURA M (2005), 'Representation of action-specific reward values in the striatum', *Science*, **310**, 1337–1340.

SARINOPOULOS I, DIXON GE, SHORT SJ, DAVIDSON RJ & NITSCHKE JB (2006), 'Brain mechanisms of expectation associated with insula and amygdala response to aversive taste: implications for placebo', *Brain, Behavior, and Immunity*, **20**, 120–132.

SAWYER AG (1975), 'Demand artifacts in laboratory experiments in consumer research', *Journal of Consumer Research*, **1**, 20–30.

SCHARF A & VOLKMER HP (2000), 'The impact of olfactory product expectations on the olfactory product experience', *Food Quality and Preference*, **11**, 497–504.

SCHIFFERSTEIN HNJ (1994), 'Contextual effects in the perception of quinine HCI/NaCI mixtures', *Chemical Senses*, **19**, 113–123.

SCHIFFERSTEIN HNJ (1995), 'Contextual shifts in hedonic judgments', *Journal of Sensory Studies*, **10**, 381–392.

SCHIFFERSTEIN HNJ (1997), 'The role of expectancy disconfirmation in food acceptability', in *Proceedings of the 26th EMAC Conference*, Warwick, UK, 20–23 May, 2019–2025.

SCHIFFERSTEIN HNJ (2001), 'Effects of product beliefs on product perception and liking', in Frewer L, Risvik E & Schifferstein H (eds.), *Food, People and Society, A European Perspective of Consumers' Food Choices*, London, Springer Verlag.

SCHIFFERSTEIN HNJ & FRIJTERS JER (1992), 'Contextual and sequential effects on judgments of sweetness intensity', *Perception & Psychophysics*, **52**(3), 243–255.

SCHIFFERSTEIN HNJ & OUDEJANS IM (1996), 'Determinants of cumulative successive contrast in saltiness intensity judgments', *Perception & Psychophysics*, **58**(5), 713–724.

SCHIFFERSTEIN HNJ, KOLE APW AND MOJET J (1999), 'Asymmetry in the disconfirmation of expectations for natural yogurt', *Appetite*, **32**, 307–329.

SCHUTZ HG & CARDELLO AV (2001), 'A labeled affective magnitude (LAM) scale for assessing food liking/disliking', *Journal of Sensory Studies*, **16**, 117–159.

SHAPIRO AK & MORRIS LA (1978), 'The placebo effect in medical and psychological therapies', in Garfield SL & Bergin AR (eds.), *Handbook of Psychotherapy and Behavior Change: an Empirical Analysis*, New York, Wiley.

SHERIF M & HOVLAND CI (1961), *Social Judgment: Assimilation and Contrast Effects in Communication and Attitude Change*, New Haven, Yale University Press.

SIRET F & ISSANCHOU S (2000), 'Traditional process: influence on sensory properties and on consumers' expectation and liking. Application to paté de campagne', *Food Quality and Preference*, **11**, 217–228.

SMITH RE (1993), 'Integrating information from advertising and trial: processes and effects on consumer response to product information', *Journal of Marketing Research*, **30**, 204–219.

SMITH RE & SWINYARD WR (1982), 'Information response models: an integrated approach', *Journal of Marketing Research*, **46**, 81–93.

SMITH RE & SWINYARD WR (1983), 'Attitude–behavior consistency: the impact of product trial versus advertising', *Journal of Marketing Research*, **20**, 257–267.

SMITH RE & SWINYARD WR (1988), 'Cognitive response to advertising and trial: belief strength, belief confidence and product curiosity', *Journal of Marketing Research*, **17**, 3–14.

SNYDER M & DEBONO KG (1985), 'Appeals to image and claims about quality: understanding the psychology of advertising', *Journal of Personality and Social Psychology*, **49**(3), 586–597.

STAYMAN DM, ALDEN DL & SMITH KH (1992), 'Some effects of schematic processing on consumer expectations and disconfirmation judgments', *Journal of Consumer Research*, **19**, 240–255.

STEENKAMP JBEM (1989), *Product Quality: An Investigation into the Concept and How it is Perceived by Consumers*, Assen NL, Van Gorcum.

STEFANI G, ROMANO D & CAVICCHI A (2006), 'Consumer expectations, liking and willingness to pay for specialty foods: do sensory characteristics tell the whole story?', *Food Quality and Preference*, **17**, 53–62.

STEVENS DA (1991), 'Individual differences in taste and smell', in Lawless HT & Klein BP (eds.), *Sensory Science and Applications in Foods*, New York, Marcel Dekker, 259–316.

SWAN JE & TRAWICK IF (1981), 'Disconfirmation of expectations and satisfaction with a retail service', *Journal of Retailing*, **57**(3), 49–67.

TOLMAN EC (1938), 'The determiners of behaviors at a choice point', *Psychology Review*, **45**, 1–45.

TOLMAN EC (1951), *Behavior and Psychological Man*, Berkeley, University of California Press.

TRAWICK IF & SWAN JE (1980), 'Inferred and perceived disconfirmation in consumer satisfaction', in Bagozzi RP (ed.), *Marketing in the 80's*, Chicago, American Marketing Association, 97–100.

TUORILA H & CARDELLO AV (2002), 'Consumer responses to an off-flavor in juice in the presence of specific health claims', *Food Quality and Preference*, **13**, 561–569.

TUORILA H, CARDELLO AV & LESHER LL (1994a), 'Antecedents and consequences of expectations related to fat-free and regular-fat foods', *Appetite*, **23**, 247–263.

TUORILA H, MEISELMAN HL, BELL R, CARDELLO AV & JOHNSON W (1994b), 'Role of sensory and cognitive information in the enhancement of certainty and liking for novel and familiar foods', *Appetite*, **23**, 231–246.

TUORILA H, MEISELMAN HL, CARDELLO AV & LESHER LL (1998a), 'Effect of expectation and the definition of product category on the acceptance of unfamiliar foods', *Food Quality and Preference*, **9**, 421–430.

TUORILA H, ANDERSSON A, MARTIKAINEN A & SOLOVAARA H (1998b), 'Effect of product formula, information and consumer characteristics on the acceptance of a new snack food', *Food Quality and Preference*, **9**, 313–320.

VAN KLEEF E, VAN TRIJP, HCM & LUNING P (2005), 'Consumer research in the early stages of new product development: a critical review of methods and techniques', *Food Quality and Preference*, **16**, 181–201.

WAGER TD, RILLING JK, SMITH EE, SOKOLIK A, CASEY KL & DAVIDSON RJ (2004), 'Placebo-induced changes in MRI in the anticipation and experience of pain', *Science*, **303**, 1162–1167.

WANSINK B, VAN ITTERSUM K & PAINTER JE (2005), 'How descriptive food names bias sensory perceptions in restaurants', *Food Quality and Preference*, **16**, 393–400.

WICKER B, KEYSERS C, PLAILLY J, ROYET J-P, GALLESE V & ROZZOLATTI G (2003), 'Both of us disgusted in my insula: the common neural basis of seeing and feeling disgust', *Neuron*, **40**, 655–664.

WIGGINS MS & BRUSTAD RJ (1996), 'Perception of anxiety and expectations of performance', *Perceptual and Motor Skills*, **83**, 1071–1074.

WOODRUFF RB, CADOTTE ER & JENKINS RL (1983), 'Modeling consumer satisfaction processes using experience-based norms', *Journal of Marketing Research*, **20**, 296–304.

ZANNA MP & COOPER J (1974), 'Dissonance and the pill: an attribution approach to studying the arousal properties of dissonance', *Journal of Personality and Social Psychology*, **29**, 703–709.

ZELLNER DA, STEWART, WF, ROZIN P & BROWN JM (1988), 'Effect of temperature and expectations on liking for beverage', *Physiology and Behavior*, **44**, 61–68.

11

Boredom and the reasons why some new food products fail

**E. P. Köster and J. Mojet, Wageningen University,
The Netherlands**

11.1 Introduction

11.1.1 Flop analysis

The reasons why so many new food products fail on the market are manifold and range from product-related causes such as food boredom to bad marketing concepts and inadequate consumer research. Systematic flop analysis is an unpopular sport in the food industry, where success has many parents, but flops are orphans. Usually, much effort is put into trying to forget a flop. Nevertheless, flop analysis is perhaps the best and most profitable way to gain insight in product development processes and thus help avoid future failures (Köster, 2004). It starts with a simple analysis of the course of the product in the market to find out whether the problem lies in the intrinsic sensory properties of the product itself, in the marketing concept, in the communication about the product, in the extrinsic properties of the product such as packaging or convenience of use or in the inadequacy of the tests on which the launching decision was based. Any combination of these causes is also possible.

11.1.2 'Product boredom'

Here, special attention will be given to the analysis of problems with conceptually sound products that are initially well accepted, well advertised and convenient in use, but that nevertheless have a very short life cycle in the market. For such products the problem obviously resides in the intrinsic properties of the product and in the adequacy of the methods to test the long-term acceptance of the product in the market. Usually, the term 'product boredom' is used to describe the phenomenon of the short life cycle, but this may be quite misleading, because there are at least two possible causes for the phenomenon. Causes differ greatly in their origin and in the steps that one has to take to avoid or remedy them. These causes are:

1. True product boredom, which results in indifference towards the product (and will simply be called product boredom from here on).
2. Slowly rising aversion, which leads to a real dislike for the product.

The origins of these phenomena, the methods to test products in order to avoid these types of problems with them and the possible remedies for marketed products that suffer from them, will be discussed in the remainder of the chapter.

11.2   Product boredom

11.2.1   Product boredom and psychological theory

Some 80 years ago, Tolman (1925) published a paper on spontaneous alternating behaviour (SAB) in rats. He had noted that rats placed a second time in the bottom part of a T-shaped maze had a very strong tendency to visit the arm of the T-maze that they had not visited the first time. A long series of experiments followed in order to determine whether SAB was due to a need for change in the motor activity of the rats or whether perceptual processes were involved. Finally, Glanzer (1953) in a series of ingenious experiments showed that, under the influence of what he called 'stimulus satiation' or loss of interest in previous visual experience, the rats were searching for new perceptual stimulation. In other words, boredom made them look for new stimulation. Around the same time Hebb (1949) launched a theory that changed the outlook of psychologists on motivation. His central theme was that each individual strived towards an optimal level of activation. This idea deviated from the older motivational theories in as far as it made activation into a normal condition of the organism and did not consider activation only as a means of re-establishing a resting equilibrium and abating the unrest created by needs or by stimulation.

Berlyne (1955, 1960) combined the idea of an optimal level of activation with his own experiments on curiosity and exploratory behaviour to formulate his theory on arousal. He distinguished two types of exploration: specific and diversive exploration. Specific exploration finds its origin in an uncertainty about the characteristics of the stimulus, which Berlyne describes as an aversive condition. The exploratory responses, which consist of prolongation or intensification of the stimulation, are directed at obtaining access to additional information to be able to reduce the uncertainty and to relieve the perceptual curiosity. Stimulus novelty, change, surprisingness, incongruity, complexity, ambiguity and indistinctness are the stimulus properties that are the main determinants of specific exploration. Berlyne calls these properties 'collative' properties since they all depend on collation or comparison of information from different past and/or present stimulus elements by the subject. These collative properties show a large amount of overlap in that they all are related to subjective uncertainty and that they all entail conflict, i.e. they all simultaneously instigate incompatible responses. Uncertainty about the stimulus situation means that the subject needs to hold a number of mutually incompatible responses in readiness. Thus, in specific exploration everything is directed at resolving the uncertainty and conflict due to the stimulation. Once this is achieved, it results in ending the specific exploratory behaviour itself.

Diversive exploration on the other hand 'has the function of introducing stimulation from any source that is "interesting" or "entertaining". It is exemplified by the various activities through which human beings seek "amusement", "diversion", or "aesthetic experience"...' (Berlyne, 1965, p. 244). It arises when the external environment of the subject is inordinately dull or monotonous, which creates an aversive condition characterised as 'boredom'. Diversive exploratory behaviour can be reinforced by the stimulation and activation that results from it and relief of boredom may be an important factor contributing to its reward value.

From a practical point of view the distinction of these two types of exploratory behaviour is rather important. Specific exploration is not concerned with pleasure or aesthetic appreciation, but with resolving puzzling stimulation and reducing uncertainty. This is hardly a moment to test new products and to expect that one gets reliable results on the durable appreciation of them. Nevertheless, this is precisely what most market researchers try to do when they test first impressions. Instead of real preference and liking data, they obtain data that are mostly based on curiosity and the desire to learn more about the product because the uncertainties caused by its novelty, surprise and incongruity are still unresolved. The responses are disguised as 'preference and liking', because the subjects simply have no choice to answer other questions and feel obliged to give an answer. That in the case of new products such first impressions do not predict later appreciation (Köster et al., 2001, 2003) is not surprising, although specific exploration is not the only reason for the lack of predictive value as will be shown below.

Diversive exploration with its attention for the pleasurable and aesthetic aspects of the product and thus for the more durable appreciation of it, sets in only after specific exploration is completed and the uncertainties are resolved. Unfortunately, usually at that point most marketers have already gone home to analyse their unstable first impression data. Instead of collecting first impression data of several hundreds (or even thousands of subjects), it would probably be better to follow the development of preference and liking in a group of 60 to 80 consumers individually (i.e. not based on the averages in so-called in-home-use tests; see below).

To return to the theory after this practical excursion, it should be noted that Berlyne assumed that exploratory behaviour is usually accompanied by increased arousal. In the case of specific exploratory behaviour an aversively high state of arousal is caused by the uncertainty or conflict implicated in the collative properties of the stimulation and the exploratory behaviour is directed at reducing the arousal. In contrast, diversive exploratory behaviour is directed at increasing the arousal from an aversively low level characterised by boredom and monotony, to the organism's optimal level by seeking stimulation from other sources. Thus, stimuli have a certain 'arousal potential' and will lead to different effects, depending on their arousing properties relative to the optimal arousal level of the organism. This means that the relationship between pleasantness of stimuli (Fig. 11.1, *y*-axis) and the arousal provoked by them (Fig. 11.1, *x*-axis) is an inverted U-curve. (See the solid curve in Fig. 11.1.) Both low and high arousal provoking stimuli can be equally unpleasant, depending on their distance to the optimal complexity level of the organism. It should be noted, however, that the relationship may not be as strictly symmetrical as suggested in the schematic example in the figure.

The addition of Dember and Earl (1957) to the theory is the introduction of the idea that the organism's optimal arousal level – or optimal level of perceived complexity as they call it – is not stable, but can be increased by exposure to stimuli that are a little more complex than the optimally preferred one, whereas exposure to stimuli with lower than optimal complexity would just be perceived as boring, but would leave the optimum level of perceived complexity of the organism unaltered. Thus, exposure to such a slightly more complex than optimal stimulus (B in Fig. 11.1) – called a 'pacer' by Dember and Earl – would cause a shift in the appreciation of all stimuli (see dotted line in Fig. 11.1), causing the originally most liked stimulus to decrease in pleasantness for the organism.

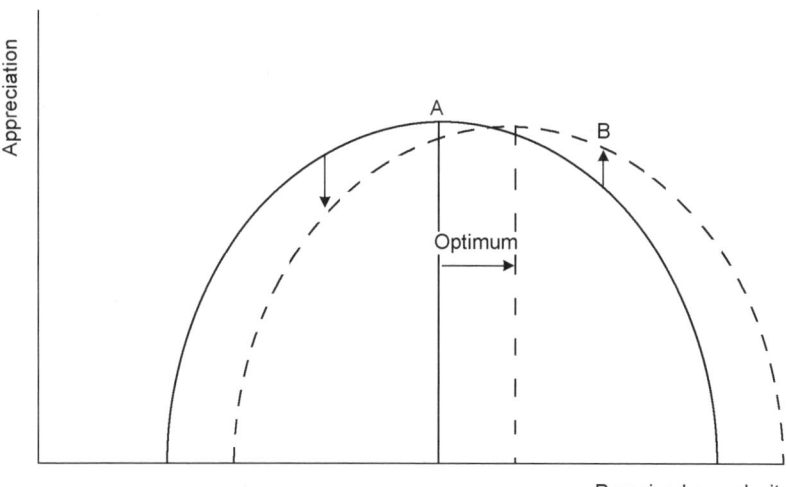

Fig. 11.1   The relationship between perceived complexity and appreciation according to Berlyne (solid line) and the shift of the curve (dotted line) and the optimum of perceived complexity (A) under the influence of the presentation of a 'pacer' (B) of a somewhat higher than optimal complexity according to Dember and Earl.

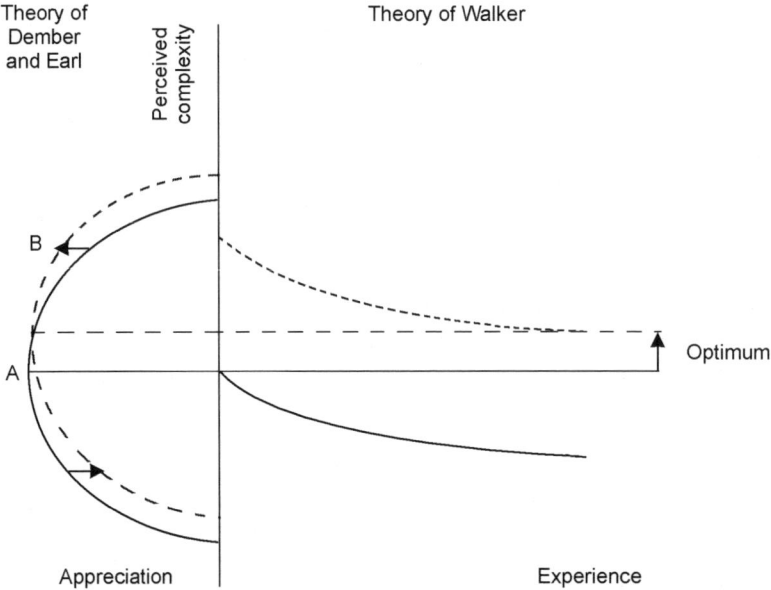

Fig. 11.2 The influence of repeated exposure to the stimulus on its perceived complexity and thus also on its appreciation according to Walker. Product boredom and mere exposure effects (Zajonc) can be explained as special cases of this theory. Their occurrence depends on the complexity of the presented stimulus relative to the optimal complexity of the subject.

This idea, which explains how our taste for more complex stimuli develops under the influence of exposure, has been successfully tested in animals for visual patterns (Dember and Earl, 1957; Dember, 1964) and in humans for different forms of art (see studies in Berlyne, 1974), for beverages (Lévy and Köster, 1999, Levy et al., 2002, 2006) and tentatively for food (Köster et al., 2003). Together with a theory formulated by Walker (1980), it provides also another strong argument against the current practices in consumer and market research that are based on choosing the product that is liked most by the consumers in a test based on first impressions. Figure 11.2 illustrates the effects of the combined actions of the theories of Berlyne, Dember and Earl, and Walker. Walker showed that as stimuli become more familiar they become less complex to the observer and that, as a consequence, the appreciation for them will change. The direction of this change will depend upon the original perceived complexity (arousal potential) of the stimulus relative to the optimal perceived complexity (arousal). For stimuli that are initially well above the optimal level (B in Fig. 11.2) and are probably too complex to be liked very much at first, the appreciation will increase with exposure, since their perceived complexity will decrease (dotted curve on the right) and will come closer to the shifted optimum (dotted straight line) that has moved in the direction of B under the influence of exposure. However, for stimuli that are originally at the level of optimal preference (A in Fig. 11.2) or below it, the appreciation will decrease with exposure, since their perceived complexity will also decrease (filled curve to the right) and will move away from the unchanged optimum (filled straight line). Thus, upon exposure, the originally most liked product of the subject (the one with the optimal complexity) will soon become too simple and lose appeal.

In the everyday practice of marketing research this means that we carefully select built-in losers when we select the product that is most liked at first sight. What we should look for is a product that, although it is a bit less liked (not too much, of course) than the most liked one, has growing potential because it is more complex than the optimally preferred one. In our view the practice of selecting on the basis of 'first impression market research' has had two deleterious effects. It produced many products that started boring people very soon and it thereby increased the volatility of the market with consumers looking constantly for other more arousing products (diversive exploration). One can only sympathise with the product developers in industry, who see that the more complex products – that they propose because they satisfy their more educated senses – are constantly rejected by the marketing people, who in their naivety select the ones that will bore us in no time. Market volatility is to a large extent the result and not the cause of failing market and consumer research.

When returning once again to the psychological theory, it should be pointed out that the combination of theories sketched in Fig. 11.2 not only explains product boredom (filled curve to the right, but also accommodates its opposite (dotted curve to the right), which is the well known and often cited 'mere exposure' theory of Zajonc (1968). This theory claims in simple terms that the more we are exposed to a stimulus, the more we will like it. Support for this theory in the food domain has come from Pliner (1982), who showed that exposure to novel stimuli resulted in the reduction of neophobia (fear for new foods) and increased liking. This might be expected also according to Berlyne, who would speak of gradual reduction of uncertainty and arousal by specific exploration. At the same time it is clear (Lévy and Köster, 1999) that exposure does not always lead to increased liking, but may also induce product boredom. This means that the theory of Zajonc can be seen as a special case of the more general theories of Walker, Berlyne, and Dember and Earl. Zajonc's theory describes the case for stimuli that have an initially higher than optimal perceived complexity (arousal) level (like B in Fig. 11.2). It does not work for stimuli that are initially at or below that optimal level. In the latter case, exposure leads to boredom and as has been argued here, this is more often than not the case for the products selected for the market by the current market and consumer research methods.

11.2.2   Methods to assess product boredom

The inadequacy of many of the current market and consumer research techniques that are based on first impressions and liking has been sufficiently discussed and it is time to talk about new methods for measuring and predicting product boredom and the development of long-term preference. In marketing and consumer research the so-called in-home-use test is sometimes proposed to solve the problem. If properly carried out and controlled, this test has the added advantage that the situations in which people consume the products are more natural. In predicting long-term product acceptance, such tests can be helpful as long as they obey the following rules:

1. The amount of product provided should be adjusted to the size of the normal consumption of the family and should permit at least seven repetitions of the normal consumption by the whole family.
2. The time allowed for the experiment should be sufficient to let people use the product at the normal frequency of consumption. When this is not possible because the normal frequency is too low (e.g. less than twice a month), it is better to use other methods (see below).
3. A diary should be provided in which the person responsible for providing the product in the family notes the quantities used and the quantities of other products of the same type bought and consumed.
4. Unexpected home visits should be made to ask some general questions (e.g. about the preparation of the product, ease of use of the packaging) and to verify in a gentle way that both the product and the diary are indeed used.
5. The use of questionnaires should be limited to the first and the last time the family consumes the food during the experiment. The first of the question-naires should just consist of a hedonic rating of the product and some factual questions about frequency of previous experience with the type of product and about the number of times per week, month or year this type of product is consumed. The second questionnaire should first ask a hedonic rating and then eventually a series of sensory questions to be answered on 'just right' scales followed by another hedonic rating and perhaps some questions about convenience if applicable.
6. One week after the experiment one can ask the participants their opinion about the product and the test and give them the possibility to be put on a list for eventual future participation in similar experiments. The response may be a reasonably good indication of any bad memories (not of the good ones). It should be calibrated against earlier experience with other successful and unsuccessful products in the same type of population, however.

Unfortunately, these criteria are seldom all met in the usual in-home-use tests that are offered by market research institutes or carried out by industrial mar-keting and consumer research groups. In most cases, the participants get insufficient quantities of product and the time allotted to them for carrying it out is insufficient to even outlive the specific exploratory phase. Furthermore, they are often confronted with very suggestive questionnaires that put them in an analytical attitude. Such an attitude is completely unnatural and makes them pay attention to all sorts of aspects that they normally never notice. The worst possible scenario is the 'group discussion in the family' that takes place when they are asked to come to a communal response. Nevertheless, in-home-use tests can be quite valuable in the prediction of long-term preference if the above six rules are respected. However, such tests are rather expensive and may be quite time consuming when products are involved that are not frequently consumed (see rule 2).

In order to reduce costs and not to lose time during product development, a number of tests were developed in The Netherlands and tested with 15 samples consisting of five products, which were each represented with a set of three 'sister' variations that had been launched together in the market simultaneously and with the same type of packaging and the same publicity. The products were delivered to the researchers without any information with regard to their market success. At the end of the experiments, the researchers stated predictions about the development of the market acceptability of the products based on the outcomes of their tests. The results showed that the methods employed were able to 'predict' (post hoc in the double blind procedure described) about 87% of the flops that were due to boredom or to slowly growing aversion, but that they could not predict the success of one of the products that, after an initially low acceptance, started to grow in the market. After their publication of the methods (Köster, 1990) these tests were regularly used in France and in Germany in confidential research and (judging by the reduced flop rates where they were introduced) seem to have contributed to the elimination of potentially unsuccessful product alternatives. The most frequently used forms of these methods are described here.

Boredom test (quick central location screening)
The boredom test is used to compare two or three versions of a new product among each other or with an existing product, usually the market leader. It is carried out with a sample of 80–120 users of the product or, if it is a totally new product, with 120–180 consumers from the target consumer group. As many subgroups of 40–60 subjects are formed as there are different stimuli that are to be tested against each other.

All subjects begin the test by rating all stimulus variations (in an order which is systematically varied over subjects) twice on a hedonic scale (pre-test). After this, each of the subgroups receives a different product variation in a monotonous series of 15 stimuli (small portions to avoid satiation), which they rate for liking with the following instruction: 'you will now receive a series of samples of very similar products. It may be that even when you cannot tell the difference you may nevertheless like some better than others. Please indicate for each of them how well you like it'. At the end of this series of 15 stimuli, each subject receives and rates all stimulus variations again twice (post-test) in the same order as the one in which they were presented to her/him during the pre-test.

Two types of results can be obtained from the test. In the first place one can compare the development of the liking over the 15 stimulus presentations in each of the subgroups who were exposed to different stimuli (see Fig 11.3). Do the average group ratings remain stable over the 15 presentations for the different stimuli? Are the changes (and the non-changes) in liking found for the group as a whole, based on consensus, or are there large individual and/or

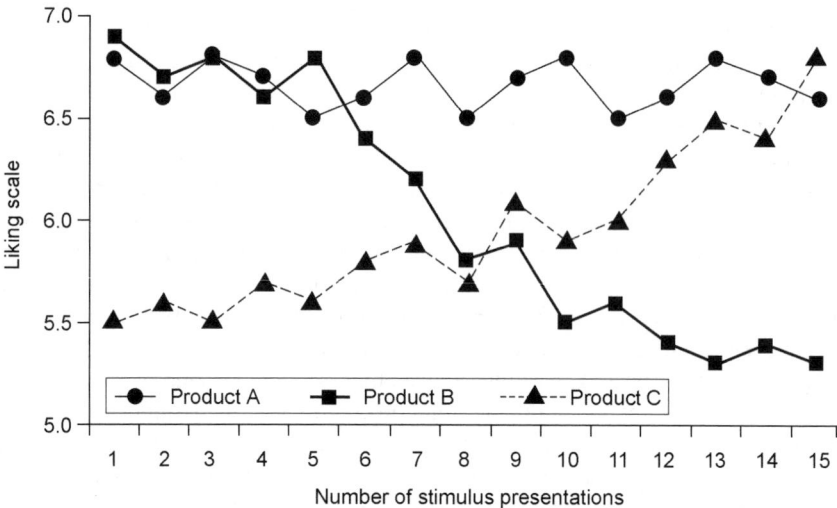

Fig. 11.3 Development of the liking expressed on a 9-point hedonic scale for a product over 15 presentations in three groups of subjects that were respectively exposed to product A, B or C.

subgroup differences (a possible indication of large differences in optimal complexity level)? Is boredom directly visible in a downward trend over the 15 presentations and do the tested products differ in this respect? Are there any signs of positive 'mere exposure' effects (an initially too complex stimulus becoming more and more liked)?

A second indication of boredom or 'mere exposure' is obtained by the comparison of the results for each of the stimulus varieties in the pre- and post-test and the extent to which this is dependent on the stimulus they have been exposed to during the monotonous series (see Fig 11.4). Do all or some of the total group ratings in the post-test go up or down relative to the pre-test? How has the relative position of the monotonously presented stimulus shifted in relation to the other stimuli and how is this compared to the subgroups that were exposed to the other stimuli (or to a non-exposed control group if one has been included)? Often the pre-post shifts of the other stimuli that take place after monotonous exposure to a stimulus tell more about the effects than the shifts in pleasantness of the monotonous stimulus itself. When these shifts are negative they can usually be interpreted as the effects of shifts in the optimal level of complexity as predicted by Dember and Earl (1957). When they are positive, they can be interpreted as diversive exploration behaviour resulting from contrast to the boredom produced by the repeatedly presented stimulus.

Extended boredom test (combined in-home-use and central location testing)
The extended boredom test was originally developed for products that one cannot present in rapid succession to subjects (e.g. cosmetic creams) or that pose problems or risks when presented in relatively large quantities in central

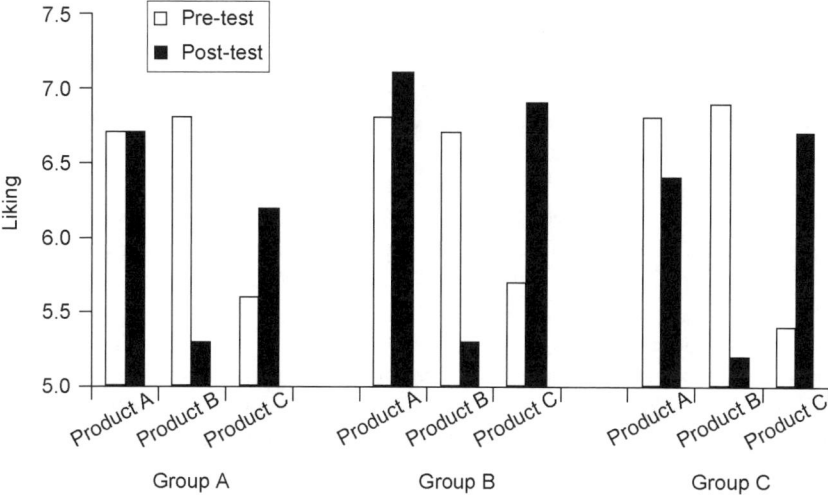

Fig. 11.4   Comparison of pre-test and post-test liking on a 9-point hedonic scale for all three products in the groups that were exposed to respectively products A, B or C.

locations to people who have to travel afterwards (e.g. alcoholic beverages). It combines the advantage of the more natural home-use with a well-controlled exposure and measurement of the effects and because of this it is now more often used than the simpler central location test even for products that can easily be used in the latter case.

As in the boredom test, as many groups of 40–60 users of the type of product as there are different samples to be tested are invited to a central location. There they perform the pre-test in the same way as described above, but instead of the presentation of the monotonous series, they receive a set of 14 carefully blinded and numbered samples with the instruction to consume or use them at home at the rate of one per day in the order in which they are numbered, starting that same day. They are also asked to rate their liking for each of 'these slightly different samples' (which are in fact identical) and then to come back after 2 weeks for another central location test (the post-test). Quite often this test is repeated once or twice more with the same people, but with randomised exposure to another product and a new post-test. Thus, one can compare exposure effects based on within-subject data and also verify whether possible shifts in optimal level persist when the same subjects are later confronted with exposure to one or more of the other stimuli. Finding the response to such a detailed question may demand a design with more subgroups, though.

The extended boredom test is of course more time consuming and more expensive, but it has the definite advantage that the exposure is not massed as in the central location test, but is more normal and spread out. Thus, risk of interference by satiation can be excluded. Although mostly used in confidential research, some of the results obtained for a more scientific goal have been published (Köster *et al.*, 2001). An experiment was carried out to verify the effects of differences in upbringing on the malleability of food preferences for different types of products in the two parts of Germany after the reunification. At that time, there were two societies that had been separated for about 40 years, and in which whole generations had grown up under very different circumstances with regard to the availability and the quality of food. About a year after the reunification, both in Potsdam (former East Germany) and in Munich, 100 housewives with families consisting of at least three persons (including children) were selected. They took part in an experiment consisting of three central location tests, separated by two 14 day in-home-use periods. Eight types of product were selected for which the original East German version still existed at the time of the experiment and could be compared with the leading brand in West Germany and when possible with another comparable European product that was not available in Germany. The product types were yoghurt (two products only), vanilla wafers, honey (two products only), sausage (two products only), gherkins, blue cheese, ketchup and apple juice. During the first central location test, the subjects rated the pleasantness of all 21 products (three products with two versions and five products with three versions). For the first in-home-use period they took all 21 products home. They were asked to acquaint themselves and their families with the products. For the second in-home-use period, they were divided into three groups of about 33 persons in both cities and they received only one of the versions of each product (the East German, West German or European product) to take home in sufficient quantity to last their family for the whole period. The composition of the groups was done on the basis of their results in the first central location session and arranged in such a way that the average appreciation of each product was as well balanced over the three subgroups as possible.

As regards the results that are of importance here, the comparison of the appreciation measured in the first and third session showed that the first impressions predicted little of the liking of the new products (European and from the other part of Germany) in the third session and that the stability of the preference for old products (East products in Potsdam and West products in Munich) differed considerably between product types, but was in general stronger in the East German than in the West German groups. Thus, there was less change of preference under the influence of exposure for products that had been scarce in the East (e.g. sausage) or for products that parents had traditionally restricted the use of by their children (e.g. ketchup), whereas products that parents had always encouraged their children to eat (e.g. yoghurt) were more easily exchanged for others in both parts of the country. This shows that there are factors other than arousal and complexity alone that influence or at least limit the development of preference. This point has also been discussed by Lévy et al. (2002, 2006) who clearly show the validity of Dember and Earl's theory for explaining the effect on liking of exposure to beverages of different perceived complexity after exclusion of other factors like the one mentioned here. It should be clear from this that boredom tests can only effectively be used to compare versions of the same type of product.

11.2.3   Remedies for products that suffer from boredom

What can one do with a boring product? Obviously one should raise its 'arousal potential'. This can be accomplished in different ways according to Berlyne (1967), who defined the arousal potential of a stimulus as the set of properties that determine the individual's level of arousal at a given moment. These properties fall into three classes: psychophysical properties (e.g. stimulus intensity, stimulus quality), ecological properties (e.g. internal changes that accompany hunger, thirst, sexual appetite, etc.) and 'collative' properties that influence the arousal level via the attention process (e.g. novelty, perceived complexity). Since the ecological properties are states of the consumer that cannot be directly manipulated by the producer of the product only the psychophysical and the collative properties remain. Furthermore the psychophysical properties that can be varied are limited almost exclusively to intensity for an existing product that retains its characteristic quality (e.g. orange juice, lasagne). Raising the intensity of the stimulation without changing its complexity, as it is sometimes tried in so-called 'tasty' products, is not an effective way to reduce boredom and may even enhance it or lead to aversion, once the shock of the first encounter is over. Thus, one should resort to perceived complexity as the collative property that remains the most important factor after the arousal raised by novelty has gone. This means that one has to add some other well-known and well-liked notes to the product, but in such low quantities that they are not perceptible in themselves and do not change the basic character of the product to which they are added. They should just make the product more intriguing, through unconscious and unresolved uncertainty. This should be tested. When presented monadically, these changes should not be noticed consciously. The changed product should still be seen as prototypical for its category.

However, when compared directly with the old one, the new product should seem more complex, i.e. seem to be more elaborated and surprising and to contain more different notes, although one cannot tell any of these notes by name (for examples see Lévy and Köster, 1999; Lévy et al., 2002, 2006). Whether such a new version of the product will be more resistant against boredom than the old one, may be checked with the (extended) boredom test.

11.3   Slowly rising aversion

11.3.1   Slowly rising aversion and psychological theory

Sometimes products that are well liked in the beginning do not become boring and uninteresting after a while, but lead to downright aversion. Something slightly puzzling or annoying in them, which seemed unimportant compared with the positive aspects at the time of the first contact, starts growing in importance and finally may render the product unbearable. In psychology this is described as a typical approach–avoidance conflict and problems like this have been studied extensively since the middle of the last century (Dollard et al., 1939; Lewin, 1951; Miller, 1959). There is general consensus over the fact that avoidance based on negative feelings grows fast over time and that approach based on positive feelings usually diminishes over time as a result of boredom or, when it increases as in the case of the mere exposure effect, grows much more slowly than the avoidance. Thus, in products as described here, there will come a moment at which the avoidance becomes stronger than the approach and aversion sets in. The speed with which this point will be reached depends both on the rate of growth of the negative feelings for the aspect that causes the avoidance and on perceived complexity of the stimulus relative to the optimal complexity, since this determines the occurrence of boredom or mere exposure effects. The presence of a negative aspect is more absolute as a cause of the phenomenon, because if there is no puzzling or slightly irritating aspect in the product, there is no approach–avoidance conflict and although the product may still become boring, it will not become aversive.

11.3.2   Methods based on the detection of slight dislike to avoid the occurrence of slowly rising aversion

If it is true that in general small irritations become stronger over exposure and positive effects diminish, many marketing and consumer research methods will have to be adjusted. In product testing procedures far too much attention is directed at positive appreciation by the subjects and small dislikes for minor aspects of the product are too often neglected. Nevertheless, the latter ones are much more important in the prediction of the final market success or failure than the positive appreciation in a first impression measurement. Furthermore, averaging over populations without sufficient attention to minorities that dislike the product (even only slightly) is also very dangerous. If in a household of five, one child starts to dislike the product, the chances are very large that the product will be replaced by another one, even if the other family members like it very much.

So far, no completely satisfactory methods have been developed that specialise in the recognition of slight dislikes. At the end of their questionnaires market researchers often ask people what they do and don't like about the product, but since consumers are usually rather unaware of the true causes of their feelings (Köster and Mojet, 2006), and because the subjects have not eaten the product in sufficient quantity to be completely satiated with it, this seldom leads to relevant responses. The following three methods have been developed to get at least some indication of the potential danger of the presence of small annoyances in products.

Aversion test
This test is based on the idea that it is possible to speed up the formation of the dislike by (over-)eating large quantities of the product in a rather short time. Often, when one eats too much of a substance one develops a temporary aversion to it. This happens much more quickly for certain products or product variations than for others and such aversions may well be the magnified reflection of the slight dislikes of the subjects.

As the test was described earlier (Köster, 1990), it has been tried out on the same five groups of three products as the boredom test (see above) and its use permitted to indicate growing aversion in two of the cases that were a flop in the market. Of these two, one was later followed up and repaired by the industry.

In applying the method, as many groups of around 36 product users as there were different products to be tested are formed. The subjects in each of these groups are first exposed to a small quantity of one of the products, which they rate for pleasantness on a line scale (dislike very much; like very much) and familiarity ('how much this product resembles the one you know'; very different; identical). After this they receive a large portion (1.5 times the average ad libitum portion eaten in a preliminary experiment with other subjects) and are asked to eat all of it and then to rate it again for liking (the time it took them to eat all is measured). After a short (3 minute) interval they receive another large portion of the same size and are asked to rate their liking for this one also and to eat as much of it as they like (the amount left over is weighed afterwards). After this, they receive again a small portion like the first one, give their last hedonic rating, indicate again how much the test product resembles their usual product and indicate – if it is not judged identical – in what way they think the presented product differs from the usual one.

The different quantitative results of this method (difference post-test minus pre-test liking, time needed to consume the first large portion, weight of the leftovers of the second large portion) are entered in a formula that gives equal weight to them and are used as an indication of the relative tendency to develop aversion of the products that have been compared. The distance in familiarity between the tested product and the subject's usual product serves as an indication of the unexpectedness of the tested product and is sometimes used as a weighting factor in the importance of the subjects' qualitative remarks about the difference.

Authenticity test

The second test that can be used to detect minor irritating or unpleasant properties of products is the so-called 'authenticity' test, which was originally developed by Mojet and Köster (1986) and published four years later (Köster, 1990; also Köster, 2003). The method is based on the idea that subtle affective reactions to products are often overshadowed and lost when observers are brought into an analytical attitude by asking them what they like about a product or even just how much they like it. Instead, the method tries to evoke and enhance purely affective reactions. It works particularly well with types of products to which people are strongly attached.

Groups of about 60–80 regular users of the type or brand of product under study are invited for a central location test. When they arrive they are told an upsetting cover story about ways in which the producers want to change their product in order to make more profit for the producing firm (e.g. new raw materials, faster production methods). They are told that this would of course be to the firm's benefit, since it can produce cheaper but still sell at the same price, but that the firm is not yet sure whether the original taste is still maintained in the new product and that they are afraid of losing customers if it is not. For this reason the subjects are asked to quickly smell and taste about 20 samples and to indicate which of these they believe to be the original, authentic and traditionally made products and which are the ones they suspect of being the products made with the new cheap method. In the series of 20 (or 21) samples, both the new product (or two new products) and the original product are each given ten (seven) times in a random sequence. After judging all stimuli, the subjects are finally asked to give an indication of their overall liking for the stimuli they had judged to be authentic traditional ones and for the new stimuli in order to verify that they indeed valued 'their' authentic product more than the cheap 'renegade'. In combination with this latter global response one can decide what is the meaning of the possible discrimination (measured with signal detection type measures) between the stimuli for each subject. As in the other tests described here, special attention should be given to bimodality in the data and to subgroups that are especially sensitive to minor negative feelings evoked by the stimuli. The authenticity test has been applied especially in areas where people are much attached to their products (beer, cigarettes, milk). In a number of cases it has been shown to be more sensitive in detecting differences than a trained panel (Wolf-Frandsen et al., 2003, 2006). Thus, it is also a good way to find the affective effects of small differences due to a change of base material or production techniques. It can be used in cases where triangular tests show no difference between products, but the consumers in the market nevertheless develop a preference for the one over the other (see Lévy and Köster, 1999).

The frequency of unacceptability method
Although this method has been developed for the assessment of acceptability in products that are not homogeneous and in which one may find occasionally unpleasant ones – although on average the quality is sufficient or even high (e.g. walnuts, assortments of chocolates) – it can perhaps also be used to determine the danger of slowly rising aversion at an early stage.

The method is very simple. Instead of asking people to rate the quality of a product on a hedonic rating scale, one gives 20–30 small samples of the type of product involved to each of about 50 consumers and asks them to try the samples and to indicate the ones they would rather not have eaten. The frequency of rejection serves as the criterion of acceptability of the product rather than the average appreciation. Thus, it was decided once that in the case of an assortment of chocolates, no single chocolate should be rejected by more than 30% of the population and that in the case of walnuts no more than 3% should be rejected (a rancid walnut loses a customer for a long period).

Perhaps it would be possible to use the same method also with for instance 30 samples of a homogeneous product and to predict long-term liking or slowly rising aversion by the development of the rejection rate over blocks of ten stimuli. As far as the present authors know, this has never been tried, but it might be a way to find small but growing irritations with the product.

Some rather similar methods have been used to measure features such as 'drinkability' of beers (Köster, 1981) or 'eatability' of snacks (Issanchou *et al.*, 1987) in quite extensive and time-consuming experiments.

11.3.3   Remedies for products with slowly rising aversion

Finding the cause of a slowly rising aversion can be a painstaking task involving much experimentation with descriptive panels interacting directly with the experts that develop the product and with a panel of consumers that are selected on the basis of their sensitivity to the aversion. In order to find out in what sensory characteristics the rejected product differs from the ones that remain well accepted, a well-trained panel must describe the many different sensory aspects in great detail. The molecule(s) that is (are) responsible for the aversion should be detected and then eliminated without ruining the whole product. The help of the experts and product developers is necessary to provide hypotheses about the aspects that might cause the aversion and to reformulate the product. New formulations should be tested with the selected consumer panel in order to verify that the true cause of the aversion has indeed been taken away. All of this is a long and cumbersome process, but sometimes it can be solved quite easily by understanding the psychological cause of the aversion.

An example of such a case was a cola drink that was strongly rejected by children. When tested in the descriptive panel, one of the points on which it differed from other colas was a slight and barely detectable hint of apple odour (sliced Golden Delicious to be precise). Since children in the Netherlands are usually restricted by their parents in the amount of cola they can drink, they develop a real craving for it, whereas, since they are encouraged to drink apple juice instead, they do not like that very much. Thus, when a cola has even only a faint odour of apple juice, it is immediately seen as false and rejected. When the responsible molecule had been found and could be eliminated from it, the cola became very well accepted by children. Unfortunately in some products it is not so easy to understand why people develop a growing aversion.

11.4   Future trends

Methods for the prediction of long-term appreciation of the products in groups of subjects that are regularly exposed to them should be further developed and be given a permanent place in the strategy leading to the launching decision. It should be realised that although first impressions are important in the first acceptance of products, they do not predict long-term preference at all and that they even lead to the selection of products with a high potential for developing boredom. Research on minor negative feelings using affective tests such as the authenticity test should be stimulated in order to see whether they provide a way to obtain a better prediction of developing aversion. At the same time a critical attitude should be developed with regard to many current practices in both sensory and consumer research. This critical attitude should also include the rather indiscriminate use of statistical methods that are based on averaging and the implicit but often false assumption that people's response behaviour only differs in degree and not in essence. Looking at differences between people in their choice behaviour with regard to food or even looking at differences of such behaviour in the same person in different situations and interpreting these differences in terms of psychological motivation theory will be more fruitful than many of the current methods in sensory and consumer research that are based on averaged sensory and hedonic data or on the sometimes rather flimsy correlations that make up preference maps and that may be built on bimodal or otherwise non-normally distributed data.

11.5   Sources of further information and advice

The relevant literature on the theories of Hebb, Berlyne, Dember and Earl, Walker and Zajonc has been mentioned in Section 11.1.2 on product boredom and psychological theory. Although this literature forms the background of much that is treated in this chapter, it does not deal specifically with food or even with food-related sensory data. In the beginning, these theories have been mainly applied in visual and auditory aesthetics (painting and music; see Berlyne, 1974). The basic literature on approach – avoidance conflicts that are at the base of slowly rising aversion has been mentioned in Section 11.3.1. A number of the basic fallacies that tend to invalidate much research in the fields of sensory and consumer research are described in a paper on food choice (Köster, 2003). A broader view of the unconscious motivational processes that influence the development of preference and food choice over time is provided in a chapter by Köster and Mojet (2006). In that chapter attention is also given to the theories on motivation and emotion of Zajonc (1980), Damasio (1994, 1999, 2003), LeDoux (1998), Kahneman (2003) and Wilson (2002), which play an essential role in understanding the determinants of decision making and food choice.

11.6   References

BERLYNE D E (1955), 'The arousal and satiation of perceptual curiosity in the rat', *J Comp Physiol Psychol*, **48**, 238–246.
BERLYNE D E (1960), *Conflict, Arousal, and Curiosity*. New York, McGraw-Hill.
BERLYNE D E (1965), *Structure and Direction in Thinking*, New York, John Wiley & Sons.
BERLYNE D E (1967), 'Arousal and reinforcement', *Nebraska Symposium on Motivation*, Lincoln, University of Nebraska Press, 1–110.
BERLYNE D E (1974), 'The new experimental aesthetics', in Berlyne D E, *Studies in the New Experimental Aesthetics: Steps Towards an Objective Psychology of Aesthetic Appreciation*, New York, John Wiley & Sons.

DAMASIO A R (1994), *Descartes Error: Emotion, Reason and the Human Brain*, New York, Avon.

DAMASIO A R (1999), *The Feeling of What Happens*, San Diego, Harcourt Inc.

DAMASIO A (2003), *Looking for Spinoza*, London, Harcourt Inc.

DEMBER W N (1964), *The Psychology of Perception*, London, Holt, Rinehart and Winston Inc, ch. 10, p. 341.

DEMBER W N and EARL R W (1957), 'Analysis of exploratory, manipulatory and curiosity behavior', *Psychol Rev*, **64**(2), 91–96.

DOLLARD J, DOOB L W, MILLER N E, MOWRER O H and SEARS R R (1939), *Frustration and Aggression*, New Haven, Yale University Press.

GLANZER M (1953), 'Stimulus satiation: an explanation of spontaneous alternation and related phenomena', *Psychol Rev*, **45**, 1–41.

HEBB D O (1949), *The Organization of Behavior*, New York, Wiley.

ISSANCHOU S, KÖSTER E P and TEERLING A (1987), 'Caractéristiques de l'árôme et preférence' (Aroma characteristics and preference), *Sci Alim*, **7**, nr hors-série VIII, 53–58.

KAHNEMAN D (2003), 'A perspective on judgement and choice: Mapping bounded rationality', *Am Psychologist*, **58**(9), 697–720.

KÖSTER E P (1981), 'Time and frequency analysis: a new approach to the measurement of some less-well-known aspects of food preferences?', in Solms J and Halls R L (eds.), *Criteria of Food Acceptance*, Foster Verlag AG/Forster Publishing Ltd., Zürich, pp. 240–252.

KÖSTER E P (1990), 'Les épreuvers hédoniques', in Strigler F (ed.), *Évaluation Sensorielle: Manuel Méthodologique*, Paris, Lavoisier Tec & Doc.

KÖSTER E P (2003), 'The psychology of food choice: some often encountered fallacies', *Food Qual Pref*, **14**, 359–373.

KÖSTER E P (2004), 'Analysis of flops: helpful for future innovations', in Bomio, H (ed.), *International SAM Seminar*, Zürich, SAM.

KÖSTER E P and MOJET J (2006), 'Theories of food choice development', in Frewer L and Van Trijp H C M (eds.), *Understanding Consumers of Food Products*, Cambridge, Woodhead Publishing, 93–124.

KÖSTER E P, RUMMEL C, KORNELSON C and BENZ K H (2001), 'Stability and change in food liking: food preferences in the two Germany's after the reunification', in Rothe M (ed.), *Flavour 2000: Perception, Release, Evaluation, Formation, Acceptance, Nutrition and Health*, Bergholz-Rehbrücke, Germany, 237–254.

KÖSTER E P, COURONNE T, LÉON F, LÉVY C and MARCELINO A S (2003), 'Repeatability in hedonic sensory measurement: a conceptual exploration', *Food Qual Pref*, **14**, 165–176.

LEDOUX J (1998), *The Emotional Brain*, 3rd edn, London, Phoenix Orion Books Ltd.

LÉVY C M and KÖSTER E P (1999), 'The relevance of initial hedonic judgements in the prediction of subtle food choices', *Food Qual Pref*, **10**, 185–200.

LÉVY C M, MacRAE A and KÖSTER E P (2002), 'The role of "collative properties" in the development of preference during exposure to an orange drink', *Appetite*, **39**, 246.

LÉVY C M, MacRAE A W and KÖSTER E P (2006), 'Perceived stimulus complexity and food preference', *Acta Psychologica*, **123**, 394–413.

LEWIN K (1951), *Field Theory in Social Science*, New York, Harper.

MILLER N E (1959), 'Liberalization of basic S-R concepts: Extensions to conflict behavior, motivation, and social learning', in Koch S (ed.), *Psychology: A Study of Science, Vol. II*, New York, McGraw-Hill.

MOJET J and KÖSTER E P (1986), 'Onderzoek naar de waardering van drie laag-alcoholische bieren' (Investigation into the appreciation of three low-alcohol beers). *Confidential Report Psychological Laboratory*, Utrecht University.

PLINER P (1982), 'The effects of mere exposure on liking for edible substance', *Appetite*, **3**, 283–290.

TOLMAN E C (1925), 'Purpose and cognition: the determiners of animal learning', *Psychol Rev*, **32**, 285–297.

WALKER E L (1980), *Psychological Complexity and Preference: a Hedgehog Theory of Behavior*, Belmont, Wadsworth.

WILSON T D (2002), *Strangers to Ourselves: Discovering the Adaptive Unconscious*, Cambridge, MA, The Belknap Press of Harvard University Press.

WOLF-FRANDSEN L, DIJKSTERHUIS, G B, BROCKHOFF P B and MARTENS M (2003), 'Subtle differences in milk: comparison of an analytical and an affective test', *Food Qual Pref*, **14**, 515–526.

WOLF-FRANDSEN L, DIJKSTERHUIS G B, BROCKHOFF P B, HOLM NIELSEN J and MARTENS M (2007), 'Feelings as a basis for discrimination between slightly different types of milk', *Food Qual Pref*, **18**, 97–105.

ZAJONC R B (1968), 'Attitudinal effects of mere exposure', *J Personality Social Psychol*, Monograph supplement, **9** (part 2), 65–74.

ZAJONC R B (1980), 'Feeling and thinking: preferences need no inferences', *Am Psychol*, **35**, 151–175.

12

SensoEmotional optimisation of food products and brands

D. Thomson, mmr Research Worldwide Ltd, UK

12.1 Using sensory characteristics to build brands

12.1.1 Sensory characteristics and liking

Much of the pleasure we experience from consuming foods and beverages derives from our enjoyment of sensory characteristics. While this simple truth has probably been recognised anecdotally, since time immemorial, it is only recently that scientific research has provided us with the formal means of identifying which sensory characteristics drive liking in a particular product category, for whom and to what extent.

The result has been sophisticated statistical models that specify the relationships between the sensory characteristics of a product, as described and measured using trained sensory panels, and liking, as evaluated by consumers (Arditti, 1997). For reasons that will become apparent shortly, the author prefers to describe these as SensoHedonic models: where 'Senso' refers to the sensory characteristics of the products in question and 'Hedonic' (from the Greek *hedone*) refers to pleasure.

The breakthrough in SensoHedonic modelling came from the recognition that even within the narrowest product category (e.g. coffee, white bread, plain white rice, lager beers), people have very different sensory preferences. Trying to capture this diversity in liking within a single SensoHedonic model is not only counterintuitive but it usually leads to worthless models with very poor predictive validity

Identifying the various 'hedonic types' among target consumers calls for segmentation tools such as cluster analysis or developments of preference mapping (MacFie and Thomson, 1984). In the author's experience of over 1000 commercial studies worldwide, somewhere between three and eight sensory preference segments will typically emerge from a single product category, reflecting the huge underlying diversity in sensory preferences.

Separate SensoHedonic models are created for each segment, showing which sensory characteristics contribute positively and which contribute negatively to liking and to what extent. The relationship between liking, the dependent variable, and sensory characteristics, the explanatory variable, is modelled using tools such as partial least squares (Martens and Naes, 1989) or other proprietary methods.

In practice, product technologists use SensoHedonic models to help them decide which sensory characteristics to change and by how much, in order to achieve a specific sensory optimum. The product is then re-formulated to deliver this optimum sensory profile. A trained panel subsequently evaluates the revised product and the new sensory profile is plugged into the SensoHedonic model(s) from where a predicted mean liking rating is obtained. After several iterations of product development, sensory profiling and SensoHedonic modelling, the technologists should have created a near optimal product, ready for submission to final confirmatory product testing with target consumers.

SensoHedonic models, provided they are constructed competently, have brought a new level of sophistication and accuracy to product optimisation. However, while this has undoubtedly been of enormous practical value, the single-minded pursuit of this end has perhaps blinded some of us (the author included) to the other, surprisingly diverse, roles of sensory characteristics in the creation and maintenance of brands.

12.1.2   A means–end chain explanation for SensoHedonic associations

In order to appreciate the 'shortfall' in our thinking, and the size of the opportunity presented by this new-found realisation, it is appropriate to consider the SensoHedonic route to enjoyment, and hence pleasure, in terms of motivational theory. Means–end chain theory, as made popular by Jonathan Gutman in his seminal publication in 1982, provides a useful template (Gutman, 1982).

The primary assumption in means–end chain theory is that the attainment of desired end-states of existence (states of being) is a powerful driving force in human behaviour (Gutman, 1982). For example, a particular individual might characteristically seek feelings of happiness, pleasure, love, appreciation, being liked and being indomitable while wishing to avoid feelings of insecurity, inferiority and foolishness. According to means–end chain theory, seeking and avoiding these various end-states will be a dominant force in this individual's life and, consequently, in determining his/her behaviour. Conversely, this individual will be motivated to behave in ways that maximise the possibility of attaining his or her desired end-states of existence. This can be represented simplistically as:

$$behaviour \rightarrow consequence \rightarrow desired\ end\text{-}state$$

This desire to achieve various end-states of existence has been variously described by others as needs (Maslow, 1943), motivations (Kelly, 1955) and values (Gutman, 1982).

In the context of consumers' choices of brands, products and services, the means–end chain model needs to be adapted in several ways. In particular, it is probable that most of us will *not* have rationalised fully the desired end-states of existence that drive our behaviour. The simple reason is that self-psychoanalysis, or this level of self-awareness, eludes most of us.

So, does this mean that consumers mindlessly follow their urges and the process of product choice is non-cognitive and without any sort of rationale? Of course not! One way of embracing rational, cognitive choice within the means–end process is to envisage the individual's desired end-states of existence as being transformed into operational goals by the context prevailing at that moment in time. This notion is best illustrated by example.

Consider choosing something for a working lunch. The individual's operational goals may be speed, convenience, availability, to feel energised, to feel full, positive nutrition and enjoyment. These goals are subsequently translated into criteria such as, no preparation, eat while still working, low fat, high sugar, high protein and a sweet rather than savoury 'taste'. Since these are positive reasons for choice, the author describes them as 'drivers'. However, choice will also be constrained by various barriers such as cost and nutritional concerns, perhaps.

Consumers perceive brands and products as having attributes. The individual will choose from amongst the options available, the category of product or the brand with the attributes that s/he believes will address his/her criteria most comprehensively.

In this hypothetical example, let's imagine that of the options available only Kellogg's Nutrigrain bar (www.nutr-grain.com) or a can of Red Bull (www.redbull.co.uk) come anywhere close to fitting the bill, and each for different reasons. Let's suppose the individual chooses and consumes the Kellogg's Nutrigrain bar. S/he must then decide whether or not the product has addressed the criteria as expected and through this, whether or not goals have been met and desired end-states attained. The consumption experience will either confirm or refute the individual's expectations thereby reinforcing or undermining her beliefs via a classic feedback loop.

This does not imply that the process is always or entirely rational and cognitive but these revisions allow for the possibility of rational, cognitive intervention in some parts of the process, depending on the scenario, the individual and the familiarity of the individual with the scenario. The revised means–end chain model is represented as follows:

$$\text{product} \rightleftharpoons \text{criteria} \rightleftharpoons \text{goals} \rightleftharpoons \text{desired end-states}$$

In the context of the SensoHedonic association between sensory characteristics and pleasure, the criterion might be a sweet taste, the source of which is the sensory characteristics (attributes) of the product. This delivers the goal of momentary enjoyment and ultimately a pleasant state of being (desired end-state):

SensoHedonic association

product	→	**criterion**	→	**goal**	→ **desired end-states**
sensory		sweet taste		enjoyment	pleasure
characteristics					

Needless to say, enjoyment and hence pleasure could be realised via other sensory characteristics such as the look and feel of the product or the pack, its taste, its smell and even its sound.

12.1.3   SensoEmotional and SensoFunctional associations

One of the values of this revised means–end chain model is that it highlights two important issues in the context of consumers' choices of brands, products and services:

- *Single end-points via multiple routes.* The foregoing example shows a mechanism through which enjoyment and hence pleasure can be attained via various sensory experiences. However, it must also be recognised that enjoyment and pleasure can be attained via non-sensory routes such as, for example, the emotional experience of interacting with a favourite or much sought-after brand or the physiological experience obtained from an increase in blood sugar.
- *Multiple end-points via a single experience.* Although sweetness can trigger enjoyment and pleasure (SensoHedonic association), it can also trigger various other consequences leading to other desired end-states.

Perhaps the most obvious and best-known example is the association that exists between sweetness and positive emotions such as love, comfort, safety and kindness. In this case, the goal might be to experience reassurance and, through this, attain feelings of confidence. The author describes this as a SensoEmotional association:

SensoEmotional association

product	→	**criterion**	→	**goal**	→ **desired end-states**
sensory		sweet taste		reassurance	confidence
characteristics				(emotional)	

This is just one example of innumerable SensoEmotional associations. Think about perfumery and fragrance!

Sensory characteristics also have functional associations. Lemon fragrance, for example, sometimes cues the belief of augmented cleaning power in household products. Likewise, consuming sweet foods and beverages can sometimes make people feel physically energised (goal), which perhaps leads them to believe that they will perform better at work or at leisure (desired end-states). In keeping with the previous nomenclature, these are described as *SensoFunctional* associations:

SensoFunctional association

product	→	**criterion**	→	**goal**	→	**desired end-states**
sensory		sweet taste		energised		perform better at
characteristics				(physical)		work or at leisure

These straightforward, yet far from trivial, examples show how one sensory characteristic (sweetness) can help deliver various goals and through these, a variety of different end-states of existence (Fig. 12.1). The key point to take from this is that all sensory characteristics potentially have hedonic, emotional and functional associations. This has significant practical implications for the manner in which product optimisation and consumer research are conducted.

For most of the past 60 years or so, by default or design, it has been assumed that sensory facilitation of desired end-states (particularly pleasure) is mediated primarily through enjoyment (as shown above). In witness to this assertion is the fact that vast sums of money are routinely spent on the sensory optimisation of products, where increased enjoyment, measured operationally as increased liking, is the primary criterion through which success or failure in research is judged.

So far, it has been shown how sensory characteristics are able to confer liking, trigger emotional reactions and communicate functionality. However, sensory characteristics also have a fourth and hitherto largely unrecognised function: adding distinctiveness to the brand's persona by adding a unique sensory signature.

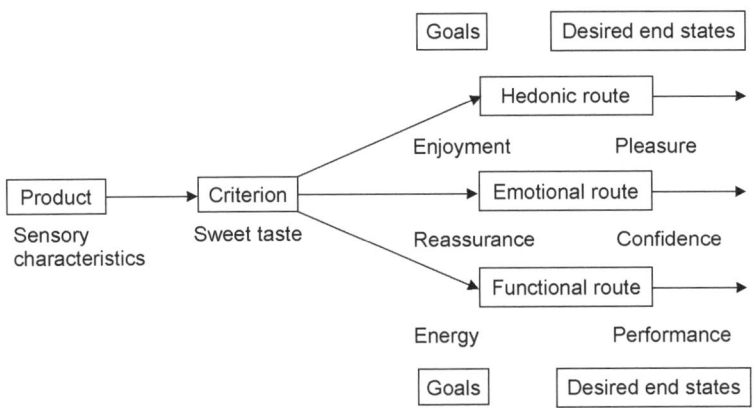

Fig. 12.1   The SensoHedonic, SensoEmotional and SensoFunctional routes to desired end-states of existence.

12.1.4   The role of sensory characteristics in brand identity

One of the key purposes of branding is to label a product or service and otherwise give it a recognisable and memorable identity that is liked and differentiates it from its competitors. Sensory characteristics play a curiously similar role in giving a product an identity, thereby making it recognisable and differentiating it from other products. In short, we recognise a product through its sensory characteristics! This, in turn, triggers a whole series of psychological and even physiological events that add to and otherwise augment the entire consumption process.

In view of the pivotal role played by sensory characteristics in giving a product its identity, it is curious that sensory characteristics have not been used more widely in branding. But they haven't: quite the contrary, in fact.

Visual imagery, the printed and spoken word and music are the primary media through which branding is normally communicated and of these two sensory channels, visual imagery, and hence the visual sense, dominates. There are a small number of brands such as Intel, Microsoft, Nokia and the BBC, for example, where the brand signature is primarily auditory, but otherwise sound is generally second to vision in the hierarchy of sensory channels used in brand communication.

The senses of touch, smell and taste currently play very minor roles in branding. This is in spite of the key role of sensory characteristics in giving a product its identity. Perhaps it is because we need to interact directly with the product, in order to touch it, smell it or taste it, which, you may think, is hardly conducive to mass communication. Or perhaps it is because of a natural reluctance among marketers to even consider sensory characteristics, because this is generally the province of product technologists and therefore less easy for them to influence and control.

Whatever the reason, it is a wasted opportunity according to Martin Lindstrom. In his book *Brand Sense*, Lindstrom (2005) describes various brands where communication has been extended beyond vision and sound to include one or more of the other senses. Examples cited by Lindstrom include the signature fragrance of Singapore Airlines and the characteristic feel of Bang and Olufsen's audio and video products. Although neither of these brands are perhaps obvious candidates for multi-sense branding, Lindstrom describes how building in an extra sensory dimension adds character and distinctiveness to the brands.

Food and drink brands, on the other hand, are very obvious candidates for sensory branding, yet the incorporation of touch, smell and taste into brand communication rarely goes much beyond the inclusion of simple sensory descriptions and claims ('Only the crumbliest, flakiest chocolate'; 'Now even cheesier'; 'More strawberry taste') or claims of sensory superiority ('Best Tasting'). Hardly very imaginative, not exactly compelling and a considerable lost opportunity!

In order to appreciate the potential role of sensory characteristics in branding, it is worth dwelling on the role of fragrance described by Lindstrom in branding Singapore Airlines. After all, transporting people around the world by air is a service rather than a product, so how can a service possibly have sensory characteristics? According to Lindstrom, Singapore Airlines has developed a distinctive fragrance that is introduced into various aspects of the travel experience; the lounge, the aircraft cabin, the refreshing towels, the fragrances worn by the cabin staff and the on-board toilets and toiletries. Needless to say, the nature of the fragrance and the degree of subtlety with which it is introduced, are both crucial.

This could be described as the 'fragrance signature' of the Singapore Airlines brand. The ubiquity of the fragrance, and the fact that it has come to be associated with an otherwise pleasing travel experience, means that this fragrance is now a key part of Singapore Airlines identity and hence its branding, in much the same way as the logo, the colours, the dress of the cabin staff, and so on. The fact that we are extremely sensitive to smells and they are extremely emotive means that fragrance can be a subtle yet powerful adjunct to the brand.

There are a few food and drink brands that have distinctive sensory signatures. Cadbury's Dairy Milk Chocolate is one such brand (www.cadbury.co.uk). All Cadbury's milk chocolate products, whether block chocolate, filled bars, boxed chocolates, Easter eggs, cakes, biscuits, ice cream, hot drinks or cold drinks, have the same unique Cadbury's chocolate taste. Those who know it, recognise it immediately. However, the sensory signature of Cadbury's chocolate is not just a defining characteristic of the product, it is also a defining characteristic of the brand. When people who enjoy Cadbury's Dairy Milk think of the brand, they immediately think of the 'taste' with all its positive aesthetic and emotional associations.

The Kettle Chips brand also has a distinctive sensory signature (www.kettlefoods.com). These potato chips/crisps are slightly more brittle, dense and tougher in texture, with a more translucent oily appearance than most brands of potato chips/crisps. In the crowded and relatively undifferentiated world of potato chips/crisps, the texture and appearance of Kettle Chips plays a very significant role in defining the brand.

Then, of course, there is Red Bull, where the distinctive flavour that characterises the product is also an integral part of the brand. There are several healthcare brands, of which Dettol (www.dettol.co.uk) and Savlon (www.savlon.co.uk) are prime examples, where the sensory signature associated with smell has become a defining characteristic of the brand as well as the product.

With some brands, the sensory signature will have emerged either by accident, by evolution or by careful nurture, over a long period of time. For other brands, it will have been designed into the brand architecture right from the outset. Either way, a distinctive sensory signature is an 'ownable' brand equity that really differentiates the brand, that is tricky for competitors to emulate and, above all, that plays a huge part in delivering the brand promise to consumers.

It is characteristic of this mode of brand augmentation, that the sensory signature is not always mentioned specifically in branding but reinforced more subtly, through brand experience (trial and repeat consumption). Whether or not this relatively understated approach makes the most of such a distinctive and valuable equity, is quite another matter! It is rather surprising, perhaps, that more brands do not strive to acquire a unique sensory signature.

12.1.5   Harnessing the four functions of sensory characteristics to make better brands

Thus far, it has been described how sensory characteristics have four main functions; labelling/identification, causing enjoyment, triggering emotional reactions and reinforcing functional beliefs. In reality, this means that the sensory characteristics of any product, by design or by default, have the potential to fulfil all four of these functions, whether or not we want them to.

There are two very important implications for marketers, product developers and product technologists. Firstly, we must always think about the sensory characteristics of a product in terms of these four functions. So, if we optimise sensory characteristics for liking, we need to concern ourselves with what it has done for the product in terms of its identity, its emotional message and its perceived functionality. These functions may be in conflict. There is no point in creating a great-tasting product and losing the product's identity. Equally, there is no point in creating a great tasting product that clashes with the emotional or functional messages associated with the brand. Whether or not it is a good idea to create a product that is emotionally optimal but hedonically compromised, for example, is something that we will come back to.

Secondly, in the past, product development and market research have focused almost exclusively on SensoHedonic facilitation of enjoyment and hence pleasure (i.e. increasing feelings of pleasantness by making things taste nice), probably because the other routes have gone largely unrecognised. On reflection, this will undoubtedly have caused a considerable degree of frustration among colleagues in marketing who may have wished, albeit unwittingly perhaps, to position their products and brands at desired end-states facilitated through SensoFunctional or SensoEmotional pathways, only to find that judgements about the potential of the product to succeed have (also unwittingly) been based on liking (or correlates of liking). In short, the wrong criteria and action standards may have been applied.

Nowhere is this more apparent than in the drinks market where the clarion call from marketing is for more 'masculine' liquids that are less sweet and otherwise more challenging. The notion of masculinity is an emotional construct, yet researchers would typically evaluate it using liking (a hedonic construct). Clearly, there's a conceptual mismatch here!

This raises two fundamental questions for market researchers. Would it be better to set liking thresholds at minimum acceptable levels rather than optimum levels and judge the success or otherwise of the liquids against constructs other than liking? If so, would it be possible to build SensoEmotional models (as opposed to SensoHedonic models mentioned earlier) to identify which sensory characteristics seem to confer masculinity, for example, on the liquids?

It is also worth considering that for any product in a particular category and market (white bread in the United Kingdom, for example), the sensory characteristics that drive liking are common across brands. Although SensoHedonic modelling would doubtless identify a number of different sensory optima, all the brands have essentially the same chance of ending up at the same place, because of the common set of sensory drivers. However, it is much less likely that the brands will share common emotional or functional equities, so factoring-in SensoEmotional or SensoFunctional optimisation creates the possibility of an optimum product with a more unique sensory signature.

Ideally, the sensory profile of a product would, of course, be optimised so that it gives the brand a unique identity (sensory signature) and takes consumers to a spectrum of desired end-states, simultaneously via all three sensory facilitated routes, without contradiction and without conflict. There are only a few examples of such products. Of these, Savlon antiseptic skin healing cream is a classic. Savlon is liked because, for most people, it has a very pleasing smell (SensoHedonic association). The smell also communicates Savlon's ability to reduce infection (SensoFunctional association) and it triggers very positive emotions, perhaps about our mothers' roles in caring for us when we were young (SensoEmotional association). It also gives the brand a unique identity.

As always, Red Bull makes an interesting case study! The brand owners are quite open about the fact that optimising the taste of the liquid was never their primary concern. Their main objectives were to give the product a unique sensory signature while communicating the functionality and emotionality of the brand. The cocktail of stimulants included in Red Bull means that it is highly functional and the strong emotional message that attaches to this functionality, and to the way that the brand is presented and marketed, means that it also carries a very powerful emotional message. For consumers, the very positive emotional and functional consequences of Red Bull are associated with the taste of the liquid and so, via a classical feedback loop, consumers come to like the taste. Very shrewd product development!

12.1.6   Bringing it all together

There are two major conclusions to draw from Section 12.1. Firstly, there is much more to sensory characteristics than merely conferring liking. Sensory characteristics also have the potential to communicate something of the emotionality and the functionality of the brand as well as adding distinctiveness to the brand's persona by adding a unique sensory signature. These are untapped resources that provide brand managers and product developers alike with significant opportunities.

Secondly, researchers (and their clients) need to be far more imaginative and daring in selecting criteria on which to judge product performance. Rather than relying on liking (or correlates of liking), it may be appropriate to set modest thresholds for liking and actually judge the product in terms of its ability to deliver the emotional and/or functional positioning of the brand. Section 12.2 focuses on the relationship between the sensory characteristics of the product and the emotional characteristics of the brand.

12.2   SensoEmotional optimisation in brand and product development

12.2.1   What is SensoEmotional optimisation?

When the sensory characteristics of a product cue an emotional reaction, this is described (by the author) as a SensoEmotional relationship, and these sensory characteristics are said to be emotionally active. That sensory characteristics can cue emotional reactions in consumers' minds is surely beyond contention! The very existence of the fragrance industry, and the fact that it has flourished for millennia, testify to the fact that the human senses provide a very direct conduit to the emotions. If further evidence is required, then simply check out the imagery in any perfumery, in any department store, or the advertising messages communicated by perfumers on screen or in print.

SensoEmotional relationships are far more powerful and far more widespread than most of us imagine. As previously mentioned, the SensoEmotional linkages between sweetness and feelings of safety, comfort and reassurance, are obvious and very well known, as is the relationship between the melt-in-the-mouth characteristics of chocolate and sensuality, for example. It is perhaps less obvious that a seemingly ordinary product such as dried pasta could be SensoEmotionally active. However, the sensory differences in shape and texture between tagliatelle and spaghetti, for example, undoubtedly cue measurably different emotional reactions, as does the colour difference between plain tagliatelle and green tagliatelle.

SensoEmotional relationships are all around us, and always have been, so what is new? The big change in recent times has been a more widespread recognition among marketers that emphasising SensoEmotional relationships can bring a powerful new dimension to branding. By featuring emotionally active sensory characteristics in brand architecture and brand communication, it is possible to cue emotions that are fundamental to the way they want consumers to perceive and interpret their brands. For example, richness, melt-in-the-mouth and smoothness are widely used to reinforce the notion of sensuousness, love, care and feelings of indulgence in chocolate branding.

Moreover, because the sensory characteristics of a product are obvious and accessible to consumers and, in some ways, much more enduring than words, music or graphical images, hitherto the mainstays of brand communication, SensoEmotional linkages seem to offer a particularly effective means of communicating emotional messages.

12.2.2   SensoEmotional optimisation strategies

Emotionally active sensory characteristics can be used to great effect to build brands in two, very powerful, yet hitherto under-utilised ways. These are the 'brand-first' and the 'product-first' SensoEmotional branding strategies.

The 'brand first' approach starts with the defining emotional characteristics of the brand and then determines which of these, if any, have SensoEmotional associations to one or more of the sensory characteristics of the product.

Assuming that 'hot' SensoEmotional associations can be established, the sensory characteristics of the product are then adjusted to deliver these emotions to maximum effect. Aligning the sensory characteristics of the product with the defining emotional characteristics of the brand creates synergy. This, in turn, elevates the 'branded totality' to a higher level so that consumers enjoy it more, and value it more.

Baileys Cream Liqueur (produced by Diageo) perhaps provides 'the' textbook example of the 'brand first' approach, where key emotional characteristics of the brand (sensuousness and deliciousness) are delivered in a very powerful way by emphasising the corresponding emotionally active (and optimal) sensory characteristics of the product (creamy rich texture and taste).

Baileys started life in the 1970s as an Irish Cream Liqueur with a brand message that communicated Irishness, along with the deliciousness and indulgence associated with cream, chocolate and whiskey. Thirty years later, the brand architecture seems to be built around sensuality delivered via the unique sensory characteristics of the Baileys liquid itself. Baileys is now one of the biggest alcoholic drinks brands in the world and is absolutely ubiquitous (www.baileys.com).

The notion of sensuality associated with the Baileys brand is, in itself, quite ethereal. However, by linking sensuality directly to the sweetness, chocolate character, creaminess, thickness and the smoothness of the liquid, for example, the sensory characteristics end up communicating sensuousness directly, and the notion of sensuality becomes much more substantial and real to consumers.

Sadly, the author can claim no part in Baileys' success and has no knowledge of Baileys brand strategy beyond that what is apparent to any interested observer. However, it seems that Baileys is an excellent example of synergistic uplift achieved by harmonising the emotional characteristics of the brand with the sensory characteristics of the product.

The 'product first' approach works in the opposite direction, by determining in the first place, whether or not any of the defining sensory characteristics of the product are emotionally active. Assuming that some of them are, then it must then be decided whether or not any of these emotions make a fitting emotional platform on which to completely (re)build or otherwise re-shape or augment the brand.

A few years ago, Mars Confectionery used the unique and distinctive sensory profile of Galaxy Milk Chocolate to breathe new life into the Galaxy brand by associating the inherent smoothness and creaminess of the chocolate, for which it was well known and much loved, with sensuality. New brand architecture and radically different imagery were then created around sensuality and underpinned with overt messages about the smoothness and creaminess of the chocolate. Prior to this very successful re-launch, the Galaxy brand had very little emotional depth, making it highly vulnerable in a category that was (and is still) heavily driven by emotion (www.mars.com).

The SensoEmotional aspects of both Baileys and Galaxy branding are constructed around the notion of sensuality and other related emotional constructs. Perhaps it should not be surprising therefore that their respective web sites show very obvious similarities in the colours, images, words and, above all sensory characteristics that are used to position the brands.

However, SensoEmotional associations are not restricted to sensuality nor are they limited to indulgent products and brands. Any emotion that lifts the spirits (i.e. causes emotional uplift) has the potential to form a SensoEmotional association, provided that the following criteria are satisfied:

- The featured emotion must be motivating to target consumers and must have the potential to cause emotional uplift.
- This emotion must also be credible in the context of the product category.
- It must not be too much of a stretch to (re-)position the brand and otherwise build this particular emotion into the architecture of the brand.
- Crucially, the emotion must complement the defining sensory characteristics of the product, and vice versa, otherwise there will be no synergistic uplift.
- The emotionally active sensory characteristics of the product must be optimised to communicate the emotion as effectively as possible.

All of the foregoing stems from the ability to identify, through research, any hot linkages that may exist between the emotional characteristics of the brand and the sensory characteristics of the product (SensoEmotional linkages), or vice versa. The research process, known as SensoEmotional profiling, is described in detail in Section 12.3.

12.3   SensoEmotional profiling

12.3.1   Research plan

In theory, SensoEmotional optimisation is relatively straightforward. Take block milk chocolate as an example. All that is required is an emotional profile of a particular brand of chocolate (or unbranded product if chocolate was being studied generically) and the corresponding sensory profile of the same chocolate. The two profiles are then overlaid to establish which sensory characteristics of chocolate are emotionally active, and which emotions are triggered by these sensory characteristics (Fig. 12.2).

In practice, it is rather more subtle than this. The process starts with groups of target consumers rather than a trained sensory panel. In this example, these would be people from all walks of life who are consumers of block milk

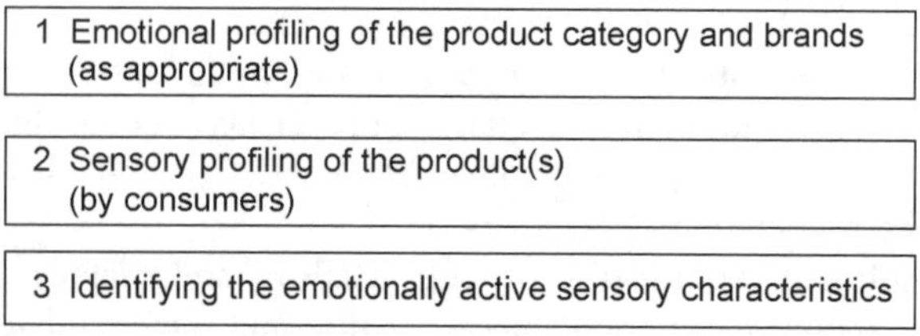

> 1 Emotional profiling of the product category and brands (as appropriate)

> 2 Sensory profiling of the product(s) (by consumers)

> 3 Identifying the emotionally active sensory characteristics

Fig. 12.2   The three stages of SensoEmotional profiling.

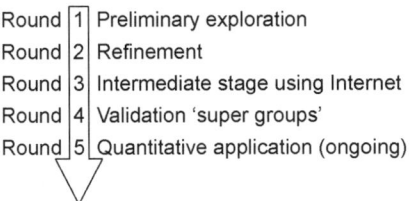

Fig. 12.3   SensoEmotional profiling research plan.

chocolate. They would be pre-recruited to attend SensoEmotional groups, each lasting about 2½ hours. Typically, there would be between six and eight SensoEmotional groups, depending on the nature of the product. These are run in a sequence, initially using different consumers in each group (Fig. 12.3). These are not the same as focus groups. They require different skills from the moderator, they have different objectives and they need to be conducted differently, so they look very different to the outside observer.

12.3.2   Round 1 – preliminary exploration

In the time interval between recruitment and attending the groups, the respondents are asked to create a scrapbook from magazines, newspapers or Internet downloads showing pictures or images that capture the emotion of milk chocolate. It is reasoned that if a pre-recruited respondent has not got the ability or the inclination to create the scrapbook, s/he generally will not show up on the night, so this is a very effective way of screening out the incapable and the uncommitted. Each group is normally split into two parts, starting with emotional probing and elicitation, followed by sensory profiling.

Emotional elicitation

Chocolate is ubiquitous, so it is likely that each of us will hold in our minds personal constructs relating to block milk chocolate. It is probable (although this can never be known for sure) that some of these constructs will be common to all of us, because individuals within a particular society or culture are potentially subject to a common set of inputs. However, the nature, depth, complexity and degree of refinement of these constructs, how they relate to each other and how they relate to tangential constructs will depend on the individual and his/her degree of immersion in the category. Idiosyncratic personal likes and dislikes are also likely to shape our personal constructs. An individual who likes a particular style of product or a particular aspect of a brand, will probably interpret it differently from someone who does not like it. Conversely, personal constructs can shape liking and opinion. The objective of the emotional elicitation phase of SensoEmotional optimisation is to access and explore the whole gamut of personal constructs associated with milk chocolate.

For some inexplicable reason, the process of emotional elicitation seems to have become elevated to the status of 'high art' within market research circles;

only open to especially gifted people, seemingly vested with mysterious powers of insight. Gladly, lesser mortals can also achieve a fair degree of success by substituting 'smoke and mirrors' with good scientific practices and common sense overlaid with a dollop of creativity.

That said, merely asking the individuals directly how they see the world of milk chocolate, would be much too simplistic. Whilst they would certainly be able to reveal something about their personal world of chocolate, largely because the emotional characteristics of chocolate are constantly being re-played to us in the media and also because chocolate is such an emotive product, direct questioning would probably fail to access deeper constructs, nuances and construct interconnectivity.

Two simple things can be done to help consumers to access and explore their deeper constructs. First of all, try using comparative techniques. As mentioned above, asking someone directly to think about milk chocolate would doubtless reveal something. However, ask the same individual to think about milk chocolate in relation to dark chocolate, white chocolate, bitter chocolate, chocolate with nuts, chocolate with raisins, chocolate with mint crisp, filled chocolate bars, boxed chocolates, liqueur chocolates, chocolate ice cream, vanilla ice cream, chocolate milk, chocolate milk shake, hot chocolate, cocoa, etc., and the nuances that differentiate them emotionally become apparent quite readily. (Why not try it yourself?) The researcher then builds on this by making other relevant comparisons.

The second 'trick of the trade' is to provide the group with pictorial images as a further aid to expressing and describing their emotions. Most naïve or unsophisticated consumers just do not have the lexicon or the experience to translate complex, vague and often shadowy emotional constructs into words. Images can really help in this respect, not just as a descriptive tool, but also as a means of encouraging interaction among group members.

Pictures and other graphics are easy to source from Internet image banks, and this can even be done interactively during the group. The images that the individuals brought with them in their scrapbooks, as an 'entry ticket' to the group, are also especially effective at stimulating constructive discussion and description. Supplementary word lists, paint colour swatches and fabric swatches also help.

Having expressed their emotions as pictures, colours or textures, it often much easier for consumers to then translate these into words. Needless to say, the moderator plays an important role in keeping the group focused, providing stimuli, overcoming any blockages and otherwise ensuring that an appropriate amount of progress is made.

The most desirable end-point at this stage would be emotional characterisations of the products and brands in words. However, words can fail consumers, so images are sometimes retained, mixed in with emotional descriptors. Ideally, this would take the form of a prioritised list of emotional descriptors, interspersed with pictorial images where appropriate.

Sensory profiling (with consumers)

The primary focus of the first two (or three) groups in the series is usually on emotional profiling. However, having taken the consumers through this very deep and detailed emotional probing process, they are now in prime condition to start thinking about the sensory characteristics that relate to these emotions.

Prior to the beginning of the research, a set of products (milk chocolates in this case) would be selected to represent some, or all, of the relevant sensory universe. It is crucial that all the products are sensorially different from each other. They may be marketplace products and/or prototypes, but all must be evaluated unbranded.

The chocolates are presented in pairs or triads (groups of three) to facilitate sensory comparison. At this stage, the sensory evaluation protocol is much more relaxed and less formal than it would be for laboratory-based sensory profiling. Elicitation of sensory descriptors is approached from two different angles. One option is to present the respondents with the chocolates, along with their own prioritised list of emotions, ask them to taste the chocolates, and then decide which, if any, of the chocolates conveys more or less of any of the emotions. It is always quite surprising how the slightly different tastes of otherwise broadly similar products can convey such different emotions (provided, of course, the respondents have been through the emotional priming in the earlier part of the group). Once the emotional differentiators have emerged, the respondents are then asked to describe which aspects of 'taste' (sensory characteristics) seem to convey these different emotions. The second approach simply reverses this process: respondents are asked to describe the sensory differences and then relate these to the emotional descriptions or images.

There are three outputs at the end of Round 1:

- Emotional descriptions – a fairly well-developed and prioritised list of emotional descriptions, interspersed and augmented (where appropriate) with pictorial images.
- Sensory descriptions – a draft sensory vocabulary, expressed in consumer terminology, of the sensory characteristics of milk chocolate.
- SensoEmotional linkages – an early indication of some of the Senso-Emotional linkages, although these are fairly tenuous at this stage.

During the course of these groups, it will become apparent that some consumers seem to have a particularly good aptitude for the process. The best of these individuals will be invited to participate in Round 4 (Validation).

12.3.3   SensoEmotional group moderator skills

SensoEmotional groups require a moderator with two particular skills: talented as sensory moderator (in a trained panel environment) and also skilled at probing and eliciting emotions and feelings, in words and pictures, from naïve consumers. This requires 'right brain' creativity as well as 'left brain' analytical skills, which is quite a rare combination.

Experience suggests that it is the talented sensory moderator aspect that is most important. This person must be highly skilled in eliciting sensory vocabularies, as would be done when developing a sensory profiling vocabulary from scratch with a trained panel. However, the moderator must also have the aptitude to do this with inexperienced consumers, which takes just a little more patience and natural talent. It is also highly desirable (although not essential) that the moderator should have experience of profiling the same product set with a trained sensory panel. Being familiar with the nuances of the sensory profile, and the words used to describe it, really helps.

Of course, the trained sensory professionals also need to be trained in the art and science of emotional probing. Some have the aptitude for this while others are just too analytical. In our experience, it is easier to train a sensory person with the right aptitude in the skills of emotional probing than vice versa. Sometimes two people are used to run the group: a group moderator to conduct the emotional probing and a sensory moderator to develop the sensory vocabulary and the emotional linkages. However, it is essential that the sensory moderator should be involved in the emotional elicitation, so that the individual is immersed fully in the emotional aspects of the product, when conducting the sensory description.

12.3.4   Round 2 – refinement

The next series of two or three groups follows essentially the same pattern as Round 1. However, during the short interlude between these two sets of groups, the emotional characterisations are captured and developed into word and picture boards. During the early stages of this next series of groups, the images that the respondents will have brought with them to the groups (remember, no images means no entry) and the image boards mentioned above, are used to get these people thinking about the emotionality of chocolate and to get them into the right mindset. This process takes about 30–45 minutes, leaving much more time to develop the sensory vocabulary and the SensoEmotional linkages.

This is where the sensory moderator's skill really makes a difference. As the moderator is developing and extending the sensory vocabulary with consumers, s/he will also be trying to reconcile this with the more formal sensory vocabulary from the trained panel. While this is not essential, it helps to build a bridge between the two vocabularies, which is often very helpful. It also keeps the sensory moderator focused and otherwise helps to engender thoroughness.

The end-point is a more refined and better-prioritised list of emotional descriptions, a well-developed and well-understood consumer sensory vocabulary and the confirmation and establishment of further SensoEmotional linkages. Again, we will have been 'talent-spotting' during the groups; looking for consumers who have the right aptitude to participate in Round 4. Across the two stages, expect to ear-mark about 10–12 people (~20% of consumers involved so far) for possible inclusion in the validation groups at the end.

12.3.5 Round 3 – intermediate developmental stage using the Internet

This stage has two purposes: pre-screening of respondents to take part in the validation groups in Round 4 and further exploration of SensoEmotional linkages. Typical SensoEmotional group recruitment procedures would be used to select about 35–40 target consumers, with the proviso that they need to participate in some preliminary exercises on the Internet before final selection. They are made aware that only about a third of them will be selected. Those people 'talent-spotted' from Rounds 1 and 2 are also included in this process.

There are four tasks, all of which are conducted remotely on the Internet.

Task 1 – emotional profiling

First of all, the recruits identify their favourite brand of chocolate, for example, from a list. They are then presented with five emotional descriptions (or five pictorial images) and asked which of the five emotions they identify most closely with their brand and conversely, which is least closely associated with their brand. This process, known as Best/Worst Scaling (Finn and Louviere, 1993; Cohen and Leopoldo, 2003) is repeated for 15–20 different combinations of various emotion descriptions or images. At the end, the recruits are shown the entire list of emotions, in alphabetical order, and asked to tick the descriptions they associate with their brand of milk chocolate in particular. They are also asked to tick any other descriptions they associate with milk chocolate brands in general, and finally any descriptions they do not understand. This helps to identify terms that seem to be redundant or misunderstood. Fairly standard procedures are used to convert best–worst scaling data into scale values; the higher the scale value, the more relevant the emotion to milk chocolate.

Task 2 – sensory profiling

Again taking their favourite brand of chocolate, the respondents are presented with five sensory descriptions and asked which of the five contributes most to their enjoyment of this chocolate, and which contributes least. Again, they would see 15–20 different combinations of five sensory characteristics. As with the emotional terms, the sensory descriptors would also be listed alphabetically and consumers would indicate which were associated with their brand in particular, milk chocolate brands in general and which, if any, they do not understand. Scale values are then calculated in order to prioritise the contribution of the various sensory characteristics to enjoyment.

Task 3 – exploring SensoEmotional linkages

Two or three of the key emotions that seem, on the basis of Rounds 1 and 2, to have strong SensoEmotional linkages are explored more formally. The process is based loosely on mean drop analysis, although the object here is to be exploratory rather than quantitative.

As an example, consider the relationship between the sensuousness of milk chocolate and 'melt-in-the mouth', which is a sensory characteristic. First of all, each respondent is asked to consider, 'To what extent does the sensuousness of milk chocolate contribute to your overall enjoyment of milk chocolate?' using the following scale:

❑ Extremely
❑ Strongly
❑ Moderately
❑ Slightly
❑ Very slightly
❑ Nothing at all

They are also given the default option:

❑ Milk chocolate isn't at all sensuous

Those who do not use the default are then asked, 'In order to improve the sensuousness of milk chocolate do you think that chocolate should melt in the mouth faster, slower or at the same rate?'

❑ Melt-in-the-mouth *much* faster
❑ Melt-in-the-mouth *slightly* faster
❑ No change
❑ Melt-in-the-mouth *slightly* slower
❑ Melt-in-the-mouth *much* slower

The general idea here is to establish whether or not those people who want milk chocolate to melt faster, for example, get more or less enjoyment from the sensuousness of chocolate, than those who want it to melt more slowly. Our purpose here is to make people think more deeply about the SensoEmotional relationships.

However, this process can also be used quantitatively to determine the sensitivity of the emotional characteristics to changes in specific sensory characteristics and the direction of change, otherwise known as the *emotional activity* of the sensory characteristic. This is a very powerful process.

Task 4 – Internet chat room discussions
Internet chat room bulletin boards are used to determine the extent to which these individuals are willing to involve themselves and, also, the extent to which they are able to articulate their thoughts and feelings about the emotionality of chocolate. There are usually about eight people per chat room, with each room managed by an experienced moderator. Although the chat room is used only for screening purposes in the context of SensoEmotional optimisation, it transpires that Internet chat rooms actually provide a really good forum for conducting certain types of qualitative research.

The primary output from Round 3 will be 12–24 respondents, depending on the circumstances, who have been screened for aptitude, interest and commitment, prior to their participation in the Round 4 validation groups. However, Tasks 1–4 also provide a significant amount of secondary information about the various SensoEmotional relationships, which is also fed into Round 4.

12.3.6   Round 4 – validation groups

These groups usually comprise six to eight people and last for up to 4 hours (usually an entire morning or an entire afternoon). Since the participants have been selected for their aptitude and they already have experience of working with the emotional and sensory characteristics of chocolate, the introductory part of the group need only be very short. The group is split into three distinct tasks.

Task 1 – validation of the emotional descriptions
First of all, the participants are presented with the emotional vocabulary, as developed in Round 2 and refined in Round 3, and this is discussed in the context of milk chocolate. The object is to make sure that the individuals comprehend the emotional descriptions and that they concur regarding the relevance of these to milk chocolate. They are provided with all the supporting imagery and words, to illustrate and describe the emotions and the key brands in the market. This group should require almost no 'professional' moderation because of the nature of the individuals involved. Their purpose is to identify any emotional descriptions that are considered to be doubtful or difficult to understand and also any gaps in the emotional landscape of milk chocolate. Any issues must be resolved, with or without the assistance of the moderator.

The final output is an agreed and prioritised list of emotional descriptions, represented using both words and images, as appropriate. The number of emotional terms depends utterly on the category of product but it typically ranges from 2 to 20.

Task 2 – validation of the sensory vocabulary
The participants are presented with a number of unbranded milk chocolates, typically four or five, depending on the product category, along with the consumer-derived sensory vocabulary. The moderator describes each of the attributes in the sensory vocabulary and may use several chocolates to highlight each attribute. This often involves comparison. Undoubtedly, this is one of the most challenging aspects for the moderator and it is one of the reasons why he or she needs to be an experienced sensory professional. The object is to identify which sensory characteristics, if any, are too difficult to understand or otherwise seem doubtful or incongruous.

Again, the final output is an agreed and prioritised list of sensory descriptors, in this case. The target is to end up with a minimum of five and a maximum of about 20 sensory characteristics, depending on the category. Remember, the object here is not necessarily to provide a deep and detailed sensory description of each product but merely to identify the primary sensory characteristics that define the category and differentiate the main brands.

Task 3 – SensoEmotional profiling
A form of 'two-dimensional' profiling is used. The process starts with the first emotional description from the prioritised list, which is again represented using words and images. Three sensory descriptions are then presented simultaneously

and the group, all working together, must identify which, if any, is most closely associated with the emotion, and which is least. This is noted and the process is repeated for three more sensory descriptors. The respondents are strongly encouraged to do this as quickly and as spontaneously as possible. This is repeated for each of the key emotions.

The process is then turned on its head. A single sensory characteristic is presented along with three emotional descriptions and the group must decide which emotion, if any, is most closely associated with the sensory term, and which is least. This is noted and the process is repeated for three more emotional descriptors. Again, this is repeated for the key sensory characteristics. Working with the group, the moderator will then collate the findings and the final associations will be dawn in by hand. These are the much-prized SensoEmotional associations.

It must be emphasised that these three tasks are well beyond naïve consumers. Those involved in the validation groups need to be preselected for aptitude and commitment and they need to be 'conditioned' into thinking about the emotionality and taste of chocolate. They also need to be well paid! It could, of course, be argued that these people are no longer representative of the consuming public and therefore the information they provide is atypical. However, experience suggests that this is not the case because the SensoEmotional associations uncovered (i.e. the main output from the process) are always 'real' when fed back to naïve consumers.

To summarise, there are three final outputs from Round 4: an emotional portfolio for milk chocolate (with supporting graphics and images), a consumer-derived sensory vocabulary for milk chocolate and, most important of all, a prioritised short-list of the emotionally active sensory characteristics of chocolate, showing which sensory characteristics cue which emotions.

12.4   Commercial applications

The outputs of SensoEmotional optimisation are of enormous commercial value. Typical applications are as follows.

12.4.1   Brand architecture

The primary objective of SensoEmotional optimisation is to identify emotionally active sensory characteristics that can be featured in brand communication to convey the corresponding emotions. There are two strategies from a marketing point of view:

- Alluding to the emotionally active sensory characteristic(s) in brand communication as a means of reinforcing or otherwise underpinning the existing emotional message (as Baileys may have done).

- Featuring the emotionally active sensory characteristic as a way of building a new emotional dimension into the brand architecture, or even as a way of creating a completely new positioning (as Galaxy has done).

12.4.2   Emotional profiling of brands

The SensoEmotional optimisation process will have identified a prioritised list of emotions for the category, along with supporting descriptions and graphics. This is a hugely valuable tool that can be used to characterise, profile and then map brands quantitatively. Although relatively new to market research, best/worst scaling (Finn and Louviere, 1993; Cohen and Leopoldo, 2003) is undoubtedly one of the best ways of doing this. The combination of superb emotional stimuli and a simple quantitative process, makes subsequent brand profiling easy to incorporate into routine category appraisals.

12.4.3   Consumer sensory profiling as part of routine product evaluation

Most product evaluation questionnaires include scales for measuring sensory magnitude (e.g. how sweet?) and sensory ideal point scales (too sweet or not sweet enough?) Often these scales will be selected and described in a fairly ad hoc manner. The process of SensoEmotional optimisation yields a refined and validated vocabulary of consumer-derived sensory terms that can be included very readily in routine product evaluation questionnaires. The quality of the sensory vocabulary means that it is usually more relevant, more easily understood by naïve consumers and much more discriminating.

12.4.4   SensoEmotional optimisation

Technologists routinely optimise the sensory characteristics of their products to maximise liking (i.e. SensoHedonic optimisation). However, if it transpires that the emotionality of the brand plays a much bigger role in the totality of product enjoyment than liking of taste, then it may be more beneficial to optimise the sensory characteristics of the product to underpin, reinforce and otherwise deliver these key emotions (i.e. SensoEmotional optimisation). Until the emergence of SensoEmotional optimisation it was not widely recognised that SensoEmotional and SensoHedonic optima may be quite different and that optimising against the wrong target could detract greatly from the totality of product enjoyment.

12.4.5   New criteria for product performance

Product performance is routinely, and often exclusively, evaluated using liking as the primary criterion of success. SensoEmotional optimisation has highlighted the fact that delivering against emotional criteria (or functional criteria) may sometimes be equally or more important than liking. This makes a very strong case for the inclusion of emotional profiling (and functional profiling) as a key part of any product test. As mentioned earlier, this calls into question any form of product testing that uncouples the sensory experience from the emotional experience. This poses serious methodological issues for researchers.

Clearly, SensoEmotional optimisation is taking brand and product research across new and challenging frontiers. The truly appalling failure rate of new product development suggests that this might be long overdue.

Details of the final outcome for SensoEmotional optimisation of milk chocolate is, of course, highly confidential. Something that is this valuable and this commercially sensitive could never exist in the public domain. However, it is feasible to reveal something about the more obvious sensory characteristics:

- Melt-in-the-mouth is indeed closely associated with sensuality. The rate of melt is crucial; too fast detracts more quickly from sensuousness than too slow. There is, of course, an optimum.
- Sweetness is closely associated with reassurance. However, the rate of decline of sweetness must be rapid (although not too rapid) because cloying-ness caused by prolonged sweetness is actually negative for adult indulgence. The term 'clean sweetness' has been coined!
- Whiteness/blackness of the brown colour of chocolate is a key determinant of adultness. If the brown colour is tinged with too much white, this diminishes adultness. Too black and it is not credible as milk chocolate. The optimum shade of brown for adultness is not optimum for sensory liking!

12.5   Sources of further information and advice

The concept of SensoEmotional Optimisation of branded products is relatively new and still evolving, as is the process of SensoEmotional profiling. Consequently, there is no formal body of literature on the subject, making this chapter one of the first comprehensive explanations of both concept and process.

Should the reader wish further explanation or advice on how to use sensory characteristics to cue emotionality, reinforce functionality or give a brand a unique sensory signature, please contact the author directly by email (d.thomson@mmr-research.com) inserting 'SensoEmotional' in the subject line.

12.6   References

ARDITTI, S. (1997). Preference mapping: a case study. *Food Quality and Preference*, **8**, 323–327.

COHEN, S.H. and LEOPOLDO, N. (2003). Measuring preference for product benefits across countries: overcoming scale usage bias with Maximum Difference Scaling. *ESOMAR 2003 Latin America Conference Proceedings*. Amsterdam: ESOMAR.

FINN, A. and LOUVIERE, J. (1993). Determining the appropriate response to evidence of public concerns: the case of food safety. *Journal of Public Policy and Marketing*, **11**, 12–25.

GUTMAN, J.A. (1982). A means–end chain model based on consumer categorisation processes. *Journal of Marketing*, **46**, 60–72.

KELLY, G.A. (1955). *The Psychology of Personal Constructs*. Volumes 1 and 2. New York: Norton.

LINDSTROM, M. (2005). *Brand Sense – How to Build Powerful Brands Through Touch, Taste, Smell, Sight and Sound*. London: Kogan Page.

MACFIE, H.J.H. and THOMSON, D.M.H. (1984). Multidimensional scaling methods. In Piggott, J.R. (Editor), *Sensory Analysis of Foods*, pp. 351–375. London: Elsevier Applied Science Publishers.

MARTENS, H. and NAES, T. (1989). *Multivariate Calibration*. Chichester: John Wiley & Sons.

MASLOW, A.H. (1943). A theory of human motivation. *Psychological Review*, **50**, 370–396.

Part III

Methods for consumer-led food product development

13

Sensory research and consumer-led food product development

H. Stone and J. L. Sidel, Tragon Corporation, USA

13.1 Introduction

Developing products is easy, developing products that appeal to consumers is less so, and developing products that appeal to a sufficient number of consumers and achieve commercial success based on specific business criteria is very difficult. The evidence for these observations can be found in various trade and/ or technical publications (e.g. *Food Technology*, *Prepared Foods*, *New Products Magazine*) since the mid-1990s. For any product category, one can easily find listings of new products introduced to the market that cannot be found in the market a year later. Not surprisingly, the number of withdrawals far exceeds the number still in the market. It is clear that companies have made and continue to make substantial investments in developing new products in the expectation that they will make a significant contribution to their respective bottom line profits and in the process, enhance that company's image and all those involved in the program. Despite the very low rate of success, about 10% or less, companies continue to make the investment because of the enormous profit and reputation opportunities expected (by the winners). Companies also have invested in workshops, in symposia, and in support of research into new ways for improving the success rate. These have included use of the Internet for both surveys and product home placements, new approaches to understanding consumer behavior including in-home video cameras, and measuring brain activity of individuals observing commercials (and products), to name a few.

To date there is little evidence that any significant breakthroughs have occurred; however, it may be too early to measure the impact of these new developments. There usually is a time delay as research results are adapted to the practicalities of the product development process and the dynamics of the marketplace. In addition, consumers are not static in their behavior regarding foods and beverages. Changes occur as a result of trying new products and by events around them. Some are easily observed and measured while others are less so. External events such as a nationally publicized food contamination situation, obesity, various diets, trans fat, acrylamide, BSE, organics, and sustainability, and concerns about diseases such as bird flu, have all had a significant effect on food and beverage choices.

With these issues discussed at length in the media and on the Internet, it makes for a complex and challenging environment for anyone contemplating investing in the process, let alone the actual introduction of a new product. It is equally daunting for the consumer who is less informed and often misinformed by individuals with vested interests or without access to the facts. Nonetheless, there also have been beneficial effects, including increased support of research into consumer behavior, awareness of the need to better understand what product and imagery variables influence preference choices, alternative approaches to managing the product development process, etc.

This brief introduction is intended as background to the discussion on why, where, and how sensory resources fit in the overall scheme of product development. As we will show, sensory information has a significant role in the product development process. Failure to obtain accurate sensory information or how and when to use sensory resources can account for many, but not all, of the problems that occur in the development process. Before describing the role of sensory and the how and when, it is useful to briefly discuss the development process from a behavioral perspective.

13.2   The product development process

Much has been written about product development from a technical perspective as well as from organizational/business perspectives. It is complex and interactive involving many different interests within and outside a company. In some instances these interests have been working together and in some instances, working at cross-purposes, and this latter situation results in costly delays and missed opportunities. In most companies, formal teams are organized and may include individuals representing purchasing, marketing, marketing research (consumer insights), production, quality control, packaging, logistics, sales, corporate, legal and, last but not least, the product development group itself. In recent times, chefs have been added to the team; although some companies have had chefs working with development teams for many years. In addition to internal resources, external groups such as advertising agencies and key suppliers (or 'partners') are included. It is interesting to note the extent to which companies have come to rely on these external groups as a way of reducing development costs and bringing a different view to the process. Suppliers become an integral part of the development team and also are often expected to contribute financially to the cost of the development efforts.

Clearly, any development effort involving so many different elements makes for unique challenges for those directly involved in the actual development effort itself but especially for the management of the process. In the past, many of these business units operated independently or only became involved in the process when a particular stage was reached. In addition, it was not unusual for information about the market, the target consumer, or some aspects of a technology to be kept from other members of the team. In some instances this was done for security reasons, or it was believed that the information was not relevant to another section's activities. Managers now recognize that this exclusivity has to be changed if meaningful progress is to be achieved in the development process and in the introduction of new products. Some companies now make use of more formalized practices in which a core team is organized with sufficient management support to ensure that the development efforts operate efficiently. The process includes scheduled timelines and information exchanges to everyone involved in the activity. These techniques have been described in various texts including Smith and Reinertsen (1997) and were adapted from other industries; e.g. airline development and construction.

Depending on one's perspective, organizational, business, brand or technology, the texts by Aaker (1991, 1996), Fuller (1994), Kotler (1999), and Smith and Reinertsen (1997) are useful guides to current product development ideas as well as the role and importance of a company's brand image. As Fuller (1994) noted, the process involves a logical and systematic sequence of events, or at least it does when described in print. Ideas are generated from many sources and tested with targeted consumers, preliminary market projections are developed, a product/marketing brief is prepared, formulation is initiated, and individuals within technology and purchasing source potential ingredients. While these activities are in progress, formulation efforts continue, safety and legal issues are investigated, prototypes are prepared and evaluated by the team, formal evaluations are made by marketing research-consumer insights group and/or sensory, reformulation and more extensive evaluations take place. All the while, additional business research is in progress to refine potential market size, production will be examining plant requirements, scheduling, raw materials sourcing and price are discussed, packaging and label requirements are being detailed, etc. At various stages products will have been evaluated using a variety of protocols; e.g. focus groups, sensory and small-scale preference tests. When formulation shifts to a pilot plant, more comprehensive sensory tests may be organized.

In principle, the project team uses this information to brief senior management for approval to proceed. Based on meeting various benchmarks, the team will receive approval to expand the scope of the effort. This could include a review of the market projections, packaging and labeling requirements, etc. All this is intended to maintain a momentum in the direction of product introduction. This is a complex multi-step process with ever-increasing budgetary implications as the program moves from an early to a later stage, with the potential increase of unanticipated events. Competitive developments, changing budgets and timetables, availability of key ingredients, production capacity issues, changing concepts, and management priority shifts all can and will conspire to facilitate and/or derail the process. In practice it is atypical for the sequence of activities to occur as planned. A company's product development budget will change, key managers change responsibilities, the concept is modified, test results are not promising, and so forth. To minimize these problems requires a strong commitment to the process from senior management and that has to be counterbalanced by the team able to provide information demonstrating progress relative to specific goals. There is no simple solution to these unanticipated developments/impediments to progress; however, it is clear that inadequate or insufficient product information should not be the issue. More will be said about this in the sections that follow.

13.3   Sensory's role in product development

Sensory should have a key role in the product development effort, beginning when the new product team is organized. As discussed previously (Stone and Sidel, 2004), sensory's role is essential to the potential success of any new product for strategic as well as for practical reasons. Strategically, sensory information defines how consumers will perceive a product in relation to how marketing plans to position the product, how the product competes with other products in that category, what language is used to communicate to the consumer, how it could be used, and so forth. Developing some of this information in the early stages keeps the development efforts focused, and serves as a baseline from which any directional changes (attitudinal as well as technically) can be mapped. From a practical perspective, sensory can identify and measure any unique characteristics such as a product sensory deficiency not previously noted. It provides additional information such as correlations with specific ingredient changes. It also enables the sensory group to be organized for any unique testing requirements, for example, testing in a non-laboratory environment, extended-use evaluation.

The first question to ask is why should sensory be a part of any new product development team vs functioning simply as a supplier of test services when requested? There are several reasons/answers. The first is about being prepared when testing is requested vs waiting several weeks (or months) to organize a panel; the second is being able to document or benchmark the initial products/ formulations for comparative purposes in the future, the third is to be able to support development efforts when different formulations are prepared. In the early stages of a program, products may be prepared by chefs and evaluated by the team but are usually not subjected to any formal evaluations. Comments are noted and recommendations made to proceed or return to preparing additional samples. Without any sensory input during this stage, the team is left with qualitative descriptions and their source (a product technologist vs the project manager). Over time, even as brief as a few days, recalling a sensory experience can be risky and this without formal documentation will lead to changes. At some point in time and as early as is reasonable, some kind of sensory analysis will be important. While a quantitative analysis could take a few days or less, the delay will be trivial compared with following a formulation sequence that takes several weeks but leads in the wrong direction. Once a project goes beyond the stage of a chef's preparation, the use of a comprehensive sensory analysis becomes even more important. Commercially available raw materials will not be the same as fresh ingredients and a processing system is not the same as a stove. These product changes could result in substantive sensory changes that need to be documented for several reasons; e.g. comparison with existing results, documenting the extent of change as a function of the specific ingredients that were used, etc. If this information is not available, there are no data to confirm whether subsequent formulations accurately reflect what was originally prepared and agreed to by the team. Also, the brand and consumer insights group will be fielding focus groups with product, but which product(s) will they be using? The results from any such focus group sessions will be significantly impacted by the product and influence subsequent directions about the products; for example, making formulation changes that may not be relevant.

A second question to ask is whether any sensory information is needed; why not obtain this information from consumers while measuring preferences and related attitudinal information? This enables two tests to be done for the price of one! Since the test will be fielded with a large number of target consumers, the results will be more likely to be believed, adding to the attractiveness of this approach. On the surface this sounds like a good idea; however, obtaining preference and sensory judgments from the same consumer leads to biased information, especially if none of the products is well liked. If the consumer likes a product, responses to that product's characteristics will probably be positive as well (the halo effect). Conversely, if a product is less liked, the responses to some (or all) of those product 'sensory' characteristics also will be lower. This latter result could lead one to assume that the specific characteristics need to be changed to improve acceptance, and this will probably be misleading.

Extensive research (with subjects qualified and screened for their sensory skills, and with typical communication skills) and the empirical evidence has shown that the language of perception is complex. A word used to represent a specific sensation will mean something different to different subjects and only after discussion and frequent samplings is it possible for a group (of subjects who also happen to be consumers) to reach a consensus about the use of a specific set of words. Even simple words such as sweet and sour can be mis- understood. As a further complication, some words have quality connotations that need to be avoided because quality is an equally, if not more, complex word. It takes several hours for a group of subjects to resolve differences in language so that to agree on what word best represents a particular sensation. In addition, consumers have varying levels of sensory skill and may or may not correctly detect or recognize a particular sensation. This will be discussed further in Section 13.4.1 on descriptive analysis. So the argument for using a single test to obtain both attitudinal and sensory information is inappropriate. It will lead to confusing information and it will compromise the sensitivity of both types of measures. It also helps to explain, in part, why so many products fail.

13.4   Sensory evaluation

In the past few decades there have been numerous texts and other publications devoted to the topic of sensory evaluation available to the sensory professional and other interested technical and marketing specialists. Space limitations preclude an extensive review and description of methods; nonetheless, a brief summary is provided here as a context for the role it plays in product development.

Sensory evaluation is a science that measures the responses of people to products as perceived by the senses, all of them, not just odor and taste. As a science it follows documented procedures and practices (Stone and Sidel, 2004), including use of qualified subjects, applying appropriate methods and, in the case of analytical tests, use of replication as a means of measuring reliability of the obtained information.

There are two types of methods that are most appropriate for product development – descriptive analysis and affective methods. Discrimination (a type of analytical test), while useful, is less so in the development process primarily because it provides limited information; i.e. determining that two products are perceived as different vs the types of differences and their magnitudes. This should not discourage one from using difference testing, but thought must be given to the usefulness of the results.

Before describing methods, some comments are warranted regarding subjects, products, and the design and analysis of tests. Failure to use qualified subjects is one of a few important challenges (and opportunities) confronting the sensory professional. The statement 'use qualified subjects' is very important. Anyone presented with a product will reply to questions about that product whether or not they understood the question. Unfortunately, one always obtains responses but the source may, in fact, not be qualified to provide that response. This can be easily demonstrated through the use of the difference test methodology. Using a qualified subject methodology begins with identification of those individuals who are average or above average in their consumption of the product to be tested or if not the actual product, a related product from that category. As a general guide, individuals who consume an average or above average amount of a product are more sensitive to differences than below average consumers.

The second and equally important step is to have the individuals participate in a series of difference tests. These tests are organized such that they represent a range of difficulty, easy to difficult, and incorporate all relevant modalities of visual, aroma, taste, etc. The basic approach relies on having each person participate in about 30–40 trials. Across this number of trials it will be possible to assess a person's sensory ability to perceive differences. Examining response patterns reveals that as many as 30% will exhibit discrimination ability at less than chance; i.e. they should not participate in sensory analytical tests of discrimination and descriptive analysis because they are not adequately sensitive to differences. Including them increases variability and decreases sensitivity, leading to incorrect conclusions. This exclusion criterion is important in a sensory test regardless of the number of subjects tested. It also has implications in larger-scale tests where consumers may be asked to respond to a range of product 'attributes' such as flavor, mouthfeel, etc. The likelihood is high that many of the responses will be random and not reflect true detection of an attribute or of a difference.

The need to have qualified subjects is essential if test results are to be believed and credible. When results do not meet expectations, they are often challenged. Demonstrating use of a systematic qualifying procedure shows the care with which the subjects were selected. Of course it is also important to use the appropriate method, as discussed in the next section. It may come as a surprise that this range of sensitivity in the general population also applies within a company, across ages and gender. Just as larger-scale consumer tests have very precise qualifying requirements in terms of demographics (age, gender, household income, brand use, etc.) so, too, should sensory tests. This is especially important with the limited numbers of subjects in most sensory tests. Because it could take a week or more to recruit individuals and another 3–4 days to complete screening, being part of a development team enables the sensory staff to anticipate panel needs vs having to react after a test request is received.

A second important sensory resource is the ability to organize and field studies that use powerful multivariable designs. With the increased power of PCs and a wide array of design software available, the possibilities for exploring the interactive effects of formulation and process variables within a single test are significantly enhanced. This minimizes reliance on and error from testing a single variable at a time. While design studies provide numerous advantages, they also require substantial knowledge about the variables of interest, and availability of time to plan and prepare products. Such planned experiments enable the sensory professional to measure the interactive effects of changing variables and better understand their effects on preferences and related attitudinal measures. These are very powerful ways of identifying combinations of variables that best predict optimal consumer liking. Further applications are described later in this section.

We now focus on the methods themselves and, as noted earlier in this discussion, the focus is on the use of descriptive analysis and preference testing. These methods will be most useful in the product development process.

13.4.1 Descriptive analysis
Descriptive analysis is a methodology that provides word descriptions (attributes) of products that also includes the intensities, the strengths for each of those attributes. There are numerous methods described in the literature

(Stone and Sidel, 2004, chapter 6; Sidel and Stone, 2006); however, all are derived from either Flavor Profile® or QDA®. The former is limited to prespecified attributes and the latter has no such restrictions; i.e. subjects must decide what attributes they want to use to represent their perceptions. There are no restrictions as to the number of them as long as the panel reaches a consensus about the words and their definitions. Other differences relate to training time, use of references, and scale use. A detailed description of the methodologies can be found in the aforementioned text and associated references; for purposes of this discussion, basic requirements are described; however, it is not intended as a plan for developing a complete sensory testing capability.

It takes about 2–3 weeks to establish a new QDA® panel. Of course, if a panel already exists, then screening and training are not required and a panel capability is available within 1 or 2 days. Starting a new panel requires that subjects are qualified based on product usage (average usage or above) and on their ability to discriminate differences among the products they will be evaluating. This process takes about 5–7 hours of an individual's time over a period of 3 or 4 days. It is expected that about 30% of those who participate will not discriminate at better than chance among the products tested. Therefore if one wants to have about 15 people available for language development and testing, then it is necessary to start with about 25 individuals. This group participates in a series of language sessions managed by a panel leader. Language development is a consensus building process; i.e. subjects develop and agree on the words they want to use to represent their perceptions and also define those words. They also practice scoring products to familiarize themselves with the products and the rating system. These activities usually take about 90 minutes per day for 5 consecutive days. During the language development process, subjects determine the product examination procedure needed to provide a realistic use experience. Actual testing relies on each subject scoring each product individually and on a repeated basis. The number of replicates can be as few as two and as many as four or more, depending on the number of products in the test, the stage of development, time constraints, and how the information will be used. The results can be easily communicated by way of plots or maps such as shown in Fig. 13.1; and these diagrams are related back to the obtained mean values along with the relevant variance measures.

13.4.2 Preference

Measuring preferences/liking is another useful tool available in support of the product development effort. There are limited methodologies available. One can directly measure preference, using the paired preference method – 'here are two coded samples, which do you prefer?'; or using the hedonic method which measures preferences indirectly by obtaining degree of liking scores ('1, dislike extremely' and '9, like extremely'). An example of this scorecard is shown in Fig. 13.2. The latter responses can be converted to preferences; however, the more useful information is based on the measures of liking. In multi-product

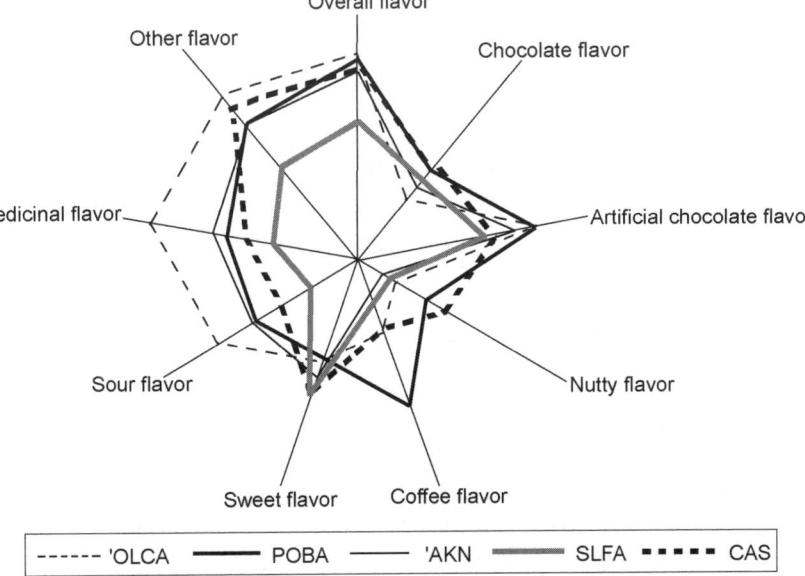

Fig. 13.1 A spider web plot of five competitive food bars for selected flavor attributes. The individual attribute intensity score (mean value) is determined by measuring the distance from the center point to that place where the product crosses the line. The visual display provides a simple but effective means for displaying product differences; reference to the definition for each attribute provides the developer with a basis for any formulation changes.

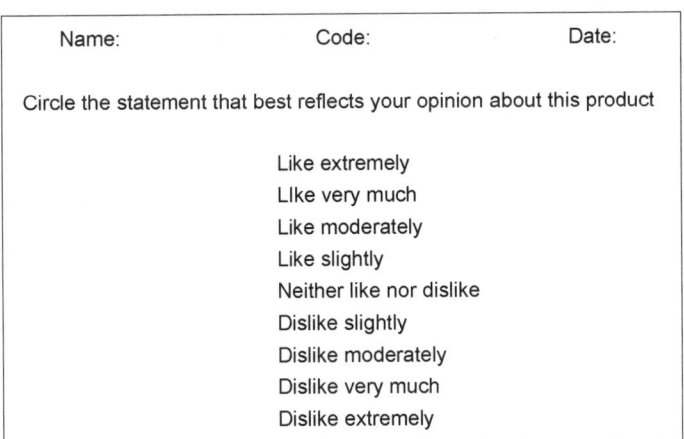

Fig. 13.2 The 9-point hedonic scale is an example of a scale that has been applied to a wide range of products in most places in the world (when appropriately translated). Effectiveness of the scale is based on providing the consumers with a context for its use prior to a test. The empirical evidence indicates the consumers use the scale as if it were equal interval. The circled statement is converted to a numerical value, '1, dislike extremely' and '9, like extremely' for computational purposes.

tests, the responses provide an ordering of liking and the magnitudes of difference provide a context for where each product is on the continuum. In many companies there is a reluctance to having a sensory preference test vs a test fielded through consumer insights/market research. This competitive separation should not occur because the important point is having rapid measures of liking in the early stages of the development process. Since the latter group usually thinks in terms of large numbers of consumers tested, field time and cost will create delays. Sensory preference tests typically involve fewer consumers (typically about 75) and the focus is product liking and not measures of purchase intent, etc., where larger numbers are needed for confidence in the conclusions obtained. This enables sensory tests to be fielded quickly, cost less, and, perhaps most important, exhibit a high correlation with larger scale tests in which preferences were measured.

For details about the methodologies and their relative benefits, the reader is referred to Sidel and Stone (2006). The value of the method rests with the usefulness of the information; a quick read of where each product is located on the liking continuum. This helps answer a frequently asked question, are the different formulations liked better than the current formulation?

13.5 Applications – opportunities

The greater value of sensory resources is demonstrated when product similarities and differences are measured by a trained panel, and preference and related attitudinal questions are measured by target consumers using the same set of products. This enables relationships to be identified between sensory and preference differences. As shown in Fig. 13.3, the sensory attribute, chalky mouthfeel, exhibits a strong inverse relation with preference; i.e. as chalky mouthfeel decreases in strength, preference increases. While single correlations have considerable surface appeal, they can be deceptive in as much as the sensory and preference worlds are multivariate; i.e. consumer behavior is complex and rarely, if ever, based on a single product dimension.

Figure 13.4 depicts selected results from a design study in which systematic changes in product viscosity and sweetener were evaluated by a trained panel and by consumers. The results for these attributes depict the impact of changes in their relative strengths on preferences. This type of information enables the technologist to formulate to a specific preference based on changes to these attributes. Once the reformulations are available, the trained panel can quickly determine whether the strengths of these attributes are in the desired direction.

As the product development process moves closer to commercialization, the testing process can be expanded to incorporate more consumer information such as specific product benefits, use situations, price/value, and so forth. In addition, it is possible to identify combinations of sensory attributes and/or formulations that result in optimal preference. For example, it will be important to measure the degree to which specific product benefits are influenced by the sensory

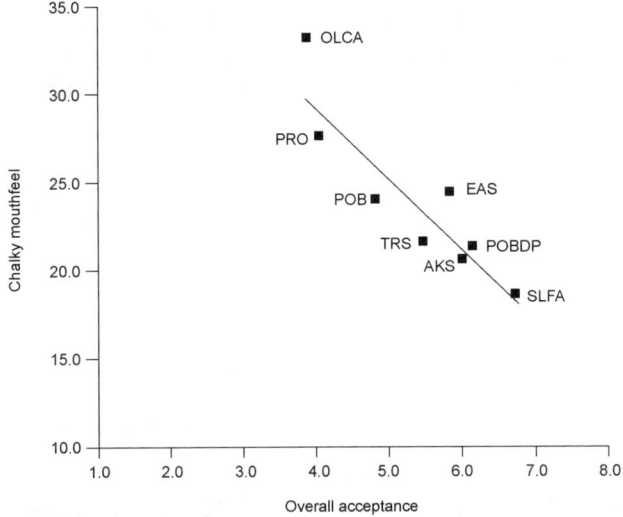

Fig. 13.3 Correlation for a single sensory attribute, chalky mouthfeel (y axis, scale from 0 – none to 60 – maximum strength), and overall acceptance (x axis 1 – dislike extremely to 9 – like extremely). The sensory data were obtained from a trained panel evaluating each product 3 times using a monadic sequential serving order. The preference scores were obtained from a target population of about 125 consumers. Products are identified by letter codes.

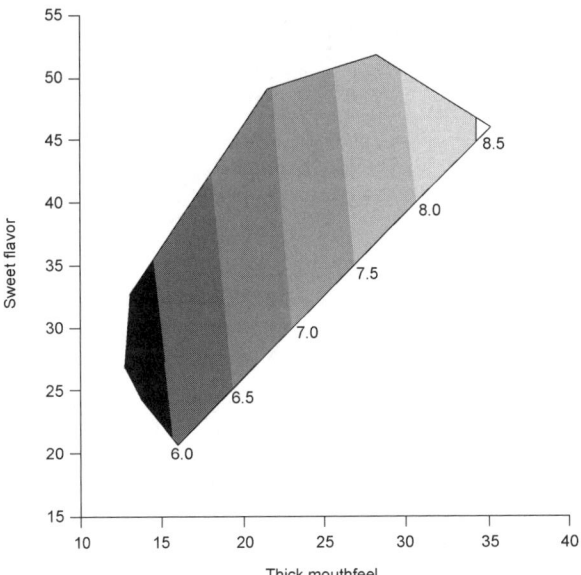

Fig. 13.4 A three-variable contour map for visualizing the simultaneous effects of these related variables of sweet flavor, thick mouthfeel and preference. The attributes were scored by a panel of 11 subjects on four occasions using a 0–60 line scale. For details about the scale see Stone and Sidel (2004).

experience. This provides a more comprehensive picture of consumer choice behavior and reduces the risk of too large a gap between the imagery and the product. Such studies involve testing a relatively large number of products, usually more than 15, but the output is significantly greater in value vs multiple tests with fewer products. Results are presented in a variety of formats, of which the most common is mapping. Mapping enables one to display a complex set of results; however, the reader is reminded that such maps are intended as a way of visualizing relationships and by themselves are not solutions.

Figure 13.5 shows a series of maps, created by presenting consumers with product descriptions and various benefits/use situations. They were asked to rate them using a scale for appropriateness from 'not at all' to 'very'. Statements that were considered most appropriate were located in the upper right quadrant.

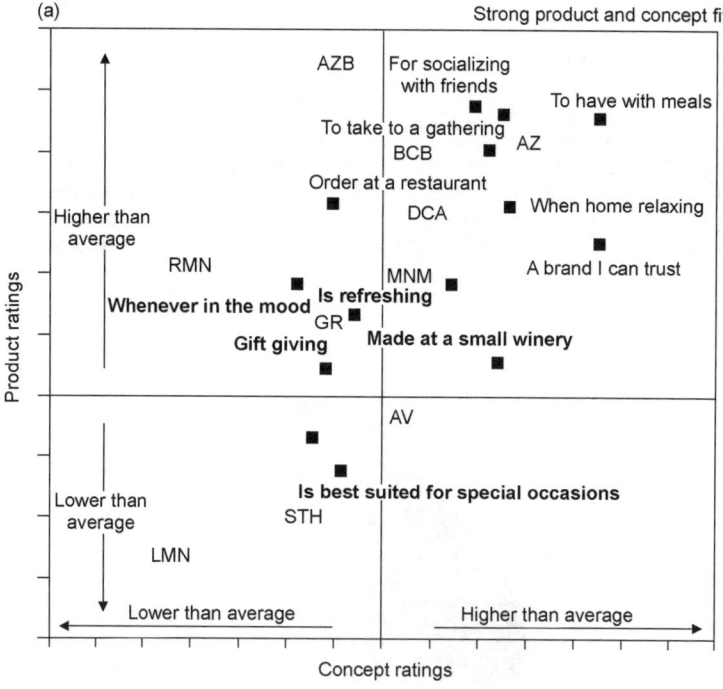

Fig. 13.5 **(a)** A map of product and concept ratings based on appropriateness measures. Letters represent selected products. Statements in bold were about 20%+ lower in appropriateness ratings. **(b)** The same map with the incorporation of the important attributes (identified through multiple regression analysis) listed on the right to provide for a more readable plot. The + and − designations refer to increasing (+) or decreasing (−) the strength of a listed attribute; for example, straw colour. The specific change is determined through reference to the data to prepare plots (see Fig. 13.1 as an example). For more detail about the interpretation, see previously cited references by the authors. **(c)** This map adds the product preferences shown italicized and underlined (evaluated without brand identification). The mapping provides direction for the location of a product in the space, and the sensory information provides product-specific direction and reference to the specific plot (shown in Fig. 13.1) focuses the formulation efforts.

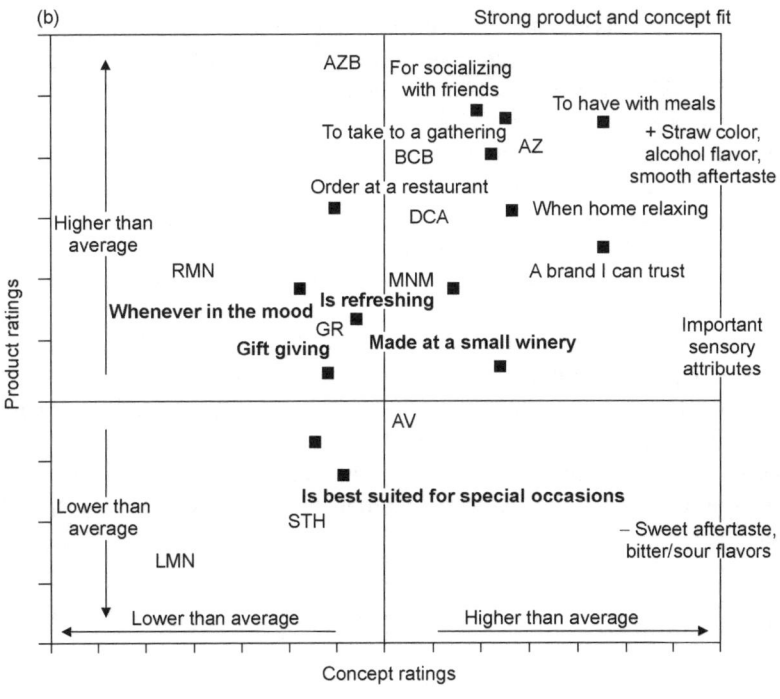

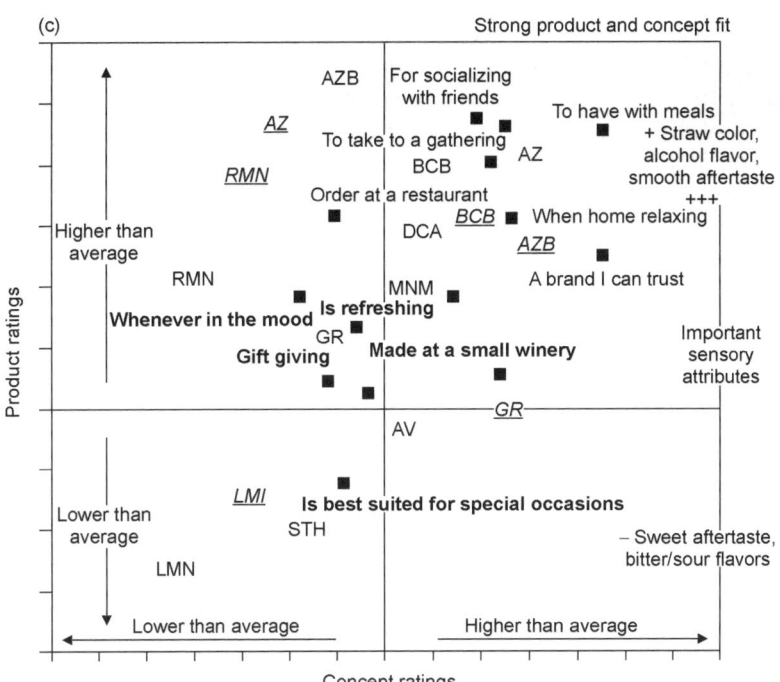

Fig. 13.5   Continued

Figure 13.5a depicts product and concept descriptions, Fig. 13.5b is the same information with the most important sensory attributes (+ and −) added, and Fig. 13.5c adds the coded product preferences and brand preferences. This enables the product development team to understand how a product is 'viewed' by the target consumer not only in terms of sensory differences but also in terms of the marketing messages. In this example, some product benefits relate well to the product sensory/use experience, and the team uses this information to bring greater focus to the market strategy and, where necessary, enables the technologists to focus their efforts on the product attributes that reinforce the marketing message. The role of the sensory information is critical in this instance. It identifies which attributes warrant change, which ones are most important, and how well that matches the various benefits and uses that will be advertised.

As previously noted, incorporating sensory resources enables a product development team to obtain useful product information not available by other means. Of specific value is the use of multivariate designs that provide a better connection between how a product is formulated and how well it links with a product's market strategy.

13.6   Conclusions

From a sensory perspective, the available methods enable one to provide substantive information about products as they are being formulated as well as how they compare to competitive products. The critical element in this process is whether the development specialist and/or the brand managers understand the potential of the sensory resources and include them in the team from the outset. As noted earlier in the discussion, all too often, sensory resources are not included because they are not recognized for their value. Sensory resources represent a significant but untapped resource. Much can be accomplished at a relatively low cost and a significant increase in value can be expected.

13.7   References

AAKER, D. A. (1991). *Managing Brand Equity*. The Free Press, New York.

AAKER, D. A. (1996). *Building Strong Brands*. The Free Press, New York.

FULLER, G. W. (1994). *New Food Product Development: From Concept to Marketplace*. CRC Press, Boca Raton, FL.

KOTLER, P. (1999). *Kotler on Marketing*. The Free Press, New York.

SIDEL, J.L. and STONE, H. (2006). Sensory science: methodology. In: *Handbook of Food Science, Technology, and Engineering*. Y. H. Hui, ed. CRC Press, Boca Raton, FL, pp. 57–124.

SMITH, P. G. and REINERTSEN, D. G. (1997). *Developing Products in Half the Time*. 2nd edn, Van Nostrand Reinhold, New York.

STONE, H. and SIDEL, J. L. (2004). *Sensory Evaluation Practices*, 3rd edn. Elsevier Academic Press, San Diego, CA.

14

Opportunity identification in new product development and innovation in food product development

E. van Kleef and H. C. M. van Trijp, Wageningen University,

14.1 Introduction

New products that deliver added consumer value contribute significantly to the success of companies. New product development (NPD) is generally recognised as the basis for profitability and growth of most companies. Additionally, innovativeness of companies has a positive impact on economic growth (Porter, 1990). Eliashberg *et al.* (1997) reported a survey among 154 senior marketing officers of US corporations: 61% of the respondents expected that 30% or more of their sales would come from new products within the next 3–5 years. This finding is consistent with the survey of 700 firms (60% industrial, 20% consumer durables and 20% consumer non-durables) of Booz *et al.* (1982) who found that over a 5-year period new products accounted for 28% of these companies' growth. Hultink and Robben (1995) reported that new products introduced in the past 5 years generated 41% of company sales and 39% of company profits. Besides these benefits, NPD offers other benefits such as the positive impact on company image, the opening up of new markets and the provision of a platform for further new products (Storey and Easingwood, 1999).

The central task in NPD is to develop those products (characteristics) that deliver desired benefits for consumers. Unfortunately, this is more easily said than done. Although the development of new products can be rewarding, it is risky as well. A great majority of new products never make it to the market and those new products that enter the marketplace face very high failure rates. Exact figures are hard to find and vary depending on the type of market (industrial versus consumer) and product (high-tech versus fast-moving consumer goods).

Moreover, criteria for the definition of success and failure differ between studies, further complicating their direct comparison. However what is clear is that failure rates have remained high over the previous decades, averaging 40% according to Griffin (1997) and 25% according to Crawford (1987). Later, Cooper (1993), a leading researcher in the field of NPD, estimated a failure rate in the order of 25–45%. A more recent study of ACNielsen (2000) showed that only one-third of all fast-moving consumer goods (FMCG) introduced in 1998 in Dutch supermarkets could be considered successful.

Since the 1960s it has become apparent that the high failure rates justified research into the underlying reasons for success and failure in new product development. Prior to the 1960s the development of new products was considered a technological linear process; new technologies and a proactive research and development (R&D) effort were believed to drive the success of new products (Poolton and Barclay, 1998). Later on it became clear that many more factors played a role. The first studies on NPD performance showed that the marketplace played a major role in stimulating the need for new and improved products. Since the pioneering studies of Booz et al. (1968), the success and failure of new products has been studied intensively. Success depends among other factors on the degree to which the new product successfully addresses identified consumer needs and at the same time exceeds competitive products. The key learning emerging from NPD performance analysis is that success is primarily determined by a unique and superior product and that the achievement of that is primarily driven by the effective marketing-R&D interfacing at the very early stages (opportunity identification) of the NPD process (e.g. Henard and Szymanski, 2001).

Two sources of risk in strategic decision-making need to be managed in the new product development process (Table 14.1). A first risk is that resources are erroneously invested in a new product that would (later) appear to be a failure in the marketplace (Type-1 error). But also, a potential successful new product (idea) may undeservedly be screened out or overlooked in the new product development process, reflecting an opportunity loss or missed market opportunities (Type-2 error).

Table 14.1   Type-1 and type-2 errors in new product development (cf. Eliasbergh *et al.*, 1997; Van Trijp, 1999; Van Kleef *et al.*, 2002)

| | | **External reference** | |
| | | Consumer acceptance of new product will be | |
		low	*high*
Internal reference	*Success*	Type-1 error: Unjustified investment	Opportunity agreement
According to company, new product will be	*No success*	Non-opportunity agreement	Type-2 error: Opportunity losses

Consumer research in the new product development process aims at reducing the decision uncertainty described in Table 14.1. Miller and Swaddling (2002) argue that the shortcomings in the current state of NPD practice can be directly or indirectly tied with consumer research (or the lack thereof) done in conjunction with NPD. The nature and application of consumer research for reduction of the two types of error in decision making differ considerably. Consumer research to reduce potentially erroneous investments (Type-1) is largely confirmative in nature, well established and to considerable extent standardised within stage-gate or innovation funnel approaches (e.g. Cooper *et al.*, 2004). This type of consumer research is nicely summarised by Ozer (1999) and essentially searches for negative consumer feedback on ideas, concepts, prototypes and as-marketed products prior to their market launch. The standardisation of these methods in formal evaluation approaches provides an objective yardstick against which management can take and justify investment and product decisions. These methods essentially reflect the *testing* of new ideas and their further development, which is very important but only part of the story.

Increasingly, is it being recognised that success in NPD depends to large extent on the quality and quantity of new product ideas that are submitted to the testing approach. This is where Type-2 error comes more important, and to prevent overlooking market opportunities, the firm needs not only to test ideas but also adopt a proactive search for new opportunities in the neighbourhood of existing market supply. This search for new and distinctive market opportunities reflects the more creative, proactive side of product development research as a complement to confirmative research. The aim is to obtain a new and stimulating perspective on product ideas from outside the company through consumer feedback. Successful innovation largely depends on the firm's ability to find a good balance between these two types of consumer research thereby balancing the reactive (prevent unjustified investments/Type-1 errors) with the proactive (prevent 'myopia' or opportunity losses/Type-2 errors) angle to consumer research.

Confirmatory consumer research has been the dominant focus in much of previous NPD research, with much less attention for consumer research at the early stages of NPD. To many, these early stages in NPD have come to be known as the 'fuzzy front-end of NPD' as they typically involve ill-defined processes, uncertainties and ad-hoc decisions (Cooper and Kleinschmidt, 1986). Consumer research to support these early decisions is not easy. An often-heard argument is that asking consumers what they want is useless, because they do not know what they want unless they see it (Ulwick, 2002). So we need more creative techniques than 'just ask consumers what they want' to help to raise the odds of success in the marketplace. Even though consumers may not always be able to express their wants, it is important to understand how they perceive products, how their needs are formed and influenced, and how they make product choices based on them. Several authors have argued that the opportunity identification stage, where product ideas are generated and initially screened, provides one of the greatest opportunities for improvement of new product success rates (Rosenau, 1988; Khurana and Rosenthal, 1998). During these early stages, the product has not yet been specified and the aim is to search for novel product ideas. Wind and Mahajan (1997) argue in their influential paper that most of the improvements of the NPD process would be most beneficial for activities dealing with the earlier stages of the NPD process.

One of the main conclusions of the many studies into new product performance is that predevelopment activities significantly improve new product success rates and are strongly correlated with financial performance (Cooper, 1988; Montoya-Weiss and Calantone, 1994). During this phase in NPD, new product concepts are generated and initially screened, prior to the actual development phase. It is a critical phase because deficiencies here result in costly problems in later stages of the NPD process. Product concepts are the basic components for NPD and concept selection decisions dictate all further development activity within a company (Roozenburg and Eekels, 1995). Cooper (1988) found that the quality of the execution of the predevelopment steps – preliminary market and technical studies, market research, business analysis and initial screening – is closely tied to financial performance. Basically, he showed that weaknesses in up-front activities seriously enlarge the chances for failure. In addition, he found that successful projects have over 1.75 times as many person-days spent on predevelopment activities, as do failures. Other authors claim as well that more time and resources should be devoted to activities that precede the actual development of products. Hise *et al.* (1989), for example, suggest that companies that use a full range of up-front activities (e.g. market definition, identifying consumer needs) have a 73% success rate compared with a 29% success rate for companies that use only a few of the up-front activities.

In this chapter we discuss the role and importance of consumer research at the early stages of NPD for the purpose of inspiration and focus in the NPD process. Despite their availability, methods and techniques for consumer research at the early stages of NPD are poorly used by companies or mostly applied in an ad-hoc manner (Mahajan and Wind, 1992; Nijssen and Lieshout, 1995; Nijssen and Frambach, 2000). There exists a bias towards the application of focus groups, surveys and the study of demographic data, which is considered to be one of the reasons for the relatively low new product success rates (Wind and Mahajan, 1997). The failure of methods to reach their full potential in NPD is perhaps the result of the limited and confused way in which they have been evaluated and made clear to potential users. Therefore, Van Kleef *et al.* (2005) developed a taxonomy of ten common methods and techniques for opportunity identification with the aim to support researchers and practitioners in identifying the most optimal methods and techniques. We first discuss the two dimensions of this taxonomy in Section 14.2.1. The first dimension is actionability of the output of methods to support marketing versus R&D. The second dimension in the taxonomy is the innovation strategy that is pursued (winning in existing well-defined markets versus building a new market through radically new products). Section 14.2.2 presents the classification scheme in which the ten methods and techniques are positioned against these two dimensions.

The key to successful new product development lies in delivering superior and distinctive need satisfaction in the marketplace. In this process, many of the consumer research methodologies put the consumer first, to ensure that the voice of the customer is being heard in the new product development process. The underlying idea is that first the relevant functions of the new product are being identified from consumer research, after which a product form is being found to satisfy the consumer needs (i.e. 'form follows function'). More recently, several scholars have expressed concern with the extent to which this linear process really identifies new product opportunities as consumers are limited in their introspection into their own need structures, in what they can articulate and their creativity in identifying new 'out of the box' options satisfying those needs. Section 14.3 addresses some of these concerns and limitations of supporting methodologies in opportunity identification. Wind and Mahajan (1997) argue strongly for the development of new approaches, which aim to overcome these problems. In this respect, we discuss in Section 14.4 the method of innovation templates recently introduced into the marketing literature by Jacob Goldenberg and colleagues. This approach turns around the 'form follows function' logic, provocatively arguing for the 'voice of the product' as a source of innovation ideas which are subsequently tested with consumers for their market relevance (i.e. function follows form). The chapter closes with a brief discussion on important issues in opportunity identification.

14.2   A typology of consumer research for opportunity identification

Consumer research for opportunity identification aims at improving the quality of initial new product ideas. Numerous consumer research methods are available to understand consumer needs and wants for product development purposes. But despite the widespread recognition of the important role that a focus on the consumer plays in NPD, most companies fail to use these methods in an appropriate manner (Nijssen and Lieshout, 1995). Product developers are still relying on gut-feel with respect to 'best practice' in NPD. Van Kleef et al. (2005) deal with the problem that it is often difficult to select a consumer research method or technique in the early product development stages, especially for product developers from a (food) technological disciplinary background. They critically reviewed and categorised ten methods and techniques that are used most frequently to uncover unmet consumer needs and wants: (1) empathic design, (2) category appraisal (including preference analysis), (3) conjoint analysis, (4) focus group, (5) free elicitation, (6) information acceleration (IA), (7) Kelly repertory grid, (8) laddering, (9) lead user technique, and (10) Zaltman metaphor elicitation technique (ZMET). Table 14.2 presents a short description of each of these methods.

Table 14.2 Ten consumer research methods and techniques for opportunity identification in new product development (methods and techniques are described more thoroughly in the original paper (Van Kleef *et al.*, 2005), including relevant references about method development and applications).

Method	Description
Category appraisal	Set of techniques (e.g. preference analysis) which visually (typically two-dimensional) shows the market structure of a product category, as perceived or preferred by consumers.
Conjoint analysis	Method in which consumer preferences are derived from experimentally varied product profiles. Statistical analysis shows the relative importance of product attributes in the overall consumer preference for a product.
Empathic design	Observational approach in which consumers are observed by product developers in their own environment. Researchers aim to find signs that refer to latent needs, frustrations with products/situations, confusion or unexpected product uses.
Focus group	Interactive group discussion technique in which consumers (8–12) discuss about a variety of products or topic led by a moderator. Results are interpreted by researchers/observers.
Free elicitation	Personal interview technique in which participants are asked to mention spontaneously all associations that they have a certain product (category).
Information acceleration	Concept-testing method for very new products, which is based on experimental design with virtual stimuli in a future purchase environment. Consumers are asked for their preferences and purchase intentions.
Kelly repertory grid	Personal interview technique in which constructs are revealed that consumers use to understand a product category. Participants are repeatedly confronted with a selection of three products and have to indicate which two are similar and different from the third.
Laddering	Personal interview technique to reveal knowledge structure of product(category). Means–end chains are elicited by evaluating multiple products, while the interviewer repeatedly probes for the reasons why a respondent gives an answer.
Lead user technique	Technique which involves innovative consumers in the product development process. These lead users are presumed to benefit the most from potential new products and therefore they have a better understanding of consumer needs and problems surrounding existing products.
Zaltman metaphor elicitation technique (ZMET)	Motivation-driven technique that analyses consumer needs by means of pictures, metaphors and emotions. Expressive collages are made based on material (pictures, stories, etc.) collected by consumers.

14.2.1 Determinants of the appropriateness of consumer research methods for opportunity identification

Despite their common objective, consumer research methods for opportunity identification differ in many respects: not only in the procedure they follow, but also in the resulting consumer needs. Fundamental differences in these methods may lead to different 'optimal' solutions to consumers' unmet needs. The choice for using a particular method or technique in the predevelopment stages is therefore not arbitrary. Our review and classification reveals that methods and techniques can meaningfully be classified according to their degree of actionability for marketing versus R&D and their ability to develop 'out of the box' ideas. We will discuss these two dimensions of the taxonomy next.

Actionability for marketing versus R&D

Applying methods is no guarantee for the actual use of their results. Information will be used only if it is perceived to be relevant for the task for which the receiver is responsible (Moenaert and Souder, 1996; Madhavan and Grover, 1998). Consumer research at the opportunity identification phase should hence deliver (1) understanding what drives consumers' decision processes and which factors influence these processes as foundation for the generation and screening of new product ideas, and (2) concrete input for the subsequent technical development stage (Rochford, 1991; Mascitelli, 2000). For that reason, it is relevant to evaluate methods on their actionability in providing critical input to both technical and marketing-related tasks in NPD. Actionability refers to the ability of information to indicate specific actions to be taken in order to achieve the desired objective (Shocker and Srinivasan, 1979).

In assessing the actionability of elicited consumer needs, we distinguish a hierarchy of concrete product characteristics that form the basis of the technical product specification to abstract consumer values (Fig. 14.1). *Product characteristics* are measurable and manipulable, and are physical properties of products under the control of technical product developers (Shocker and Srinivasan, 1979; Myers and Shocker, 1981). These characteristics are also referred to as 'tangible'. *Product attributes* are those technical features/characteristics as they are perceived by the consumer and potentially expressed in consumer terminology. These perceived product attributes (either intrinsic or extrinsic) also form the substrate from which the consumer infers and interprets the *benefits*

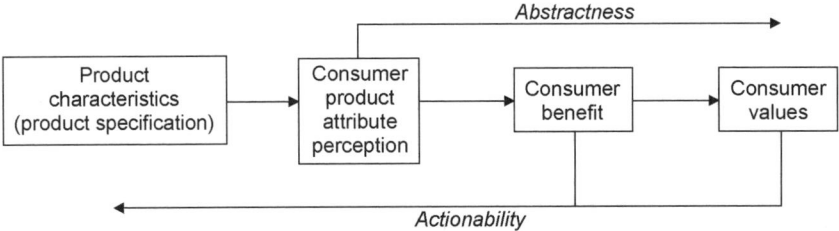

Fig. 14.1   Actionability and abstractness of provided information.

that the product is expected to deliver (e.g. creaminess). The key characteristic of these benefits is that they reflect what the product does for the consumer. Benefits are pleasant consequences of consuming a product. Different products can deliver the same benefit, which implies that benefits are not product specific. Benefits differ from attributes in that people *receive* benefits whereas products *have* attributes (Myers and Shocker, 1981; Gutman, 1982). Examples of benefits include 'health', 'good taste' and 'convenience'. Finally, values represent important beliefs about oneself and the perception of oneself by others. They are either 'instrumental' (preferred modes of conduct such as honesty and courage) or 'terminal' (preferred end-states of being such as freedom and living an exciting life) (Rockeach, 1973). It is important to note that the relationship between consumer benefits and product characteristics is not unique: the number of product characteristics is far greater than the number of attributes and benefits. Multiple product characteristics can satisfy a product attribute and multiple attribute combinations can provide the consumer one particular benefit. This is called the 'reverse mapping problem', which indicates the lack of one-to-one translation possibility in NPD (Kaul and Rao, 1995). The more abstract the consumer needs that are elicited, the less actionable a method is for technical product development (Fig. 14.1).

Technical product developers have the task of merging knowledge of what consumers want with knowledge of what is (technologically) possible. Product developers need to know how abstract benefits (e.g. enhancing my health condition) translate into specific, concrete characteristics sought from desirable alternatives (e.g. the specific health-promoting ingredients in a food). Methods that indicate which product attributes and characteristics consumers use to infer the presence of desired consequences permits clearer specifications for product development. Marketing-oriented tasks involve the creative phase of finding new product ideas. When consumer needs are linked too early to product characteristics, it may kill the creativity in finding really new product ideas. The more abstract consumer needs are, the more freedom in creativity is left. Information about which benefits consumers are seeking in a particular product enlarges the solution space and prevents thinking within the box of current product delivery. In this way, it can serve as a source of inspiration. Inspiration refers to becoming motivated because of new insights and possibilities being revealed that individuals would not have recognised on their own (Thrash and Elliot, 2003). Additionally, it may create a shared understanding and team spirit in the development group (Slater and Narver, 2000).

Newness of product considered
The innovation strategy pursued is the second dimension of our taxonomy outlining the appropriateness of methods. Optimisation of products requires other information about consumer needs and wishes than the development of really innovative products. The optimisation of products is a continuously needed activity to keep up with competitors and stay cost-efficient. Besides optimisation, companies can focus on developing really new products anticipating consumers'

future needs and desires. Consumers may have needs that they are not aware of, often referred to as 'latent needs'. Consumers do not ask for the fulfilment of these needs and may not have the ability to articulate them. This is because products, which could probably fulfil them, do not yet exist. Identifying and understanding such 'latent needs' is of crucial importance, since these needs, if they were fulfilled, would delight and surprise the consumer (Griffin and Hauser, 1993). Moreover, novel solutions to people's latent needs can differentiate a product from its competitors and make consumers more loyal (Oliver *et al.*, 1997). Depending on the new product being considered (winning in existing markets versus developing new markets), different methods are more or less appropriate to support product developers in their task.

14.2.2   A typology of consumer research

Based on the two dimensions discussed above, Fig. 14.2 provides a mapping of the ten methods for consumer research in the early stages of new product development. The horizontal dimension ('newness') describes the extent to which the method supports the development of really new products versus supporting incremental product optimisation. The vertical dimension ('action-ability') distinguishes methods that primarily support marketing versus the technical product development function.

Methods are mapped onto these quadrants on the basis of their distinguishing features. In particular, the following criteria are used. First, we define the method's basic approach to need elicitation, which can originate either from responses to concrete products or from reflection on consumer motivation. Second, task differences in methods can be responsible for differences in elicited consumer needs. Finally, need actionability is taken into account, which is the ability of need information to indicate specific actions to be taken to achieve the desired objective in NPD.

All methods situated at the left hand side of Fig. 14.2 (i.e. focus group, free elicitation, Kelly repertory grid, laddering, category appraisal and conjoint analysis) are particularly appropriate for incremental new products; products that are repositionings or updated versions of existing products. These methods have in common that they tend to elicit consumer needs from providing consumers with product examples (i.e. product-driven) and primarily from familiar stimuli. Product-driven methods confront consumers with products as cues to start the identification of needs and wants. Consequently, they provide insights that are limited by the particular product(s) included in the study – that is, they elicit consumer needs within an existing framework of what is already available on the market. Consumers can generally give reliable judgements about new products that are relatively similar to familiar products. Hence, the advantage of these methods lies in their capacity to capture current needs and desires and optimise existing products accordingly. However, their limitation lies in the fact that it is difficult to elicit unfulfilled needs by analysing preferences for products currently existing in the marketplace. Although they can provide clues as to

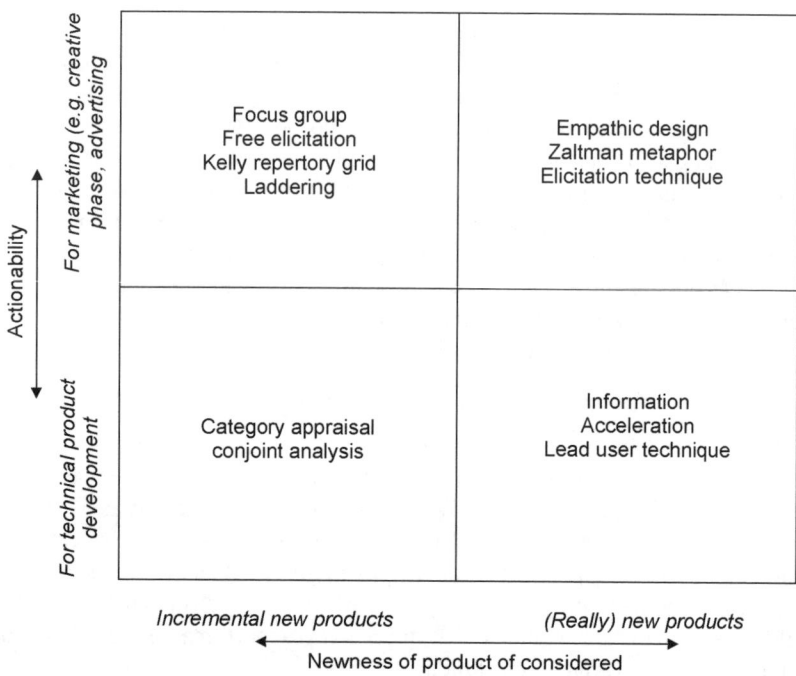

Fig. 14.2 Recommended consumer research methods for opportunity identification based on newness of product considered and actionability for technical product development or marketing (Van Kleef *et al.*, 2005).

which benefits people are seeking in the near future, these approaches primarily refer to consumer needs that are widely understood by competitors in a market as well. A risk of relying on them is that they are likely to give companies only 'me-too' ideas, which hardly excite the consumer. Category appraisal and conjoint analysis are highly actionable for technical product development, because they allow product developers to understand how consumer needs interrelate and translate to the 'physical' domain of product characteristics. Laddering, Kelly repertory grid, free elicitation and focus group are more appropriate for marketing purposes, as they reveal more abstract consumer needs and values. These more abstract needs and values are closer to what drives consumer choice behaviour. However, they are too abstract and allow too many degrees of freedom for unambiguous translation into product design.

The right hand side of Fig. 14.2 involves methods more appropriate for (radically) new products and thinking outside the box. Really new products are particularly risky to develop, but at the same time, yield the highest long-term financial gains if they succeed in the marketplace. However, this type of 'out-of-the-box' product is much more difficult to evaluate by the consumer because they do not fall into any established current category and probably combines several technologies not currently available together (Eliashberg *et al.*, 1997). Simply asking consumers what they want is not likely to elicit unfulfilled needs, because consumers tend to mention needs that are already catered for in the marketplace. As a result, highly complex, radically new products pose special challenges to consumer research. When considering radically new products, consumers have to make major modifications to their choice processes (e.g. Goldenberg *et al.*, 1999a). In particular, consumers need to change their behaviour in order to adopt the product. Hence, the major difficulty in conducting consumer research is the consumer's lack of experience with the product. Confronting consumers with unfamiliar products (e.g. a really new concept) may lead to unrealistic situations and information that has limited predictive validity. After all, for new products consumers have less information in their memory to guide them and expressions of preference are often constructed at the time that the respondent is asked to give a judgement. As a result, consumers may change their opinion by the time the product is introduced.

Two groups of methods can be distinguished on the basis of their action-ability. The lead user technique and information acceleration both try to access consumers' unspoken and latent needs, but with a clear link to physical 'solutions' against those needs. Information acceleration explicitly takes into account that consumers might not have the level of product knowledge that is necessary for judging new products. By creating a simulated future environment, respondents are guided in understanding what a new product can do for them. The lead user technique uses a sample of consumers whose present needs are expected to become general in the marketplace months or years in the future. Moreover, lead users may have developed solutions to problems encountered with existing problems. However, relying on lead users can also have its risks. Their needs many be of limited appeal, perhaps applicable only to other lead users (Ulwick, 2002). ZMET and the empathic design technique are also appropriate for really new products. They are both need-driven in that they focus on understanding consumer problems or motivations. They specifically focus on the more latent non-articulated needs and hence provide detailed insight into what really drives consumer behaviour. This information is highly actionable for marketing purposes (e.g. communication strategy). However, as a downside, this abstract insight requires additional methods for translation into actual physical product design.

14.3   Opportunity identification: some concerns and limitations of supporting methodologies

Lack of consumer relevance and poor application of consumer research at the early stages of the new product development process have been identified as key determinants of failure in new product development. Several scholars have voiced some concerns with the use of consumer research for opportunity identification, arguing that consumers may not always be the best resource in this more creative and proactive side of new product development. The first argument concerns the use of 'ordinary' consumers who may have difficulty expressing their future needs (Section 14.3.1). The possible downside from this is that many NPD consumer research methods may tend to be biased toward solutions to consumers' current problems and thereby provide relevant input only for continuous, not breakthrough, innovation (Wind and Mahajan, 1997). It is well known that the unique relative advantage of a new product is a major predictor of its success (Cooper, 1993). NPD consumer research methods should generate such out-of-the-box ideas, which demands breaking out of obvious thinking by searching outside the scope of current product delivery in the marketplace. Coming up with innovative concepts hence requires creativity and divergence in thinking of product developers. Methods to support this creative phase in NPD are increasingly criticised (Section 14.3.2).

14.3.1   Ordinary consumers as a starting point in opportunity identification

In recent years several authors have also pointed toward potentially serious limitations of using 'average' or a random set of consumers as the starting point of NPD idea generation. It is argued that being market and consumer-oriented has incorrectly been confused with being *consumer-driven* (e.g. Zaltman, 2003; Narver *et al.*, 2004; Van Kleef *et al.*, 2005). Such consumer-driven approaches in which the consumer is, to a large degree, the origin of new product ideas, would assume that (1) consumers are able to identify and articulate unfulfilled needs, and (2) consumer are able to identify and articulate potential solutions to these unfulfilled needs almost as a 'recipe' for new products (e.g. Van Trijp and Steenkamp, 2005). Several authors (e.g. Slater and Narver, 1998, 1999; Connor, 1999; Reid and De Brentani, 2004) have argued that consumers find it difficult to articulate their needs for products that do not yet exist. As consumers are limited by their current experiences and environment, their input is believed to inhibit new ideas (e.g. Lilien *et al.*, 2002). Most consumer research methods work well in understanding consumer preferences among existing products, but are less appropriate in identifying future needs that consumers cannot yet articulate. As a result, companies may fail in picking up emerging markets and consumer needs. Also, it has been argued that this approach may lead to 'me-too' products rather than real innovations.

Several researchers have suggested alternative approaches to overcome this problem of ordinary consumers not being able to articulate their future product needs. For example, the lead user approach (Von Hippel, 1986) involves more advanced consumers rather than ordinary consumers when it comes to need and solution identification. Lead users are selected on their characteristic of recognising a particular product need much earlier than average consumers. They may have even developed their own products to solve their problems with existing products (Von Hippel and Katz, 2002). The ZMET was developed to overcome similar problems with traditional methods of consumer research (Zaltman, 2003). In this approach, researchers attempt to tap into the unconscious level of consumers' thinking by including unconventional inquiry techniques (for an application of this technique, see Christensen and Olson, 2002). The basic premise of the 'empathic design' method is that the richest information on consumer needs can be acquired by observing consumers in their own surroundings (Leonard and Rayport, 1997).

14.3.2   Generating innovative product concepts

Many new product development processes find their start in well-known idea-generating techniques such as brainstorming. The underlying assumption of these techniques is that generating ideas is most productive when conducted in an unrestricted fashion. For example, brainstorming encourages creativity by creating an atmosphere in which there is deferral of judgements, which stimulates participants to generate a large number and wide variety of ideas. It is believed that the more ideas produced, the greater the probability that an original idea will emerge (e.g. Baker and Hart, 1999). Increasingly, however, there is evidence (some of it summarised in Goldenberg and Mazursky, 2002) that these open and unrestricted creativity tasks are not all that effective and efficient in the new product development process. Their perceived success tends to be based on the 'illusion of group productivity' and the usual variety of discussions held within a brainstorming group tends to interfere with a person's ability to work in a productive way (Nijstad *et al.*, 2003; Kerr and Tindale, 2004). More specifically, true group productivity is undermined because other people are talking; the pressure is not always on an individual.

Recent research in marketing has challenged the basic assumption that good new ideas would have to originate from consumer research. Goldenberg *et al.* (1999b) argue that much of the innovation potential resides in the structure of products currently in the marketplace. Hence, firms should listen to the 'voice of the product' and unlock the innovation potential from those product structures using systematic approaches. Once these new product ideas have been extracted, they should be submitted to early consumer research to explore their potential in terms of consumer relevance. Goldenberg and colleagues developed a set of templates – regularities in the emergence of successful innovations – to systematically explore directions for innovation and new product ideas. Because of the potential of this approach for the new product development process, we will discuss the approach in more detail in Section 14.4.

14.4   Goldenberg's innovation template approach

The innovation templates of Goldenberg and colleagues are a set of systematic operators that help to transform the product from an earlier version to a new version. The basic thinking behind the approach is that over time, market changes leave traces in product configurations that can be identified as product-based trends. Those trends, crystallised as templates, provide the skeleton from which new successful future product ideas can be generated. Inspired by the work of the Russian engineer Altschuller (1986), Goldenberg and colleagues (e.g. Goldenberg *et al.*, 1999a,b, 2000, 2003; Goldenberg and Mazursky, 2002) identified six underlying mechanisms ('templates') that, when applied to the initial structure of an existing product, lead to a creative output by analysis of historical product changes.

The innovation templates approach builds on three major premises: (1) structured creativity, (2) restricted scope, and (3) function follows form. Together, these three principles define a structured ideation approach (see next section), which is claimed to enhance NPD success. Empirical studies on the contribution of innovation templates in NPD success are scarce, the only evidence coming from two studies by Goldenberg and colleagues. Goldenberg *et al.* (2001) show evidence that templates significantly distinguish successful from failed new products in the marketplace, and hence are better able to identify product ideas that capture consumers' future needs. The rationale behind the success of the templates is that by their very nature of being systematic variations on existing products, they facilitate consumer evaluation by providing a sense of familiarity to an otherwise unexpected and novel product idea (i.e. 'an optimal level of newness').

14.4.1   Three key premises underlying the innovation templates approach

Three key premises form the basis of the innovation templates approach (Goldenberg and Mazursky, 2002, p. 41), each of which challenges some conventional thinking in idea generation:

- Structured creativity: several identified, universal templates underlie product evolution and these can be exploited to predict new candidate products.
- Restricted scope principle: channelling thinking along predefined inventive routes makes people more productive in idea generation.
- Function follows form: enhancing the recognition of innovative ideas by applying an unusual sequence: first new configurations for a product are proposed, for which potential consumer appeal is inferred afterwards (rather than 'form follows function' as usually applied in more traditional NPD approaches).

Structured creativity
Creativity is pursued along a very structured process. As a first step, the essential elements of the existing product, both its physical components and its attributes, such as colour and shape are being listed, including those available in the product's immediate environment such as type of consumer using the product or outside temperature. Then, these components and attributes are manipulated in a very structured way following one or more of the templates in order to come up with a new product configuration. The six basic template operators can be characterised as follows (see also Goldenberg *et al.*, 1999b; Goldenberg and Mazursky, 2002).

The *attribute dependency* template is based on finding two independent variables (i.e. a change in one does not cause a change in the other) and creating a new dependency between them. Take, for example, a standard mayonnaise product. There is no dependent relation between an ingredient (i.e. mustard) and an external situation (such as 'region of origin'). By introducing a new dependency to previously independent product attributes (the procedure applied by the attribute dependency template), a product developer can come up with a Dijon mayonnaise, in which a mayonnaise ingredient (i.e. mustard) is related to the region of origin of the ingredient (Dijon).

The *replacement* template is based on the replacement of an essential component of the product by something in the immediate environment of the product that can fulfil the same necessary function. An example is a keyboard of a portable computer which transforms mechanical energy (from the user's fingers) to charge the battery. In this example, the battery in the product (an essential component) is replaced by a more beneficial system that draws on the user in providing the necessary energy. An example in food products could be the first product which replaced sugar by an artificial sweetener which has the same function (sweetening of product) including other advantages, such as lowering the amount of calories of the product. Another example is a mayonnaise from which oil is removed and replaced by plant sterols. The plant sterols have the same function (giving structure and taste to product) including a new health advantage, which is lowering the serum cholesterol.

The *component control* template involves creating a new link between a component in the internal environment of the product and a component in its external environment. An example is 3M's Post-It notes, which can repeatedly be attached and removed from a table. Compared with ordinary notes, a new link is created between the note and the table, leading to new benefits.

In the *displacement* template, an essential component is removed including its associated function. An often mentioned example is the first Sony Walkman, where the recording device is removed from the cassette player, making it feasible to develop a smaller product that can be carried around.

In the *division* template, a product component is split in two and each new component is made responsible for a new function. For example, the ingredients of a strong washing powder are split to produce two products, one regular and one strong for highly soiled laundry.

Finally, the *multiplication* template involves making one or more copies of an existing product component and alters them in some important way. For example, think of the opening of a ketchup jar which is copied. This second opening made it possible to precisely dose the amount of ketchup.

Restricted scope
Rather than broadening the scope to enhance creativity, the restrictive scope principle states that inventive productivity is enhanced when the search area of an issue will be limited/restricted. In contrast to idea generation techniques that lack a structured framework, within the template approach the number of

variables under consideration is being limited through the application of innovation templates and this is believed to increase effectiveness and efficiency in creativity.

Function follows form

The application of the template operators leads to new (virtual) product forms, which are then examined for consumer relevance and benefits. This examination for consumer relevant benefits can be done by experts or by means of consumer concept evaluation studies. Goldenberg *et al.* (2001) argue that this unusual sequence of steps (i.e. 'function follows form') in obtaining new product ideas is beneficial for the idea generation process itself. People are more likely to make creative discoveries when they analyse novel forms and then assess what consumer benefits they might possess, rather than when they try to create an optimal form solely on the basis of consumer-desired benefits (Finke *et al.*, 1992). This contrasts with the traditional sequence of steps, also known in design literature as 'form follows function' (e.g. Krishnan and Ulrich, 2001). Form is the set of product characteristics that make up the product. Function is the set of consumer-desired benefits that would be fulfilled by the product form. Take, for example, the quality function deployment approach, which starts by identifying consumer-desired product attributes (Hauser and Clausing, 1988) and then systematically translates these attributes to measurable product characteristics.

14.4.2   Templates and market success

Goldenberg and Mazursky (2002) argue that templates carry codes for the evolution of successful new products and that they can be exploited to generate a competitive advantage based on minimal *a priori* market information. Besides their claimed advantages in the idea generation process, Goldenberg *et al.* (2001) conclude also, based on two studies, that templates significantly distinguish successful from failed new products in the marketplace. In the first study, a set of successful and unsuccessful products (from three categories: kitchen devices, garden tools and car devices) was collected from the Israeli patent office. A product was considered a failure when it was either totally rejected by the market or when its introduction was cancelled because of poor test market results. Each product was classified according to the templates by judges trained in template identification. The predictive power of the templates was assessed by a logistic regression analysis which indicated that a high proportion (88.6%) of the failures and successes could be predicted by the model. Of all 41 included successful products, 36 products were predicted to be successful based on their template structure. Similarly, of all 29 included failed products, 26 products were predicted to be a failure based on their lack of template structure. The second study likewise assessed the predictive power of template variables, but also included other variables such as source-idea determinants and project-level determinants. A set of 127 detailed cases of successful and unsuccessful consumer products was collected (70 successes and 57 failures) from three different books using the same criteria of success and failure as in the first study. Two of these books described a large set of product failures, including some of the classic cases such as New Coke (Adler and Houghton, 1997; McMath and Forbes, 1998). One book describes 50 well-known successful inventions, such as Post-It notes, disposable nappies (diapers) and the Swiss army knife (Freeman and Golden, 1997). Interjudge agreement in classification of product templates was high ($\alpha = 0.89$). A logistic regression analysis indicated that most (81.9%) of the failures and successes can be predicted by the template and idea-source variables. Based on these results, the authors concluded that products that follow the template structure have a greater likelihood of success.

14.5   Conclusions

In the numerous studies of new product performance over the years, consensus has developed that understanding consumer needs is of greatest strategic value, especially in the early stages of the NPD process. During these early stages, the product has not yet been specified and the aim is to search for novel product ideas. As indicated in the introduction of this chapter, product success rates continue to be extremely low and little improvement can be seen over time. Consequently, a consumer orientation in NPD will continue to be extremely important in the future. The challenge is to continuously generate new knowledge about consumer needs and how to best satisfy them (Slater and Narver, 1998; Narver and Slater, 2000). Listening to current consumers of a product may lead companies to miss significant opportunities for innovation.

Of special importance in the design and improvement of consumer research methods for NPD is the ability to provide guidance in the development of really new products and not just line extensions and incremental improvements to existing products. The difficulties that consumers have with expressing their needs and evaluating the potential of new products do not imply that consumer research should be left out. It does, however, pose special challenges to consumer research. This kind of research is the most challenging, but ultimately may yield the greatest payoff. In particular, this kind of research should support the elicitation of latent and emergent needs. Wind and Mahajan (1997) argue that most consumer research methods focus on continuous innovations in predictable markets. Although this kind of research may provide valuable input in the NPD process, consumer research should become proactive and focus on overcoming the problems of ordinary consumers having difficulty expressing their future needs. In particular, Wind and Mahajan (1997) argue that new research approaches are required that avoid consumers' short-term and current experience bias and enable them to identify their true needs and wants as they may evolve under future scenarios.

Ideally, consumer research for opportunity identification is carried out on a continuous basis. It is not just enough to be able to describe the current state of the market in detail. The consumer's own circumstances may have changed or what used to be a valuable benefit is no longer so important. Competitors' offerings change as well, so it is not safe for a company to assume that it understands consumers' value perceptions for very long (Miller and Swaddling, 2002). An early understanding of changes in consumer behaviour makes it possible to anticipate market opportunities and respond before competitors do. In this way, consumer research helps to expand the time horizon of innovation. Rather than be executed on an ad-hoc basis with a short-term focus, it should strongly and coherently be embedded in the total business process. This allows for systematic learning and anticipating on developments rather than only reacting to them (Hughes and Chafin, 1996).

14.6   Sources of further information and advice

JONGEN, W.M.F. and MEULENBERG M.T.G. (Eds) (2005). *Innovation in Food Production Systems: Product Quality and Consumer Acceptance*, 2nd edition. Wageningen: Wageningen Press.

GOLDENBERG, J. and MAZURSKY, D. (2002). *Creativity in Product Innovation*. Cambridge: Cambridge University Press.

URBAN, G.L. and HAUSER, J.R. (1993). *Design and Marketing of New Products*, 2nd edition. Englewood Cliffs, NJ: Prentice-Hall.

14.7   References

ACNIELSEN (2000). *Product introducties: de feiten op een rij.* ACNielsen Research.

ADLER, B. and HOUGHTON, J. (1997*). America's Stupidest Business Decisions.* New York: William Morrow and Company Inc.

ALTSCHULLER, G.S. (1986). *To Find an Idea: Introduction to the Theory of Solving Problems of Inventions.* Novosibirsk, USSR: Nauka.

BAKER, M. and HART, S. (1999). *Product Strategy and Management.* London: Prentice-Hall Europe.

BOOZ, ALLEN AND HAMILTON (1968). *Management of New Products.* Chicago: Booz, Allen and Hamilton Inc.

BOOZ, ALLEN AND HAMILTON (1982). *New Product Management for the 1980s.* New York: Booz, Allen and Hamilton Inc.

CHRISTENSEN, G.L. and OLSON, J.C. (2002). Mapping consumers' mental models with ZMET. *Psychology & Marketing*, **19**(6), 477–502.

CONNOR, T. (1999) Customer-led and market-oriented: a matter of balance. *Strategic Management Journal*, **20**, 1157–1163.

COOPER R.G. (1988). Predevelopment activities determine new product success. *Industrial Marketing Management*, **17**, 237–47.

COOPER, R.G. (1993). *Winning at New Products: Accelerating the Process from Idea to Launch.* Reading, MA: Addison-Wesley.

COOPER, R.G. and KLEINSCHMIDT, E.J. (1986). An investigation into the new product process: steps, deficiencies, and impact. *Journal of Product and Innovation Management*, **3**(2), 71–85.

COOPER, R.G., EDGETT, S.J. and KLEINSCHMIDT, E.J. (2004). Benchmarking best NPD practices – I. *Research Technology Management*, **47**(1), 31–43.

CRAWFORD, M.C. (1987). *New Products Management*, 2nd edn. Homewood, IL: Irwin.

ELIASHBERG, J., LILIEN, G.L. and RAO, V.R. (1997). Minimizing technological oversights: a marketing research perspective. In Garud, R., Nayyar, P.R., Shapira, Z.B. (Eds), *Technological Innovation: Oversights and Foresights* (pp. 214–230). New York: Cambridge University Press.

FINKE, R.A., WORLD, T.B. and SMITH, S.M. (1992). *Creative Cognition*. Cambridge, MA: MIT Press.

FREEMAN, C. and GOLDEN, B. (1997). *Why Didn't I Think of That?* New York: John Wiley & Sons.

GOLDENBERG, J. and MAZURSKY, D. (2000). First we throw dust in the air, then we claim we can't see: navigating the creativity storm. *Creativity and Innovation Management*, **9**(2), 131–143.

GOLDENBERG, J. and MAZURSKY, D. (2002). *Creativity in Product Innovation*. Cambridge: Cambridge University Press.

GOLDENBERG, J., LEHMANN, D.R. and MAZURSKY, D. (1999a). The primacy of the idea itself as a predictor of new product success. *MSI working paper*, Report No. 99-110.

GOLDENBERG, J., MAZURSKY, D. and SOLOMON, S. (1999b). Toward identifying the inventive templates of new products: a channeled ideation approach. *Journal of Marketing Research*, **36**(May), 200–210.

GOLDENBERG, J., LEHMANN, D.R. and MAZURSKY, D. (2001). The idea itself and the circumstances of its emergence as predictors of new product success. *Management Science*, **47**(1), 69–84.

GOLDENBERG, J., HOROWITZ, R., LEVAV, A. and MAZURSKY, D. (2003). Finding your innovation sweetspot. *Harvard Business Review*, March, 3-11.

GRIFFIN, A. (1997). PDMA research on new product development practices: updating trends and benchmarking best practices. *Journal of Product Innovation Management*, **14**(6), 429–458.

GRIFFIN, A. and HAUSER, J.R. (1993). The voice of the customer. *Marketing Science*, **12**(1), 1–27.

GUTMAN, J. (1982). A means–end chain model based on consumer categorization processes. *Journal of Marketing*, **46**(Spring), 60–72.

HAUSER, J.R. and CLAUSING, D. (1988). The house of quality. *Harvard Business Review*, May–June, 63–73.

HENARD, D.H. and SZYMANSKI, D.M. (2001). Why some new products are more successful than others. *Journal of Marketing Research*, **38**(August), 362–375.

HISE, R.T., O'NEAL, L., McNEAL, J.U. and PARASURAMAN, A. (1989). The effect of product design activities on commercial success levels of new industrial products. *Journal of Product Innovation Management*, **6**(1), 43–50.

HUGHES, G.D. and CHAFIN, D.C. (1996). Turning new product development into a continuous learning process. *Journal of Product Innovation Management*, **13**(2), 89–104.

HULTINK, E.J. and ROBBEN H.S.J. (1995). Measuring new product success: the difference that time perspective makes. *Journal of Product Innovation Management*, **12**(5), 392–405.

KAUL, A. and RAO, V. R. (1995). Research for product positioning and design decisions: an integrative review. *International Journal of Research in Marketing*, **12**, 293–320.

KERR, N.L. and TINDALE, R.S. (2004). Group performance and decision making. *Annual Review of Psychology*, **55**, 623–655.

KHURANA, A. and ROSENTHAL, S.R. (1998). Towards holistic 'front ends' in new product development. *Journal of Product Innovation Management*, **15**(1), 57–74.

KRISHNAN, V. and ULRICH, K.T. (2001). Product development decisions: a review of the literature. *Management Science*, **47**(1), 1–21.

LEONARD, D. and RAYPORT, J.F. (1997). Spark innovation through empathic design. *Harvard Business Review,* December, 102–113.

LILIEN, G.L., MORRISON, P.D., SEARLS, K., SONNACK, M. and VON HIPPEL, E. (2002) Performance assessment of the lead user idea-generation process for new product development. *Management Science*, **48**(8), 1042–1059.

MADHAVAN, R. and GROVER, R. (1998). From embedded knowledge to embodied knowledge: new product development as knowledge management. *Journal of Marketing*, **62**(October), 1–12.

MAHAJAN, V. and WIND, J. (1992). New product models: practice shortcomings and desired improvements. *Journal of Product Innovation Management*, **9**(2), 128–139.

MASCITELLI, R. (2000). From experience: harnessing tacit knowledge to achieve breakthrough innovation. *Journal of Product Innovation Management*, **17**(3), 179–193.

MCMATH, R.M. and FORBES, T. (1998). *What Were They Thinking?* New York: Times Business–Random House.

MILLER, C. and SWADDLING, D.C. (2002). Focusing NPD research on customer-perceived value. In: Belliveau, P., Griffin, A., Somermeyer, S. (Eds), *The PDMA Toolbook for New Product Development* (pp. 87–114). New York: John Wiley & Sons.

MOENAERT, R.K. and SOUDER, W.E. (1996). Context and antecedents of information utility at the R&D/marketing interface. *Marketing Science*, **42**(11), 1592–1610.

MONTOYA-WEISS, M. and CALANTONE, R. (1994). Determinants of new product performance: a review and meta-analysis. *Journal of Product Innovation Management*, **11**(5), 397–417.

MYERS, J.H. and SHOCKER, A.D. (1981). The nature of product-related attributes. In: Sheth, J.N. (Ed.), *Research in Marketing Vol. 5* (pp. 211–236), Greenwich, CT: JAI Press.

NARVER, J.C. and SLATER, S.F. (2000). The positive effect of a market orientation on business profitability – a balanced replication. *Journal of Business Research*, **48**(1), 69–73.

NARVER, J.C., SLATER, S.F. and MacLACHLAN, D.L. (2004). Responsive and pro-active market orientation and new-product success. *Journal of Product Innovation Management*, **21**(5), 334–347.

NIJSSEN, E.J. and FRAMBACH, R.T. (2000). Determinants of the adoption of new product development tools by industrial firms. *Industrial Marketing Management*, **29**, 121–131.

NIJSSEN, E.J. and LIESHOUT, K.F.M. (1995). Awareness, use and effectiveness of models and methods for new product development. *European Journal of Marketing*, **29**(10), 27–44.

NIJSTAD, B.A., STROEBE, W. and LODEWIJKX, H.F.M. (2003). Production blocking and idea generation: does blocking interfere with cognitive processes? *Journal of Experimental Social Psychology*, **39**, 531–548.

OLIVER, R.L., RUST, R.T. and VARKI, S. (1997). Customer delight: foundations, findings and managerial insight. *Journal of Retailing*, **73**(3), 311–335.

OZER, M. (1999). A survey of new product evaluation models. *Journal of Product Innovation Management*, **16**(1), 77–94.

POOLTON, J. and BARCLAY, I. (1998). New product development from past research to future applications. *Industrial Marketing Management*, **27**, 197–212.

PORTER, M.E. (1990). *The Competitive Advantage of Nations*. New York: Free Press.

REID, S.E. and DE BRENTANI, U. (2004). The fuzzy front end of new product development for discontinous innovations: a theoretical model. *Journal of Product Innovation Management*, **21**, 170–184.

ROCHFORD, L. (1991). Generating and screening new product ideas. *Industrial Marketing Management*, **20**, 287–296.

ROKEACH, M.J. (1973). *The Nature of Human Values*. New York: The Free Press.

ROOZENBURG, N.F.M. and EEKELS, J. (1995). *Product Design: Fundamentals and Methods*. Chichester: Wiley & Sons.

ROSENAU, M.D. (1988). From experience. Faster new product development. *Journal of Product Innovation Management*, **5**(2), 150–153.

SHOCKER, A.D. and SRINIVASAN, V. (1979). Multiattribute approaches for product concept evaluation and generation: a critical review. *Journal of Marketing Research*, **16**(May), 159–180.

SLATER, S.F. and NARVER, J.C. (1998). Customer-led and market-oriented: let's not confuse the two. *Strategic Management Journal*, **19**(10), 1001–1006.

SLATER, S.F. and NARVER, J.C. (1999). Market-oriented is more than being customer-led. *Strategic Management Journal*, **20**(12), 1165–1168.

SLATER, S.F. and NARVER, J.C. (2000). Intelligence generation and superior customer value. *Journal of the Academy of Marketing Science*, **28**(1), 120–127.

STOREY, C. and EASINGWOOD, C. (1999). Types of new product performance: evidence from the consumer financial services sector. *Journal of Business Research*, **46**(2), 193–203.

THRASH, T.M. and ELLIOT, A.J. (2003). Inspiration as a psychological construct. *Journal of Personality and Social Psychology*, **84**(4), 871–889.

ULWICK, A.W. (2002). Turn customer input into innovation. *Havard Business Review*, January, 92–97.

VAN KLEEF, E., VAN TRIJP, H.C.M., LUNING, P. and JONGEN, W. (2002). Consumer-oriented functional food development: how well do functional disciplines reflect the 'voice of the consumer'?, *Trends in Food Science & Technology*, **13**, 93–101.

VAN KLEEF, E., VAN TRIJP, H.C.M. and LUNING, P. (2005). Consumer research in the early stages of new product development: a critical review of method and techniques. *Food Quality and Preference*, **16**(3), 181–201.

VAN TRIJP, J.C.M. (1999). Consumer behaviour: Inspiration for innovation. Inaugural Speech, delivered on 25 March, Wageningen University (in Dutch).

VAN TRIJP, J.C.M. and STEENKAMP, J.E.B.M. (2005). Consumer-oriented new product development: an update on principles and practice. In: Jongen, W.M.F., Meulenberg M.T.G. (Eds), *Innovation in Food Production Systems: Product Quality and Consumer Acceptance*, 2nd edition. Wageningen: Wageningen Press.

VON HIPPEL, E. (1986). Lead users: a source of novel product concepts. *Management Science*, **32**(7), 791–805.

VON HIPPEL, E. and KATZ, R. (2002). Shifting innovation to users via toolkits. *Management Science*, **48**(7), 821–833.

WIND, J. and MAHAJAN, V. (1997). Issues and opportunities in new product development: an introduction to a special issue. *Journal of Marketing Research*, **34**(February), 1–12.

ZALTMAN, G. (2003). *How Customers Think: Essential Insights into the Mind of the Market*. Cambridge, MA: Harvard Business Press.

15

Consumer-driven concept development and innovation in food product development

H. Moskowitz, Moskowitz Jacobs Inc., USA

15.1 Chapter summary

Concepts are critical as blueprints for product design. This chapter presents a systematic approach to concept development using experimental design, and the migration of the systematic approach to the Internet. The topics range from how one goes about conceptualizing the process, through to study execution and data interpretation. The chapter then deals with two case histories. The first case history presents the concept development approach as the precursor to system-atized innovation, showing how elements from three product categories can be mixed/matched to generate an entirely new-to-the-world product category combining features of donuts, cookies and chocolate candy. The second case history shows how Internet-enabled concept development can lead to the creation of databases of concept ideas. The case history shows two examples: large-scale sets of elements for 70 different food/beverage/lifestyle categories (the so-called InnovAidOnline[TM].Net database) and cross-category concept studies of product features, brands, emotional benefits (the so-called It![®] databases for beverages, foods, healthful foods, and fast foods).

15.2 Importance of concepts as blueprints for product design

In corporate product development a great deal of interest has focused on the creation of new product concepts which have the promise to achieve high levels of consumer acceptance. The reason for such importance is clear to anyone who has worked in a company, but not necessarily as clear to those without such experience. Concepts constitute blueprints for products. No one in a company who suffers from limited time and resources wants to commit those resources to a project with little payout. At the onset of a development project where there is no concrete proof of product acceptance: only the concept and trend information stand between potential success and potential failure (Fuller, 1994).

Concept research is a branch of marketing research. The specific goal of concept research is to quantify responses to concepts. Concepts, in turn, constitute statements about the product. The concept may be a description of the product features (so-called product concept) or a set of reasons why the consumer should purchase the product (so-called positioning concept). For the most part marketers have dispensed with pure concept and positioning concepts when they test new ideas, preferring to work with concepts that combine both features. The goal is to create an idea that will please the consumer, and thus lead to purchase. The product developer, however, needs concepts to guide the up-front creation of the new product, and thus pays far more attention to product concepts than to positioning concepts. It is the product concept that provides part of the development blueprint.

Concept research as a discipline is primarily applied research, so that there is little in the way of training young students to do concept research in the universities. If there is any such training it occurs in the business schools, where the emphasis is on knowing the 'tools' with which to do the concept research study, and not on the rules for writing good concepts. These rules are left to the individual creative at an advertising agency, or assumed (without much justification) to be part and parcel of the business person's capabilities. Certainly concept research has not found much interest in the curricula of other schools and departments, except marketing and marketing research in business schools. A recent book on the topic by the author and two colleagues (Moskowitz *et al.*, 2005a) attempted to remedy this lack of information about concept development.

15.3   The need for faster and better concept design and innovation

With today's ever-increasing requirements on companies to market products that are better, to get the products to the market faster, and to make the products cheaper, it is vital for a company to have a source of new product ideas. These ideas need to be tested and optimized. A poor-scoring idea may actually be very profitable in the marketplace but miss out because the concept scores poorly. In contrast, a high-scoring idea may turn out to be a very poor market performer, but is introduced because somehow the concept scored well in a test, leading to expensive, time-consuming development.

During the past 40 years consumer researchers have increasingly availed themselves of higher-level methods for concept tests in order to aid the

---

Introducing **Orchard**

Desserts...

Now you can savor the taste of homemade treats anytime!

There's nothing quite like the satisfying taste of homemade snacks
and desserts. But sometimes it's hard to find the time to make them.
Now you can savor the delicious taste of homemade treats anytime,
with new Orchard Desserts.

Orchard Desserts combines real fruit with a crumb topping for a
homemade taste. Just sprinkle the crumb topping over the fruit and enjoy.
Simply pop one in the microwave for a tasty, warm treat anytime.
And because they don't need to be refrigerated they are great for on-the-go.

Choose from three deliciously distinctive varieties:
Cranberry, Apple, or Blueberry.

Comes in a package of three convenient 4 oz. portions for $2.39

Fig. 15.1   Example of a complete concept.

development and marketing functions. These methods fall into two categories –
methods that predict 'success rates' and methods that 'diagnose features which
drive acceptance'. A word about the former, prediction of success rates, is in
order. Success rates in concept testing mean the ability to estimate the
'volume' or 'sales' of a product from the concept. This topic will not concern
us here because it is typically left to marketers, who must factor in concept
scores (acceptance), estimated distribution, price, and the competitive frame.
This chapter is concerned with the latter issue because it deals with the
development function, whose goal is to create better concepts and afterwards
better products.

Diagnosing the features of concepts that drive acceptance means understanding what is important in the concept. A concept, such as that shown in Fig.
15.1, may comprise a number of different elements. Precisely knowing which of
the elements 'drives' acceptance is not easy. One might present this concept to a
respondent and ask the respondent to 'circle the elements or phrases that are
important'. This rather simple method is surprising effective – consumers do
have an idea about what works for them and what does not. Of course, the
method is based upon introspection and can be only as good as the ability of the
respondent to look inside his or her own criteria for judging.

15.4   Systematic exploration of concepts by experimental design

Experimental design refers to the systematic layout of combinations of variables (see Box *et al.*, 1978). The layouts in the case of concepts are test concepts or test vignettes. The combinations are not just combined at random, willy-nilly, to make up a group of test concepts. That strategy of random combinations simply will not work, although to many individuals the systematic approach is often seen as being 'random', i.e. that somehow the research 'randomly puts together' the elements. Nothing could be further from the truth.

If we look at the material of a concept, we can see that there is a set of silos or buckets, overarching variables. One of these might be brand name, another might be the way the product looks, and another might be special ingredients added to the product. These silos are general groups of elements.

The components of the silo are the elements or attributes. (Experimental psychologists are likely to call these individual phrases 'elements', whereas market researchers are likely to call them 'attributes'.) A sense of the different categories and attributes for cookies appears in Table 15.1. These elements, divided into their proper silos or buckets or categories, come from a variety of sources – brainstorming sessions, analysis of what the competitors are doing/ saying (competitive intelligence; Moskowitz *et al.*, 2002a), previous ideas tested for this same product, lead users (von Hippel, 1986), etc. What needs to be remembered is that there are systematically classified raw materials or elements, which will be put together according to a plan.

The foregoing set of concept elements constitutes the raw material for experimental design. The design itself constitutes a set of combinations, predetermined in such a way that the elements appear independently of each other. Furthermore, the experimental design specifies that the same element appears against different backgrounds. Table 15.2 shows an example of the experimental design. Note that each element either appears or does not appear in a specific concept, that each concept comprises a limited number of elements, and that the same element appears in several combinations. When the respondent evaluates these concepts, and rates the concepts on an attribute, the researcher can then relate the presence/absence of the concept elements to the rating, or transform the rating (e.g. ratings of 1–6 are transformed to 0 to reflect low interest in the concept; ratings of 7–9 are transformed to 100 to reflect high interest in the concept). The experimental design ensures that the researcher can relate the presence/absence of the concept elements to the consumer ratings (Moskowitz and Martin, 1993).

Experimentally designed concepts are beginning to gain acceptance in the business community, and especially in the food and beverage businesses. The reason for this acceptance becomes obvious once we realize that the companies all compete with each other for 'share of wallet' (purchase dollars) and 'share of stomach' (relative amount eaten). The food and beverage industries are not growing rapidly, so most of the competition tends to be based on taking a 'share'

Table 15.1   Silos and elements

Silo A: Appearance and texture
A1   Soft and chewy ... just like homemade
A2   Crisp and crunchy ... perfect for dunking
A3   Soft and chunky ... for an extra special treat
A4   Oversized chunks of dark chocolate to sink your teeth into
A5   Bite size for a quick indulgence
A6   Jumbo size ... for when you want a little extra

Silo B: Primary ingredients
B1   Real creamery butter for a rich, indulgent taste
B2   Made with canola oil which helps lower blood cholesterol levels
B3   For a healthy source of protein ... made with unpasteurized egg whites
B4   Made with unprocessed whole grain flour ... keeping all the goodness in
B5   Made with only the freshest ingredients ... eggs, milk, butter
B6   Sweetened with natural fructose for a healthy indulgence

Silo C: Special ingredients
C1   Calcium enriched for strong bones
C2   Low carb ... when you're counting carbs and looking for a great snack
C3   With added iron and isoflavones ... a cookie that not only tastes good but is good for you
C4   0 grams trans fat and cholesterol free ... a heart-friendly cookie
C5   A high-fiber cookie that boosts your energy level and leaves you feeling full
C6   With no trans fats or preservatives ... for a healthy snack you can feel good about giving your kids

Silo D: Flavors and tastes
D1   Comes in spicy flavors ... cinnamon, nutmeg and allspice
D2   Comes in cool, citrus flavors ... orange, lemon and lime, perfect for a lady's afternoon tea
D3   Dark Belgian chocolate, Swiss milk chocolate and bittersweet chocolate ... simply irresistible
D4   Rich and creamy peanut butter ... for those who love an old favorite
D5   Vanilla flavored ... a traditional favorite
D6   Oatmeal ... for old-fashioned goodness and packed with nutrition

Silo E: Packaging
E1   All cookies are individually wrapped for freshness
E2   Comes with a stay fresh inside wrapper ... just twist and seal
E3   Comes in resealable bags ... take out only what you want
E4   Available in variety packs ... three of your favorite varieties in one box
E5   Available in decorative tins ... the perfect gift idea
E6   Comes in a crush-proof box ... no more broken cookies

Silo F: Size or location in store
F1   In the bakery section of your supermarket ... always fresh
F2   In the frozen foods section of your supermarket ... just thaw and serve
F3   In the frozen foods section of the supermarket ... bake and serve ... hot from the oven
F4   In the refrigerated section of your supermarket ... just thaw and serve
F5   In the refrigerated section of your supermarket ... just bake and serve
F6   Sold in bulk ... enough to please a crowd

Table 15.2   Experimental design

Concept	A	B	C	D
1	1	0	3	1
2	2	1	3	0
3	2	2	3	2
4	3	0	0	3
5	0	3	2	1
6	3	3	2	1
7	3	1	0	0
8	0	1	2	2
9	1	1	1	2
10	3	3	3	3
11	2	0	2	2
12	3	2	1	0
13	0	1	1	1
14	1	2	2	0
15	2	0	0	1
16	2	3	1	3
17	0	2	0	3
18	1	3	0	2
19	1	0	3	3
20	0	2	1	0

or proportion of the current market from the current competitors. Since the market is not growing, it is important to create ideas that attract consumers, so they take their money and spend it with one's company, not with the competitor company. Product and positioning concepts play an extremely important role in exciting and persuading consumers to buy; afterwards it is up to the product and the distribution systems to maintain this momentum.

Conjoint analysis is the name given by business researchers to this type of experimental design. Conjoint research itself enjoys a long and venerable history, beginning in the 1960s with the pioneering work of Luce and Tukey (1964). At that time the marketing and business communities were beginning to recognize the importance of systematic exploration of stimuli through experimental design, but the focus was on product rather than concept. Indeed, in those early days the notion of systematically varying concepts was at least 5–10 years away, would come from the business rather than the technical communities, and would take at least another 15–20 years to enter the food industry (in the early 1980s). Wittink and his colleagues have chronicled the emergence of conjoint analysis in the business world, among both practitioners and academics (Cattin and Wittink, 1982; Wittink and Cattin, 1989; Wittink *et al.*, 1994).

The basic premise of conjoint analysis, like all such analyses based upon experimental design, is that the responses to systematically varied combinations would, if properly analyzed, show the part-worth contribution of the components. In conjoint analysis the focus was on the contribution of discrete entities that were either absent or present in the concept. In contrast, food scientists just beginning to be introduced to experimental design were more familiar with so-called continuous variables (amount of starch, length of heating, degree of mixing, amount of added chemical). We will see two examples of the conjoint analysis, executed on the Internet, later on in the chapter. For now the important thing to keep in mind is that concept research has evolved from simple concept evaluation ('How am I doing?') to a more sophisticated analysis of great importance to the product developer in the food world ('What specific statements in this product or positioning concept drive consumers to be interested in the product, or feel that the product is appropriate for a specific end use?').

15.5   Consumer research venues and the Internet

Consumer research can be expensive and difficult. Although interviews themselves appear to be 'easy' and 'well orchestrated', especially when done by professional interviewers, the truth is that good consumer research is neither as easy nor as straightforward as one might believe it to be. The issues are 'Are we testing the right stimuli?', all the way to 'Are we testing the stimuli right?'.

The Internet has forever altered the nature of consumer research, making it easier, faster, and less prone to biases, but of course introducing possible biases of its own. Beginning in the early 1990s, the Internet gathered speed and became less 'idealistic', and by the late 1990s more commercially driven (Mosley-Matchett, 1998). Market researchers were quick to recognize the value of Internet-based research, first because of the lowered costs (respondents could be interviewed at home), then because of the increased convenience (respondents could be interviewed at any time they wanted), and finally because the Internet promised 'rich-media' (one could test full text, but pictures and video with sound were quickly coming up). Both Dahan and Srinivasan (2000) and Pawle and Cooper (2001) recognized and elaborated on the opportunity about a half decade before this writing (2007), but the progress since then can only be described as accelerating beyond even their vision.

Concept evaluation turned out to be one of the great opportunities for Internet-based consumer research. It took very little effort to create programs that would present concepts to consumers, and gather reactions. The computer had to be able to accept and show high-resolution concepts, although even that requirement did not stop researchers from showing simple text ideas. The researcher needed to have a method by which to acquire data (such as ratings of the concept on a set of scales, e.g. 2–15 such scales), and a way to randomize the concepts so that if there were two or more concepts to test, the respondents would see these test concepts in different order. This is simply good research policy – there tends to be a bias in consumer responses to the first test stimulus, whether this is product or concept.

By 1999, a number of companies were working on conjoint analysis and experimental design using the Internet to present the test stimuli and acquire the data. Rather than having to show a respondent a set of concept boards representing different variations of the same test stimulus, as researchers did when working in a central location, the interview was conducted on-screen. The computer program presented the respondent with the systematically varied concepts, and acquired the data. Most programs pre-created the concepts and sent them to the respondent, over the Internet, one at a time. Some more advanced versions used 'overlays', and a schematic, sending down the overlays and schematic ahead of time, and having a small remote-controlled program order the local computer to put together the concepts according to the experimental design.

The early days of Internet-based consumer research were filled with issues such as the representativeness of the consumer population, an issue which, since the 1930s, has plagued every new research venue. In the 1930s the *Literary Digest* had conducted a poll for president, and reported quite proudly that Alf Landon had been elected, only to discover the next morning to its chagrin that the sample had been biased (they used telephones, in the possession of only the well-to-do in those days; *Literary Digest*, 1936). The humiliating miscall chastened consumer researchers for the next seven decades and almost all consumer researchers would be sensitized to the need for proper representation. The Internet recalled some of these problems, especially since it, like the telephone, represented a venue open only to those with up-to-date technology.

Today, in the middle of the decade, in 2007, issues of consumer representativeness on the Internet have become a thing of the past, especially for the United States, Western Europe, and the developed countries of Asia and the Pacific. No one is particularly worried about access; indeed as ESOMAR reports, more and more companies are working on the Internet, and a great deal of research has migrated from traditional venues to the Internet. Some companies such as General Mills Inc. and Procter and Gamble proudly announce that a majority of their concept testing takes place on the Internet.

Of course there are the ever-present issues of whether the respondent who participates on the Internet is really the target sample, whether the respondent on the Internet is 'paying attention' (i.e. are the data really valid), and whether interviewing the respondent in a different location (home rather than a central location) changes the respondent's point of view, leading to different answers, and invalidating the norms.

15.5.1 Experimental design and concepts for food moves to the Internet

With the increasing popularity of the Internet, let's look at the progression of concept testing methods used on the Internet. We will look quickly at three methods, and how they are implemented on the Internet. The methods are benefit screening, full concept testing and then experimental design (conjoint analysis) which will occupy us for the rest of the chapter.

Benefit screening
Here the respondent is given a list of product features, and asked to rate each one on interest, and possibly on other attributes as well, such as uniqueness,

believability, and degree to which the benefit or idea does a specific 'job'. The benefits tend to be simple, single-minded ideas rather than complete vignettes. The goal of benefit screening is to identify what simple ideas 'work'. Researchers find benefit screening easy to do, with the outcome being a set of ideas that appear to work or at least to have promise. For the food and beverage industry the benefits are typically expressed in factual language, rather than in more emotional, connotative language. The research objective is quite simple – discover what works and what does not work. By no means can the benefit screening be construed as anything more than a sorting of potential ideas. On the Internet such a screening takes about 4–10 minutes. The benefits can be presented one at a time or as a grid. Grid-based presentation makes the respondent task much easier; the respondent need only read the benefit and rate it on a set of attributes by checking off the appropriate locations on the grid. The task can, however, become repetitive and boring.

Full concept testing with the concepts as 'gestalts'
We saw an example of a full concept in Fig. 15.1, dealing with Orchard Desserts. Companies are quite accustomed by now to testing dozens, if not hundreds or even thousands, of these complete concepts in a systemized way. The concepts are not systematically varied, but the testing is designed to push through a lot of these complete concepts (really small vignettes), get scores for the concepts on a number of different measures and then estimate the chances of success. Prior to the Internet such full concepts might be tested in person (e.g. one on one interview at a mall), discussed in focus groups, presented to large numbers of pre-recruited respondents in a central location hall (e.g. Acupoll®, 2005), read over the telephone, or sent home to the respondents, either with product, or in a book of other concepts.

Given the attention paid to budgets, it is not hard to see why so much concept testing (or con-screen) has migrated to the Internet. The ability to obtain rapid feedback on from one to dozens of concepts by Internet-based evaluation is leading many companies to shift their con-screen research to the Internet. A few years ago when the Internet was just beginning to become a mainstream research venue the critics would claim that one could simply not get the tailored sample of respondents through the Internet in the way one could by mail panel, phone, etc. Today, however, many companies feature special panels of hard-to-get respondents, with a lot of these panels set up with con-screen in mind. Companies working with such specialized groups of respondents can now test many concepts in an efficient manner, simply by creating the concepts and sending the panellist an e-mail to participate in the latest study.

15.5.2   Conjoint analysis over the Internet
Traditionally, conjoint analysis (really all experimental designs using concepts) was reserved for very expensive, high-profile projects. The researcher had to have extensive training, often lasting a half-day or more, about how to administer

the study. Typically each respondent would spend about 30–60 minutes with the interviewer, being shown different combinations and asked either to rate the combination or choose between pairs of combinations. The experience was often unpleasant, and for the most part the research dealt with very expensive items such as cars, refrigerators, hotel design, credit card design, and the like. The high cost of research precluded the conjoint approach from being used for most food and beverage products, unless the issue was one that had legal ramifications.

The Internet ended this high cost of conjoint analysis. Almost at the outset of the Internet boom in the late 1990s researchers began to port their conjoint studies to the web because it was fairly simple to create the web pages representing the different combinations, send these to respondents sitting in front of a computer at home or at work, acquire the data and process the results. The length of the interview was still long, although the fact that the respondent was in familiar surroundings, proceeding at his own pace, dropped the field cost considerably. Various companies began to offer conjoint analysis over the Internet. Such early movers as Greenfield Online, Sawtooth Software, and Moskowitz Jacobs Inc. developed these conjoint products to take advantage of the inexpensive and easier fielding offered by the Internet.

15.5.3   The Internet promotes self-reliance

As a consumer research procedure that developed as a practice rather than as a science, concept testing was never taught as a discipline in school. One of the key consequences of emerging as a practice is the nature of research implementation. For many years concept testing was handed over to research companies or corporate practitioners (in-house market research functions) to execute as a 'test.' Students were not trained in concept research because there was no corpus knowledge about the field that could be inserted into a school curriculum. In a sense the field had to develop as it matured, but without solid academic underpinnings. One could not go to the library to read about concept testing, except perhaps as part of a text on marketing research, and even in that happy situation the attention paid to concept testing was scant. Consumer research thus consigned concept research to a practice, not a science.

The explosion of Internet-based research changed the face and the structure of concept research. All of a sudden tools became available that allowed anyone to test concepts by creating their own questionnaire. Despite the plaints from professional researchers that somehow amateurs would not be able to test concepts properly, these Internet-tools democratized concept research, and made everyone aware of how straightforward it was to evaluate concepts for their own particular needs. This was a particularly important evolutionary step since there was no cadre of students who had been trained to a high level of expertise in concept research, and who could take the field forward. The Internet replaced that generation of students, deeply skilled, by a wide swath of individuals who could use the Internet tools. Concept research moved out of the hands of the skilled professional into everyone's domain.

One of the developments emerging from the Internet was the so-called self-authoring research tools, just alluded to. Self-authoring means that anyone can become a researcher by following an Internet-enabled template. Thus for conjoint analysis the self-authoring part means that anyone can do previously hard-to-execute conjoint studies, simply by following a point and click regimen that is driven by the program. The results of these developments were a set of self-authoring conjoint programs that allowed almost anyone to create the study, launch it painlessly on the Internet, gather ratings, and then have the data automatically analyzed (Moskowitz *et al.*, 2001). This development would prove to be very important in the development of research skills among those groups which hitherto had little budget for conjoint analyses and other hitherto expensive procedures.

15.5.4   The six steps for Internet-enabled conjoint analysis

Prior to the two case histories (on invention machines and genomics-inspired databases of mind-sets) we need only review the relatively simple steps needed to develop, launch, run, and analyze the conjoint analysis. By so doing, it will become clearer to the reader how the two case histories were executed. We can list the six steps and their rationale as follows.

Step 1 – Select design and create the elements, as well as define the rating scale
Here the researcher first identifies the architecture of the concept (how many 'silos' or 'buckets' of elements, and how many elements in each bucket). Figure

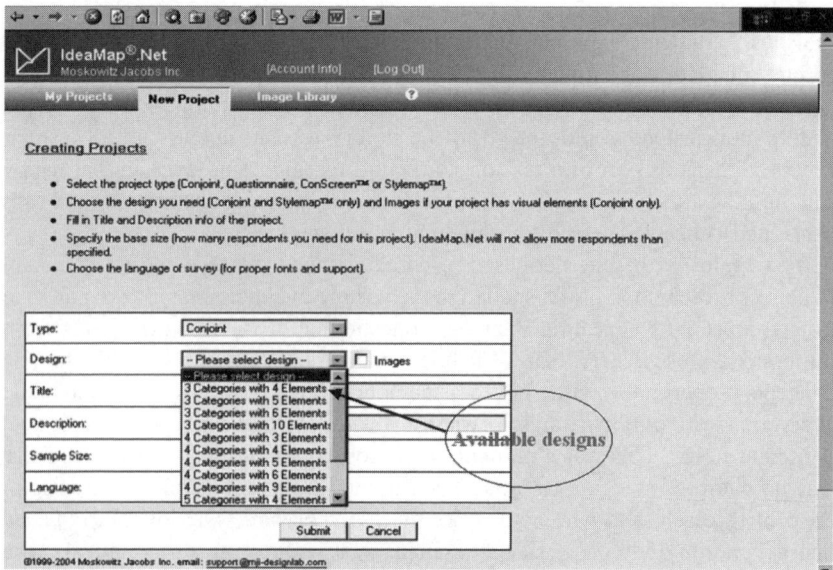

Fig. 15.2   Select the experimental design.

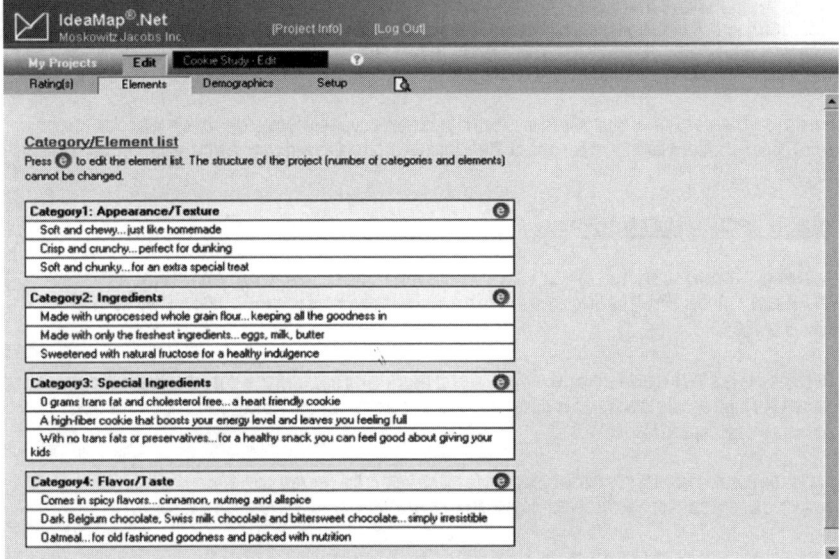

Fig. 15.3   Screen shot of part of the template which the user completes by typing in the elements.

15.2 shows the selection of different experimental designs. For simplicity, we choose a six categories, six element design without any pictures. Figure 15.3 shows the spreadsheet template, wherein the researcher simply types in the elements. If there are pictures, the first category would be reserved for these pictures (as an element of the concept), and the user would be prompted to add pictures from a set of pictures. The set is under the user's control.

Step 2 – Launch the study on the Internet and invite respondents to participate
This is very straightforward. The computer program used here (IdeaMap®.Net) generates its own URL, once the program is launched. The user simply distributes this URL to panellists, either from a standing panel, or from an ad or pop-up on the website. Respondents who click on the URL, or receive an invitation, go straight to the study. Figure 15.4 shows a typical invitation letter. It is important to keep in mind that the invitation must be 'zippy' and attractive to the reader. If the invitation is boring the respondent will probably delete the invite letter. Response rates will vary – from low as 0.5% to as high as 30%. A lot has to do with the incentives and the way the letter is worded. Notice that the letter in Fig. 15.4 tries to invite the respondent to get involved and tells them about the conditions to win a prize. It is important at this stage to offer them an incentive; very few respondents today are interested in volunteering their time if there is no promise of a reward. The reward shown in Fig. 15.4 is a chance at a sweepstake; other rewards are donations to a charity or actual cash/gift awards.

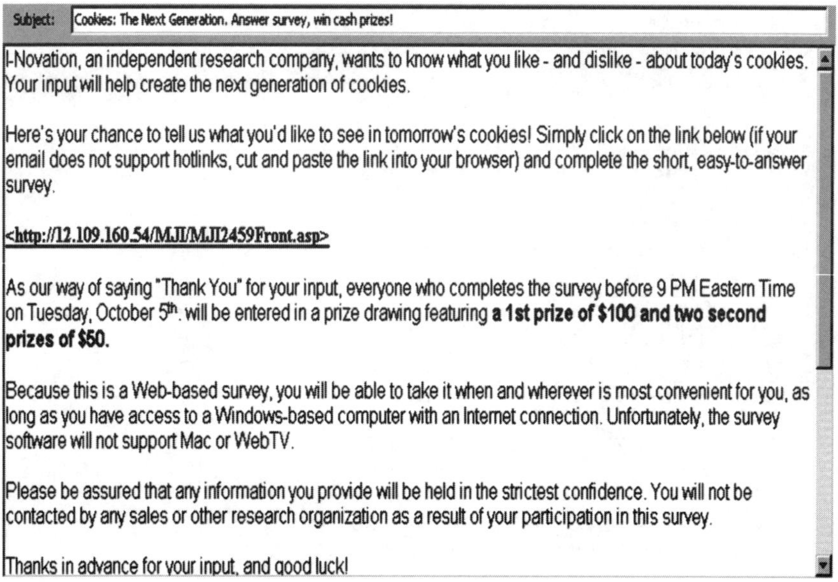

Subject: Cookies: The Next Generation. Answer survey, win cash prizes!

I-Novation, an independent research company, wants to know what you like - and dislike - about today's cookies. Your input will help create the next generation of cookies.

Here's your chance to tell us what you'd like to see in tomorrow's cookies! Simply click on the link below (if your email does not support hotlinks, cut and paste the link into your browser) and complete the short, easy-to-answer survey.

<http://12.109.160.54/MJI/MJI2459Front.asp>

As our way of saying "Thank You" for your input, everyone who completes the survey before 9 PM Eastern Time on Tuesday, October 5th, will be entered in a prize drawing featuring **a 1st prize of $100 and two second prizes of $50.**

Because this is a Web-based survey, you will be able to take it when and wherever is most convenient for you, as long as you have access to a Windows-based computer with an Internet connection. Unfortunately, the survey software will not support Mac or WebTV.

Please be assured that any information you provide will be held in the strictest confidence. You will not be contacted by any sales or other research organization as a result of your participation in this survey.

Thanks in advance for your input, and good luck!

Fig. 15.4 Invite letter.

Step 3 – Orient respondents
The respondents who click on the link embedded in the invitation are directed to the website, where the interview begins immediately. The respondents are shown an orientation page (Fig. 15.5) which tells them about the purpose of the study.

Step 4 – Interview
The test stimuli comprise either systematically varied concepts for conjoint analysis (see Fig. 15.6) or a fixed concept. In either case the respondent rates each concept in the study on a set of attributes, in an interview lasting 12–18 minutes. This length of interview promotes acquisition of data without being unduly long, which would generate a lot of incomplete interviews (MacElroy, 2000). For conjoint analysis, which will require a dozen to several dozen concepts, the recommendation is to have one or at most two rating attributes; beyond that number of ratings the respondent stops attending and begins to get bored. For con-screen, which comprises evaluations of one or two concepts, the practice is to ask many questions about the few concepts. These two practices follow conventional market research thinking. When researchers work with many concepts that are systematically varied, the insights come from the pattern of responses to these concepts, so the respondent only has to give a little information. The systematic variation of concepts adds the rest; the researcher need only discover the pattern relating the presence/absence of elements to the ratings and he knows the 'key' to the data. In contrast, when the researcher opts for the apparently simpler concept screening, the more information that the

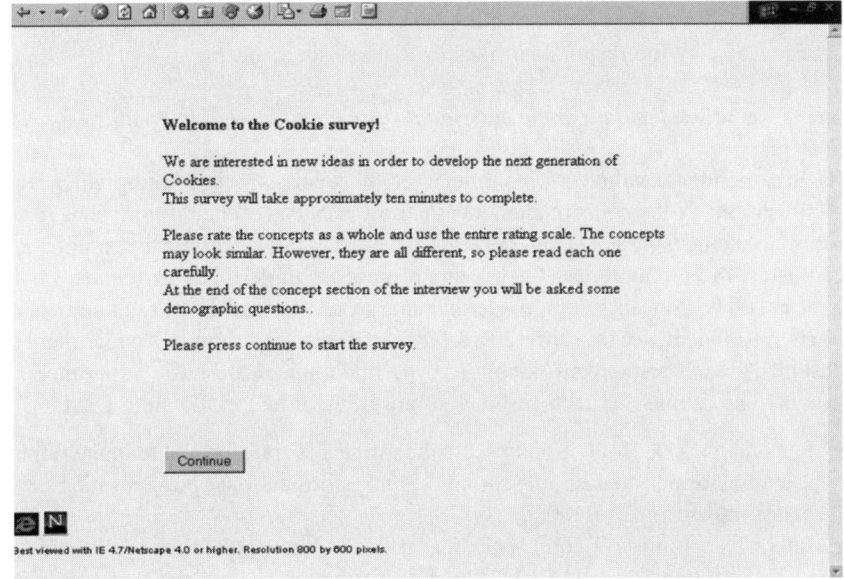

Fig. 15.5   Orientation page.

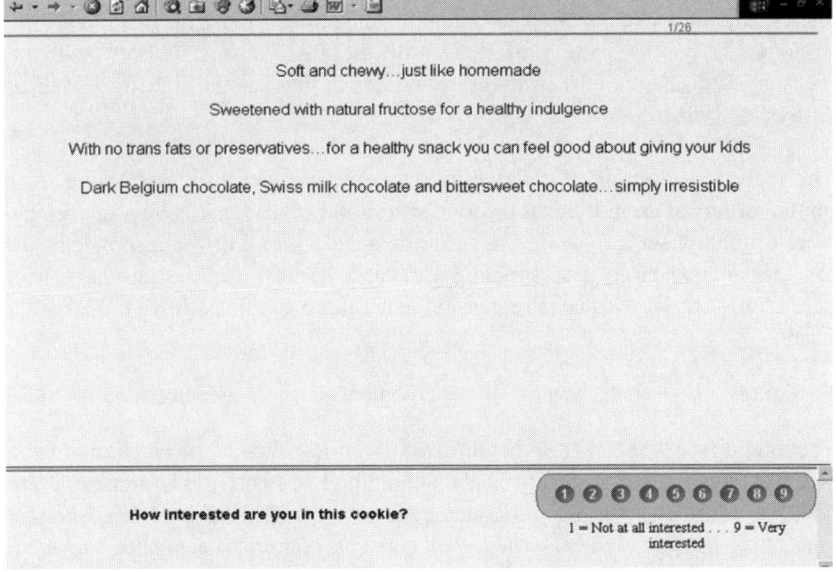

Fig. 15.6   Example of a test concept.

respondent gives when evaluating the concept, the more likely it will be that the data reveal why the respondent likes or dislikes the concept.

Step 5 – Analyze the data at the individual respondent level, and create summary data files

Each respondent evaluates a complete set of concepts, representing sufficient combinations of the concept elements to allow analysis of the ratings. Thus for a set of 36 elements (e.g. six categories, six elements per category) the respondent evaluates 48 combinations. Each element appears three times in the set of 48 concepts. The concepts comprise either one or no element from each category. Each respondent sees a different set of combinations, but the experimental design for each respondent is simply a permutation of the basic experimental design. The ratings for each respondent are transformed in a simple fashion:

- Persuasion scale: for a 9-point scale the rating 1 is transformed to 0, the rating 2 is transformed to 12.5, the rating 3 is transformed to a 25, up a rating of 9 being transformed to 100.
- Interest scale: for a 9-point scale the ratings 1–6 are transformed to 0, ratings of 7–9 are transformed to 100. This transformation follows the convention of consumer research, which looks at membership in the acceptance class rather than rating on a scale.

Step 6 - Summary data files

Each respondent generates his or her own data vector comprising the additive constant and utilities. These data vectors can be averaged across all respondents (total panel), against prespecified respondents (subgroups, e.g. males/females; ages, etc.), and even treated as input to clustering programs that divide the respondents by the pattern of their utilities (e.g. k-means cluster, with the measure of distance between two respondents defined as the statistic 1-Pearson R; Systat, 1998).

The analysis uses straightforward regression (ordinary least squares), done either on the ratings of an individual respondent (valid because that is how the designs were originally set up), or on the entire data set with all of the respondents put together in one large data stream (also valid because the designs have been permuted). The least-squares regression generates a simple additive model of the form:

$$\text{Rating} = k_0 + k_1(\text{Element \#1}) + k_2(\text{Element \#2}) \ldots k_{36}(\text{Element \#36}) \quad 15.1$$

The foregoing equation states that the overall rating (whether the persuasion value or the interest value) comes from the summation of two types of numbers. The additive constant k_0 is the estimated rating that would be obtained were the constant to have no elements. Clearly all concepts comprised elements, so that the additive constant is an estimated parameter which can be used as a benchmark to show basic level of acceptance of product or service, without any communication.

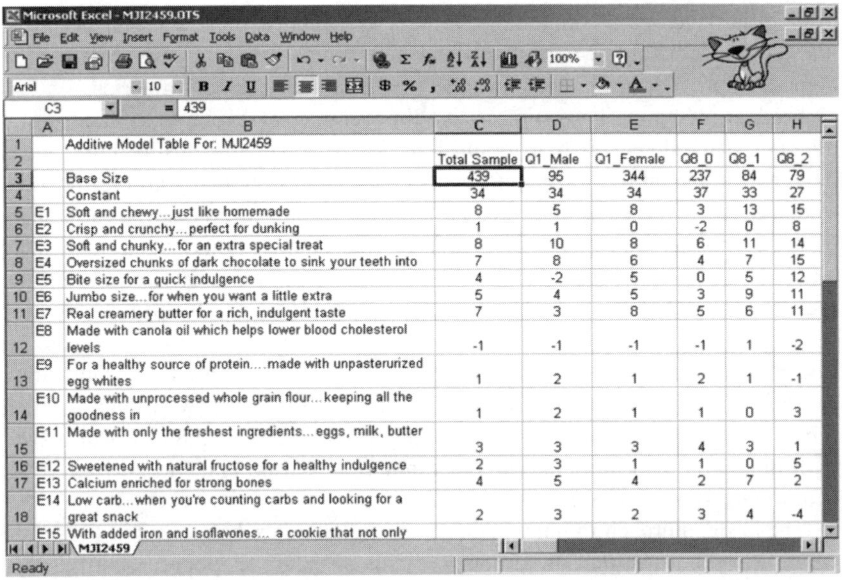

	Additive Model Table For: MJI2459	Total Sample	Q1_Male	Q1_Female	Q8_0	Q8_1	Q8_2
	Base Size	439	95	344	237	84	79
	Constant	34	34	34	37	33	27
E1	Soft and chewy...just like homemade	8	5	8	3	13	15
E2	Crisp and crunchy...perfect for dunking	1	1	0	-2	0	8
E3	Soft and chunky...for an extra special treat	8	10	8	6	11	14
E4	Oversized chunks of dark chocolate to sink your teeth into	7	8	6	4	7	15
E5	Bite size for a quick indulgence	4	-2	5	0	5	12
E6	Jumbo size...for when you want a little extra	5	4	5	3	9	11
E7	Real creamery butter for a rich, indulgent taste	7	3	8	5	6	11
E8	Made with canola oil which helps lower blood cholesterol levels	-1	-1	-1	-1	1	-2
E9	For a healthy source of protein....made with unpasteurized egg whites	1	2	1	2	1	-1
E10	Made with unprocessed whole grain flour... keeping all the goodness in	1	2	1	1	0	3
E11	Made with only the freshest ingredients...eggs, milk, butter	3	3	3	4	3	1
E12	Sweetened with natural fructose for a healthy indulgence	2	3	1	1	0	5
E13	Calcium enriched for strong bones	4	5	4	2	7	2
E14	Low carb...when you're counting carbs and looking for a great snack	2	3	2	3	4	-4
E15	With added iron and isoflavones... a cookie that not only						

Fig. 15.7 Screen shot of data from the cookie study, showing Excel-ready format of element × subgroup.

Consistency of response has continued to be an issue in concept research, especially over the Internet. When the researcher creates the individual level model, he or she can assess the consistency of any individual's responses by computing the goodness of fit of an individual's persuasion model to the ratings. This measure is the goodness-of-fit statistic, the multiple R^2 of the equation. Typically more than 80% of the individual models generate an R^2 statistic exceeding 0.65 (Moskowitz et al., 2002b).

Since the additive constant and the 36 utilities (for this example) are saved at the individual respondent level, it becomes possible to aggregate the data for total panel (all respondents), for defined subgroups as stated in the self-profiling classification, and for concept–response segments defined by clustering the utilities. For these studies the segments are defined by the method of k-means clustering, with respondents in the same segment showing similar patterns of utilities. Figure 15.7 shows an example of the data as the computer program places it in an Excel-ready format.

15.6   Application 1: creating a product-concept 'innovation machine' through mixing/matching

Product innovation in the food industry has become an increasingly popular topic over the past years. The increased competition among manufacturers means that it is no longer sufficient for a company to create and market a product to a willing consumer audience. As soon as a viable market for a new product is developed, competitors, armed with reports of new products on the market, immediately swoop in to determine whether this new product can be copied, improved, or changed and the new customer segment diverted.

Product invention/innovation is like the weather; everyone talks about it but no one really does much to control it. There are many business processes devoted to innovation in the food industry, and certainly many articles devoted to the topic with titles as alarming as 'innovate or die.' Certainly the early stage of development, the so-called fuzzy front end, is where innovation could take place (Khurana and Rosenthal, 1997). The question is simply 'how' – not the metrics, but rather the actual 'nitty gritty'. The first case history shows how developers can use concept evaluation on the Internet in order to create an invention machine. The key to the approach is that invention and innovation comprises the recombination of small pieces of ideas into new 'wholes,' which represent the product ideas. The conjoint approach discussed above works well with this organizing principle of recombination (Ewald and Moskowitz, 2001, Moskowitz et al., 2005b).

The basic approach to innovation on the Internet is schematized in Fig. 15.8, which simply says that in order to innovate, the developer identifies the components of several different products, and recombines these components into new combinations. The original components come from winning or at least strong performing elements in different products, which are then combined in a new study. The combination generates the new idea.

The steps in the innovation process follow the sequence discussed above for executing the conjoint study on the Internet. The steps and the results follow.

Step 1 – Develop concept architecture
The architecture comprises concepts with six silos, each having six elements. The silos and an example from the cookie study appear in Table 15.3. It is important to note here that each of the six silos is 'actionable,' so that the concepts are essentially product concepts. The first set of studies comprised 36 elements chosen for cookies, the second comprised a parallel 36 elements

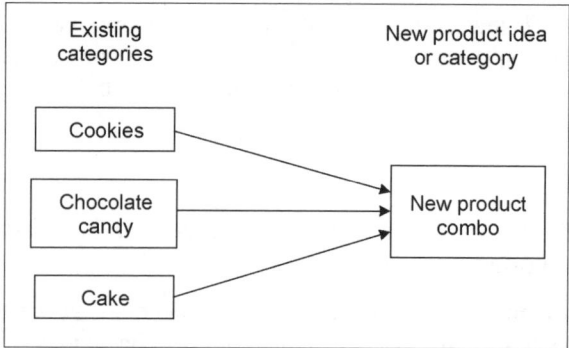

Fig. 15.8 Innovation/invention of new product concepts by recombining ideas from different products into a new product.

Table 15.3   Architecture of concepts underlying the conjoint analysis (cookies)

Silo A: Appearance and texture
Soft and chewy ... just like homemade

Silo B: Primary ingredients
Real creamery butter for a rich, indulgent taste

Silo C: Special ingredients
Calcium enriched for strong bones

Silo D: Flavors and tastes
Comes in spicy flavors ... cinnamon, nutmeg and allspice

Silo E: Packaging
All cookies are individually wrapped for freshness

Silo F: Size or location in store
In the bakery section of your supermarket ... always fresh

chosen for chocolate candy, and the third comprised yet another parallel 36 elements chosen for donuts.

Step 2 – Run the first set of three studies, one for each product, to identify winning elements
The ingoing strategy was to run an initial set of three studies that would identify the elements that did well for each of the three products. Table 15.4 shows the utility value for some of the cookie elements. The same type of table can be generated for chocolate candy and for donuts. With a table of element utilities the product developer rapidly recognizes what elements contribute to acceptance versus detract from, for the total panel, for gender, and even for segments. Segments are groups of consumers with similar mind-sets, who respond to the same type of elements.

A developer looking at the data in Table 15.4 might draw the following conclusions:

- The additive constant for cookies is 34 for the total panel, meaning that without any other information other than the product is a cookie, approximately 34% of the 439 respondents who participated would rate the idea as interesting (6–9). The remaining 66% of the respondents would have rated the cookie as 1–6 (not interesting).
- It is not only the predisposition, but also the specific messaging and features that make a difference. Putting in the element *A dark Belgian chocolate, Swiss milk chocolate and bittersweet chocolate ... simply irresistible* adds 11% respondents to the interested group.
- Gender does not make much of a difference. Certainly there are gender differences, but for the most part winners among men remain winners among women; the same pattern holds for losing elements.
- There are radical differences among consumers, but these have to be

Table 15.4 How the different elements for cookies performed for the total panel, gender, and four concept-response (or so-called latent) segments. The numbers in the body of the table are interest utilities

| | | Total | Gender | | Concept response segment | | | |
			Male	Female	S1 Indulgers	S2 Health oriented	S3 Classics	S4 Chocolate loves
	Base size	439	95	344	226	59	89	65
Silo	Constant	34	34	34	34	24	33	46
D	Dark Belgian chocolate, Swiss milk chocolate and bittersweet chocolate … simply irresistible	11	10	11	12	9	4	18
A	Soft and chunky … for an extra special treat	8	10	8	10	1	14	1
A	Soft and chewy … just like homemade	8	5	8	10	-2	14	-1
A	Oversized chunks of dark chocolate to sink your teeth into	7	8	6	12	-3	-3	10
B	Made with unprocessed whole grain flour … keeping all the goodness in	1	2	1	2	12	-10	4
B	For a healthy source of protein … made with unpasteurized egg whites	1	2	1	-1	23	-9	2
F	In the frozen foods section of your supermarket … just thaw and serve	1	-2	2	5	7	-12	-1
D	Comes in cool, citrus flavors … orange, lemon and lime, perfect for a lady's afternoon tea	-9	-10	-9	-23	10	5	2

extracted by looking at the pattern of utilities for the different elements. Table 15.4 can give only a small taste about the richness of such segment differences, which lie at the heart of successful product development. A few words are in order about the segments.

a Typically there is more than one segment, but the precise number of segments is a research issue.

b These segments are 'latent' in the data; they are not explicit but must be extracted by looking at the patterns of the utilities (Vermunt and Magidson, 2000; Vigneau *et al.*, 2001).

c Segments begin as clusters of like-minded respondents. It is the job of the researcher to assign names to these segments. Usually the names are assigned on the basis of the elements that do best for the segment. Generally, but not always, the elements that perform best tend to have a common theme. In contrast, the elements that do worst differ dramatically even within the same segment. One is reminded of Tolstoy's famous opening line in Anna Karenina '*Happy families are all alike; every unhappy family is unhappy in its own way*'.

d The additive constant differs by segment, as does the size of the winning element. A strong concept may be created in two ways – by having a strong additive constant (strong basic interest in the idea) + moderate/ strong elements, or by having a low additive constant (modest basic interest) but very strong performing elements. The implications of these two patterns supporting a strong concept are important to bring forward in development.

e Segments are of different size. Statistical analyses do not force the segments to be equal. For cookies more than half of the respondents belong to Segment 1, which has been labelled (by the author) as Indulgers.

f Which elements to choose as most promising with which to go forward is a matter of research proclivity and point of view. Research is an aid to judgment and development, not as a replacement. It is neither necessary nor perhaps even wise simply to choose the winning elements. It is probably better to form an idea of what opportunities present themselves from the pattern of utilities for the three studies, and then move on with a judicious selection of elements.

Step 3 – Mix and match promising components from the three data sets to generate a fourth set of elements, and then identify what wins
This third step is essentially like the previous two steps. The only exception is that the elements have been taken from different products, recombined, and then given to the consumer respondents. Table 15.5 shows the selection of elements for the first three silos.

The process is quite simple, following the steps described above. The only unusual thing is that the respondents are not told that the concept elements come from a graft of different product ideas from different products, but merely that they are evaluating a new dessert product. To the degree that the respondents can

Table 15.5 Two examples from each product (cookie, donut, and chocolate candy) for the first three product categories

	Silo 1 – Appearance and texture
Cookies	Soft and chewy … just like homemade
Cookies	Crisp and crunch … perfect for dunking
Donuts	A glazed snack that will melt in your mouth
Donuts	A yeast raised snack shaped into a twist. Puts a new twist on an old favorite
Choc Candy	A creamy milk chocolate snack with a soft, chewy center for a satisfying experience
Choc Candy	A creamy chocolate snack with a crunchy, nougat center
	Silo 2 – Primary ingredients
Cookies	Real creamery butter for a rich indulgent taste
Cookies	Made with canola oil which helps lower blood cholesterol levels
Donuts	We only use Kosher ingredients
Donuts	Sweetened with natural fructose for a healthy indulgence
Choc Candy	Made with the finest Belgian chocolate … for the discerning chocolatier
Choc Candy	Made with the finest Swiss chocolate
	Silo 3 – Special ingredients and additives
Cookies	Calcium enriched for strong bones
Cookies	Low carb … When you're counting carbs and looking for a great snack
Donuts	Made with no trans fat to fit into your healthy lifestyle
Donuts	A high-fiber snack that boosts your energy level and leaves you feeling full
Choc Candy	With soy isoflavones … nutrition at its best
Choc Candy	With 4 grams of fiber … A great way to add fiber to your diet

be shielded from knowing any more than those few facts, they will have a more open mind. Figure 15.9 shows the page, and how the origin of the product is disguised.

Step 4 – Run the study and identify promising elements from the total panel, and from key subgroups that may emerge from segmentation
Segments constitute respondents with different response (i.e. utility) profiles, and may reflect different basic mind-sets in the respondent population (Green and Krieger, 1991; Wells, 1975). Four segments emerged from the study. It would be onerous to list and read all of the different subgroups and elements. Instead, Table 15.6 shows promising elements for Segments 1 and 3. All of these promising elements score well. They come from different products and different silos. They do not yet provide a concept; they simply provide some additional insights about the mind-set of the group.

Step 5 – Searching for synergisms and suppressions
A unique aspect of the Internet is that it allows researchers to work with a large number of respondents. Combine this scope of research with a technique that

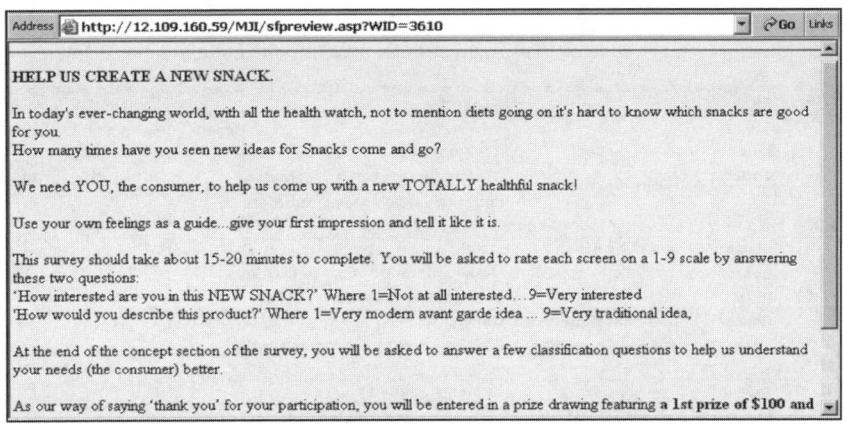

Fig. 15.9 Orientation page for the snack concept that was created by grafting together elements from cookies, chocolate candy and donuts.

systematically permutes the different basic concept design to a family of isomorphic combinations, and the researcher is given the ability to identify synergism and suppressions (Moskowitz and Gofman, 2004). A synergism is a combination of elements that performs far better than would be expected on the basis of its linear sum of utilities. Thus a combination can be said to be synergistic when one of the elements scores, say, +3, the second element scores +6, and in

Table 15.6 Promising elements for Segments 1 and 3 from the 'fourth study' combining elements from chocolate candy, cookies and donuts

File Edit View Insert Format Tools Data Window Help Adobe PDF

A	B	C	D	BE	BF	BG	BH
			Tot	S1	S2	S3	S4
		Base Size	335	76	72	145	42
		Additive Constant	38	36	59	29	39
		Promising elements - segment 1					
A6	Candy	A creamy chocolate snack with a crunchy, nougat center	6	18	-5	6	5
A5	Candy	A creamy milk chocolate snack with a soft, chewy center for a satisfying experience	6	10	-4	11	-1
E5	Candy	Wrapped in aluminum foil...stays fresh longer	3	10	4	0	-2
A2	Cookies	Crisp and crunchy...perfect for dunking	3	10	-1	4	-8
E4	Donuts	Comes in a crush-proof box	1	9	1	-1	-4
A1	Cookies	Soft and chewy...just like homemade	5	9	-4	10	-4
A3	Donuts	A glazed snack that will melt in your mouth	2	8	-7	6	-6
		Promising elements - segment 3					
B5	Candy	Made with the finest Belgian chocolate...for the discerning chocolatier	9	2	-8	18	19
B6	Candy	Made with the finest Swiss chocolate	5	1	-17	17	10
B1	Cookies	Real creamery butter for a rich, indulgent taste	3	-2	-10	11	6
A5	Candy	A creamy milk chocolate snack with a soft, chewy center for a satisfying experience	6	10	-4	11	-1
D5	Candy	Filled with crunchy cookie chocolate chips	4	1	-12	10	13
B4	Donuts	Sweetened with natural fructose for a healthy indulgence	1	-7	-10	10	4
A1	Cookies	Soft and chewy...just like homemade	5	9	-4	10	-4
B2	Cookies	Made with canola oil which helps lower blood cholesterol levels	-1	-13	-9	10	-1

Table 15.7   Pairs of concept elements from different silos that significantly interact with each other. The magnitude of their interaction is given by the column labeled Inter.

Pair	First	Second	First	Second	Inter.
D5E1	Filled with crunchy cookie chocolate chips	Each snack is individually wrapped to preserve freshness	5	1	13
B3C5	We only use Kosher ingredients	With Soy isoflavones...nutrition at its best	-4	-8	12
B2C3	Made with canola oil which helps lower blood cholesterol levels	Made with no trans fat to fit into your healthy lifestyle	2	2	11
B5D1	Made with the finest Belgian chocolate...for the discerning chocolatier	Comes in spicy flavors...cinnamon, nutmeg and allspice	13	-2	-10
B6C2	Made with the finest Swiss chocolate	Low carb...when you're counting carbs and looking for a great snack	11	5	-12
C2E1	Low carb...when you're counting carbs and looking for a great snack	Each snack is individually wrapped to preserve freshness	5	1	-13

those happy combinations where the two elements appear together and are treated as a third element the combination scores +13. This is simply an example.

The search for synergisms is a simple statistical task, albeit a big one. The data set comprises all of the linear terms. The researcher creates all of the pair-wise interaction terms as well. That is a simple matter of multiplying columns. For our new product comprising six categories and six elements per category there are 36 basic terms in the model (contribution of each element), and 540 pair-wise combinations. This large number is simply the result of counting all of the pairs of elements from the pairs of silos. There are 15 distinct pairs of silos; there are 36 distinct pairs of elements for any pair of silos, and thus there are 15 × 36 or 540 pairs. Only a very few of these pairs are sufficiently strong statistically to add substantial predictability.

We interpret the synergisms and suppressions in Table 15.7 quite easily. The first column, labelled 'Pair,' shows the combination of concept elements that are found to interact significantly, above and beyond the contribution of their components. There are only six of these out of a possible 540, or a little more than 1%. The column marked 'First' gives the text and the utility of the first element; the column marked 'Second' in turn gives the text and the utility of the second element. Finally, the column marked 'Inter' shows the additional utility that is added (or subtracted) when those two elements appear together in the same concept. As we can see, some pairs of elements strongly interact to generate an additional 13 points, whereas other pairs interact so that their mixture is far worse (13 points lower) than we might expect.

Searching for 'what works with what' scenarios
Common sense tells us that when we create a product concept, some ideas ought to go well with each other, whereas other ideas should not go well with each

other. Yet, these ideas do not have to synergize. How do we find out what ideas 'work' with each other, so that they fit together? The respondent cannot easily make every possible comparison between pairs of elements – this would mean 540 comparisons, which is more than most respondents are willing to tolerate, even if they are paid. By the time a respondent has evaluated the first 200 combinations more than likely they are totally turned off by the project, and have switched to 'automatic pilot', no longer paying attention.

Recent developments in conjoint measurement provide a method by which to identify what elements go together in concepts. The approach is quite simple, following these steps, which we will illustrate with the data on the new hybrid product.

1. Lay out the raw data matrix, so that each row comprises the 36 columns (one per product feature) and the 37th column (corresponding to the 9-point rating). With 335 respondents, and this particular experimental design comprising 48 combinations per respondent, the data matrix comprises 16 080 rows ($48 \times 335 = 16\ 080$).
2. Create a new variable called interest, which takes on the value '100' if the rating was 7–9, and takes on the value '0' if the rating was 1–6.
3. Create a second new variable called 'bylocation', which takes on the value '1' if variable F for that particular concept is F1 (*In the bakery section of your supermarket*), takes on the value '2' if variable F for that concept is F2 (*In the frozen foods section of your supermarket – just thaw and serve*), etc. When the concept does not have an element from category F (location), then the variable 'bylocation' takes on the value '0'. It is clear that with six alternatives for variable F and a true zero, the new variable 'bylocation' takes on seven distinct values, and that knowing the particular value of 'bylocation' automatically tells us where in the store the concept says the product will be merchandised.
4. Sort the large matrix of 16 080 rows by the variable 'bylocation', from low to high. This sort will put all of the concepts with no element F first, and then put all of the concepts with element F1 second, etc.
5. Look at 'slices' of data in this matrix; first concentrating on all of the concepts having element F1. The independent variables are A1–A6, B1–B6, C1–C6, D1–D6, and E1–E6. The dependent variable is interest, as it was before. The variables F1–F6 no longer act as independent variables because in any dataset set being analyzed here there is only one level of F (e.g. F1) that is present in the concept.
6. Since there are six levels of F (F1–F6), there are six corresponding models, each having 30 independent variables, and interest as the dependent variable.
7. We interpret any of the six models as showing what are the contributions of the remaining 30 elements, if the location in the store (way the product is merchandised, silo F) is held constant. The additive constant is the esti-mated acceptance of that concept (with constant level of F, e.g. F6).

Table 15.8 What elements work with the six different elements in silo F (merchandising)

	Available through the Internet and delivered to your door	Located in the gourmet food section of all fine department stores	In the bakery section of your supermarket	Find it in the Snack aisle of your local supermarket	In the frozen foods section of your supermarket ...just thaw and serve	In the refrigerated section of your supermarket ...just thaw and serve
	F6	F5	F1	F3	F2	F4
Additive constant	9	20	31	33	30	39
Constant + Avg. Util.	17	25	34	36	33	39
B5 Made with the finest Belgian chocolate...for the discerning chocolatier	18	10	17	13	10	10
A5 A creamy milk chocolate snack with a soft, chewy center for a satisfying experience	17	7	9	16	11	2
B6 Made with the finest Swiss chocolate	10	14	8	10	10	11
D4 A rich, pure buttery taste...makes your mouth water in anticipation	14	13	9	5	8	0
A1 Soft and chewy...just like homemade	13	10	10	7	1	7
A6 A creamy chocolate snack with a crunchy, nougat center	16	10	5	8	1	5
A high-fiber snack that						

8. Let's interpret the results for two of the more interesting merchandising ideas. F1 (*Available through the Internet and delivered to your door*), versus F6 (*In the refrigerator section of your supermarket – just thaw and serve.*) These are two very different ideas. Their additive constants are quite different (the Internet delivery has an additive constant of 9 so without any elements this is a very poor idea; the refrigerated section of the supermarket has an additive constant of 39 so it is a very good idea).

9. Table 15.8 shows a subset of elements, namely those that do well. However, even among this small subset we see that the same element performs differently. For example, let's look at element A5 (*A creamy milk chocolate snack with a soft, chewy center for a satisfying experience*). This element scored moderately well for the total panel (+6), as we can see from Table 15.6. However, the scenario analysis tells us a different story altogether. The element scores very well when paired with concepts about Internet delivery (+18), perhaps because it promises a good sensory experience. The element scores only +2 (indifferent) when paired with the refrigerated section in a supermarket.

10. Elements that perform well for one merchandising section may perform just as well for another section (e.g. B6, *Made with the finest Swiss chocolate*).

15.7   Application 2: Sourcebooks for concept ideas: InnovAidOnline™.Net and It!® databases

The introduction to this chapter presented the approaches to systematic concept development and the Internet, concentrating on conjoint analysis as it is currently being implemented through the IdeaMap®.Net system. If the reader is so inclined, a good exercise is to look for databases that provide the user with ideas to incorporate into the conjoint analysis. At the time of writing, a quick Google® search lists only databases of concept elements that the user can incorporate into a product development exercise. The two databases are the InnovAidOnline™.Net database and the It!® database (Beckley and Moskowitz, 2002; Moskowitz *et al.*, 2005c). Both of these have been developed with the view that there needs to be a science underlying concept development and that with the emergence of the web and the subsequent empowerment of R&D it is vital to create a public database of concept elements. Only with such a database can concept development grow from an art to a science.

For the past 50 years most researchers working with concepts conceived of their job as an art, not a science. Although there is a science of food and drink, the science comprises facts about the physical features of products. There is a science dealing with the consumer aspect (known, not surprisingly, as consumer science), but this science deals with the nature of decision making, not with the materials on which the decisions are made. Journal after journal discusses criteria on which people decide about foods, whether responses to statements about the food, the actual taste of the food, or even the packaging.

What is missing, however, is a fact-based list about what features of a concept really 'work' in the mind of the consumer, or even a list of statements that one might use for one's concept research. It is generally left as an exercise for the person doing the concept development to come up with the ideas for the concept. More often than not that person throws up their hands in frustration, not even knowing where to begin. What are the relevant elements of a concept? What should be used? What works (or has worked in the past)? Are there any segments? What are the rules of the product category, and do these rules transcend single product categories, applying to a whole range of products at the same time? All of these are valid questions, all are issues that need to be addressed, and all are worthy of a database that a concept developer can consult.

15.7.1   The InnovAidOnline™.Net database

We begin with a database of food ideas. The notion of the database was conceived as a way that product developers could structure the way they thought about developing a product concept. Indeed, the cookie data from the first case history was taken from that database. The notion is quite simple – structure a set of elements, one per product, in the same format, and allow this in an open format so it is available to anyone. The result is a database of product features, all compatible with each other. Those who are involved with creating food

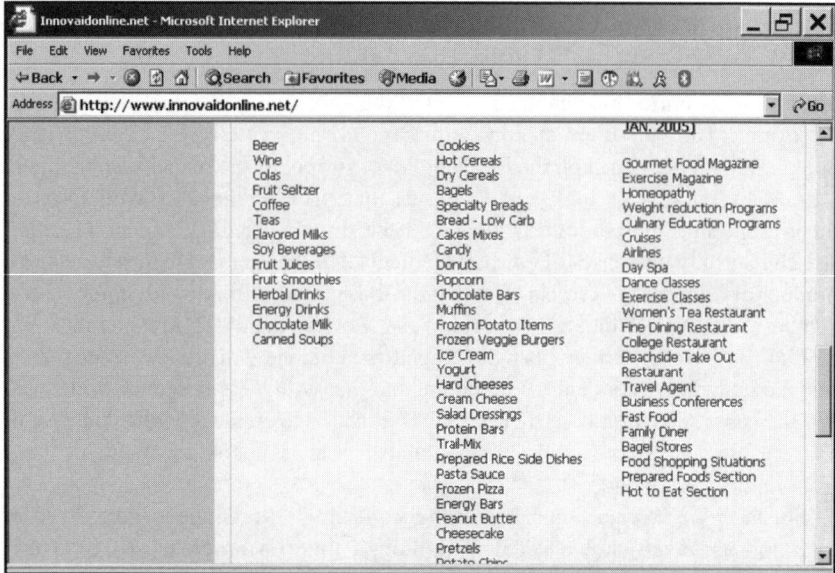

Fig. 15.10 Screen shot of the InnovAidOnline^{TM}.Net database list.

concepts can use these elements on a complimentary basis; those who wish to avail themselves of the IdeaMap®.Net tool can do so as well. In a sense the InnovAidOnline™.Net database follows today's convention of open-source technology.

Only one figure and one table are needed to illustrate the premise of InnovAidOnline™.Net. We see the list of available element databases shown in Fig. 15.10, and a comparison of two element sets in Table 15.9. The element databases are currently being expanded to consumer electronics and to financial services, each with its own set of relevant product/service categories, architecture, and elements. In all cases the organizing principle is six categories × six elements per category.

Figure 15.10 shows the available databases, reachable by going to www.innovaidonline.net. The different available data sets of elements are available by downloading. Table 15.9 shows the text for two of the element databases: pasta sauce and energy drink. Note that the element databases do not have any utilities attached to them; the actual response data is left to the user. Moving down Table 15.9 there are the six categories that we have already seen. Following the 36 elements we see four questions of a problem-solution nature, designed to have respondents complete open-end questions, and by so doing provide even more new ideas. The approach of generating new ideas by asking respondents how they solve current problems was suggested by Professor Yoram (Jerry) Wind at Wharton. The database concludes with questions about what particular factors are important for the respondent's choice, and what brands and flavors are currently being used.

Table 15.9   Two representative databases of elements and other relevant questions, one from food, one from beverage, representing the elements available in the database. The databases are taken from the food/beverage/lifestyle InnovAidOnline.Net database

Pasta sauce	Energy drinks
Appearance and texture	
A1 Smooth and creamy ... coats your pasta evenly	Smooth and thick ... so satisfying
A2 Light and thin ... doesn't weigh you down	Silky smooth ... glides down easy
A3 Thick and creamy texture ... so smooth in your mouth	Light and crisp ... sends a shiver down your spine
A4 Hearty and rich ... for a taste your family will love	No carbonation ... smooth and inviting
A5 Dark and rich tasting	Bubbly ... lightly carbonated
A6 Chunky, stick-to-your-ribs goodness	Decadently creamy ... excite your tastebuds
Primary ingredients	
B1 Made with extra virgin olive oil to help maintain a healthy heart	Made with 100% pure spring water
B2 Made with canola oil ... helps lower blood cholesterol levels	Made with oxygenated water ... the latest trend in Hollywood
B3 100% organic ... healthy eating that tastes great	Made with 100% organic green tea from China
B4 Seasoned to taste the red and green peppers	Made with 100% freshly bottled water from the south of France
B5 Made with only the freshest vine-ripened tomatoes	Made with 90% spring water and sweetened with 10% fruit juice
B6 We use only freshly grown herbs and spices for an unforgettable taste	No artificial sweeteners ... sweetened with nature's natural ingredients
Special ingredients	
C1 A low-fat food ... perfect for your healthy lifestyle	With vitamins A, C and E and potassium ... feel good knowing you're living healthy
C2 Extra fiber added to make your favorite pasta sauce even healthier	Added calcium ... to give your bones a boost
C3 All natural ... no MSG, artificial colors, flavors or sweeteners	Contains Guar Gum ... helps maintain healthy cholesterol and glucose levels
C4 No salt for those on low sodium diet	Zero fat and 12 net carbs ... the push you need
C5 Reduced carb ... 75% less carbs than traditional pasta sauce	Contains the essential nutrient choline ... shown to improve memory loss
C6 Low in saturated fat	Loaded with electrolytes ... for a quick boost
Flavors and tastes	
D1 Hot peppers ... added for those who like it hot	Slightly tangy ... outrageously refreshing
D2 Made with extra basil ... for a homemade taste	Infused with Lemon Lime extracts ... a traditional favorite

Table 15.9   Continued

	Pasta sauce	Energy drinks
D3	Made with pesto and basil ... for a different kind of sauce	Flavored with ginseng and honey ... helps reduce stress
D4	Flavored with sausage and peppers ... versatile, not just for pasta	Infused with Boysenberry ... healthy never tasted so good
D5	Made with spinach and cheese ... a healthy combination	With the pure taste of cranberries ... straight from Nantucket
D6	Full of fresh vegetables ... tastes good and is good for you	Peachy delicious

Packaging

	Pasta sauce	Energy drinks
E1	Comes in recyclable glass jars ... help do your part with the environment	Packaged with your needs in mind ... a thermal barrier helps your juice stay colder longer
E2	Comes in glass jars ... see all the goodness inside	Packaged in a carton with aluminum foil on the inside ... to preserve freshness
E3	Comes in plastic containers ... easy to handle	Aluminum can with a large opening ... makes it easier to gulp
E4	Vacuum sealed for added freshness	Packaged in 100% recyclable paper ... take it with you on your workout
E5	Comes in glass bottles perfect for canning ... use again and again	Packaged in a lightweight container with a resealable plastic pop-up lid ... a good choice for people on the go
E6	Comes in a foil pouch ... just tear and serve	Packed in a shatterproof plastic container with a straw attached ... a good choice for people on the go

Size store location

	Pasta sauce	Energy drinks
F1	Find it in the produce section of your supermarket	In the refrigerator of your local convenience, gas and drug stores
F2	Conveniently located on the grocery aisle shelf	Chilled for your enjoyment ... look in the refrigerated section of your local convenience store
F3	Find it in the gourmet section of your local supermarket	Can be found in a 30 pack at most club stores nationwide
F4	Available in single serve size ... when you're cooking for one	Available in a convenient 8-pack ... in the beverage section of your local supermarket
F5	Available in variety packs ... because your taste changes	Now available in easy to carry 8 oz. bottles ... in the refrigerator section of your supermarket
F6	Find it in the refrigerated section of your supermarket ... fresh, homemade taste	Available at your local deli

Problem plus solution

	Pasta sauce	Energy drinks
1	You buy Pasta Sauce and use it as a BASE INGREDIENT. When you get home, SPECIFICALLY, what other ingredients do you add to the Pasta Sauce to jazz it up and make it your own style?	You are going on a run and want to bring a few bottles of your favorite Energy Drink with you. How do you take it along on your run and keep it cold?

Table 15.9   Continued

	Pasta sauce	Energy drinks
2	Other than serving over pasta, SPECIFICALLY, what other ways do you use PASTA SAUCE?	You love Energy Drinks, but sometimes the caffeine in them is just too much for your system, what do you do to them so you don't overload on caffeine.
3	From the many varieties and brands available to you, SPECIFICALLY, what determines which PASTA SAUCE you buy?	You have a blender at home. How would you jazz up an energy drink to make it special?
4	If you were asked to design new packaging for a PASTA SAUCE which would be catchy to the eye and entice consumers to buy it, SPECIFICALLY, what would you do?	You love feeling healthy and drink ENERGY DRINKS often. You would like to incorporate ENERGY DRINKS into your LIFE in other WAYS. How would you incorporate them into your FOOD?

New flavors

1	The manufacturer is always trying to figure out new flavors of a PASTA SAUCE that will stick around for a while. SPECIFICALLY, what new flavors would you tell them you'd like to see that are not already available?	The manufacturer is always trying to figure out new flavors that will stick around for a while. What flavors would you like to see available that are not already available in an ENERGY DRINK?

Attribute importance for choice and eating

1	Amount of calories	Amount of calories
2	Amount of fat	Amount of carbohydrates
3	Appearance	Amount of fat
4	Aroma	Amount of caffeine
5	Brand	Appearance
6	Healthy ingredients	Aroma
7	Packaging/storage	Brand
8	Preserved for longer freshness	Flavor
9	Price	Amount of electrolytes
10	Size	Healthy ingredients
11	Taste	Packaging/storage
12	None of the above	Preserved for longer freshness
13	Other	Price
14	I do not eat jarred pasta sauce	Size/quantity
15		Special ingredients
16		Taste
17		None of the above
18		Other
19		I do not consume Energy Drinks

Types of products eaten

1	Alfredo sauce	Sugar free
2	All natural, no preservatives added	Caffeine free
3	Chunky tomato sauce with vegetables	Those made with 100% organic materials

Table 15.9 Continued

	Pasta sauce	Energy drinks
4	Homemade pasta sauce	Those 100% water based
5	Low sodium pasta sauce	Those made with fruit juice and added ingredients
6	Newburgh sauce	An Energy Drink you make yourself at home using your blender or juicer
7	Plain tomato sauce	None of the above
8	Ready-to-serve from a specialty store purchased in a container which keeps the sauce hot	Other
9	Ready-to-serve pasta sauce from a can which you heat	I do not consume Energy Drinks
10	Ready-to-serve pasta sauce from a jar which you heat	
11	Tomato sauce with herbs and spices	
12	Tomato sauce with meat added	
13	None of the above	
14	Other	
15	I do not eat jarred pasta sauce	

Flavors purchased and eaten most often

	Pasta sauce	Energy drinks
1	Creamy Alfredo	Boysenberry
2	Marinara	Cherry
3	Mushroom and garlic	Cranberry
4	Roasted garlic and onion	Ginsing and Honey
5	Sweet pepper and garlic	Grape
6	Tomato basil	Grapefruit
7	Pesto	Guava
8	None of the above	Kiwi
9	Other	Lemon Lime
10	I do not eat jarred pasta sauce	Mango
11		Melon
12		Orange
13		Peach
14		Pear
15		Pineapple
16		Pomegranate
17		Raspberry
18		Strawberry
19		Tangerine
20		Watermelon
21		None of the above
22		Other
23		I do not consume energy drinks

Influencers to eat this product

	Pasta sauce	Energy drinks
1	Advertising	Advertising
2	Aroma	Aroma
3	Associations with family or friends	Associations with family or friends
4	Brand	Brand
5	Environment and social situation	Environment and social situation

Table 15.9   Continued

	Pasta sauce	Energy drinks
6	General mood or how you feel	General mood or how you feel
7	Packaging	Packaging
8	Product appearance	Product appearance
9	Portion size	Portion size
10	State of being, i.e. relaxed, on the go	State of being, i.e. relaxed, on the go
11	Stress level	Stress level
12	Taste	Taste
13	Texture	Texture
14	Weather	Weather
15	Season	Season
16	None of the above	None of the above
17	Other	Other
18	I do not eat jarred pasta sauce	I do not consume energy drinks

Brands eaten in the past year

	Pasta sauce	Energy drinks
1	Barilla	Amp
2	Bertolli	Chase 5 hour
3	Classico	Dark Dog
4	Del Monte	Hansen Diet Red
5	Francesco Rinaldi	Hansen Purified Water
6	Hunts	Juicy Lady
7	Muir Glen Organic	Mad Croc
8	Prego	Omega Blast
9	Ragu	Red Bull
10	None of the above	Red Bull ... sugar free
11	Other	Sobe
12	I do not eat jarred pasta sauce	Sports
13		Rhino's
14		XS
15		None of the above
16		Other
17		I do not consume energy drinks

Armed with this information the enterprising developer can create his or her own concepts, or using IdeaMap®.Net identify the utilities of the concepts, the synergies among concepts, and the scenarios in concepts where elements work well or poorly together. The InnovAidOnline™.Net database presents a first in databases that have been collected to help users create new food and beverage concepts.

15.8   Going beyond product features as determiners of consumer choice – the It!® databases

This chapter has principally concentrated on the concrete aspects of concepts; what the product developer needs to know. Yet there is a whole world beyond the specifics. Concepts, properly configured, present vignettes to consumers and the reactions to those concepts in turn provide insight into the consumer mind. This thinking motivated the development of an integrated set of databases to understand consumer choice. The databases, called the It!® studies, were originally developed to understand why people desire food (the Crave It! studies), but in the past four years the databases have expanded to deal with beverages, the dining situation, and beyond foods to the shopping experience, as well as to insurance and social issues (Beckley and Moskowitz, 2002; Moskowitz et al., 2005c).

The It! databases are constructed in a specific fashion, to comprise multiple product studies (usually 20–40), each study designed to have four silos, each with nine elements, and a total of 60 'totally permuted' combinations per respondent, along with an extensive classification. Thus the It! Studies constitute already-executed studies about the consumer response to different product categories.

The ongoing organizing principle surrounding It! is that the consumer choice is dictated by a number of factors, including product sensory features, emotional statements, brands, other product features, and other general phrases of relevance to the product. A single It! study might deal with only one food, such as pizza (see Table 15.10 for the Eurocrave UK Pizza Study). However, the integrated It! databases would deal with 20–40 such foods shown on the 'wall' for the German Eurocrave study (see Fig. 15.11).

In practice the It! database is constructed and analyzed in the following choreographed manner:

1. The general topic is selected. This topic applies to all of the studies within a single database. For instance in the Eurocrave study the topic was 'what features make a food craveable' to consumers? That is, what makes a food highly desired so that consumers think about it even when the food is not there?
2. Develop the basic structure of the elements for any study. The It! databases go beyond the product features to emotions and brands. Thus if we look at Table 15.10 (for Eurocrave Pizza, UK), we see the following four silos, each comprising nine elements. The silos are divided into the following topics:
   a Silo 1 Description of the product
   b Silo 2 Situation and imagination
   c Silo 3 Emotions
   d Silo 4 Brands, outlets or advertising slogans
3. What are the foods, and who are the respondents? For Eurocrave the database encompassed three countries (France, Germany, United Kingdom), with 19–20 relevant foods/beverages for each country. The studies are created using the self-authoring Internet tool discussed above (Moskowitz et al., 2001).
4. Invite respondents to participate. The respondents are not invited to a particular study, but rather to a 'wall' which acts as a clearing-house. The wall presents the available studies. For example, Fig. 15.11 shows the 19 products

Table 15.10 Sample results from the Eurocrave™ 2002 database for pizza (UK results). The table shows how the 36 elements score for total panel and some selected subgroups

	Total	Female	Male	Slightly hungry	Very hungry	Seg 1	Seg 2
Base size	119	90	29	48	25	69	50
Constant	35	36	32	26	40	26	48
Silo #1 – Description of the product (from simple to romanced)							
A1 Thin crust pizza with a layer of tomato sauce and cheese	−2	−2	−2	3	0	7	−13
A2 Pizza with a crust so crispy you have to be careful when you eat it	−4	−4	−2	−3	−3	3	−13
A3 Soft and gooey slices of pizza with cheese	15	20	1	22	15	25	2
A4 Pizza with a lot of tomato sauce, ham and cheese	1	1	2	−4	21	10	−12
A5 Pizza with a filled or twisted crust	10	10	9	9	17	17	0
A6 Pizza with a crust that doubles as breadstick	6	6	6	3	13	9	2
A7 Pizza with a thick crust and a rich topping	7	9	−1	13	4	15	−4
A8 Vegetarian pizza with vegetables and cheese	−11	−9	−16	−5	−25	10	−40
A9 Pizza with fish and seafood	−26	−30	−12	−23	−37	−22	−31
Silo #2 – Situation and imagination							
B1 Pizza is great for parties	−1	−1	−3	−4	1	−3	0
B2 With a glass of lemonade, beer or wine	−3	−6	7	−6	−5	−6	1
B3 With a fresh mixed salad	−2	−1	−6	−2	−4	2	−8
B4 Premium quality … that great traditional taste	−1	−1	−1	−1	−3	−3	3
B5 You can just savor it, when you think about it	1	2	0	4	4	1	3
B6 100% natural, fresh and carefully prepared	3	4	1	3	−1	1	6
B7 With all the toppings and accompaniments you want	10	12	5	9	16	10	10
B8 You can imagine the taste as you walk in the door	2	1	2	−2	4	0	4
B9 So tasty you have to lick your lips twice after each bite	4	4	1	5	7	3	5

Table 15.10 Continued

	Total	Female	Male	Slightly hungry	Very hungry	Seg 1	Seg 2
Silo #3 – Emotions							
C1 Quick and fun ... eating alone doesn't have to be boring	−1	−2	5	1	2	−2	1
C2 When you think about it, you have to have it ... and once you have it if you want more	5	5	7	5	13	4	7
C3 It fills you up – just when you need it	4	4	2	4	5	4	3
C4 Cheers you up	−3	−3	−4	−3	−3	−4	−2
C5 Celebrate special occasions with pizza – escape from the routine	2	2	1	3	5	1	2
C6 A joy for your senses ... seeing, smelling, tasting	2	2	3	4	10	2	3
C7 A real experience ... shared with friends or family	3	2	6	1	11	1	5
C8 Pure enjoyment	2	1	3	1	6	0	3
C9 It satisfies your hunger	1	1	1	4	4	0	4
Silo #4 – Brands, outlets, advertising slogans							
D1 At Pizza Hut	6	5	10	6	11	7	5
D2 At the Italian restaurant with a traditional pizza oven	8	9	7	11	15	8	8
D3 From Pizza Express	2	3	−2	6	3	4	−1
D4 From McCain	−6	−7	−2	−2	−8	−4	−7
D5 From Dominos	−1	−3	3	−4	2	−2	0
D6 From Findus	−8	−8	−8	−4	−12	−9	−7
D7 Made fresh, especially for you	5	7	−1	2	13	5	5
D8 Simply the best pizza for me	0	2	−6	1	1	−1	1
D9 With highest quality and standards that you trust	1	2	−3	−3	3	2	−1

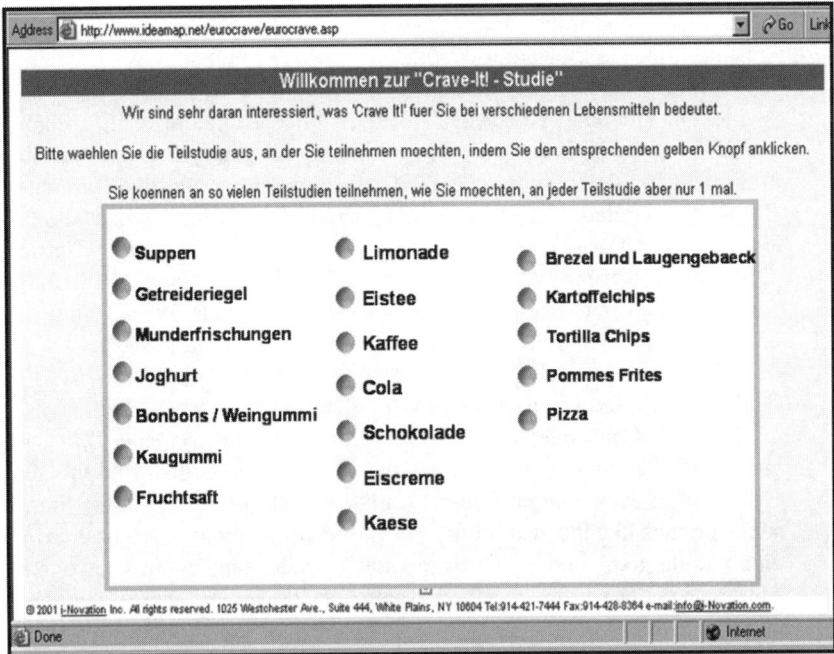

Willkommen zur "Crave-It! - Studie"

Wir sind sehr daran interessiert, was 'Crave It!' fuer Sie bei verschiedenen Lebensmitteln bedeutet.

Bitte waehlen Sie die Teilstudie aus, an der Sie teilnehmen moechten, indem Sie den entsprechenden gelben Knopf anklicken.

Sie koennen an so vielen Teilstudien teilnehmen, wie Sie moechten, an jeder Teilstudie aber nur 1 mal.

Suppen   Limonade   Brezel und Laugengebaeck

Getreideriegel   Eistee   Kartoffelchips

Munderfrischungen   Kaffee   Tortilla Chips

Joghurt   Cola   Pommes Frites

Bonbons / Weingummi   Schokolade   Pizza

Kaugummi   Eiscreme

Fruchtsaft   Kaese

© 2001 i-Novation Inc. All rights reserved. 1025 Westchester Ave., Suite 444, White Plains, NY 10604 Tel:914-421-7444 Fax:914-428-8364 e-mail:info@i-Novation.com.

Done   Internet

Fig. 15.11 Wall for the German Eurocrave™, showing the 19 studies. Respondents are led to this wall, and instructed to choose a food that interests them by clicking on the proper button. Afterwards the program leads the respondent to the proper study.

available in the German Eurocrave. Typically the respondent chooses the particular food which he or she finds interesting. There are some quotas that one sets, so that an overly popular food such as pizza does not get all of the respondents. The wall strategy for Internet testing allows the researcher to measure the distribution of interest in a product category; respondents going to a wall to participate behaviorally show that they are interested in the particular food. As one might expect, pizza is generally the most popular food.

5. Develop the database for each product, using the individual-level presentation of concepts, and subsequent modeling of the ratings. Although a lot of the traditional focus has been on single or 'one-off' studies, a potential area of opportunity is a database, as discussed above with respect to InnovAidOnline™.Net. The It!® studies go further – they work across products, across countries, and across time. The utilities emerging from the interest models have absolute meaning and fall on a ratio scale. This means that one can compare these utilities across silos within a study, across studies within a particular It! Database, across databases, and across uses. Thus, taking a look at the Eurocrave database for pizza, run in three countries, the user can learn a great deal about what drives responses to a pizza. Table 15.10 shows the data for Eurocrave pizza for the United Kingdom. Let's look at some of these results from the UK, and compare them to the results for

Eurocrave pizza for France and for Germany, respectively. We are looking at the results from the viewpoint of a database, not a single study.

a The additive constant (a measure of basic interest in pizza, without any elements). The additive constant for the United Kingdom is 35, meaning that 35% of the UK respondents would likely rate this pizza as interesting (rating 7–9 on the 9-point scale). We can compare this additive constant of 35 in the United Kingdom to a constant of 39 for France (French Eurocrave database), and a constant of 30 for Germany (German Eurocrave). German respondents are less interested in pizza (without any elements) than are French respondents (30 versus 39, meaning that almost one third more respondents in France would rate the idea of pizza as 7–9, without any elements).

b The analysis can extend to the concept elements themselves. Let's look at element A1. Consumers differ in their reaction to this element *'Thin crust pizza with a layer of tomato sauce and cheese'*. The British find this idea irrelevant or even a slight turn-off (utility = −2), whereas both the French and Germans like the idea (utility = +6). It is precisely this type of information that the food industry needs to know to understand country to country, culture to culture differences, and to anticipate the future. There is simply no library of this type of information available, nor could such a library be developed unless there were the tools and the vision for the database.

c Emotional elements appear to do modestly and similarly across the three countries. Emotions are becoming a 'hot topic' for food concept development, but the exact nature of what works in these more ethereal ideas is not clear. The data from Eurocrave suggests that most of the emotion elements do not perform particularly well across the three countries. For example, the element *'Celebrate special occasions with pizza – escape from the routine'* generates a utility of +2 for the United Kingdom and France, and a +1 for Germany. However, if the emotional element brings in somewhat of a sensor expectation the utilities are higher. For example, the emotional element *'When you think about it, you have to have it . . . and once you have it you want more'* which promises a more sensory reward has a utility of +5 for the United Kingdom and France, and +4 for Germany.

6. Work with defined subgroups. These subgroups are either based on standard geo-demographics (e.g. age, income, market), on self-profiling (e.g. self-reported hunger at the time of taking the interview), or on segmentation emerging from the pattern of the utilities (concept-response segmentation).

15.9   Conclusions

How should the reader tie a nice bow around the information presented here, with the bow comprehending all of the facets? We might reach some closure by keeping a few organizing principles in mind when we deal with the consumer end of food and beverage ideas.

1. It is time to found a science of concept studies. Look at the curricula of business schools, departments of food science, and allied disciplines. Their curricula are filled with substantive areas of knowledge – whether these are statistics, consumer behaviour, sensory analysis, and the like. There is no separate science of concept studies. There is, of course, always lip-service for concept testing, but in the main the topics of concept testing are treated as a very small section in research methods, or product development processes. The opportunity exists to found a new science for concept studies, starting with the discovery of needs, continuing with their incorporation into consumer-oriented vignettes, and finishing with optimization and concept databases to know what works and what does not.

2. The notion of a concept is not well defined. As we have seen in this chapter, concepts range from simple descriptions of the product (product concepts) all the way to concepts that incorporate emotional elements, brand names, and the like. Since there is no science of concept development, and varied practices surrounding concept testing, the definition of what is a concept, and more importantly what is a good concept, is not at all clear. To some a concept comprises a fully formed paragraph, perhaps 150 words, written in literary style, with proper sentences, connectives, with a problem definition and the product as a solution. To others a concept comprises a set of phrases strung together, with the consumer putting in the connectives at a subconscious level while reading the concept.

3. We deal here with a body of practice. Because science did not necessarily reach the world of business before the practice of concept testing gained currency among management, there is no body of science. Our body of practice is just that – different methods, espoused by corporate professionals with motives varying from the desire to do the right thing, to doing the thing right, to those who might be kindly described as simply doing the test. Bodies of practice comprehend methods, not necessarily methods that have emerged from rational, considered opinion and test. An accepted method for concept research might have easily sprung fully formed, Minerva-like from the head of a corporate Zeus challenged to measure concepts for a new product.

4. It is high time to bring concept research into the product development realm. Most concept research began in marketing or the advertising agency and has remained there. For the most part R&D product development, including sensory research, has avoided the 'soft science' of concepts in favor of the hard science of actual products and processes. Indeed, many R&D scientists are proud of their splendid isolation, and their lack of involvement with the communication of ideas to the consumer. Traditional R&D generally wants concepts that have promise and can demonstrate some type of ROI (return on investment). These concepts are their marching orders; what they are missing, however, is the recognition that R&D must now become part of the group that deals with concepts, and that gives the marching orders. The days of passive obedience are finished – it is time for R&D in general, and sensory

researchers in particular, to expand their horizons into the very early stages of development, the so-called fuzzy front end.

5. Testing has migrated to the computer. The organizing topic of this chapter is concept evaluation and optimization on the Internet. A retrospective on the consumer research industry suggests an increasing proportion of concept testing to be migrating to the Internet. One reads almost monthly articles on the increase in Internet testing in such popular newsletters as *Inside Research* (Lawrence Gold) and *Research Business Reports*. Furthermore, from personal experience by the author the migration to the Internet is all but complete. In the late 1990s research on the Internet was still new. Questions asked at that time pertained to the representativeness of the respondent population (Moskowitz *et al.*, 2000) and the believability of the data. Those days are gone; today's questions are price, speed, and capability. The Internet execution of most studies is simply an accepted fact, courtesy of *homo economicus regnant* – low prices, speed, power, and simplicity drive out almost anything in their way.

6. Experimental design is becoming more popular. Experimental design, also known as conjoint measurement, has become increasingly popular, for the simple reasons that it has become cheap, easy to do, and basically it works. With today's technology it is the lack of powerful methods that holds back research, not the complexity of the method. Most complexity is being hidden by pull-down menus, automatic programs that execute the study with one or two clicks, and simple analysis that boils down complex data sets to a few easy-to-understand charts. Systematic exploration works, and researchers are simply becoming less afraid, less intimidated.

7. A big opportunity exists in innovation. A lot of research on the business process in product development deals with the different activities that the product development enterprise must follow in order to conceptualize, develop, commercialize, and then market a product. Many of the steps have been laid out by Cooper and his associates in their creation of the Stage Gate[TM] process (Cooper, 1993). These processes deal with the so-called fuzzy front end, that start of the business process where the possibilities are many, and the product has not yet been codified. This chapter shows that concept research can be used to formalize the invention methods at this fuzzy front end. The big opportunity in innovation thus consists of a structured invention process, underpinned by concept research using experimental design on the one hand, and systematic grafting of different types of product idea into one composite on the other.

8. Another big opportunity exists in databases. Perhaps the biggest opportunity of all comes from the opportunity to access an ideas database. There is no current science of consumer development, and therefore no corpus of knowledge to which the developer can turn. Development each time has to be from the start, perhaps helped a bit by one's experience. Armed with a database, the developer has the benefit of previous knowledge that has been created, digested, and systematized. The database provides at once a practical

tool for those who seek applications, and for the others who seek a science. The outlines of such a database remain, of course, for the next generation of researchers, a daunting challenge but at once a glorious opportunity.

15.10   References

ACUPOLL® (2005), http://www.acupoll.com/homenglish/index.html

BECKLEY J and MOSKOWITZ H R (2002), *Data Basing the Consumer Mind: The Crave It!*®, Drink It!®, Buy It!® and Healthy You! Databases, Anaheim, Institute of Food Technologists.

BOX G E P, HUNTER J and HUNTER S (1978), *Statistics For Experimenters*, New York, John Wiley & Sons.

CATTIN P and WITTINK D R (1982), 'Commercial use of conjoint analysis: a survey', *Journal of Marketing*, **46**, 44–53.

COOPER R G (1993), *Winning at New Products: Accelerating the Process from Idea to Launch* (2nd edn), Boston, MA, Addison Wesley Publishing Co.

DAHAN E and SRINIVASAN V (2000), 'The predictive power of Internet-based product concept testing using visual depiction and animation', *Product Innovation Management*, **17**(2), 99–109.

EWALD J and MOSKOWITZ H R (2001), 'Always on – bringing market research down to the development engineer, closer to the customer, and into the vortex of product development', *Proceedings of the 54th ESOMAR Congress*, Rome.

FULLER G W (1994), *New Food Product Development: From Concept to Marketplace*, Boca Raton, FL, CRC Press.

GREEN P E and KRIEGER A M (1991), 'Segmenting markets with conjoint analysis', *Journal of Marketing*, **55**, 20–31.

HIPPEL, E. VON (1986), 'Lead users: source of novel product concepts', *Management Science*, **32**, 791– 805.

KHURANA A and ROSENTHAL S R (1997), 'Integrating the fuzzy front end of new product development', *Management Review*, **38**(2), 103–120.

Literary Digest (1936), http://www.historymatters.gmu.edu/d/5168 (retrieved 6 July 2005).

LUCE R D and TUKEY J W (1964), 'Conjoint analysis: a new form of fundamental measurement', *Journal of Mathematical Psychology*, **1**, 1–36.

MACELROY B (2000), 'Variables influencing dropout rates in Web-based surveys', *Quirks Marketing Research Review*, www.quirks.com; paper 0605.

MOSLEY-MATCHETT J D (1998), 'Leverage the Internet's research capabilities', *Marketing News*, **32**, April, 13.

MOSKOWITZ H R and GOFMAN A (2004), System and method for performing conjoint analysis. Provisional patent application, 60/538,787, filed 23 January 2004.

MOSKOWITZ H R and MARTIN D G (1993), 'How computer aided design and presentation of concepts speeds up the product development process', *Proceedings of the 46th ESOMAR Congress*, Copenhagen, Denmark, pp. 405–419.

MOSKOWITZ H R, GOFMAN A, TUNGATURTHY P, MANCHAIAH M and COHEN D (2000), 'Research, politics and the Internet can mix: considerations, experiences, trials, tribulations in adapting conjoint measurement to optimizing a political platform as if it were a consumer product', *Proceedings of ESOMAR Conference: Net Effects*, Dublin, pp. 109–130.

MOSKOWITZ H R, GOFMAN A, KATZ R, ITTY B, MANCHAIAH M and MA Z (2001), 'Rapid, inexpensive, actionable concept generation and optimization – the use and promise of self-authoring conjoint analysis for the foodservice industry', *Foodservice Technology*, **1**, 149–168.

MOSKOWITZ H R, ITTY B, SHAND A and KRIEGER B (2002a), 'Understanding the consumer mind through a concept category appraisal: toothpaste', *Canadian Journal of Market Research*, **20**, 3–15.

MOSKOWITZ H R, BECKLEY J, MASCUCH T, ADAMS J, SENDROS A and KEELING C. (2002b), 'Establishing data validity in conjoint: Experiences with Internet-based 'mega-studies', *Journal of Online Research*, http://www.ijor.org/ijor_archives/articles/establishing_data_validity_in_conjoint.pdf

MOSKOWITZ H, PORRETTA S and SILCHER M (2005), *Concept Research in Food Product Design and Development*, Iowa, Blackwell Professional.

MOSKOWITZ H R, REISNER M, KRIEGER B and OKSENDAL K O (2005b), 'Steps towards a consumer-driven concept innovation machine for 'ordinary' product categories in their later lifecycle', *Proceeding of the First Innovate! Conference*, ESOMAR (World Society of Market Research), Paris and Amsterdam.

MOSKOWITZ H R, GERMAN B and SAGUY I S (2005), 'Unveiling health attitudes and creating good-for-you foods: The genomics metaphor and consumer innovative web-based technologies', *CRC Critical Reviews in Nutrition and Food Science*, **45**(3), 265–191.

PAWLE J S and COOPER P (2001), 'Using web research technology to accelerate innovation', *Proceedings of Net Effects*, Barcelona, European Society of Market Research, pp. 11–30.

SYSTAT (1998), 'Systat, the system for statistics', Evanston Systat Division of SPSS.

VERMUNT J K and MAGIDSON J (2000), 'Latent class cluster analysis', in McCutcheon A L and Hayenaars J A (Eds), *Advances in Latent Class Analysis*, Cambridge, MA: Cambridge University Press.

VIGNEAU E, QANNARI E M, PUNTER P H and KNOOPS P (2001), 'Segmentation of a panel of consumers using clustering of variables around latent directions of preference', *Food Quality and Preference*, **12** (5–7), 359–363.

WELLS W D (1975), 'Psychographics, a critical review', *Journal of Marketing Research*, **12**, 196–213.

WITTINK D R and CATTIN P (1989), 'Commercial use of conjoint analysis: an update', *Journal of Marketing*, **53**, 91–96.

WITTINK D R, VRIENS M and BURHENNE W (1994), 'Commercial use of conjoint analysis in Europe: results and critical reflections', *International Journal of Research in Marketing*, **11**, 41–52.

16

Consumer testing of food products using children

R. Popper and J. J. Kroll, Peryam & Kroll Research Corporation, USA

16.1 Introduction

Nearly one-third of the world's population is under the age of 15. In the United States alone, there are over 50 million consumers in that age group, accounting for a youth market in excess of $300 billion, with food and beverage representing as much as 60% of that market.

The size of the business opportunity has resulted in a highly competitive environment for food manufacturers as they try to gain and maintain their share of the youth market. While parents serve an important gatekeeper function in determining which products are purchased, the 'pester' power and influence of even young children on food purchase decisions continues to rise. Today's children have more choices and more influence over their parents' purchase decisions than ever before. Children are also making many more of their own purchase decisions, with ever greater sums of money under their direct disposal. Therefore, no manufacturer can succeed in the youth marketplace today without optimizing products for their consumer target – the child. By successfully appealing to children, marketers also stand a chance of building long-term brand loyalty, as early exposure to products and brands may form the basis for brand choices later in life.

Successfully developing and optimizing foods and beverages for children requires involving children in the product development process. Children's needs and wants differ from those of adults, and product development must be guided by insights into what uniquely motivates and appeals to them. For new products, input from children may be needed at every stage of product development, from early idea exploration and prototype screening on through product and package optimization, as well as advertising copy development. For products undergoing reformulation, either for the purpose of cost reduction or product improvement, research may be needed to confirm that the product change is not detrimental to acceptability or does indeed improve the product.

In conducting research with children, it is important that the methods employed take into consideration the physical, emotional, and cognitive development of the children being asked to participate in the study. The assumption that methods appropriate for adults can be used with little or no modification in a study involving children is almost certain to lead to disappointing results. Furthermore, it is important to consider the specific age of the child in deciding what methods to employ, since motor skills, language skills, and reasoning abilities develop rapidly during childhood.

The primary focus of this chapter is on methods appropriate for quantitative testing of product acceptability with children who are between the ages of 6 and 12. Methods for testing pre-school-age children are also briefly considered. Infants and toddlers require methods grounded in behavioral observation, which are outside the scope of this chapter. Children who are above the age of 12 are capable of using many of the same research tools as adults. The chapter begins with a review of what is known about children's sensory perception and how food preferences develop, areas that can have important implications for product design as well as testing methodology. The chapter then reviews the techniques for quantitative consumer testing and provides examples of the types of questions and scales appropriate for testing with children of different ages.

16.2   Sensory perception: sensitivity and perceived intensity

Are children more or less sensitive than adults to taste or olfactory stimulation? Studies of taste thresholds present a confusing picture, with some studies suggesting that children as young as 5–7 years of age have similar detection thresholds as adults, others finding that children of this age have poorer sensitivity than adults (Guinard, 2001). Various methodological differences among the studies have been suggested as the source of these inconsistencies (James et al., 1997). In their study, James et al. tested 8–9-year-old children and young adults of both sexes under the same experimental conditions, determining detection thresholds for sucrose, sodium chloride, citric acid, and caffeine. A two-alternative forced-choice procedure was used, in which respondents had to indicate which sample 'tasted stronger' (one of the samples contained only water, the other the taste stimulus). The fact that the thresholds did not differ across two replications was indicative of the reliability of the procedure. Girls had detection thresholds similar to the adults; boys, on the other hand, proved to be somewhat less sensitive than adults, especially in the case of citric acid.

Greater differences in sensitivity between children and adults are evident as the task increases in complexity. Oram *et al.* (2001) compared 8–9-year-old children to adults in terms of their ability to recognize a particular taste (e.g. sweet) in a mixture of two tastes (e.g. sweet and sour). While children were able to correctly identify a taste as sweet, sour, or salty when it was the only taste present, they performed markedly poorer than adults in correctly identifying the components of a binary taste mixture. Children tended to recognize only one of the components, the adults recognized both. These differences could be due to differences in taste perception, in cognitive ability (e.g. ability to separately process two sensations), or in response strategy (e.g. children may have focused on the more intense or more appealing taste quality).

Children and young adults tend not to differ much with respect to olfactory thresholds (Lehrner *et al.*, 1999). However, while children's olfactory sensitivity is very similar to that of adults, children are less likely than young adults to correctly identify (name) an odor or to recognize it as one that had been presented earlier in the experiment (Lehrner *et al.*, 1999; Lehrner and Walla, 2002). The fact that children perform less well at identifying odors than young adults is not surprising. Children also perform less well than young adults in picture identification (Cain *et al.*, 1995). Children have the olfactory sensitivity but lack odor-specific knowledge, which accumulates slowly over time. Odors may be unfamiliar to children; and, even if familiar, may not have been become strongly associated with verbal descriptors. Semantic encoding (that is, the association of smells with words) is a key component in odor memory (Rabin and Cain, 1984), especially in children (Lehrner *et al.*, 1999; Lumeng *et al.*, 2005). Cain *et al.* showed that when given the opportunity to learn odor names in a paired association task, children improve quickly in performance, although the learning curve is much attenuated when the odors are novel to begin with.

Another measure of sensory sensitivity, separate from detection threshold, is supra-threshold intensity perception. Compared with adults, children appear to perform less well at ranking the sweetness of beverages varying in amount of sucrose (Kimmel *et al.*, 1994; De Graaf and Zandstra, 1999; Liem *et al.*, 2004a). The rate at which perceived intensity changes with stimulus concentration may also differ between children and adults. Zandstra and De Graaf (1998) varied the concentration of sucrose in orange drink and found that for children aged 6–12 sweetness increased less rapidly as a function of sucrose than it did for adolescents or adults. Using a category scaling procedure, the children rated the low sucrose concentrations as sweeter than did the adults, but rated the high sucrose concentrations as less sweet. In a second study, De Graaf and Zandstra (1999) replicated these results for orange drink, but found that in water the sweetness functions for adults and children were similar (as was also reported by James *et al.*, 2003). Thus, food context may affect children and adults differently. This conclusion is supported by James *et al.* (1999), who used the method of magnitude estimation to scale sweetness, and by Temple *et al.* (2002), who used a computer-based time-intensity scaling procedure. Both studies found that sweetness increased in a similar fashion for both adults and children in response to increases of sucrose in water, but increased more slowly in orange drink and some other food applications. In the case of these more complex stimuli, children may have found it difficult to attend exclusively to sweetness and may have been influenced in their sweetness ratings by other sensory characteristics.

While sweetness is the sensory characteristic most frequently studied with children, a number of studies have been reported on supra-threshold sourness perception. These studies indicate that adults and children rate sourness similarly in response to variations in citric acid, both in orange drink (Zandstra and De Graaf, 1998) and in gelatin dessert (Liem and Mannella, 2003).

Much remains unknown with regard to differences between adults and children in sensory perception, especially in modalities other than olfaction and taste (for the few examples of research on texture perception in children, see Szczesniak, 1972; Oram, 1998; Narain, 2005). Adults and children show many similarities in sensory sensitivity, both in terms of detection thresholds and super-threshold intensity perception. In cases where differences have been reported, it is important to consider the possible sources for the findings. Any determination of threshold sensitivity or supra-threshold perception in children is complicated by the fact that measuring sensory perception in children is difficult – differences between adults and children in sensory perception may reflect, in part at least, differences in how children interpret the question they are asked and how they use the scales on which the research is based. Liem *et al.* (2004a) provide an example of how important the question is to the experimental outcome. In their study, 4-year-old children were not able to reliably dis-criminate the intensity of beverages differing in sweetness, using either a two-alternative forced choice or a ranking procedure. However, when asked to state their preference, the children consistently preferred the sweeter of two for-mulations in a pair and were able to rank order beverages from least to most preferred. Obviously, the children could indeed discriminate among the sweetness levels, but did not understand the intensity scaling instructions.

A number of the studies on intensity scaling cited above included control conditions in which children and adults rate simple visual stimuli for appearance characteristics (such as darkness or length) using the scaling procedures used to rate taste intensity. Children and adults usually perform identically on scaling such visual characteristics, suggesting that performance differences in percep-tion of taste intensity are not solely a result of the scaling methodology. Whether the observed differences in taste sensitivity reflect maturational differences in the sensory systems of children and adults or developmental differences in attention and cognition has yet to be fully sorted out. The fact that differences between children and adults are more likely to reveal themselves with complex rather than simple taste stimuli (e.g. with taste mixtures and real foods rather than simple aqueous solutions) suggests that higher mental processing plays at least some part in accounting for the age difference in performance.

One practical implication of these findings is that to the extent to which children and adults differ in their sensory perception, children may not notice changes in product formulation to the same degree as adults. Ingredient substitutions and reformulations such as a reduction in sodium or a removal of trans fatty acids need to be tested with children in order to determine whether children notice a difference. Of course, whether a difference in sensory perception is of a magnitude to result in a difference in food acceptability is a separate, and usually more critical, question.

16.3   The origin of food preferences

Humans are born with an innate liking for sweet and an aversion to bitter, as has been shown by studies of reflexive facial expressions and food intake in newborns (see Birch, 1999, for a review). These genetic predispositions make evolutionary sense, since sweet foods (e.g. certain fruits) tend to be nutritious and high in energy, whereas bitter foods can be poisonous. Sour tastes are also rejected by newborns. A genetic predisposition towards liking of salt has not been so clearly established; newborns are indifferent to salt, but infants at 4 months show a liking for moderate levels of salt, possibly the result of a natural maturation process (Beauchamp et al., 1986).

Humans are also genetically predisposed to avoid unfamiliar foods (Birch, 1999). In human evolution, this food neophobia served a protective function, since unfamiliar foods could cause illness or death. However, an infant's early experience with foods begins to counteract this neophobic response. While basic tastes such as sweetness and bitterness are intrinsically pleasant and unpleasant, the preferences for specific foods are largely learned, and the diversity of world cuisines attests to the role of culture and environment in shaping food preferences and overcoming the neophobic response (Rozin, 1984).

Many food preferences are learned based on the positive physiological consequences that follow consumption, such as feeling of satiety (Birch et al., 1999). However preferences are acquired even in the absence of such positive physiological reinforcement, based on repeated exposure to a food alone. Zajonc (1968) was the first to identify the 'mere exposure effect' across a variety of stimulus domains, demonstrating that simply the repeated exposure to a stimulus, such as a sound or shape, can enhance liking. The role of early experience in food preferences has been demonstrated in a variety of studies. What the mother eats during pregnancy and lactation can affect an infant's flavor preferences (Birch, 1999), because of flavor cues contained in amniotic fluid or breast milk. Beauchamp and Moran (1984) showed that infants who were routinely fed sweetened water during the first months of life showed a greater preference for sweetened water at 2 years of age. Even bitter-tasting foods are subject to early learning effects. Protein hydrosolate infant formula (recommended for infants that do not tolerate cow's milk) tastes bitter as well as sour and is not well accepted by infants. Mennella and Beauchamp (2005) showed that exposure to these formulas starting shortly after birth leads to greater acceptance (as measured by intake) in infants aged 5–11 months. Early experience with protein hydrosolate formulas also has consequences for food preferences later during childhood. According to Liem and Menella (2002), children who experienced protein hydrosolate formulas early in life showed a preference at age 4–7 for higher levels of citric acid in juice (i.e. preferred a more sour taste) than did infants with no such experience.

The effect of exposure seems to be proportional to the amount of repetition. In a study conducted by Birch and Marlin (1982), 2-year-old children were exposed to initially unfamiliar foods with varying frequency (from zero to 20 times) over a period of several weeks. Preference for foods at the end of the study was almost perfectly correlated with exposure frequency: the more frequently a child was exposed to the food, the more the child liked it.

Food preference is also subject to social influences. Birch (1980) showed that 4-year-old children were influenced in their preferences by their peers who sat next to them during lunch. For example, children that initially did not like a vegetable grew to like it after repeatedly observing a peer consume that vegetable. This change in preference was relatively long lasting, persisting for several weeks and in the absence of the peers. In general (see Birch, 1999), older children are more effective role models than younger ones, mothers are more effective than strangers, and, for older pre-schoolers, adult heroes are more effective than ordinary adults.

There are several implications – for marketers, product developers, and researchers alike – of the way food preferences are acquired. Children's neo-phobia is likely to affect their willingness and response to novel foods in a research setting. Loewen and Pliner (2000) developed a questionnaire for assessing individual differences in neophobia in children that might be a useful attitudinal measurement for helping to explain children's response to novel foods in test situations.

Marketers and product developers need to balance children's desire for novelty with their propensity to prefer the familiar. The success of green colored ketchup in the United States, which combines a familiar flavor with an unfamiliar color, is a good example of this principle. In a laboratory context, Pliner and Stallberg-White (2000) demonstrated a similar phenomenon, showing that 10–12-year-old children were far more willing to try an unfamiliar chip when it was combined with a familiar dip than when it was presented with an unfamiliar dip.

The effect of repeated exposure on food acceptance among children suggests that product sampling and other strategies that encourage repeat consumption may help build acceptance of a new product. The learning effect also has implications for how to test novel foods with children. The limited exposure typical in taste tests may lead product development to underestimate the potential for unfamiliar flavors or textures. Finally, the demonstrated effects of social role models on food acceptance provides marketers with a continued reason to attempt to leverage peer influence through advertising or grass-roots marketing campaigns.

16.4   Difference between children and adults in food preferences

Perhaps the most widely documented difference in preference between children and adults is that children prefer a greater intensity of sweetness than adults. The heightened preference for sweetness among children was first reported by Desor *et al.* (1975) and was subsequently confirmed by Desor and Beauchamp (1986) in a longitudinal study, in which the same respondents were tested at two points in time – at age 11–15 and ten years later. The heightened preference for sweetness, which was reflected in the percentage of respondents who chose the sweetest of four sucrose concentration, decreased from the first test to the second, demonstrating that this preference declines with age. Using ranking or scaling methods instead of choice procedures, a number of other studies have also found that the optimal level of sweetness is higher for children than adults (Zandstra and De Graaf, 1998; De Graaf and Zandstra, 1999; Liem *et al.*, 2004a). Children also prefer higher levels of salt than adults, as has been demonstrated both with 9–11-year-old children (Desor *et al.*, 1975) as well as pre-schoolers (Beauchamp and Cowart, 1990). The reasons for this heightened preference for sweet or salty are not fully understood, but are believed to be rooted in development rather than the result of environmental influences.

Recent work by Liem and colleagues has revealed a segment of children with a preference for extreme sour tastes. Across a number of studies (Liem and Mennella, 2003; Liem *et al.*, 2004b), the researchers found that about one-third of children aged 5–12 showed a preference for an extreme sour taste, in contrast to adults as well as other children who preferred lower levels of sourness. All children, regardless of their preference, were equally able to discriminate among different levels of sourness, and showed the same level of discrimination as adults. Thus, the preference for extreme sour cannot be attributed to differences in sensory perception. The 'sour-loving' children did differ from the other children with respect to a number of other factors: they were less neophobic, had a greater preference for bright colors, tended to experience a wider variety of fruits, and tended to like other sour foods (sour candies, lemons), behaviors that may be examples of 'thrill seeking' among these children.

Whether the preference for extreme sour extends to preferences for other extreme tastes has not been investigated so far, and no longitudinal studies have yet been conducted to determine how this sour preference changes with age. It is also unclear whether the preference for extreme sour is the result or the cause of the observed differences in food habits.

Adults and children differ not only in their preferred intensity for certain basic tastes. Moncrieff (1966) studied olfactory preferences among adults and children and concluded that children aged 10–14 prefer fruity over floral smells, whereas the opposite was true for adults. With respect to food texture, children seem to prefer simple, smooth textures (Urbick, 2002) and, in bread products, dislike crunchy or chewy textures, especially at a young age where chewing efficiency is lower (Narain, 2005). The preference for simple textures may be an example of children's preference for foods with low 'complexity' (Ringel, 2005).

Children and adults may also differ in the relative importance they place on the appearance, taste and texture in assessing overall acceptability. Moskowitz (1994) cites a case study on ice cream in which children were found to place the same importance on appearance, flavor and texture, contrary to adults, who placed more emphasis on flavor and texture than appearance. Tuorila-Ollikainen *et al.* (1984) found that children put more emphasis on sweetness in soft drinks than on any other attribute, consistent with the demonstrated liking for sweet taste among children. Comparing children of different ages with respect to liking of meats, Rose *et al.* (2004a,b) found that taste and smell were of predominant importance to older children (aged 10–11), whereas texture and mouthfeel characteristics were more likely to influence acceptability in the younger children (aged 6–7), as Chambers and Bowers (1993) showed was the case for adults.

In summary, the results of published research (as well as those of unpublished industrial studies) demonstrate the importance of optimizing products for children based on their distinct preferences. A product that is optimal for children is likely to differ from one that is optimal for adults – and may also differ from one that adults *think* would be optimal for children (Moskowitz, 1994).

16.5   Research methods for testing children

The research task and measurement technique employed in a study involving children must be age appropriate, since language, cognitive, and motor skills vary significantly by age. The Swiss psychologist Jean Piaget is well known for his description of the stages of a child's cognitive and linguistic development. For example, Piaget distinguishes the 'pre-operational' stage (aged 2–6) from the 'concrete-operational' stage (aged 7–12). In the pre-operational stage, children are more likely to focus on a single aspect of a stimulus, whereas concrete-operational children have the ability to perceive stimuli multidimensionally.

Gollick (2002) describes some of the limitations of children that may affect their ability to answer questions in a research study. Very young children, for example, have difficulty with concept formation (e.g. sweetness) and classification (e.g. like/dislike). Even when they understand the principles, their attention span may limit their ability to perform the task. For example, 3½-year-old children may understand a sorting task, but only about half the children may have the attention span to remember the assignment and therefore successfully complete the task.

'Seriation,' the ability to rank things in order of magnitude, is not fully mastered until age 7, according to Gollick, and this has implications for the reliability of any scaling results from younger children. In addition, children have limited memory skills, which may affect their ability to remember a succession of flavors presented for evaluation in a sensory test. Gollick also notes the difficulty that children 6 and under have in attending to more than one aspect of a stimulus at one time, as Piaget's theory suggests. As a consequence, young children may attend to only one dimension of a food, unlike older children, who may be able to base their reaction on a simultaneous consideration of multiple aspects.

Unfortunately, the child's age is far from being a perfect predictor of a child's ability to participate in research. There is tremendous variation in skills among children of the same age. Gollick's experience with cognitive testing has shown that the age at which 10% of children can master a particular task, compared with the age at which 90% of children can do so, varies by as much as 4 years. Thus, assumptions regarding what a particular age group can do are often going to be true only approximately, and researchers need to take into account the considerable variation in children's abilities, even at similar ages. As Chambers (2005) has pointed out, particular caution is needed with respect to children in the 'cusp' years, i.e. those that are transitioning from one developmental stage to another. For example, children age 6–7 are at the border of the pre-operational and the concrete-operational stages of development, and this age group is likely to be quite variable with respect to their cognitive abilities.

Guinard (2001) has summarized published studies with regard to children's abilities to perform a variety of sensory testing methods at different ages (see also ASTM, 2003). As one would expect, the younger the age group, the more challenging it is to devise valid, reliable test methods. These reviews conclude that children 2-3 years old are capable of expressing preference between two choices, but not much else. Children aged 4–5 are, in addition, are capable of performing attribute-based paired comparison tasks ('which is sweeter?'), ranking products in terms of preference, and rating products on simple hedonic scales. Children aged 6–7 are capable of more complex scaling tasks (e.g. intensity) and performing certain discrimination tests (e.g. triangle tests). By age 8, children are capable of performing virtually any kind of standard sensory test. By this age, children are also capable of self-administering many tests with only occasional assistance from the interviewer or experimenter. At younger ages, one-on-one interviews are usually required.

When products are expected to appeal to a wide age range, it is often convenient to test older children (8 and above), who are subject to fewer limitations regarding the appropriate research technique and who can self-administer the tests (which is less costly and time consuming than one-on-one interviewing). However, when the core target age for the product is specifically younger children, it may not be appropriate to focus exclusively on the older age group.

16.6   Hedonic testing with children

16.6.1   Hedonic scales

Being able to determine the level of a child's liking for a product has obvious importance to product development. A number of hedonic scales for children have been proposed, some using pictures (often faces), some using words, and

Fig. 16.1   Examples of pictorial hedonic scales for children.

some a combination of pictures and words (see ASTM, 2003). Three examples of pictorial scales, two of which are also verbally anchored, are shown in Fig. 16.1.

Kroll (1990) introduced a verbal liking scale for testing children (see Table 16.1), which has become known as the Peryam & Kroll (P&K) or the super good/super bad scale. The scale is similar to the traditional 9-point hedonic scale, except that the verbal anchors associated with the scale are more child-friendly – instead of using terms such as 'like extremely' and 'dislike extremely,' for example, it employs the terms 'super good' and 'super bad.' Testing children in the range of 5–10 years old, Kroll compared several scale variations, including the traditional 9-point scale, the 9-point P&K scale, and a 9-point face scale (similar to the one shown at the bottom of Fig. 16.1). Kroll also tested 7-point versions of these same three scales. Scales were compared on the basis of their ability to discriminate between two beverages (which paired

Table 16.1   The traditional adult hedonic scale and the P&K hedonic scale for children

Traditional adult hedonic scale	P&K hedonic scale for children
Like extremely	Super good
Like very much	Really good
Like moderately	Good
Like slightly	Just a little good
Neither like nor dislike	Maybe good or maybe bad
Dislike slightly	Just a little bad
Dislike moderately	Bad
Dislike very much	Really bad
Dislike extremely	Super bad

preference tests showed were differentially preferred). All scales found a significant difference in liking between the two beverages, but with the P&K scale the difference was more highly significant than for the face scale and the traditional scale. Across scale types, the 9-point scales discriminated better than the 7-point scales, even among 5–7-year-old children.

These results challenged two prevailing assumptions, namely that face scales were superior to verbal scales when testing children, and that shorter scales were better than longer scales. Spaeth *et al.* (1992), working with 8–10-year-old children, confirmed Kroll's conclusions in a study comparing 3-point, 5-point, and 9-point versions of three scale types: the traditional hedonic scale, a face scale without verbal anchors, and a box scale verbally anchored only at the end-points. The authors concluded that children do not use face scales better than purely verbally anchored scales and do not use short scales better than longer ones. Also, anchoring the 9-point scale only at the end-points yields the same results as the traditional hedonic scale.

Face scales may actually be detrimental by introducing unintended bias or confusing the child. A face intended to show a degree of 'dislike' can be interpreted by a child as conveying anger, a face intended to show 'liking' may suggest 'happiness.' Children may choose the 'happy' face because they like it better, rather than because it represents their opinion about the food they taste. Cooper (2002) has found that the eyes and the mouth are particularly important to the interpretation of the facial expression and are more likely than other elements to lead to misinterpretations of the scale unless carefully chosen. She also indicates that there are cultural differences with respect to the interpretation of facial expression. Certain expressions are appropriate in some cultures, but not in others.

While children 8–10-years-old can effectively use verbal 9-point scales, their ratings, compared with adults, are often higher. Figure 16.2 compares the distribution of ratings for adults and children rating the same products, with children using the 9-point P&K scale, and adults using the standard hedonic scale. The children most frequently responded with the top most category of the scale, 'super good,' unlike the adults, who responded less positively. Other studies show that younger children will give somewhat higher ratings than older children using the same super good/super bad scale.

There are several factors that affect children's use of scales. Scale length is an obvious one. With fewer response choices (e.g. a 5-point hedonic scale), responses are more likely to be crowded in the upper end of the scale range (Crawford *et al.*, 2003). Crawford *et al.* (2005) investigated other scale factors which may affect scale usage among 8–14-year-olds. Using a 7-point hedonic scale, the authors concluded that a vertical scale orientation leads to more positive responses than a horizontal orientation, and that a horizontal scale with the positive end on the left leads to higher ratings than one with the negative end on the left. Similar scale effects have been reported for adults (Friedman and Amoo, 1999; Sauerhoff *et al.*, 2005). These findings suggest caution is needed when comparing findings across studies that have used ostensibly identical, but differently formatted hedonic scales.

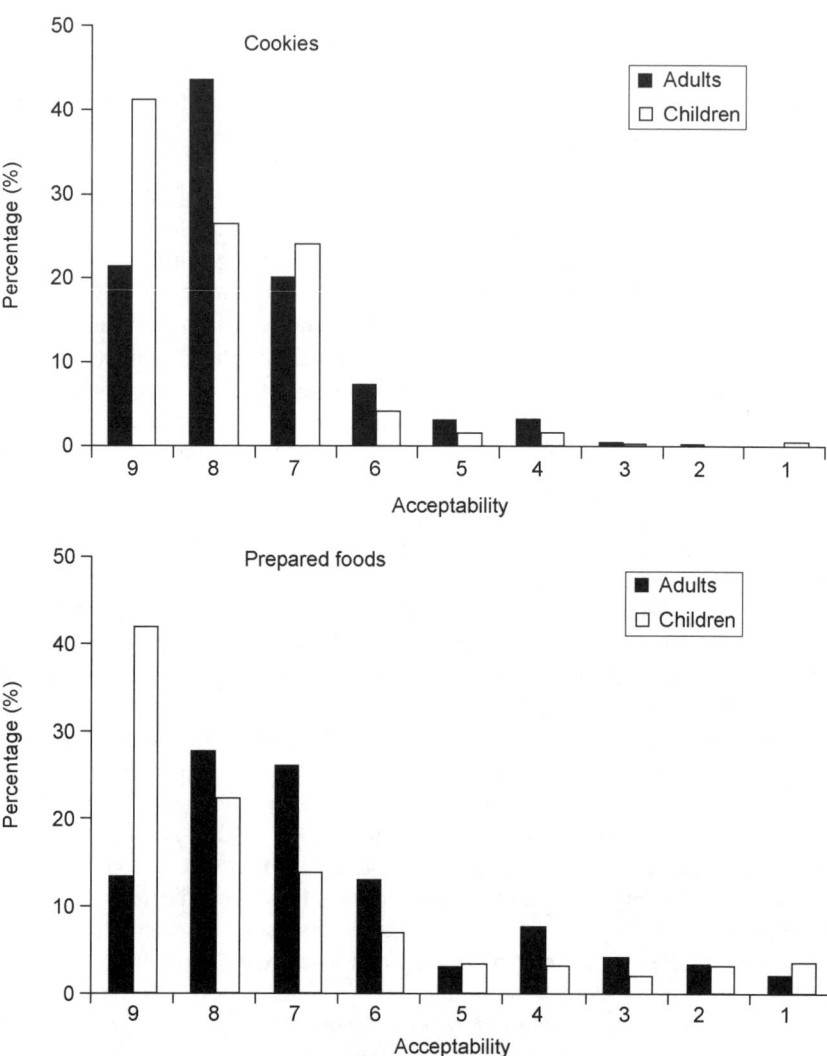

Fig. 16.2   Distribution of hedonic ratings by adults and children (8–12-years-old) who evaluated acceptability on a 9-point scale. For cookies: $N = 220$ per age group, four samples. For prepared foods: $N = 125$ per age group, nine samples. For adults: $9 =$ like extremely; for children, $9 =$ super good.

The reason that children favor the positive end of the scale more than adults is not clear. Perhaps, unlike adults, children do not feel a need to 'hedge their bets,' by reserving the scale end-points for future, yet-to-be tasted products. Perhaps children lack the frame of reference that adults have developed over the years, or perhaps children are simply easier to please. Whatever the reason, the tendency for children to favor the positive end of any hedonic scale may result in a 'ceiling effect,' especially with highly liked foods. When such ceiling effects are a concern, some researchers have found that a preference question may provide better discrimination than scaled liking. In general, however, children's hedonic ratings discriminate very effectively among products. A recent unpublished review by Peryam and Kroll of 14 studies involving five product categories (including such popular categories as cookies and pasta), with children and adults rating the same products, found that children's overall liking ratings (using the super good/super bad scale) were as likely to show significant differences as were the adults' ratings (albeit not always favoring the same products). In fact, in the six cases where significant differences were found with one group and not the other, it was the children that showed significant differences, not the adults.

16.6.2   Hedonic scale structure

One potential concern with any verbally anchored hedonic scale is how the choice of verbal scale anchors affects the psychological spacing between the scale points. The development of the standard 9-point hedonic scale was supported by extensive psychometric research (Peryam & Girardot, 1952; Jones *et al.*, 1955), which provided the basis for the selection of the phrases used to anchor the nine scale points. On the basis of an analysis of the ratings of different words or phrases, these early investigators were able to select phrases for anchoring the nine scale points that were approximately equally spaced psychologically. As Lawless and Heymann (1998) point out, the equal interval property is important, since the analysis of hedonic scale data almost always involves the assignment of numerical values to the responses and the application of parametric statistics, which assume equal interval spacing. Crawford *et al.* (2005) used a semantic analysis approach with children similar to the one Jones *et al.* used with adults, in order to investigate the psychological spacing of hedonic phrases, including six of the phrases from the standard adult scale. While the authors do not compare children's perceptions to those of adults, the agreement between the children's perceptual scale values and those reported by Jones *et al.* for the corresponding six phrases is very high (linear correlation $r = 0.98$), indicating similar semantic spacing of those phrases by children and adults.

Studies such as Jones *et al.* base their conclusion about the equal interval properties of the scale on studies of word perception. These studies leave open the question of whether the equal interval nature of the scale is preserved when respondents actually use the scale to rate stimuli (such as foods).

To investigate this question and to compare scale usage by children and adults, an analysis was conducted for the purposes of this chapter using a modified correspondence analysis approach. The data comprised ratings of 15 crackers, collected over the course of three test sessions. Overall liking ratings were collected from adults ($N = 200$) using the standard nine-point liking scale. Children (aged 9–12, $N = 200$) rated overall liking using the 9-point P&K (super good/super bad) scale. In addition to rating overall liking, both adults and children rated the products on a number of other liking dimensions (e.g. appearance, flavor) and rated several sensory attributes using intensity scales.

Multivariate mapping techniques are often used in sensory analysis and consumer research to map products spanned by 'attributes' (hedonic or intensity scales). In the present case, the perceptual mapping techniques are used to learn about differences between children and adults in scale usage, in particular in their use of an overall liking scale.

Greenacre (1984, pp. 169–184) describes how correspondence analysis can be used to analyze rating data, as opposed to frequencies (the typical domain of correspondence analysis). A related technique, dual scaling, has been described by Nishisato (1980). Only recently has this approach been applied to sensory and market research data (Abdi and Valentin, 2007; Chrea et al., 2004, 2005a; Torres and van de Velden, 2007). The technique is well suited for the present purpose, because it allows for the determination of the perceptual distance between scale points.

The correspondence analysis included all rating scale results (liking and intensity), though only the results for overall liking are reported here. Figure 16.3a shows the product map, in which products that are perceived similarly are positioned close together and products that are perceived very differently are far apart. The map shows product projections for the children, the adults, and for the consensus between the two, which represents a weighted average of the ratings of adults and children. The consensus projections of the scale points for overall liking are also shown.

Children and adults appear to rate the liking of products similarly in some cases, but not others (the interpretation of the map in terms of product attributes is beyond the scope of this chapter). Adults rated all products somewhat lower than the children (in the map, liking increases from left to right and the adult means are positioned to the left of those for the children), consistent with the age effect on liking ratings discussed above. In addition to this systematic shift, there are some differences between adults and children in the relative liking of certain products (such as M, N, and G). For example, product N is the best liked product for children, but not for adults.

Figure 16.3b shows the separate projections of the overall liking scale points for adults and children. The scale points for the standard adult liking scale and the P&K super good/super bad scale project very similarly, and are spaced at approximately equal intervals, except for some slight compression in the middle of the scale. These results provide strong evidence that the two overall liking scales are highly similar measurement instruments and are both approximately equal interval scales.

16.6.3   Hedonic testing with pre-school-age children

The hedonic scales described so far are appropriate for testing children 5–12 years of age. Within that range, younger children (aged 5–7) will require assistance in completing the test, if for no reason other than that their reading ability is

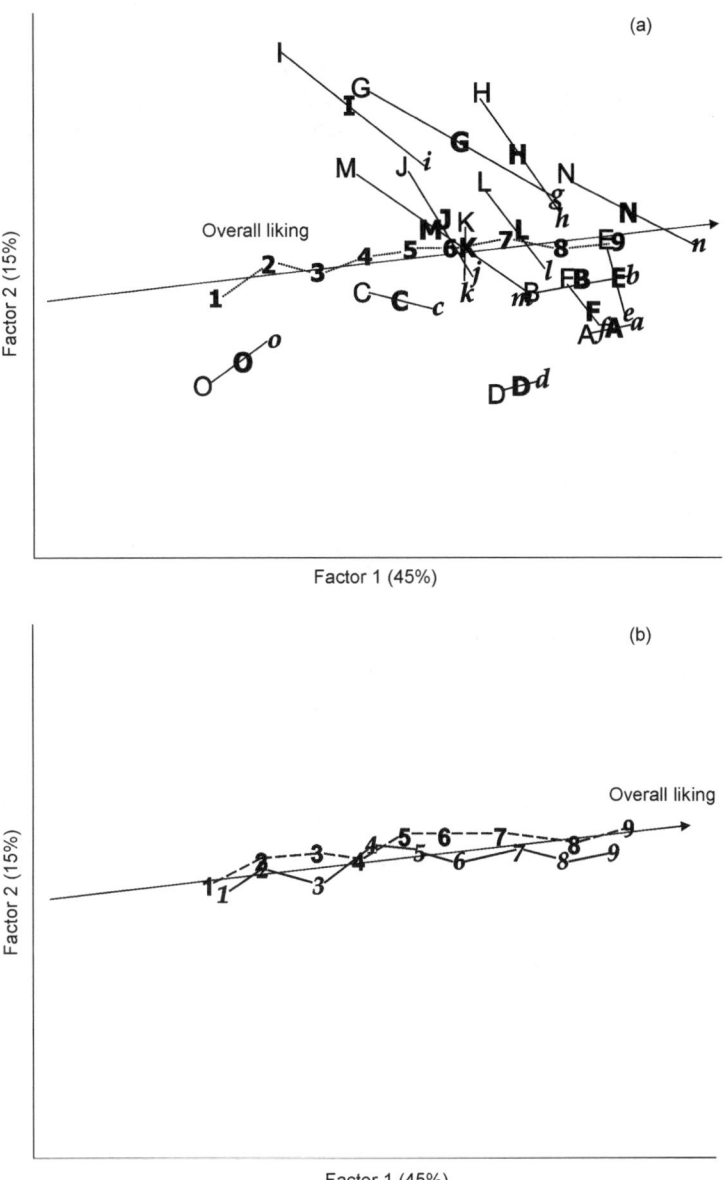

Fig. 16.3 Hedonic scale structure for children and adults as determined by correspondence analysis. (a) Products (letters) are projected in the map according to the adults' (upper case) or children's (lower case) ratings of overall liking and other hedonic and sensory attributes (only overall liking is shown). The consensus of the children and the adults (bold capital letters) is located between each pair of projections. The numbers represent the consensus projections of the nine scale points on the overall liking scale. (b) Separate projections of the overall liking scale points for adults (dashed line, traditional nine point hedonic scale) and children (solid line, P&K hedonic scale for children), indicating similar scale structure for both age groups.

limited. As an alternative to one-on-one interviews (which are costly and time-consuming to conduct), it is sometimes possible to test 5–7-year-old children using a classroom-style group administration, provided the questionnaire is short and the test moderator guides the children through the questionnaire one question at a time.

But what methods are suitable for determining liking among children aged 3–5? Birch (1979, 1980) has successfully used a 'ranking by elimination' procedure, a variation of the traditional ranking procedure. According to this procedure, children first taste a number of samples and then choose their favorite. This sample is then set aside, and children re-taste the remaining samples, once again indicating their favorite. This process continues until all samples have been ranked. While used extensively in academic research by Birch and others, the use of this method in industrial applications appears to be rare.

Kimmel et al. (1994) concluded that 4–5-year-olds were able to use a 7-point face scale. Chen et al. (1996) found that 3-year-old children were able to use a 3-point scale, 4-year-olds a 5-point scale, and children 5 years old a 7-point scale. On the other hand, Léon et al. (1999) found low repeatability among 4–5-year-olds using three different methods, including a simple binary classification (like and dislike).

Recently, Popper et al. (2002) compared two different methods for measuring liking among pre-schoolers, the ranking by elimination procedure and a 5-point bifurcated scale. In the latter procedure, the child was first asked if the sample was 'good' or 'bad,' and, depending on the answer, was then asked whether the sample was 'really good' (or 'really bad') or 'just a little good' (or 'just a little bad'). If the child had trouble committing to whether the sample was good or bad, the answer was recorded as 'neither.'

Pre-school-age children are pre-literate and must by necessity be interviewed one-on-one. Typically, research personnel (usually female) serve as inter-viewers. Popper et al. included two interviewer conditions: in one condition, the child's mother was the interviewer, in the other condition, the interview was conducted by a female researcher (P&K staff person) unfamiliar to the child. When interviewed by the researcher, Mom was not present in the room with her child.

Both the ranking procedure and the bifurcated scale resulted in significant differences among the samples tested. Greater discrimination, using either procedure, was obtained when the child was interviewed by the mother than by the unfamiliar researcher (see Fig. 16.4). When Mom did the interviewing, the average ratings for the three formulations spanned a larger range than when the researcher did the interviewing, and the differences were more likely to be statistically significant. This effect varied by age – the benefit of Mom as interviewer was evident at ages 3 and 4, but was largely absent by age 5, where Mom and the P&K interviewer gave very similar results. The authors caution that using the mother as interviewer may not be preferable in all situations, especially when there is a risk that the mother could introduce her own biases about the products. In the study, the samples the child evaluated all looked the

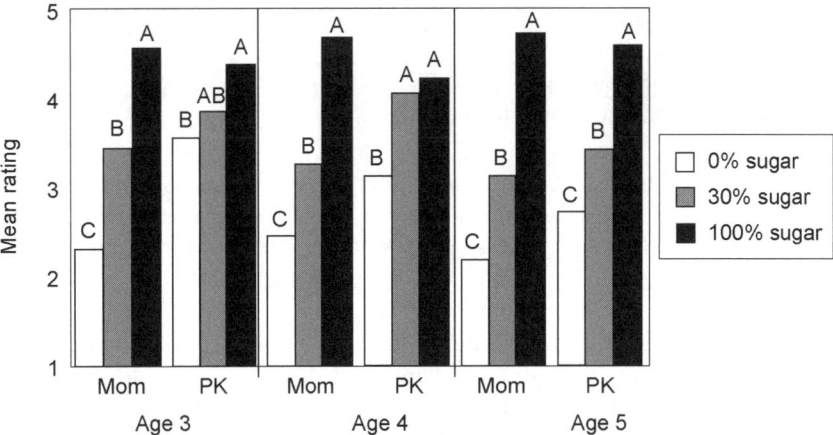

Fig. 16.4   Effect of interviewer (Mother vs Peryam & Kroll researcher) on children's liking ratings (5 = 'really good') for powdered orange drinks differing in sugar concentration. Means sharing a common letter are not significantly different from one another ($p < 0.05$). From Popper *et al.* (2002).

same (and the mother did not taste them). In situations when the appearance of the samples might suggest differences in nutrient content or when brand information is provided as part of the test, the mother's role as the interviewer would need to be carefully assessed.

16.7   Use of intensity and just-about-right scales

The research cited in Section 16.2 on intensity scaling with children demonstrates that children are capable of using intensity scales, although their ratings may differ from those of adults. In these studies, children typically become familiar with the scales using some practice tasks (e.g. scaling visual stimuli) prior to using them to rate foods. Also, these studies usually focus only on one or at most a few sensory attributes at a time.

Swaney-Stueve (2002) undertook the challenge of attempting to train children in descriptive analysis. Using six brands of peanut butter and a ballot including 14 or more attributes, she demonstrated that it is possible to train children as young as 9 years old to reliably discriminate among products using intensity ratings. The results of the 9-year-olds did not differ much from those obtained with children aged 13–14 and 16–18, although the results from the children and teen panels did differ somewhat from those obtained with two adult panels. In another example of the use of descriptive analysis with children, Narain (2005) showed that even 4-year-olds could be trained to use texture attributes to distinguish among breads. These studies demonstrate that children are capable, given appropriate training, to perform sensory tasks that are cognitively quite demanding.

In consumer tests with children, intensity scales are sometimes included in addition to hedonic scales. Intensity scales can help provide product developers information on how children perceive the differences among the products included in the test. By comparing their ratings with those of adults, these scales may also provide some insight into how children use terms such as 'sweet,' 'crunchy,' etc. In most consumer tests, children receive no or only minimal explanation of the sensory terms. For some attributes, such as sweetness, the assumption that children (typically aged 8–12) understand the meaning of the sensory characteristic they are being asked to scale is probably warranted. For other attributes, this assumption may be tenuous. The research on odor recognition and identification cited in Section 16.2 suggests that flavor concept formation and flavor naming is a learning process that extends into adulthood. Little research has been published that would tell product developers for which sensory attributes (and for which types of products) children are capable of generating meaningful intensity scaling results (in the absence of training), or what words to use to refer to different sensory characteristics (e.g. do children understand the meaning of 'mouth coating'?). Examples from Peryam and Kroll's research suggests that some attributes can indeed be scaled quite successfully, as was shown in the study on crackers described earlier. Figure 16.5 shows the correlation between children and adult ratings on three attribute intensity questions, using 9-point intensity scales. The agreement between children and adults was quite high.

Information regarding the preferred level of a sensory attribute (e.g. sweetness) is often obtained from children using just-about-right scales, following the common practice of using such scales in research with adults (see Chapter 17). The adult version of the just-about-right scale is usually a 5-point scale, in which the middle category is labeled 'just about right' and other scale points are labeled, for example, 'too weak' and 'much too weak' on one side and 'too strong' and 'much too strong' on the other. For use with children, many researchers shorten the scale to three points, 'too weak', 'just about right,' and 'too strong'. Again, little research has been published on children's use of such scales, or their benefits and limitations. In Peryam and Kroll's experience, just-about-right scales can provide meaningful results with children, although as in the case of intensity scales, careful consideration must be given to the types of attributes children are asked to evaluate. For example, children appear to be able to use just-about-right scales to flag product issues regarding appearance, size, or visual amount of an ingredient, as well as for basic tastes and simple texture attributes. In the case of flavor (other than basic tastes) and more complex appearance and texture terms, just-about-right scales tend to be less informative with children, most likely because of the difficulty children have understanding these attributes. Scaling these attributes often results in a high percentage of 'just right' responses, especially among younger children (aged 6–7), compared with adults using the same 3-point scale.

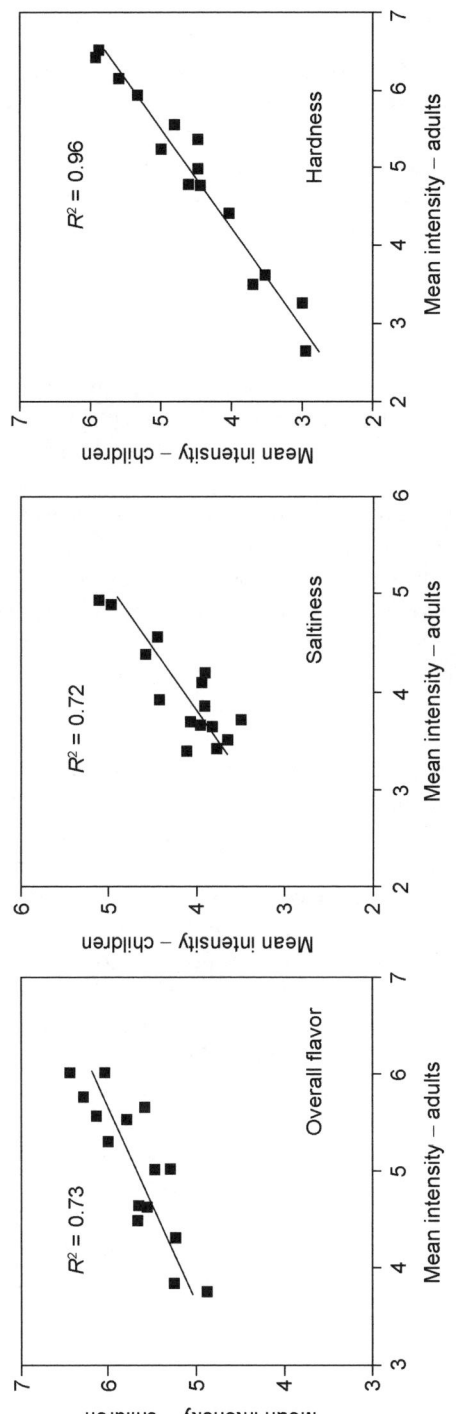

Fig. 16.5 Ratings of sensory intensity of 15 crackers by adults and 9–12-year-old children.

16.8   Future trends

Given the importance to the food industry of conducting research with children there is need for more research in several areas. At the level of basic science, much still needs to be learned about children's chemosensory abilities and perceptions. The basic tastes and olfaction have been more extensively studied than other sense modalities, such as texture, yet even in the area of basic tastes, too few studies have involved complex foods, as opposed to test solutions or simple beverages.

For applied testing, the scaling methods most appropriate for children will continue to require investigation, especially as testing technology evolves to various forms of computerized testing. The potential of leveraging technology to improve children's comprehension and use of scales in consumer tests remains to be explored. The use of the Internet for research with children is also likely to grow. Cooper (2005) has reported on the use of a children's Internet panel for qualitative research.

While methods for hedonic scaling have been widely investigated, the benefits and limitations of using intensity and just-about-right scales with children in the context of consumer research requires more investigation, leading to guidelines for when to use such scales and how to refer to different sensory characteristics in a way easily understood by children.

Several other testing parameters are in need of further study, such as time of day. According to Gollick (2002), depending on the time of day, a child's IQ score on a standardized test can vary by as much as 30 points. Urbick (2002) advocates conducting consumer tests with children in the morning, when kids are most alert, and avoiding after-school hours, when children are mentally tired and need unstructured playtime and a chance to be physically active. Time of day may also affect children's reactions for other reasons. Some foods are more appropriate at some times of the day than others. Children at an early age have acquired a sense of time appropriateness for different foods (Birch et al., 1984) and the effect of this awareness on test results remains to be better understood.

Food products developed for children are unique in many respects – their formulation, their packaging, and their messaging. Sensory researchers are prone to think about the product itself, but for children the success of a new product may depend not just on its taste and texture, but on how it handles, its play value, the image it projects, and how successfully all these aspects are integrated. It is likely that food developers will need to pay increasing attention to the overall 'product concept fit,' which may mean introducing advertising language and brand information into product research with children more often than is done today (if it is included at all). Little research has been conducted to understand how children process such information in the context of evaluating food and how such information impacts their liking ratings (Bahn, 1989; Allison et al., 2004).

16.9   Sources of further information and advice

Birch (1999) provides a comprehensive review of research on the development of food preferences in children and the differences between children and adults in sensory sensitivity. Sensory research methods for use with children are reviewed by Popper and Kroll (2005), Guinard (2001) and Resurreccion (1998). A guide on testing with children, developed in 2003 by ASTM Committee E-18 (ASTM E-2299-03), provides a great deal of practical advice on how to conduct effective testing with children and includes recommendations regarding the age appropriateness of various sensory research methods. Journals such as *Food Quality and Preference, Appetite,* the *Journal of Sensory Studies,* and *Chemical Senses* are good sources for information for the latest research findings on children's sensory perception, food preferences, and the associated research methods. The biannual Pangborn Sensory Science Symposium provides an important forum for researchers from academia and industry to share findings regarding conducting research with children.

16.10   References

ABDI H & VALENTIN, D (2007), 'Multiple correspondence analysis', in Salkind N *Encyclopedia of Measurement and Statistics.* Thousand Oaks, CA, Sage.

ALLISON A-M A, GUALTIERI T & CRAIG-PETSINGER D (2004), 'Are young teens influenced by increased product description detail and branding during consumer testing?', *Food Qual Preference,* **15**(7/8), 819–829.

ASTM (2003), *ASTM Standard Guide for Sensory Evaluation of Products by Children,* E-2299-03, West Conshohocken, PA, ASTM International.

BAHN K D (1989), 'Cognitively and perceptually based judgments in children's brand discriminations and preferences', *J Bus Psych,* **4**(2), 183–197.

BEAUCHAMP G K & COWART B J (1990), 'Preference for high salt concentrations among children', *Dev Psychol,* **26**(4), 539–545.

BEAUCHAMP G K & MORAN M (1984), 'Acceptance of sweet and salty tastes in 2-year-old children', *Appetite,* **5**, 291–305.

BEAUCHAMP G K, COWART B J & MORAN M (1986), 'Developmental changes in salt acceptability in humans', *Dev Psychobiol,* **19**, 75–83.

BIRCH L L (1979) 'Dimensions of preschool children's food preference', *J Nutr Educ,* 11(2), 77–80.

BIRCH L L (1980), 'Effect of peer models' food choices and eating behaviors on preschoolers' food preferences', *Child Develop,* **51**, 489–496.

BIRCH L L (1999), 'Development of food preferences', *Annu. Rev Nutr,* **19**, 41–62.

BIRCH L L & MARLIN D W (1982), 'I don't like it; I never tried it: effects of exposure on two-year-old children's food preferences', *Appetite,* **3**, 353–360.

BIRCH L L, BILLMAN J & RICHARDS S S (1984), 'Time of day influences food acceptability', *Appetite,* **5**, 109–116.

BIRCH L L, FISHER J O & GRIMM-THOMAS K (1999), 'Children and food', in Siegal M and Peterson C, *Children's Understanding of Biology and Health,* Cambridge, Cambridge University Press.

CAIN W S, STEVENS J C, NICKOU C M, GILES A, JOHNSTON I & GARCIA-MEDINA M R (1995), 'Life-span development of odor identification, learning, and olfactory sensitivity', *Perception*, **24**, 1457–1472.

CHAMBERS E (2005), 'Commentary: sensory research with children', *J Sensory Studies*, **20**, 90–92.

CHAMBERS E & BOWERS J R (1993), 'Consumer perception of sensory qualities in muscle foods', *Food Technol*, **47**, 116–134.

CHEN A W, RESURRECCION A V A & PAGUIO L P (1996), 'Age appropriate hedonic scales to measure food preferences of young children', *J Sensory Studies*, **11**, 141–163.

CHREA C, VALENTIN D, SULMONT-ROSSÉ C, LY MAI H, HOANG NGUYEN D & ABDI H (2004), 'Culture and odor categorization: agreement between cultures depends upon the odors', *Food Qual Pref*, **15**, 669–679.

CHREA C, VALENTIN D, SULMONT-ROSSÉ C, NGUYEN-HOAN D & ABDI, H (2005), 'Semantic, typicality, and odor representation: a cross-cultural study', *Chem Senses*, **30**, 37–49.

COOPER H (2002), 'Designing successful diagnostic scales for children', presented at Ann. Mtg., Inst. of Food Technologists, Anaheim, Calif., 15–19 June.

COOPER H R (2005), 'Taking directional sensory research on-line – experiences with the creative kids panel', presented at the 6th Pangborn Sensory Science Symposium, Harrogate, UK, 7–11 Aug.

CRAWFORD C, WARD C & SIMPSON N (2003), 'Do scales grow with children?', presented at the 5th Pangborn Sensory Science Symposium, Boston, 20–23 July.

CRAWFORD C, SIMONS C & WARD C (2005), 'When kids don't like "dislike"; creating a new hedonic scale for children', presented at the 6th Pangborn Sensory Science Symposium, Harrogate, UK, 7–11 Aug.

DE GRAAF C & ZANDSTRA E H (1999), 'Sweetness intensity and pleasantness in children, adolescents, and adults', *Physiol Behav*, **67**(4), 513–520.

DESOR J A & BEAUCHAMP G K (1986), 'Longitudinal changes in sweet preferences in humans', *Physiol Behav*, **39**(5), 639–641.

DESOR J, GREENE L & MALLER O (1975), 'Preferences for sweet and salty in 9- to 15-year-old and adult humans', *Science*, **190**, 686–687.

FRIEDMAN H H & AMOO T (1999), 'Rating the rating scales', *J Marketing Mgt*, **9**(3), 114–123.

GOLLICK M (2002), 'Asking kids questions: possible pitfalls', presented at Ann. Mtg., Inst. of Food Technologists, Anaheim, Calif., 15–19 June.

GREENACRE M (1984), *Theory and Applications of Correspondence Analysis*, London, Academic Press.

GUINARD J X (2001), 'Sensory and consumer testing with children', *Trends Food Sci Technol*, **11**, 273–283.

JAMES C E, LAING D G & ORAM N (1997), 'A comparison of the ability of 8–9-year-old children and adults to detect taste stimuli', *Physiol Behav*, **62**(1), 193–197.

JAMES C E, LAING D G, ORAM N & HUTCHINSON I (1999), 'Perception of sweetness in simple and complex taste stimuli by adults and children', *Chem Senses*, **24**, 281–287.

JAMES C E, LAING D G, JINKS A L, ORAM N & HUTCHINSON I (2003), 'Taste response functions of adults and children using different rating scales', *Food Qual Preference*, **15**, 77–82.

JONES L V, PERYAM D R & THURSTONE L L (1955), 'Development of a scale for the measuring soldiers' food preferences', *Food Res*, **20**, 512–520.

KIMMEL S A, SIGMAN-GRANT M & GUINARD J X (1994), 'Sensory testing with young children', *Food Technol*, **48**(3), 92–99.

KROLL B J (1990), 'Evaluating rating scales for sensory testing with children', *Food Technol*, **44**(11), 78–80, 82, 84, 86.

LAWLESS H T & HEYMANN H (1998), *Sensory Evaluation of Food*, New York, Chapman & Hall.

LEHRNER J & WALLA P (2002), 'Development of odor naming and odor memory from childhood to young adulthood', in Rouby C, Schaal B, Dubois D, Gervais R and Holley A, *Olfaction, Taste and Cognition*, Cambridge, Cambridge University Press, pp. 278–289.

LEHRNER J P, GLÜCK J & LASKA M (1999), 'Odor identification, consistency of label use, olfactory threshold and their relationships to odor memory over the human lifespan', *Chem Senses*, **24**, 337–346.

LÉON F, COURONNE T, MARCUZ M C & KÖSTER E P (1999), 'Measuring food liking in children: a comparison of non-verbal methods', *Food Qual Preference*, **10**, 93–100.

LIEM D G & MENNELLA J A (2002), 'Sweet and sour preferences during childhood: role of early experiences', *Dev Psychobiol*, **41**, 388–395.

LIEM D G & MENNELLA J A (2003), 'Heightened sour preferences during childhood', *Chem Senses*, **28**, 173–180

LIEM D G, MARS M & DE GRAAF C (2004a), 'Consistency of sensory testing with 4- and 5-year-old children', *Food Qual Preference*, **15**, 541–548.

LIEM D G, WESTERBEEK A, WOLTERINK S, KOK F J & DE GRAAF C (2004b), 'Sour taste preferences of children relate to preference for novel and intense stimuli', *Chem Senses*, **29**(8), 713–720.

LOEWEN R & PLINER P (2000), 'The Food Situations Questionnaire: a measure of children's willingness to try novel foods in stimulating and non-stimulating situations', *Appetite*, **35**, 239–250.

LUMENG J C, ZUCKERMAN M D, CARDINAL T & KACIROTI N (2005), 'The association between flavor labeling and flavor recall ability in children', *Chem Senses*, **30**(7), 565–574.

MENNELLA J A & BEAUCHAMP G K (2005), 'Understanding the origin of flavor preferences', *Chem Senses*, **30** (suppl 1), i242–i243.

MONCRIEFF R W (1966), *Odour Preferences*, New York, Wiley.

MOSKOWITZ H R (1994), *Food Concepts and Products: Just-in-time Development*, Trumbull, CT, Food and Nutrition Press.

NARAIN C (2005), 'Texture preference in children', presented at the 6th Pangborn Sensory Science Symposium, Harrogate, UK, 7–11 Aug.

NISHISATO S (1980), *Analysis of Categorical Data: Duel Scaling and its Applications*, Toronto, University of Toronto Press.

ORAM N (1998), 'Association of perceptual feel and general descriptors with food categories by 8–11 year olds and adults', *J Texture Studies*, **29**, 669–680.

ORAM N, LAING D G, FREEMAN M H & HUTCHINSON I (2001), 'Analysis of taste mixtures by adults and children', *Dev Psychobiol*, **38**, 67–77.

PERYAM D & GIRARDOT N F (1952), 'Advanced taste test methods', *Food Eng*, **24**(7), 58–61, 194.

PLINER P & STALLBERG-WHITE C (2000), '¤Pass the ketchup, please¤: familiar flavors increase children's willingness to taste novel foods', *Appetite*, **34**, 95–103.

POPPER R & KROLL J J (2005), 'Conducting sensory research with children', *J Sensory Studies,* **20**, 75–87.

POPPER R, SCHRAIDT M & KROLL B J (2002), 'Testing with pre-school children: the effect of the interviewer', presented at Ann. Mtg., Inst. of Food Technologists, Anaheim, CA, 15–19 June.

RABIN M D & CAIN W S (1984), 'Odor recognition: familiarity, identifiability, and encoding consistency', *J Exp Psychol: Learning, Memory Cognition*, **10**, 316–325.

RESURRECCION A V A (1998), *Consumer Sensory Testing for Product Development*, Gaithersburg, MD, Aspen.

RINGEL C (2005), 'The change of the optimal complexity level after extended exposure – comparison of elderly people, young adults and children', presented at the 6th Pangborn Sensory Science Symposium, Harrogate, UK, 7–11 Aug.

ROSE G, LAING D G, ORAM N & HUTCHINSON I (2004a), 'Sensory profiling by children aged 6–7 and 10–11 years. Part 1: a descriptor approach', *Food Qual Preference*, **15**, 585–596.

ROSE G, LAING D G, ORAM N & HUTCHINSON I (2004b), 'Sensory profiling by children aged 6–7 and 10–11 years. Part 2: a modality approach', *Food Qual Preference*, **15**, 597–606.

ROZIN P (1984), 'The acquisition of food habits and preferences', in J D Matarazzo, S M Weiss, J A Herd and N E Miller, *A Handbook of Health Enhancement and Disease Prevention*, New York, Wiley, pp. 590–607.

SAUERHOFF K, DEGNAN D & CRAIG-PETSINGER D (2005) 'The impact of scale orientation and labels on consumer response in an online environment', presented at the 6th Pangborn Sensory Science Symposium, Harrogate, UK, 7–11 Aug.

SPAETH E E, CHAMBERS E C IV & SCHWENKE J R (1992), 'A comparison of acceptability scaling methods for use with children', in Wu L S and Gelinas A D, *Product Testing with Consumers for Research Guidance: Special Consumer Groups, Second Volume ASTM STP 1155*, Philadelphia, Am Soc Testing Materials, 65–77.

SWANEY-STUEVE M (2002), 'Can children perform descriptive analysis?', presented at Ann. Mtg., Inst. of Food Technologists, Anaheim, CA, 15–19 June.

SZCZESNIAK A S (1972), 'Consumer awareness of and attitudes to food texture', *J Texture Studies*, **3**, 206–217.

TEMPLE E C, LAING D G, HUTCHINSON I & JINKS A L (2002), 'Temporal perception of sweetness by adults and children using computerized time-intensity measures', *Chem Senses*, **27**, 729–737.

TORRES, A & VAN DE VELDEN M (2007), 'Perceptual mapping of multiple variable batteries by plotting supplementary variables in correspondence analysis of rating data', *Food Qual Preference*, **18**(1), 121–129.

TUORILA-OLLIKAINEN H, MAHLAMAKI-DULTANEN S & KURKELA R (1984), 'Relative importance of color, fruity flavor, and sweetness in the overall liking of soft drinks', *J Food Sci*, **49**, 1598–1600.

URBICK B (2002), 'Kids have great taste: an update to sensory work with children', presented at Ann. Mtg., Inst. of Food Technologists, Anaheim, CA, 15–19 June.

ZANDSTRA E H & DE GRAAF C (1998), 'Sensory perception and pleasantness of orange beverages from childhood to old age', *Food Qual Preference*, **9**(1/2), 5–12.

ZAJONC R B (1968), 'Attitudinal effects of mere exposure', *J Personality Social Psychol*, **9**(Part 2), 1–27.

17

The use of just-about-right (JAR) scales in food product development and reformulation

L. Rothman, Kraft Foods, USA

17.1 Introduction to JAR scales

Food products are brought to market for a variety of reasons. New products may be developed in response to unmet consumer needs, competitive pressures, or to create category 'news.' Existing products may be reformulated, reworked or repackaged to better delight consumers, remove costs or replace unavailable ingredients or obsolete processes. No matter the underlying reasons, successful food product development requires a thorough understanding of the benefit(s) that consumers expect from the product. Whether the benefit of a particular food product is functional ('it fills me up') or emotional ('it satisfies me'), delivering that benefit is based on incorporating the right attributes (product parameters) at the right levels.

Product developers have a number of tools available to help them create or revise products to delight consumers. The broad area of quantitative consumer research is one such tool, with many different applications and techniques for execution. Quantitative consumer research typically involves obtaining consumer responses through data collection methods pertaining to the products of interest. Generally hedonic responses are obtained, either through rating scales of product acceptability or through choice behavior (preference). However, it is often, though not always the case (see Section 17.5) that merely obtaining acceptability (in the form of overall liking or preference ratings) does not enable further product development or refinement. This is because acceptability scores or preference ratings, in and of themselves, convey only the hedonic or choice status of a product. While these scores can be compared across products or time, and supplemented with liking or preference of specific attributes, liking scales and preference ratings alone do not assist the developer in understanding how product attributes need be altered to improve product acceptability or relative preference.

There are several techniques available to the product developer to pinpoint attributes that need adjustment. The incorporation of just-about-right (JAR) scales into quantitative consumer testing is one such technique. This chapter will discuss the definition, construction, analysis and interpretation of JAR scales and resulting data with an emphasis on the applications of penalty analysis. Controversies and problems associated with JAR scale usage will be presented. Future trends and sources of additional information will also be discussed.

17.2   Defining JAR scales

JAR scales are bipolar labeled attribute scales with a center point labeled 'just about right' or 'just right' (see Section 17.3.1). The end-points are anchored with labels that represent levels of the attribute that deviate from a respondent's theoretical ideal point, in opposite directions. The purpose of using JAR scales in quantitative consumer testing is to obtain information about whether to increase or decrease the attribute levels under consideration. The assumptions made here include (1) the respondent understands the attribute term, (2) the respondent can consider this attribute independently of other attributes, (3) the respondent judges the attribute intensity based on sensory perception rather than cognitive processing and (4) the respondent can anticipate the direction of his or her 'ideal' attribute level relative to the existing attribute level.

A simple example of a JAR scale is the first example in Fig. 17.1. A consumer's task in using this scale is to rate the level of sweetness appropriateness using the distance from his or her theoretical ideal point as a basis for judgment. Thus the consumer task is twofold, to determine the intensity of sweetness and to determine where that intensity lies with respect to his or her ideal level. If the perceived intensity is less than ideal, he or she will rate the product as 'not sweet enough', choosing the point on the scale appropriate to the degree of the 'mismatch' between intensity and ideal. Similarly, if the perceived intensity is greater than ideal, he or she will rate the product as 'too sweet', again choosing the point on the scale appropriate to the degree of mismatch.

17.3   JAR scale construction

While it is not the purpose of this chapter to detail the mechanics of JAR scale construction (see ASTM, 2008), a few comparisons to other scale types will be mentioned. Similar to other scale types, one can construct JAR scales as category or continuous line scales. If constructing continuous line scales, the number of points is theoretically infinite, although as a practical matter, few

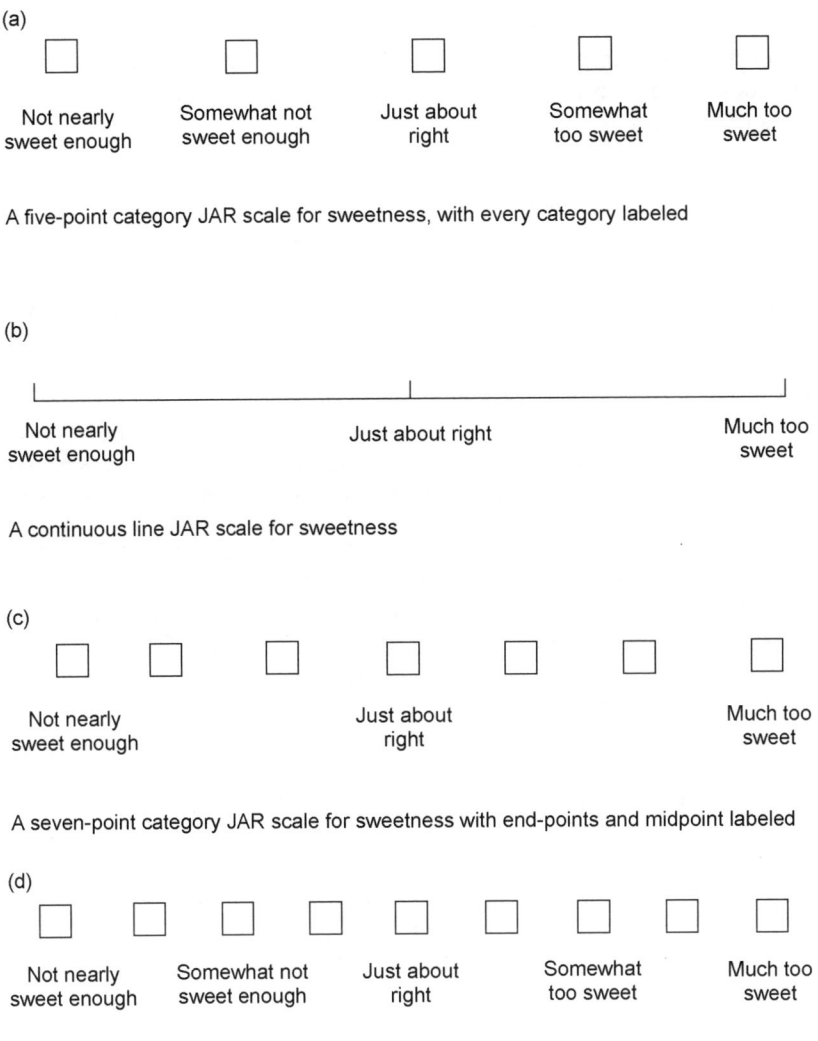

(a)

Not nearly
sweet enough

Somewhat not
sweet enough

Just about
right

Somewhat
too sweet

Much too
sweet

A five-point category JAR scale for sweetness, with every category labeled

(b)

Not nearly
sweet enough

Just about right

Much too
sweet

A continuous line JAR scale for sweetness

(c)

Not nearly
sweet enough

Just about
right

Much too
sweet

A seven-point category JAR scale for sweetness with end-points and midpoint labeled

(d)

Not nearly
sweet enough

Somewhat not
sweet enough

Just about
right

Somewhat
too sweet

Much too
sweet

A nine-point category JAR scale for sweetness with a combination of categories and midpoint labeled

Fig. 17.1   Examples of JAR scales.

researchers use more than 100-point scales (an exception is magnitude estimation scales, in which respondents can use numbers as high as they wish when rating product attributes; this author, however, has not seen magnitude estimation ratings used with JAR scales). The number of scale points must be odd, owing to the just-about-right or just right center point. Common numbers of scale points for category JAR scales typically range from three to nine points, with the unifying principles that (1) the center point is anchored with 'just about right' or 'just right', (2) there are an equal number of points on either side of the center point, and (3) the anchors on either side of just-about-right are symmetrical, both in number and placement. For category scales, researchers can choose to label only the end-points, every category, or a combination of end-points and categories. (See Fig. 17.1 for other examples of JAR scales.)

17.3.1   JAR scale midpoint

There are two choices for the mid-point of a JAR scale 'just about right' or 'just right'. Some researchers may use 'just right' as the midpoint as a means of counteracting a tendency for respondents to use the center of the scale (in order to avoid the scale end-points); the attribute intensity must therefore exactly match the respondent's theoretical ideal intensity to use the scale center point. Other researchers may feel that the choice of 'just right' represents too strong a commitment for this category to be of much use (Lawless and Heymann, 1998), and label the midpoint 'just about right.' As a practical matter, this author has not seen much difference in scale usage owing to differences between the two midpoints.

17.3.2   Constructing scales for attributes of interest

JAR scale usage requires the construction of bipolar scales for attributes of interest, which may prove difficult, particularly when looking for semantic opposites. While 'too thick' may be paired with 'too thin' as semantic opposites (in many, but probably not all cases), 'too sweet' has no semantic opposite. One way around these problems is to label the scale with the attribute name ('sweetness') and the end-points simply with 'much too weak' or 'not nearly enough' and 'much too strong' or 'much too many.' Another problem arises when one of the scale end-points is difficult to relate back to the food. Consider an attribute such as 'chewiness', which in consumer language may be defined as the degree of mastication required prior to swallowing. While the 'too much' end of the scale may make sense ('much too chewy'), the 'not enough' end of the scale may be difficult to interpret ('not nearly chewy enough'). This may be compounded by the fact that food products can be hard and chewy or soft and chewy.

It is not possible to construct JAR scales for attributes that have negative connotations, such as 'off taste' or 'artificial' taste.' It may also be difficult for respondents to use JAR scales to rate attributes that are part of a food product's flavor profile, but may be considered negative, even though they play a large role in the flavor profile. For example, a consumer may not rate coffee or chocolate as 'not bitter enough', even though the flavor lacks depth and balance.

17.3.3   JAR scales and consumer language

Similar to other attribute scaling techniques, JAR scales rely on consumer language (attributes on the consumer ballot) to communicate opportunities for product improvement to the product developer. The difficulties involved in constructing attribute scales for use by consumers have been discussed previously (Stone and Sidel, 1993; Moskowitz, 2003b). As Moskowitz (1994) so eloquently puts it, 'consumers may recognize that "something" is wrong with the product, although they may be unable to articulate the problem.' While this is an important point that applies to all consumer attribute-scaling tasks, the direct nature of perceived actionability of JAR scales makes the consumer language problem that much more prominent. Thus an extra measure of caution should be used when constructing, analyzing, and interpreting data from these scales.

17.4   Controversies

There is substantial controversy surrounding the use of JAR scales in consumer research, beyond the psychological issues common to scaling tasks in general (Lawless and Heymann, 1998).

17.4.1   The psychological task

Some researchers feel that the simultaneous consideration of both attribute intensity and appropriateness is difficult psychologically (Moskowitz, 2001). Others (Moskowitz, 2003b) consider the task to be three-fold: to 'evaluate the product, compare the product to an internal "ideal" product, and then report what characteristics are out of kilter.' Still others discuss the conflict between sensory liking and other influences, such as health related concerns (Bower and Boyd, 2003). For example, a consumer may like the saltiness of a given product, but may feel compelled to rate it as 'too salty' because he or she is on a reduced salt diet, and has been taught that salt is 'bad.'

17.4.2   'Never enough' attributes

The use of JAR scales may be problematic when considering certain ingredient additions or flavor characteristics. Such ingredients commonly include toppings or inclusions. There may never be enough sausage on the pizza, chocolate chips in the cookie, or nuts in the ice cream. Although these ingredients may always be judged as insufficient in quantity, further increases may not increase liking, and may actually serve to decrease liking. There are also certain flavor or texture attributes with such a positive halo that consumers will always say they want them increased. The cake may never have enough 'chocolate flavor' and the Alfredo sauce may never be 'creamy' enough. In other words, it may never be possible to reach the theoretical 'ideal 'level of that attribute. More likely is that at some point, further increases may lead to products that are actually less well liked (Moskowitz, 2003b).

17.4.3   The impact of JAR scale inclusion on other attributes

There has been recent evidence that inclusion of JAR scales on a consumer questionnaire may impact overall liking ratings, whereas other attribute scales do not (Popper et al., 2004), although more recent research shows that the effect may be product or attribute specific (Popper *et al.*, 2005) or may not occur at all (van Trip *et al.*, 2007). Clearly more research is needed in this area.

17.4.4   The interdependence of food attributes

Another problem arising from JAR scale usage concerns the interdependence of food product attributes. While the *amount* of a single ingredient can be varied, the *impact* of that ingredient is rarely limited to the attribute in question. In other words, if the JAR data indicate that the product is not sweet enough, increasing the sweetness may impact subsequent JAR ratings of sourness, saltiness, and other flavor attributes in addition to sweetness. In fact, it has been demonstrated that products that have been optimized using JAR scales did not produce the same optimal product as those optimized using hedonic scales (Epler *et al.*, 1998; Popper *et al.*, 1995) or that predicted optimal products did not generate estimated attribute levels at 'just right' when creating models for product optimization (Moskowitz, 1994, 2001). Thus, to accept the results of JAR scale ratings at face value, while expedient, may not be prudent. However a recent study (van Trip *et al.*, 2007), found similar deviations from ideal and directions for product improvement among three methods tested: the 'conventional' method using attribute intensities and overall liking to estimate ideal points, the 'just about right' method using overall liking and JAR scales, and the 'variant' method, using overall liking, attribute intensity and ideal point scales. The authors conclude that none of the methods enjoys an advantage in terms of predictive validity. However, the development direction provided from the different methods could not be validated, as the experimental samples were commercially available products, rather than the results of systematic experimentation. Clearly, more research is needed. As a practical matter, researchers that use JAR scales should work with product developers to help them understand opportunities and limitations associated with their use. Pilot studies should demonstrate the usefulness of collecting such data in a given product category.

17.4.5   How much should attributes be changed?

JAR scale data alone do not tell the researcher the degree to which the attribute(s) should be increased or decreased. While it is tempting to think that a larger percentage of ratings on the 'too much' side of the scale translates to a larger required attribute decrease than a smaller percentage, that is not always the case. The ingredients that deliver those attributes may react differently to changes in intensity within the food matrix; additionally, the JAR scale itself may differ by attribute in sensitivity to formulation changes by attribute (Moskowitz, 2004). Finally, it is possible that formulation changes resulting from data in the not 'just about right' parts of the scales would alienate those respondents that currently view the attributes as 'just right.' In other words, the 'cure' could be worse than the 'disease.'

Given these cautions, one may question the value of JAR scale inclusion at all. It is, however, this author's contention that the inclusion of JAR scales in consumer testing can be quite useful *in certain situations and with proper analysis.*

17.5   Appropriate uses of JAR scales

As mentioned earlier, data obtained from JAR scales are only as good as the appropriateness of the attribute terms chosen. It is *always* preferable to rely on sound experimental design and analysis (Gacula and Singh, 1984; Resurreccion, 1998) for product development direction, which obviates the need for consumer language. However, the reality is that product optimization (in the true statistical sense) is not always practiced. There is the popular (though often fallacious) belief that systematic experimentation requires inordinate amounts of resources (people, time and testing dollars). Product developers may feel that it restricts their 'artistry' in product creation, as they must follow charts that tell them what ingredients to put into which samples and at what levels (although much of the artistry comes from choosing the ingredients and their ranges in the first place). There may be a company practice of 'formulation on the fly,' with management demand to 'get something into consumer test.' Or perhaps the systematic variation does not cover all the aspects of the product. For example, the flavor components may be varied systematically, but the texture system may be fixed. Or the flavor and texture may be systematically varied, but the color or size may be fixed. Whatever its genesis, the reality is that consumer researchers are often confronted with products that have not been systematically varied or optimized. It is in these cases that JAR scales can often be used effectively and efficiently, *provided that they are analyzed and interpreted properly.*

17.6   Analysis and interpretation of JAR scales

There are a myriad of analyses for JAR scale data, ranging from simple tabulations and charts, to extremely complex statistical manipulations, some of which require special software. For a thorough review of JAR scale analysis methods, refer to the ASTM document mentioned earlier.

17.6.1   What type of data do JAR scales generate?
A first step prior to the analysis of any data is to determine the nature of the scales being used. At first glance, JAR scales would appear to be interval. Consider, however, that the direction to the developer is the same whether 100% of respondents rate a product 'much too sweet' or 'somewhat too sweet.' It is difficult to direct the decreases or increases in ingredients in any finite way as one could with properly designed optimization experiments, because without systematic experimentation there is no link between attribute intensity and ingredient level. Additionally, the tendency with many of the methods of

analysis is to collapse the points on either side of 'just about right,' essentially treating the data as categorical, rather than interval. Although some researchers (while advising caution), discuss the treatment of 7 or 9-point JAR scales as interval data (Lawless and Heymann, 1998), given the lack of precise product refinement direction, it is generally best to think of JAR scales as generating category data.

The keys to prudent JAR scale analysis and interpretation include (1) clearly articulating the objective of the analysis, (2) collecting other data to relate to the JAR scale data (if that is an objective), (3) becoming aware of the pros and cons of the analytical technique(s) under consideration both theoretically, via appropriate statistician input, and practically, through the use of comparative or pilot studies, (4) obtaining the proper software, and (5) interpreting the results in light of all other relevant information about the product and product category.

17.6.2   Types of JAR scale data analysis
While not meant to be exhaustive, a list of common statistical methods for JAR scale analysis, grouped by objective is as follows:

- What does the JAR distribution look like?
  - Graphical data display
  - Graphical scaling
- Does the pattern of JAR scale responses meet an established norm?
  - Percentage difference from norm/percentage difference from just right
- How do the JAR scale distributions compare with those of the theoretical ideal product?
  - Thurstonian ideal point modeling
- Do the distributions of JAR scores differ among products?
  - Chi square
  - Stuart Maxwell
  - McNemar
- How do the JAR scores relate to overall liking or other hedonic/preference measures?
  - Chi square
  - Correlation and regression analyses
  - Penalty analysis
    - o Graphical
    - o Total penalty
    - o Opportunity analysis
    - o Penalty vs purchase intent
    - o Penalties by preference segment.

For an extensive list of JAR scale analysis methods, refer to ASTM (2008). As listed above, typical objectives include (1) to gain an understanding of the JAR distribution itself, (2) to determine if the pattern of JAR responses meets an established norm, (3) to compare the JAR scale distribution to those of a theoretical ideal product, (4) to determine if the distributions of JAR scores differ among products, and (5) to determine how the JAR scores relate to a hedonic or other commonly used benchmark measure such as overall liking, preference, or purchase intent. Objectives that fall into the first four categories involve analysis of just JAR scale data, while those that fall into category 5 require relating JAR scale to other data.

Understanding the JAR distribution
In examining the JAR distributions, the researcher looks to see whether there are obvious attribute skews or whether the preponderance of scores are in the 'just-about-right' categories. While an obvious attribute skew (60% 'just about right', 40% 'not sweet enough') may be thought to present a clear improvement opportunity (make the product sweeter), there are additional considerations prior to undertaking such a step, including the amount of sweetness increase desired, the impact on the other product attributes and the effect of the sweetness increase on those already rating the sweetness 'just about right.' Often attribute skews provide unclear direction (40% 'just about right', 25% 'not sweet enough', 35% 'too sweet'). That is why these simple examinations rarely supply sufficient information to establish a future course of action.

Do the JAR scales meet established norms?
The underlying assumption of these methods is that the greater the percentage of 'just-about-right' responses, the better the product. While typical percentages of acceptable 'just-about-right' responses may vary with the product category, the goal is to minimize the percentage of respondents rating the attributes on either side of 'just-about-right.' Although attribute manipulation via JAR scale data and true product optimization do not always result in the same formulation (Epler *et al.*, 1998; Moskowitz, 1994), in the absence of optimization data, the assumption above is a reasonable starting point. If, however, a cognitive, rather than a sensory, response has been obtained (recall the earlier example of a consumer rating a product 'too salty' because she was on a reduced salt diet), reacting to the JAR scale skews may lead to erroneous action. If one assumes that the response *is* a true sensory response (rather than a cognitive one) and on visual inspection of the data, a large number of respondents indicate that a product is 'too salty' (what is a 'large number' will depend on company norms and the judgment of the researcher), the next step would be to lower the saltiness. As previously stated, the JAR data do not indicate *how much* to lower the saltiness, nor do they guarantee that the product with lower saltiness will be more acceptable than the prior product. It could also be the case that approximately equal numbers of respondents rated the product 'too salty' and 'not salty enough.' This could be evidence of consumer segmentation, where some consumers want saltier products and others less salty products. Using JAR scales alone, it is impossible to determine what action should be taken because the impact of the attribute skews on acceptance measures is not known.

In discussing whether the JAR data meet established norms, we have been considering the *distribution* of JAR scale ratings. It is not advisable to average the JAR scale points into a mean JAR rating (Meilgaard *et al.*, 1999). The mean rating of 'just about right' could be achieved by 100% of respondents rating the product 'just about right' or by 50% of respondents rating the product 'too salty' and the other 50% rating the product 'not salty enough.' While an extreme example, taking a mean rating for a bipolar scale is not advisable as important information may be lost. Similarly statistical methods that compare mean JAR scale ratings (*t* tests or ANOVA) have limited usefulness.

How do the JAR scale distributions compare with that of the theoretical ideal product?
Thurstonian ideal point modeling (Ennis, 2003) allows the researcher to compare multiple products with a theoretical ideal product by comparing the JAR distributions of the products with the estimated probabilistic JAR distribution of the ideal product. This technique is useful because it allows the researcher to compare several products with a theoretical ideal at the same time. However, there is no guarantee that the theoretical ideal JAR distribution would in reality yield the best product. And products with identical JAR profiles may not be identical, or even similar. Products could vary in attributes not evaluated using JAR scales such as flavor type or character. (Two products that have the same JAR distributions for 'strength of cheese flavor', for example, could still vary in their degree of cheddar-type, Italian-type, or Swiss-type flavors.) Hence, a finding that two products had similar distributions as the 'ideal' product could be misleading. Additionally, this technique requires special software.

Are there differences in distributions between JAR scale ratings from different products?
There are a number of techniques that tell the researcher whether the JAR scale distributions from two or more products are different. While these techniques accomplish that goal, the usefulness of such information may be limited (as discussed above). It has been suggested (Meilgaarde *et al.*, 1999) that targeting the JAR scale distribution of a successful product may be a viable strategy, although again, there is no guarantee that two sets of identical JAR scale ratings mean that products are similar. If, however, a product is being modified from a reference product, information that the two sets of JAR ratings are similar may help assess whether the reformulated product is close enough (in the eyes of the consumer) to the reference product.

How do JAR scale ratings relate to hedonic or preference data?
Interpretive difficulties, as discussed in Sections 17.4.2, 17.4.4 and 17.5, underscore the need for researchers to link the JAR ratings to the respondent's level of liking (Popper and Kroll, 2005). These types of analyses involve the relationship between JAR scales and hedonic (liking) data. These techniques assist the

researcher in determining whether attributes rated on one or the other sides of 'just about right' are associated with lower hedonic or preference ratings.

There are a number of regression methods that attempt to relate JAR scales to hedonic data. Hedonic scales are generally treated as unipolar, while JAR scales are bipolar. While it is true that the higher end of the hedonic scale is anchored with 'liking' terms and the lower end of the scale is anchored with 'disliking' terms, the scale is referred to as a 'degree of liking' scale (Lawless and Heymann, 1998), with mean values widely reported. The most desired hedonic ratings are at the high end of the scale, whereas generally, the most desired JAR ratings are in the middle of the scale, with hedonic ratings falling off on either side of 'just about right' (an inverted U shape). Therefore it is prudent to recode each JAR scale into two scales, one that ranges from 'not enough' to 'just about right' and the other that ranges from 'just about right' to 'too much.' In this way, the researcher can determine which side of 'just about right' had the most impact on overall liking. Scales are not collapsed, as regression assumes interval data.

Regression methods can be used *within* a product, using the raw data, to understand consumer behavior, that is, when consumers rated a product as 'too much' or 'not enough', whether hedonic ratings fell or rose compared with ratings of 'just about right', or they may be used *across products,* to aid in understanding what happens to hedonic scores across a product category, as JAR ratings change. When used to relate hedonic scores to JAR ratings across products, mean JAR ratings are often used, although mean ratings result in loss of important JAR scale distribution information and must be used with extreme caution, *if at all*.

17.7   Introduction to penalty analysis

Penalty analysis is a commonly used technique to relate JAR scale to hedonic data, specifically, to understand the degree to which each side of the JAR scale is associated with lower hedonic ratings. There are several variations to penalty analysis, each with its pros and cons. All require the same initial steps; each JAR scale is collapsed into three categories, 'too weak,' 'just about right' and 'too strong.' The mean hedonic ratings are calculated for those consumers that rated the attribute 'too weak' and 'too strong.' Those means (referred to as 'the mean of the too weaks' and 'the mean of the too strongs') are compared with either the mean of those consumers rating the product as 'just about right' ('the mean of the JARs') or with the overall product mean (the 'grand mean'). The methods of calculating a penalty is as follows:

Total number of consumers: 100

JAR scale distribution for flavor strength:
   Much too weak – 5
   Somewhat too weak – 5
   Just about right – 60

Somewhat too strong – 20
Much too strong – 10

Number of consumers rating the product just about right for flavor strength:	60
Overall liking among those 60 consumers:	7.5
Number of consumers rating the product as too strong in flavor:	30 (20 + 10)
Overall liking among those 30 consumers:	6.6
Difference in overall liking between these two groups of consumers:	7.5 – 6.6 = 0.9
Penalty for flavor too strong:	0.9

17.7.1   The mean of the JARs or the grand mean?

It has been recently (Plaehn *et al.*, 2006) demonstrated that, in certain situations, the grand mean has a greater potential to show results that would make the researcher draw an erroneous conclusion. Therefore, the mean of the JARs is recommended when using penalty analysis. The justification is as follows.

Underlying the usefulness of penalty analysis to provide product reformulation direction is the assumption that the maximum hedonic score will occur at the 'just-about-right' point. One method of determining the degree to which a hedonic score can be increased is to subtract the hedonic mean over all respondents (i.e. the grand mean) from the hedonic mean of those respondents who rated the product on one side of the 'just-about-right' point for that attribute ('the mean of the too weaks' for example). The more negative the resulting value, the greater the opportunity to increase in hedonic score by adjusting the product on that attribute. When JAR distributions are nearly symmetric, this method is appropriate to use. However, near-symmetric JAR distributions may be indications that few product adjustments are needed. When JAR distributions are skewed in one direction however, the overall mean has already been influenced to a great degree by those respondents who rated the product as having too much or too little of the given attribute. In other words, a larger proportion of the respondents are 'double-counted' in the penalty calculation because they are represented not only in the group mean (i.e. the hedonic mean of respondents in the group that rated the product as having too much or too little of an attribute) but also in the overall hedonic mean. This has the effect of minimizing the size of the penalty, especially in cases where there are large skews.

This effect may lead the product developer to conclude that adjusting the product attribute will not increase the hedonic score, when in fact it would. The recommended solution is to use the hedonic mean of respondents who rated the product as 'just-about-right' (i.e. the 'mean of the JARs') instead of the overall or grand hedonic mean, as the base in the calculation of penalties. The JAR hedonic mean is typically larger than the grand mean and represents a truer estimate of a product's potential hedonic score. This is particularly true when

JAR distributions are skewed. Product reformulation based on 'the mean of the JARs' instead of the grand mean is therefore based on larger penalties and provides a more realistic view of the degree of potential, based on adjustment of that attribute.

17.7.2   The penalties
It is nearly always the case that the mean hedonic ratings from respondents that rated the product as 'just about right' will be higher than those from respondents that rated the product as 'too strong' or 'too weak.' The difference between the mean hedonic ratings from respondents that rated the product 'just about right' and those that rated the product 'too strong' or 'too weak' are the penalties, that is, the lower rating associated with being on that side of 'just about right'. These penalties are generally not independent, as consumers may rate several attributes as not 'just about right.' Additionally, these penalties in no way guarantee that the hedonic score will increase by the amount of the penalty (or at all) when the not 'just about right' skews are corrected. They merely suggest whether or not consumers that found that product to be not 'just about right' for a given attribute rated the product lower in liking than consumers that found the product 'just about right' for that attribute.

17.7.3   The cut-off percentage
It is advisable to have a skew cut-off percentage, a maximum percentage of 'too weak' or 'too strong' that does not require a penalty calculation because it is too small, typically between 10 and 20%. The cut-off percentage will depend upon category or product norms, the stage of product research being conducted (larger skews may be acceptable in earlier stages of research, while smaller skews desired during later stages of research), the sample size of the study, and comfort level of the researcher. In larger studies, smaller skews may be considered to provide a meaningful source of consumer dissatisfaction. If, for example, 5% of consumers rated the attribute as 'too weak' and 30% as 'too strong' in a study with sample size of 100, the penalty for being 'too strong' (the difference in means between the 'mean of the JARs 'and the 'mean of the too strongs') would be calculated, but the penalty for 'too weak' would not. Table 17.1 provides a list of theoretical attribute penalties along with their associated distribution size. The 'penalty' for any attribute can be tested to determine whether they it represents a statistically significant decrease by using any number of techniques, including ANOVA and bootstrapping (Shao and Tu, 1996).

17.7.4   Reacting to the penalties
Many researchers stop at this point, using the information to make future development decisions. If, for example, there is a large penalty associated with being 'too strong', the developer may decide to decrease the ingredient responsible for that attribute. If, on the other hand, there is no substantial penalty

Table 17.1   Theoretical attribute skews, penalties and total penalties

Attribute	Skew (%)	Penalty	Total penalty
Too sweet	35	0.2	7
Not sweet enough	35	0.8	28
Too thick	50	1.0	50
Not thick enough	10	na	na
Too light	25	0.1	2.5
Too dark	30	0.3	9

Rank order of total penalties

Attribute	Total penalty
Too thick	50
Not sweet enough	28
Too dark	9
Too sweet	7
Too light	2.5

associated with being 'too strong', the developer may decide not to take action, because adjusting the attribute may not increase the hedonic score. However, there is no guarantee either way. In the first case, the developer may decrease the attribute, without increasing the hedonic score; in the second case, the developer may miss the potential to raise the hedonic score by not decreasing the attribute. In the first case where the developer decreases the attribute, the change may not result in a higher hedonic score for a number of reasons: (1) the attribute was not decreased enough, (2) the attribute was decreased too much, and now the attribute is rated 'too weak' by those who formerly rated it 'too strong', (3) the change in the attribute caused those who had rated the attribute 'just about right' to rate the attribute 'too weak', (4) the change in the attribute caused product changes in other attributes, and (5) respondents initially misjudged the attribute because of interference/confusion with other attributes. In the example above, where there was a large skew for 'too sweet' but with no substantial penalty, even a small increase in liking among a large consumer group could raise the overall level of liking. In this case, the developer could have 'missed the boat.' As mentioned earlier, penalty analysis alone does not consider what may happen to those consumers in the 'just about right' group when the attribute level is changed. They may move into another category, now finding the attribute 'too strong' or 'too weak' depending on the change. That is one reason why the previously mentioned cut-off for percentages is essential.

17.7.5   The utility of penalty analysis
Despite the caveats associated with JAR scales and with penalty analysis in particular, the practicality of penalty analysis is that (in the absence of true

optimization data) it helps to prioritize which attributes need to be fixed. Consider the list in Table 17.1. It can be seen that the largest penalty (1.0) was received for being 'too thick' and this penalty was associated with a large proportion of consumers (50%). The penalty for 'too light,' on the other hand is considerably smaller (0.1), with fewer consumers (25%) associated with this penalty. This information could also assist in deciding which, if either, side of an attribute scale to 'fix' when distribution data are bimodal. Notice that the penalty for being 'not sweet enough' is 4× that for being 'too sweet,' although the absolute skews are the same. The product developer could then prioritize what to fix based on the relative penalties and skews, his or her knowledge of the food system and the probable interactions of the ingredients. Cost and timing implications could also be considered.

17.7.6   Additions to penalty analysis

Some researchers prefer to summarize the data from the attribute skews and associated penalties into total penalties, whereby each penalty is multiplied by the relevant skew and shown in rank order (Table 17.1). The total penalties can be graphed on simple line scales to show their relative magnitude. In this way, the penalties can be prioritized. Some researchers take issue with the use of total penalty, because the total penalty value is not unique – it can be arrived at via differing combinations of attribute skew and associated penalties. While calculation of total penalties may aid in prioritizing which attributes need changing, it may oversimplify understanding of the product system and its relationship to consumers.

There are analyses that attempt to balance the results from penalty analysis with other information, such as attribute hedonics or with the potential effect on those in the 'just about right' categories. Opportunity analysis (see Fig. 17.2) looks at the number of consumers that like vs dislike each attribute and graphically demonstrates the risks associated with changing the level of any attributes in response to penalty analysis. Other analyses take the entire distribution into account and calculate the upside potential from changing the attribute in the desired direction vs the downside potential from those formerly in the 'just about right' category.

17.7.7   Graphical techniques for penalty analysis

Another technique is to plot the respondent percentages (x-axis) vs the penalties (y-axis). Such a plot is shown in Fig. 17.3. Clearly, the desired response is to have attributes located in the lower left hand corner of the plot, low percentage of consumers rating and low penalty given (Fig. 17.4). Alternate desirable plots would be to have greater than cut-off percentage skews from 'just-about-right' but with no associated penalties, or skews from 'just about right' below the cut-off percentage, even with large associated penalties. A product with many attributes on the upper right hand corner of the plot has many problems (Fig. 17.5). For the 'never enough' attributes, such as pepperoni on pizza or chocolate

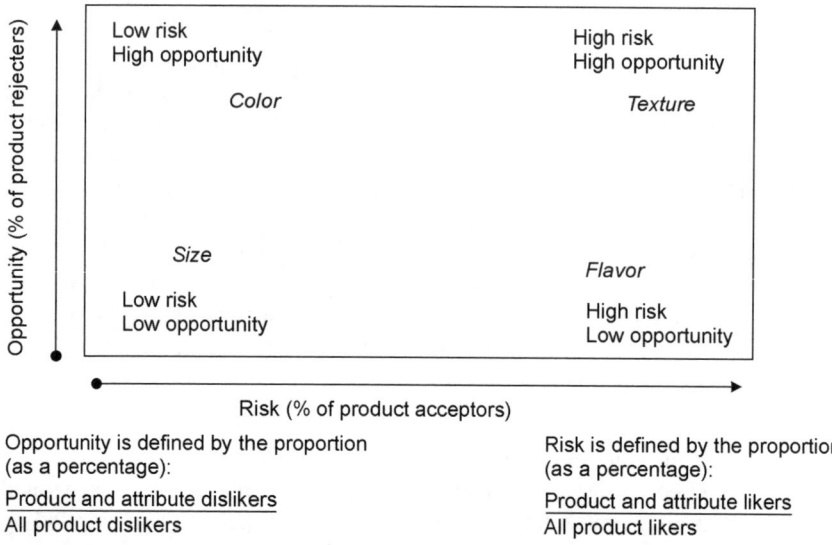

Fig. 17.2   Opportunity analysis example.

chips in cookies, the attributes tend to be located as presented in Fig. 17.6, a high percentage of consumers rating, but very low penalties. For attributes that may be cognitively, rather than sensorily, rated, a 'positive penalty' may emerge – that is consumers may rate the product higher when the attribute is rated 'too strong.' This may occur with products rated 'too sweet' or 'too salty.' In fact, a

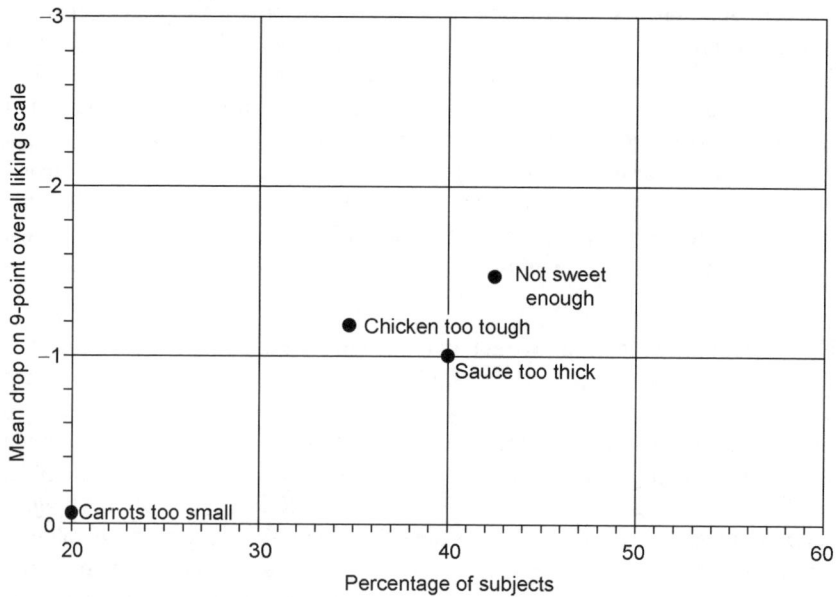

Fig. 17.3   A typical penalty plot.

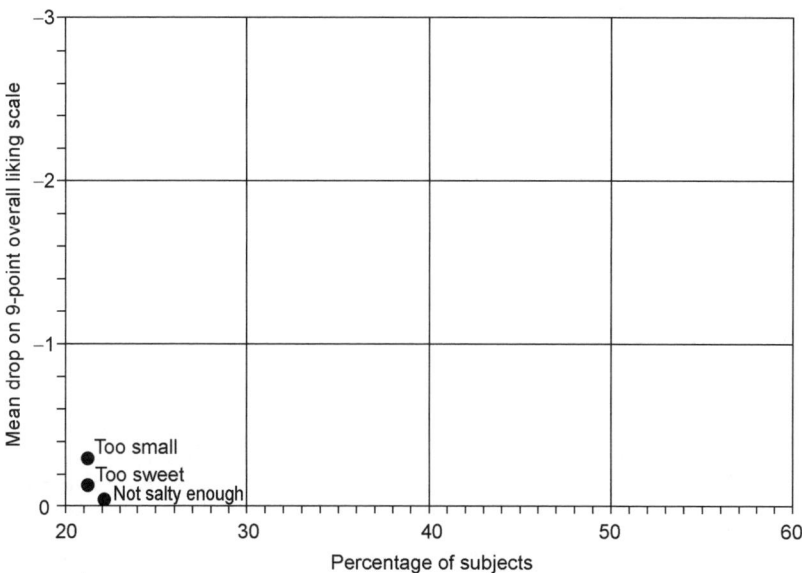

Fig. 17.4   A good penalty plot.

'positive penalty' may be evidence of cognitive-based, rather than sensory-based rating. Figure 17.7 shows positive penalties for 'too sweet' and 'too salty,' indicating that consumer that rated the products 'too sweet' or 'too salty' liked it more than consumers that rated the product 'just about right' for sweetness or saltiness.

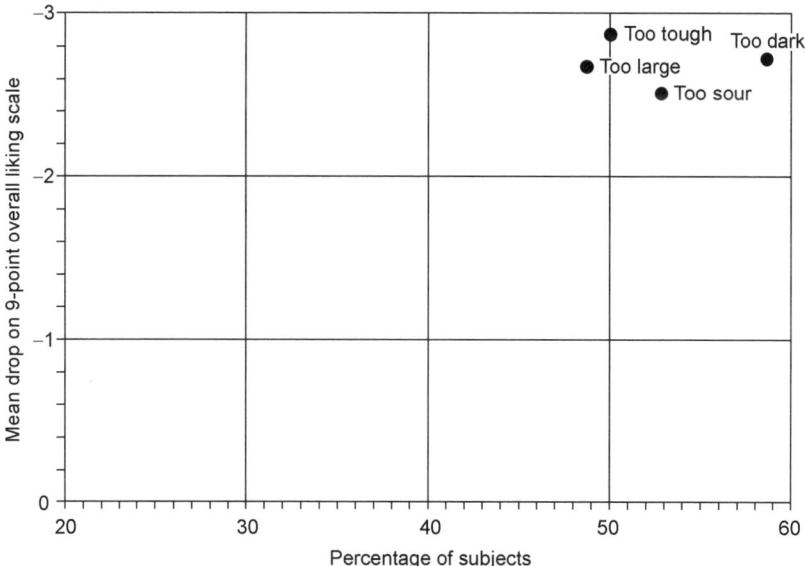

Fig. 17.5   A bad penalty plot.

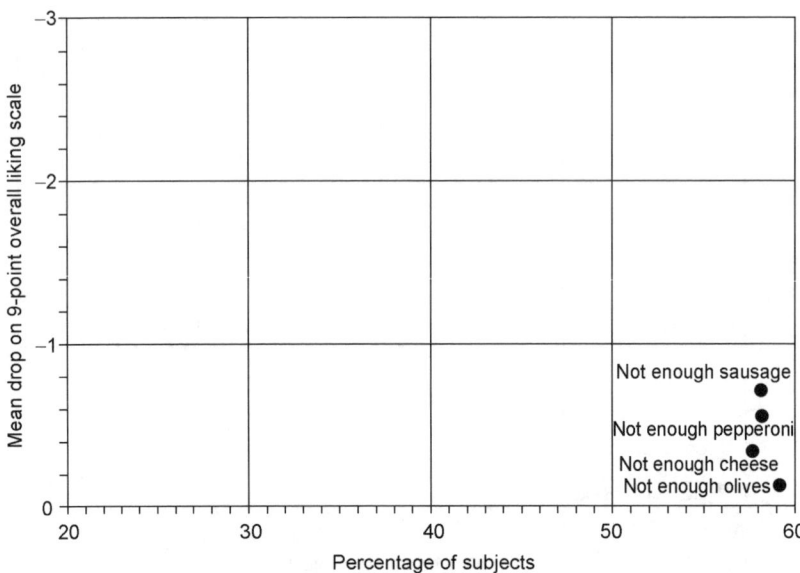

Fig. 17.6 'Never enough' attributes.

Figure 17.8 shows a problematic penalty plot; the penalties for 'too hot/spicy' and 'not hot/spicy' are similar. Depending on the attribute, this may be evidence of consumer segmentation. In spicy/hot foods, this type of response is typical, with consumers differing in their desire for spice/heat. In such a case, it is difficult, if not impossible to achieve a level of heat to please everyone; hence

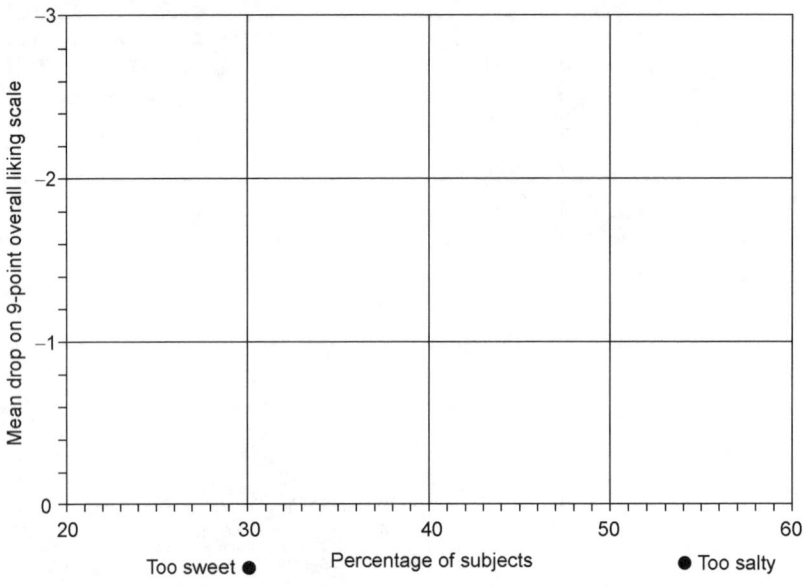

Fig. 17.7 Example of positive penalties.

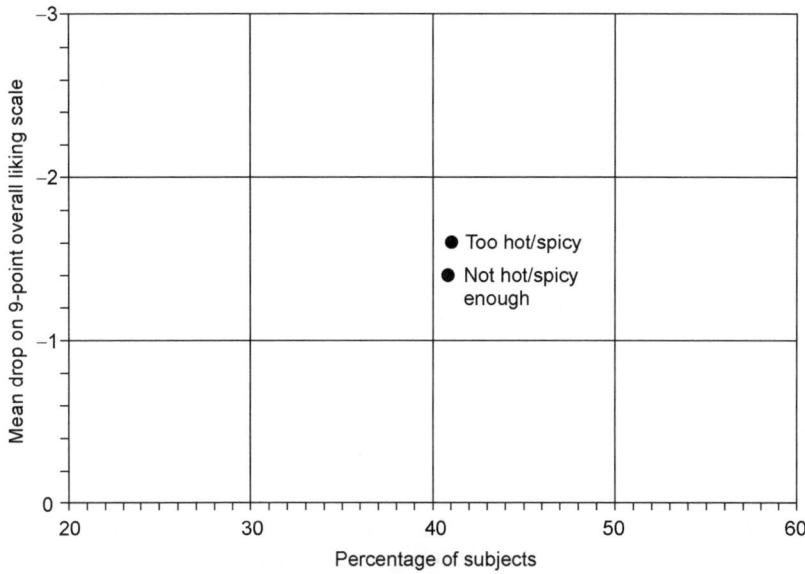

Fig. 17.8   A bimodal penalty plot.

the development of mild, medium, and hot products. In the case of Fig. 17.8, the solution would be to develop two products, one even milder and one even hotter than the product illustrated. 'Cherry flavor' in Fig. 17.9 shows a similar pattern, but resulting from a different issue. In this case, the 'cherry flavor' is artificial and medicinal tasting, although the product is a candy. Respondents were

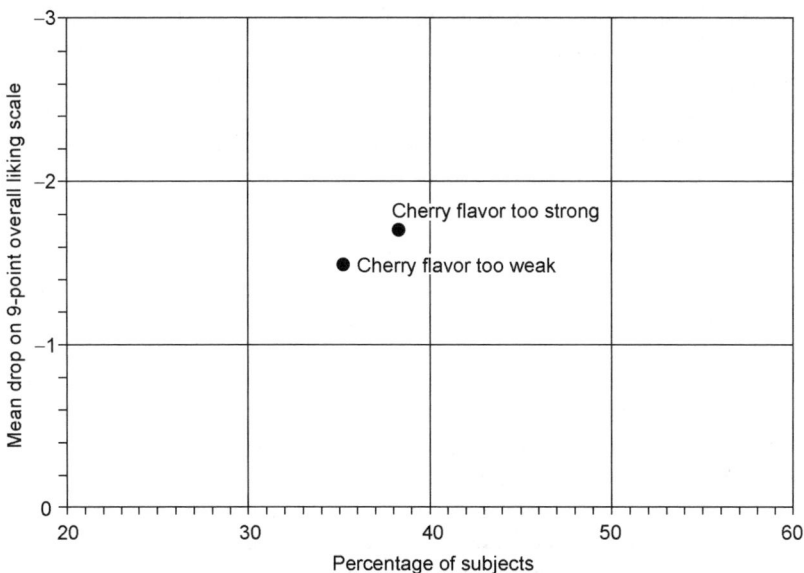

Fig. 17.9   A bimodal penalty plot – unclear attribute.

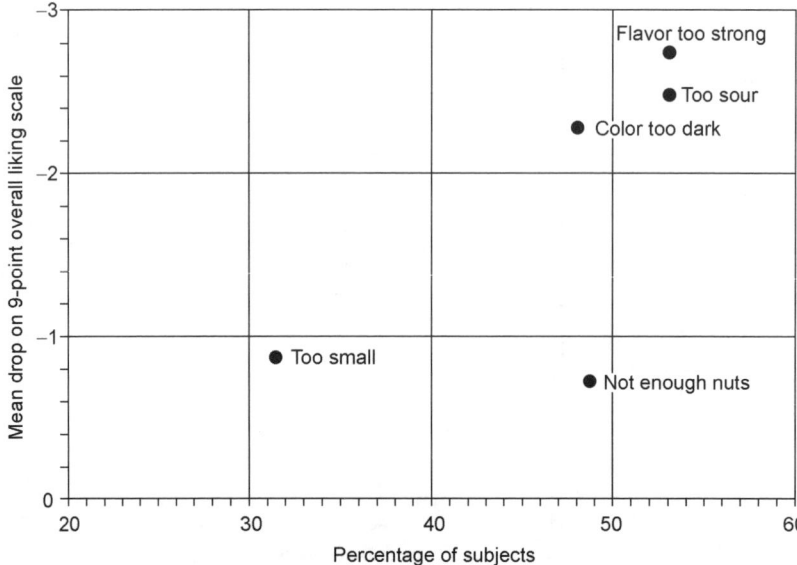

Fig. 17.10   Penalty plot before reformulation.

reacting either to the fact that the artificial cherry taste was too strong or the natural candy cherry taste was too weak. In this case, the hedonic rating for the cherry flavor itself, if included on the questionnaire, could assist in explaining the data.

Penalty plots such as these can assist with a number of other product development situations. The plots in Figs 17.10 and 17.11 show penalties received before and after product reformulation. Notice that the penalties that remain have moved from the upper right to the lower left hand section of the graph, indicating that further reformulation may not result in further increases in hedonic ratings. Also notice that, despite the amount of nuts being increased, the skew and penalty (while small) do not change.

17.7.8   Penalty plots from purchase intent data

Penalty plots can also be generated from purchase intent data. Such plots are generated similarly to penalty plots generated using overall liking data. The only difference is that instead of calculating overall liking means within each JAR category, the percentage of respondents that will purchase the product (using top or top two box purchase intent data) within each JAR category is used. These plots may provide different information from that obtained using overall liking data. Consider Figs 17.12 and 17.13. Figure 17.12 is a penalty plot based on overall liking; the direction for increasing overall liking would be to increase the flavor strength of the chicken broth, and decrease the saltiness. Figure 17.13 is a penalty plot based on purchase intent for the same product; the direction for increasing purchase intent is to increase the size of the vegetables and the

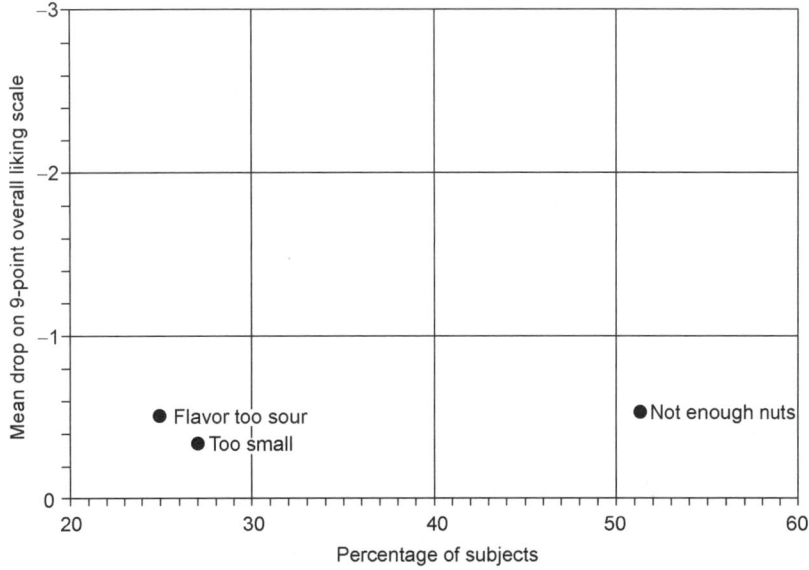

Fig. 17.11   Penalty plot after reformulation.

portion size. The attributes that will increase liking may not be the same as those that will increase purchase intent.

17.7.9   Penalty plots and preference data

Penalty plots can also help developers in understanding differences in product preference, and can aid in understanding the basis for consumer segmentation and product substitutability. Figures 17.14–17.17 illustrate such an example. Consider two competitive products, one with a mild flavor profile, the other with a strong flavor profile. Penalty plots can be generated for each product among those consumers that preferred each product and among those consumers that did not prefer each product.

Figure 17.14 illustrates the penalty plot for the strong product among those who preferred it. The penalties all lie in the lower left hand corner, indicating that consumers who prefer this product are largely satisfied with its attribute profile. Figure 17.15 illustrates the penalty plot for the strong product among those who did not prefer it. Notice the penalties in the upper right hand corner. Clearly, the consumers that did not prefer this product objected to its strong flavors. Figure 17.16 illustrates the penalty plot for the milder product among those who preferred it. Again, the penalties are located in the lower left hand corner, indicating that this product appears to be well suited for the consumers that prefer it. Figure 17.17 illustrates the penalty plot for the milder product among consumers that did not prefer this product. Note that many of the penalties remain in the lower left hand corner of the plot. This illustrates the phenomenon of 'product substitutability.' The stronger product could not be

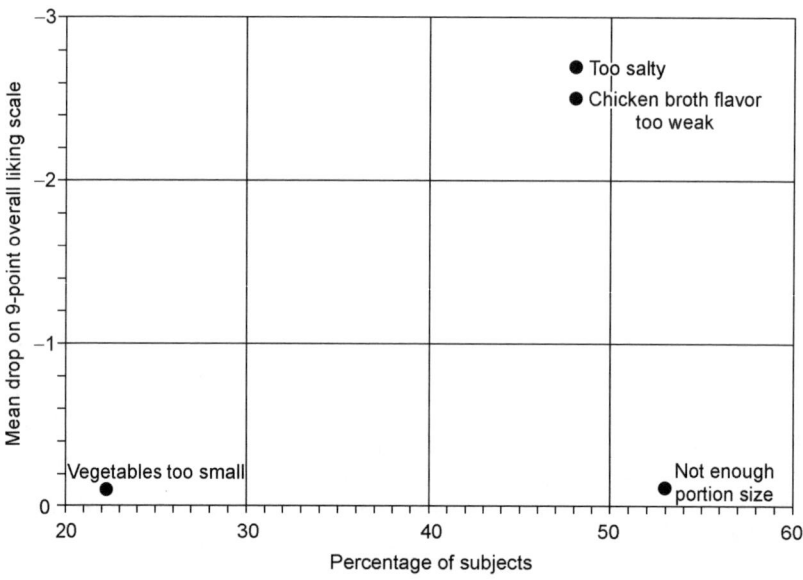

Fig. 17.12 Penalty plot based on overall liking.

substituted for the milder product as the stronger product is polarizing; however, the milder product could be substituted for the stronger product even among those that prefer the stronger product. This goes beyond the idea of consumer segmentation, in which different groups of consumers prefer products with

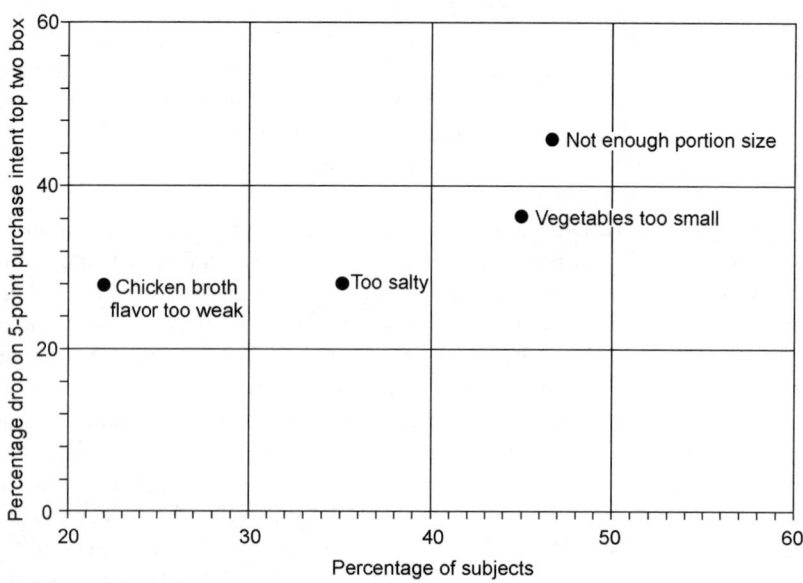

Fig. 17.13 Penalty plot based on purchase intent.

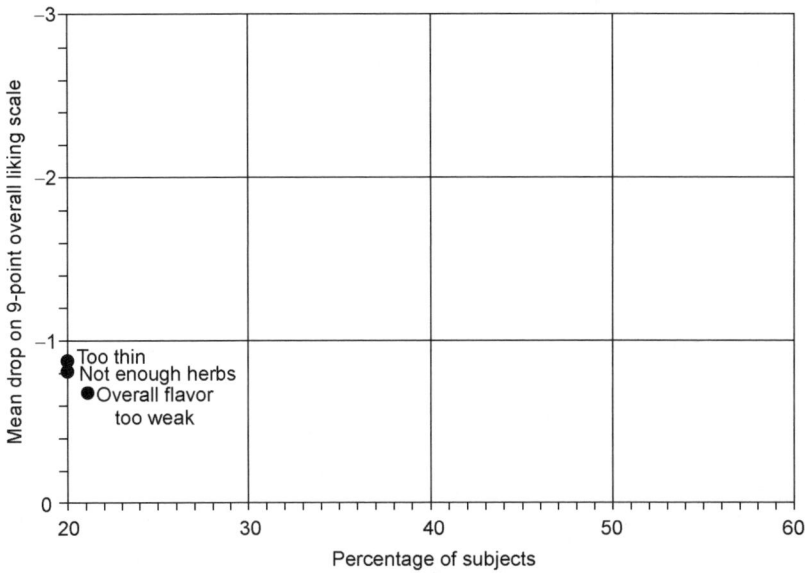

Fig. 17.14   Penalty plot – strong product among preferrers.

different profiles, toward the idea that some products, while not representing the consumer 'ideal' may be able to span the space between groups of consumers that prefer different attribute intensities.

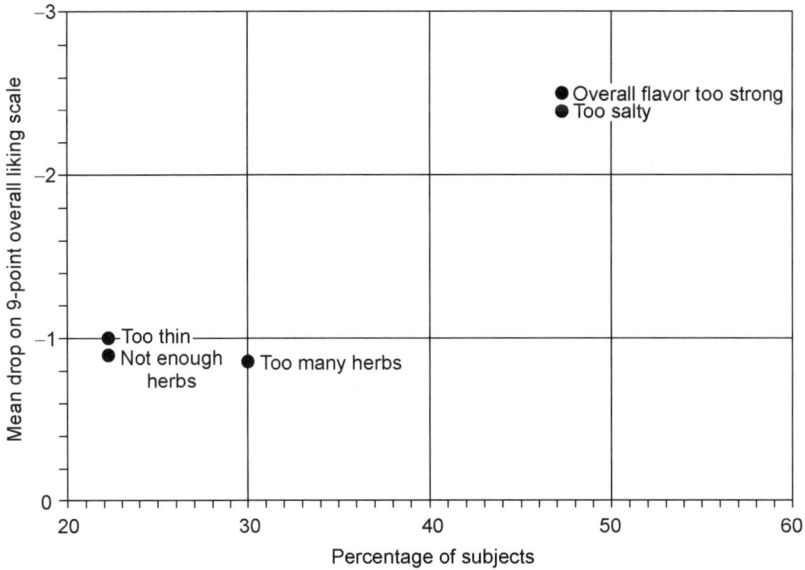

Fig. 17.15   Penalty plot – strong product among non-preferrers.

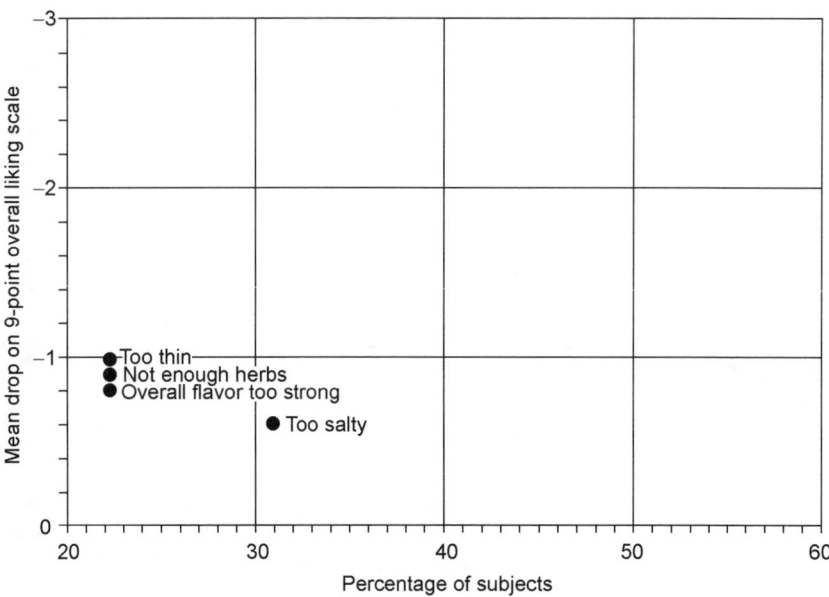

Fig. 17.16   Penalty plot – weak product among preferrers.

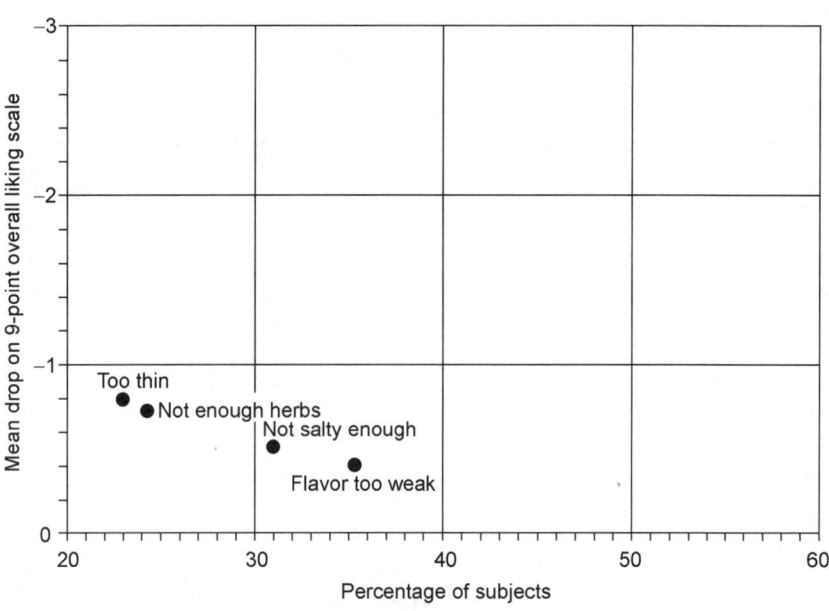

Fig. 17.17   Penalty plot – weak product among non-preferrers.

17.8   Alternatives to JAR scales

As mentioned previously, properly designed optimization experiments completely avoid the problems associated not just with JAR scales, but also with consumer language in its entirety. In the absence of optimization experiments, sensory panel data of attribute intensities can be used to link hedonic information from consumer panels without using JAR scales (Moskowitz, 1985). Another technique involves 'ideal scaling' (van Trip *et al.*, 2007), which involves separating out the hedonic response to the attribute from its intensity determination. The respondent may first be asked how well he or she likes the intensity of an attribute on a hedonic scale. Then the respondent rates the attribute intensity on a unipolar intensity scale, followed by a rating of their 'ideal 'intensity rating for the attribute on the same type of scale. It is assumed that the further the distance between the attribute intensity rating and the attribute 'ideal' intensity rating, the larger the change that must be made to adjust the attribute level. The attribute liking question is asked in order to gage the importance of adjusting the attribute level. While this latter method may remove the psychological difficulty of simultaneously rating intensity and appropriateness associated the JAR scales, many of the difficulties associated with JAR scales (the actual degree of change required, the issue of 'never enough' attributes, the impact of changing certain attributes on other attributes, etc.) remain.

17.9   Future trends

As discussed, some of the barriers to the use of JAR scales include (1) the difficulties related to consumer language, (2) the inability to relate attribute skews to levels of desired attribute change, and (3) the problems associated with trying to optimize a product with distinct consumer attribute intensity segments. These problems can all be circumvented by the use of systematic experimental design and product optimization. Properly designed optimization experiments obviate the need for consumer language, and provide a direct relationship between attribute levels and desired intensity. Models can be built that incorporate the attribute levels of distinct consumer segments. As product developers and consumer researchers become increasingly aware of its benefits, the use of systematic product development will increase. However, the nature of experimental design itself may sometimes preclude its use, especially when the entirety of product attributes is being considered.

The number of experimental prototypes required to optimize complex products may tax the abilities of developers to create them, not to mention the resources required to execute the experiments. Moskowitz (2003a) presents an approach to synthesize useful direction from product testing when the stimuli are not systematically varied, albeit for a relatively simple food system (margarine). However, the models poorly fitted the data (which the author points out is typical in studies with a limited number of products). What is required are systems for whole product optimization that preserve its advantages, while reducing the very large numbers of products that must be created. Until and unless such systems are created, product optimization may continue to be used 'piecemeal' (keep the texture system constant but alter the flavor, or optimize the main dish but keep the side dish constant). Such experiments often fail to account for the interactive nature of product ingredients and meal components. And JAR scales will be used to help developers in the all-too-common cases where products have not been systematically created.

17.10   Sources of further information and advice

Several major textbooks published on sensory evaluation (Meilgaard *et al.*, 1999; Lawless and Heymann, 1998; Sidel and Stone, 1993) contain sections pertaining to JAR scales, though few, if any, mention penalty analysis. As mentioned, ASTM International Committee E-18 (www.astminternational.org) has a task group (E.18.04.26) working toward publishing a comprehensive document on JAR scales and their analysis is to be published in 2008. University professors have published or presented JAR scale analysis methods at major conferences such as Pangborn (www.pangborn2005.com) and Sensometrics (www.sensometrics.org) and there have been round table discussions at such conferences as well. At least yearly, a lively discussion takes place on the Yahoo e-group www.sensory.org. Trade group publications, such as those from the Institute of Food Technologists Committee on Sensory Evaluation, and consultants' newsletters such as *ChemoSense* from the Centre for ChemoSensory Research or the Institute for Perception (www.ifpress.com) are also available.

17.11   Acknowledgements

Many thanks to Mary Schraidt of Peryam and Kroll for construction of the penalty plots. The data represented, while within the realm of possibilities, are entirely theoretical.

17.12   References

ASTM (2008), *Standard Guide for Benefits and Risks Associated with the Use of Just About Right Scales in Consumer*, ASTM International Committee E-18 (in press).
BOWER JA and BOYD IA (2003), 'Effects of health concerns and consumption patterns on measures of sweetness by hedonic and just-about-right scales', *Journal of Sensory Studies*, **18**(3), 235–248.
ENNIS DM (2003), 'Just-about-right scales', *IF Press*, **6**(3), 2–3.
EPLER S, CHAMBERS E and KEMP K (1998), 'Hedonic scales are a better predictor than just-

about-right scales of optimal sweetness in lemonade', *Journal of Sensory Studies*, **13**(2), 191–197.

GACULA, JR. MC and SINGH J (1984), *Statistical Methods in Food and Consumer Research*, Orlando, Academic Press

LAWLESS HT and HEYMANN H (1998), *Sensory Evaluation of Food*, New York, Chapman & Hall.

MEILGAARD M, CIVILLE GV and CARR BT (1999), *Sensory Evaluation Techniques*, Boca Raton, FL, CRC Press.

MOSKOWITZ HR (1985), *New Directions for Product Testing and Sensory Analysis of Foods*, Westport, CT, Food & Nutrition Press.

MOSKOWITZ HR (1994), *Food Concepts and Products Just-in-time Development*, Trumbull, CT, Food & Nutrition Press.

MOSKOWITZ HR (2001), 'Sensory directionals for pizza: a deeper analysis', *Journal of Sensory Studies*, **16**(6), 583–600.

MOSKOWITZ HR (2003a), 'Category appraisal within a limited product range', *Journal of Sensory Studies*, **18**(3), 182–198.

MOSKOWITZ HR (2003b), 'The just-about-right scale – do panelists know their ideal point', in Moskowitz HR, Munoz AM and Gacula MC Jr., *Viewpoints and Controversies in Sensory Science and Consumer Product Testing*, Trumbull, CT, Food & Nutrition Press.

MOSKOWITZ HR (2004), 'Just about right (JAR) directionality and the wandering sensory unit. In Data analysis workshop: getting the most out of just-about-right-data. *Food Quality and Preference*, **15**, 891–899.

PLAEHN D, LUNDAHL D, STUCKY G and HORNE J (2006), InsightsNow Personal Communication.

POPPER R and KROLL DR (2005), 'Just-about-right scales in consumer research', *Chemo Sense*, **7**, 1–6.

POPPER R, CHAITON P and ENNIS D (1995), 'Taste test vs. ad-lib consumption based measures of product acceptability'. Presented at the second Pangborn Sensory Science Symposium, University of California, Davis, 30 July–3 August.

POPPER R, ROSENSTOCK W, SCHRAIDT, M and KROLL BJ (2004), 'The effect of attribute questions on overall liking ratings', *Food Quality and Preference*, **15**, 853–858.

POPPER R, SCHRAIDT M and KROLL BJ (2005), 'When do attribute ratings affect overall liking ratings'. Presented at the Sixth Pangborn Sensory Sciences Symposium, Harrogate International Center, York, 7–11 August.

RESURRECCION A (1998), *Consumer Sensory Testing for Product Development*, Gaithersburg, MD, Aspen Publishers, Inc.

SHAO J and TU D (1996), *The Jackknife and Bootstrap*, New York, NY, Springer.

STONE H and SIDEL JL (1993), *Sensory Evaluation Practices*, San Diego, CA, Academic Press.

VAN TRIP H, PUNTER P, MICKARTZ F and FRUITHOF L (2007), 'The quest for the ideal product: comparing different methods and approaches', *Journal of Food Quality and Preference*, in press.

18

Conducting difference testing and preference trials properly for food product development

M. O'Mahony, University of California, Davis, USA

18.1 Introduction: the role of difference and preference testing

18.1.1 The goals of difference and preference testing

The food product developer might be called upon to copy a product that is doing well in the market, so as to produce a 'me too' product to steal some market share. He/she may have to modify the product because of changes in the ingredients, processing or packaging. If developing a new product, he/she may have used one or more of a range of sophisticated techniques to decide what type of product will produce maximum consumer acceptance. There are a large number of product optimisation techniques available, but these are beyond the scope of this chapter. This chapter is concerned with what a food product developer can get from difference tests and preference tests. For convenience these will be divided into three main areas, each representing a different goal, which we will call consumer preference and acceptance resting, Sensory Evaluation I and Sensory Evaluation II (O'Mahony, 1995a).

18.1.2 Consumer preference and acceptance testing

This is really the 'bottom line'. Will the product succeed in the marketplace? Probably the best way for the food product developer to determine whether a product will be accepted by the consumers is to test market that product. If the food company owns or has access to a chain of supermarkets, it could test-market the product and monitor consumer purchasing behaviour via courtesy cards or shop credit cards. If the demographics were recorded on such cards, the company would have instant feedback on which products were selling well, to whom and for how long. Measuring behaviour is really the best thing but the only problem is that not every company has such instant access to the marketplace.

The next best thing is to ask the consumers whether they like the product, which product is preferred, and whether they would buy the product. The rather naive hope is that the answers to these questions predict consumer purchasing behaviour. In most circumstances, it is the best we can do, and it is in reference to these in general questions that we will discuss preference testing.

Who do we ask? Obviously we ask the potential customer. There is no point getting preference data from people who are not going to be potential customers. If a product was designed to be sold in England, it would not be sensible to test it on Japanese consumers. We take a sample of potential customers and infer from our sample to the population, using the appropriate statistics. At this point it is as well to be aware of the concept of statistical significance. We might find a significant preference for a given product but this only means that it is not a chance event. What is usually forgotten is to consider the strength of the preference. If one product is preferred over another it is more important to know what proportion of consumers prefer that product. If the proportion is 95% it is obviously an important preference. If the proportion is 60% the preference is only slight and different business decisions might be made. Either proportion might be significant, depending on the sample size. Also, in choosing the sample size, it is as well to remember that the population we are sampling from may run into millions. We want a representative sample, so we should be thinking in terms of hundreds. A group of 30 willing friends is not really a representative sample. Finally, the products used in a preference test should be easily distinguishable. One cannot logically have a real preference if one cannot tell two products apart. In such a situation, a consumer might give a preference response but it will not represent a real preference.

18.1.3   Sensory Evaluation I: analytical sensory testing

Here we are using the human sensory system as a set of instruments to measure the various physical and chemical properties of the food. In this case the product developer takes a sample of food to the laboratory and has it analysed using a variety of instruments, one of which is a taste panel.

A gas chromatograph detects volatile chemicals but so does a nose. Both may be used as volatile detectors. The nose and the gas chromatograph have different sensitivities, so they may be used in parallel. A nose may detect what a gas chromatograph never sees. As far as a gas chromatograph is concerned we try to buy the best model that we can. We need only one gas chromatograph. Logically we need only one panellist's nose. But humans can have bad days and so we use more than one panellist, more than one nose, as a failsafe. If one panellist is having a bad day, the others can make up for his or her poor performance.

Unlike consumer preference and acceptance testing, we are sampling food products, not people.

For most people, Sensory Evaluation I, or analytical sensory testing, means descriptive analysis. Here the various attributes of the food are rated using some type of scaling procedure. The goal is to get a profile of the attribute strengths of various food products. Because scaling is designed to measure differences between attributes or stimuli that are easily discriminated, it is assumed that as far as the panelists are concerned, the differences in perceived attribute strengths are relatively large and easy to discriminate. If they were not, if they were confusable, then difference tests would be more appropriate. In this case, the attribute profiles would be in terms of the degree of difference between foods in terms of their various attributes. Difference tests are designed to measure the degree of difference that is perceived between two very similar, confusable stimuli.

Whose sensory systems do we use? We select judges who perform the best. We may train them so as to increase their discrimination skills. We may place them under controlled conditions so that there are no distractions. We would use the most sensitive and reliable methods. The measurements we make may not correspond to natural eating conditions, but that is not the point. Here, the goal is to obtain analytical profiles comparing the various sensory characteristics of the foods. For this we can use any experimental conditions that are appropriate. If we need to place the judge under red light, wearing a nose clip, then so be it. This will not correspond to normal eating conditions but it might make it easier for the judge to concentrate on assessing the taste properties of the food, while not being distracted by visual or olfactory cues.

18.1.4   Sensory Evaluation II: measuring consumer perception

Here the goal is to measure consumer perception. With difference tests this would mean seeing whether consumers can tell the difference between two foods under ordinary eating conditions. So once again, we want to test our target consumers. This is not the job for trained panellists. As with consumer preference and acceptance testing, a sample of consumers is tested and inferences made about the population. The demographics here may not be as complicated as with preference testing. For some products, there are probably two types of consumers, regular consumers and casual consumers. Regular consumers are likely to be more discriminating than casual consumers. A consumer who chews three or four packets of a particular gum per day will be more discriminating than one who chews gum maybe once a month.

Because we are dealing with ordinary consumers, who have no special training, we want to make their measurement task as simple as possible. We would not ask them to give us a detailed description in terms of the differences in attribute strengths between the products. That is the job of the Sensory Evaluation I panellist. The task is simply to see whether ordinary consumers can distinguish between foods under ordinary eating conditions.

The main problem, here, is to reproduce ordinary eating conditions. This may not be easy, so we do the best we can. We try to get as close to ordinary conditions as we can manage. For example, if we were dealing with hamburger patties, which are to be eaten in a hamburger with accompanying tomato ketchup and mustard, then that would be the way we would serve them. If our consumer always drank a beer while eating the hamburger, that would be quite easy to provide. If the hamburgers were eaten sitting on a couch at home, while watching TV, it might not be so easy to reproduce such an environment. But we do the best we can and use our common sense. Naturally, using nose clips or red lights would not be suitable here.

18.2   Difference tests

18.2.1   Response bias

One of the main problems in difference testing is response bias (O'Mahony, 1992, 1995b; O'Mahony and Rousseau, 2002). For difference tests there are two main types of response bias that must be considered. These refer to the beta and the tau criteria. First, let's consider the beta criterion. Imagine you are presented with a cake, to which has been added a slight amount of sweetener. The amount is so small that it is difficult to tell whether it is there or not. Imagine you are asked whether it tastes sweet or not. Your response will depend on two things. It will depend on whether your taste receptors were sensitive enough to pick up the sweetness. It will also depend on whether the sweetness that you perceived was sufficiently sweet for you to feel confident enough to report that it was sweet. Another way of saying this is that it depends on where you draw the line between 'sweetness' and 'unsweetness'. Does the sensation you got from the cake fall on the sweet side of the line or does it fall on the unsweet side? You might detect a very slight suggestion of sweetness but it may not be strong enough for you to want to report that it is sweet. In other words, it may not be strong enough to fall on the sweet side of the line. The line has a technical name; it is called the beta criterion.

Where you draw the line is a cognitive process. If you are feeling confident that day, you might draw the line in such a position that the cake will be on the sweet side of the line and you will declare that the cake is sweet. If you are feeling cautious, you might draw the line so that the cake falls on the unsweet side of the line and you will respond that the cake is not sweet. Where you draw the line affects your response; this is why it is called response bias. There are various ways of getting round the problem of response bias. One is to use signal detection theory (Green and Swets, 1966; Macmillan and Creelman, 1991) but the most common method in food science is to use one of the many forced choice methods (Lawless and Heymann, 1998). You simply present two cakes. One will have the added sweetener and the other will not. Your task is to indicate which cake is the sweet one. To answer this question, you have to adjust your line, your beta criterion, such that the sweet cake is on the sweet side of the line and the unsweet cake is not. Naturally, the instructions must state clearly that one cake *will* be sweet and the other will not. They cannot state that one *may* be sweet; this does not guarantee that the beta criterion line will be drawn in the correct position.

The second type of criterion is the tau criterion. This is used to answer a similar question. Are two stimuli different enough to be called different? Here the criterion can be visualised as a perceptual yardstick. If two stimuli are further apart than the length of the yardstick, they will be reported as different. If not, they will be regarded as the same. Again, a test must arrange that the yardstick, the tau criterion, is of such a length that if a slight difference is perceived, it will be reported. A test such as the triangle test manages to do this. The judges are told that two stimuli are the same while one is different. Therefore they must adjust their tau criterion so as to include two of the stimuli within the range of their yardstick while the 'different' one falls outside the range.

So the forced choice tests set up a situation where the beta criterion is in such a position or the tau criterion is of such a length, that sensation differences, if perceived, will be reported. Most of the common forced choice difference tests do this so that problems of response bias can be forgotten. It is interesting that the people who designed the forced choice tests had no conception of response bias. They merely set up the tests so that if a judge said he/she could discriminate between two foods, he/she was required to prove it by successfully performing the appropriate difference tests. The fact that most of these tests just happened to eliminate response bias was a piece of luck.

18.2.2   The tools of the trade

A detailed description of the many types of difference test is beyond the scope of this chapter. Accordingly, only a few will be mentioned and this will be in terms of how they cope with response bias. Triangle and duo–trio tests need to stabilise the tau criterion because they involve judgements of whether products are the same or different. What food scientists tend to call the paired comparison test, but psychologists tend to call the 2-AFC (two-alternative forced choice), needs to stabilise the beta criterion, because it involves judgements of whether a sensation has crossed a line (sweet/unsweet) or not. An extension of the 2-AFC is the 3-AFC. Instead of telling the judge that one cake will be sweeter and one will not, the judge is told that one cake will be sweeter while two will not, or vice versa.

Having stabilised the appropriate criteria, it is useful to know the cognitive strategy that the judge uses when performing the test. For the triangle and duo–trio tests, the judge compares the sensations from each of the three stimuli and picks the one that is most different. This has been called the comparison of distances strategy (Lee and O'Mahony, 2004; O'Mahony *et al.*, 1994). For the 2-AFC or 3-AFC tests, the judge merely picks the sensation that is the strongest (or weakest). This has been called the skimming strategy.

As a result of signal detection theory/Thurstonian modelling, fundamental measures of difference, such as d', can easily be obtained from tables, by noting the proportion of tests that were performed correctly (Ennis, 1993). Because it is a fundamental measure, the same d' value should be obtained, other things being equal, regardless of the test method used. However, to be able to compute d' it is necessary to know the cognitive strategy that is being used by the judge. Measures of d' and Thurstonian modelling are discussed in more detail in Chapter 19.

There are various measures of degree of difference. One measure is the proportion of tests the judge gets correct. Yet, this is a measure of performance; it is not a fundamental measure. By this is meant that it is a measure of how well a person performs that specific test but it is not a fundamental measure of how well the judge can discriminate. A measure of performance, like the proportion of tests correct, is test-specific. If a judge gets five out of ten triangle tests correct there is no way of knowing that he/she can discriminate better than if he/she got eight out of ten 2-AFCs correct, because the chance probabilities are different. The tests cannot be compared. There would also be no way of resolving the fact that a person who got five out of ten triangles correct was more discriminating than a person who got the same proportion of 3-AFCs correct, even though the chance probabilities for the two tests were the same. Only a fundamental measure such as d' can resolve such issues.

John Brown's R-index (Brown, 1974; O'Mahony, 1992), like d', is a measure of how well a judge can distinguish between two stimuli. Yet, it is also a measure of performance; it is not a fundamental measure like d'. The R-index obtained from ranking will be higher than the measure obtained from rating (O'Mahony *et al.*, 1980; Ishii *et al.*, 1992). Yet, Brown wanted a measure that was free of the assumptions that are necessary for the computation of d' and the R-index has the advantage of simplicity. It also converts easily to d' (Elliot, 1964; Ennis, 1993) if the usual assumptions for d' are upheld and if the cognitive process used in the test involves the beta criterion and not the tau criterion (Green and Swets, 1966).

Although the most important measure is one of degree of difference, it is also important to determine whether performance on the test was due to actual discrimination or merely to chance. For the usual forced choice tests, the statistic generally chosen is the binomial test. However, this was designed for situations involving the tossing of coins or dice, where the probability in getting a certain result (heads, six) is constant. Thus, if a coin is tossed three times, we can say that it is equivalent to tossing three coins once. We can combine coins and replicate tosses. However, with people, we cannot do the same thing. We cannot say that testing three people once is equivalent to testing one person three times. Thus, if 20 people each perform three duo–trio tests, we cannot combine people and replicate tests and treat the data as if it were a sample of 60 tests. This is because, unlike coins and dice, the probability of getting a particular result from different people is not constant. People have different sensitivities. This adds variance to the situation and the problem of this added variance is called the problem of overdispersion. There are several ways of dealing with over-dispersion (Brockhoff and Schlich, 1998; Kunert and Meyners, 1999; Kunert,

2001; Brockhoff, 2003) and one is to use a beta-binomial analysis (Harries and Smith, 1982; Bi and Ennis, 1998, 1999a,b; Ennis and Bi, 1998; Bi *et al.*, 2000). However, to do this, each judge must perform more than one difference test so that the degree of overdispersion can estimated and the appropriate corrections made. So, if the data from several judges are to be combined, each judge needs to perform more than one difference test.

The same–different test (Pfaffmann, 1954; Peryam, 1958; ASTM, 1968) would also seem to be useful. However, it does not control response bias. Merely asking whether two products are the same or different does not force the tau criterion to be of the appropriate length. It may be too long, so that all pairs of stimuli are reported as the same. It may be too short, so that all pairs of stimuli are reported as different. Merely counting the proportion of tests performed correctly is not a true measure of how well the judge can discriminate between the two stimuli. However, a signal detection/Thurstonian modelling analysis provides a measure of d' (O'Mahony and Rousseau, 2002), which is unaffected by variation in the tau criterion. The analysis is somewhat complex and involves an iterative equation but it is easily performed with the appropriate software (e.g. IFPrograms, Institute for Perception, Richmond, Virginia).

The tests mentioned so far are comparative. They involve presenting more than one stimulus at a time. Yet, in some situations it is better to present stimuli singly and ask a question such as, 'Is this sweet?' Of course, such a question elicits response bias; the response depends on the beta criterion. Yet, by using a suitable signal detection analysis, it is also possible to compute a value of d' which is free of response bias (Green and Swets, 1966).

So, which test do we use? To make this decision we have to consider several things. We have to decide whether the goal of the measurement is analytical (Sensory Evaluation I) using 'in-house' taste panellists or whether the goal is to measure consumer perception (Sensory Evaluation II) using a sample of target consumers. Then we need to do some preliminary testing with the product. How does the product affect us? Is it fairly bland or does it have a strong taste or aftertaste? Does it have an irritating trigeminal effect, like products containing capsaicin? This will affect the number of samples that we can put into our difference test.

If the product is bland we can use one of the forced choice tests, involving more than one sample. If the product is strong, we may have to present samples one at a time and use a suitable signal detection analysis. How many tests can we perform in one session? Would rinsing between stimuli or between tests increase this number? Such questions can be answered by a little preliminary testing and then the appropriate difference test can be designed. In designing the test, we have to be aware of the possibility of response bias. What is our strategy for dealing with beta or tau criterion variation? For this, it is important to understand something about the models and theories associated with difference testing. However, it is a better approach than trying to fit different products into the same routine test all time. You would not test a hot chilli pepper in the same way that you test a tomato.

18.2.3   What is the best test to use for Sensory Evaluation I?

Here, we can use the notion of statistical power. A more powerful statistical test is better at rejecting the null hypothesis; it will reject the null hypothesis using a smaller sample size. Obviously, we want our testing to be as efficient as possible, so we want to be able to make a decision based on statistically significant data, with as small a sample of tests as possible. This will be more efficient, take less time and accordingly cost less money.

Using Thurstonian modelling, Ennis noted that for a given d', the most powerful statistical test was the 3-AFC (Ennis, 1990, 1993). Very close in power was the 2-AFC. The triangle and duo–trio tests were not as powerful and so would not be suitably efficient for our purpose. However, in reality, because different tests cause different perturbations in the mouth and make different demands on memory, the d' that they elicit will not be exactly the same.

Sequential sensitivity analysis (O'Mahony and Odbert, 1985; O'Mahony and Goldstein, 1987; Vié and O'Mahony, 1989; Tedja et al., 1994) is a model for predicting how sequences of tasting in a difference test affect sensitivity. It predicts that the 2-AFC has more favourable sequences of tasting than the 3-AFC and will thus elicit higher d' values, which was later confirmed (Rousseau and O'Mahony, 1997; Dessirier and O'Mahony, 1999). The 2-AFC having two rather than three stimuli, would also be expected to have a memory advantage (Rousseau et al., 1998; Lau et al., 2004) With the 2-AFC, the product that is tasted second is compared with the memory of the product tasted first. With the 3-AFC, the product that is tasted last has to be compared with the memories of two products tasted beforehand. The time since tasting the first product is longer than with the 2-AFC, allowing more chance for the memory to be distorted. Also, the memory traces of the first two products could interfere with each other and cause more memory distortions. In fact, it seems that interference is the more important effect here (Lau et al., 2004) These two effects render the d' from the 2-AFC to be greater than that of the 3-AFC, which causes it to be a more powerful test (Rousseau and O'Mahony, 1997). So where the 3-AFC wins on central processing, the 2-AFC more than compensates for this with its advantages from the sequence of tasting and the effects of forgetting. So, in summary, the 2-AFC is the most efficient and powerful test and is thus the most suitable for Sensory Evaluation I.

The 2-AFC method requires the experimenter to be able to explain to the judge the attribute that has to be attended to (indicate which sample is sweeter, spicier, crunchier, etc.). Because these are Sensory Evaluation I panellists, this should present no difficulty, because the panellists would have been trained regarding the attributes of the foods under consideration. Should there be any difficulty in this regard, then the use of 'warm-up' should provide an answer to this problem.

Warm-up is a phenomenon that has been noted in psychology (Heron, 1928; Ammons, 1947; Thune, 1950) and is associated with skilled behaviour. It was first mentioned with regards to taste by Pfaffmann (1954) and has been the subject of study (O'Mahony et al., 1988; Thieme and O'Mahony, 1990;

Dacremont *et al.*, 2000; Mata-Garcia *et al.*, 2007). For example, before a game of tennis, the players will warm up. They will hit the ball back and forth over the net, so as to fine-tune the messages travelling from the brain to the muscles, to ensure that the ball goes in the direction intended. In the same way, tasting is a skilled behaviour and so can also warm up. Old and experienced technicians used to say that when performing difference tests, it was not worth collecting data until two or three tests had already been completed. They were referring to the fact that over the first two or three tests, the judge warmed up. However, rather than waste tests like this, it is better to go through a warm-up procedure. This involves the judge tasting alternately the two stimuli to be discriminated, while knowing which is which. At first, the two stimuli will feel identical but after a few tastes, the difference will begin to appear. It would seem that warm-up is some form of signal search procedure. At first the brain scans the inputs from both products, which are pretty well identical. However, there are some slight differences and when the brain locates them, it focuses on these and attenuates the irrelevant input that is the same. The sensation that the judge experiences when this happens is that at first the two stimuli taste the same and then suddenly the difference between the stimuli appears. It is rather like an old valve radio. When the radio is switched on, the sound does not appear until valves warm up. Then, suddenly, the sound appears. In warm-up, at first, the two stimuli appear identical and then suddenly they appear different.

Once the judge has warmed up, it is a simple matter to ask him/her to describe the sensation difference between the two products. This description can then be used to indicate the target product for the 2-AFC. It does not matter what the description is as long as it has meaning for the judge in describing the difference between the products. The judge may say that one product is 'fresher', 'punchier' or 'crazier' than the other. Then the appropriate question for the 2-AFC would be to ask which product is 'fresher', 'punchier' or 'crazier' than the other. If the judge cannot describe the difference between the products, they could be designated 'X' and 'Y' and the judge asked in the 2-AFC to identify the 'X' product or the 'Y' product. If after a suitable amount of warm-up, the judges report that they still cannot 'find' the difference, then the testing starts regardless. Presumably the judges cannot tell the difference between the products or have not realized that they can tell the difference and proceed to get the tests correct despite believing they are only guessing.

Warm-up is a sensible routine procedure, which can be used for Sensory Evaluation I and certainly should be used in sensory psychophysics. It is merely a way of getting judges to focus on the relevant attributes. Of course, warm-up would not be suitable for Sensory Evaluation II. Nobody warms up before tasting a meal. It would also have the effect of making the consumer over-sensitive.

There are some other tricks that can be used for increasing sensitivity for Sensory Evaluation I. One of these is bilateral tasting. This involves placing a stimulus on one side of the mouth and dipping the tongue into it. Then, on swallowing or expectorating this stimulus, a second stimulus is immediately introduced into the other side of mouth and the tongue dipped into that. This technique is useful for stimuli with strong tastes or after tastes or stimuli that have a strong trigeminal effect. It has the advantage of using the most powerful 2-AFC without a lengthy rinse in between tasting, during which the memory of the stimulus would be distorted. In fact, sometimes the two stimuli can be tasted simultaneously. This was found possible with toothpaste, where one often has to wait approximately 20 minutes for the effects of the first toothpaste to wear off before assessing the second toothpaste. Naturally, this allows a golden opportunity for forgetting. In fact, it was found that the two toothpastes could be tested simultaneously, with each paste on a separate toothbrush, brushing separate sides of the mouth. As long as the teeth were slightly separated, a simultaneous judgement could be made of the relative irritation effect of each toothpaste. This was found to be highly efficient, because it utilised the more powerful 2-AFC and introduced no delay between tasting the two stimuli. In this way, two stimuli could be compared efficiently before having to wait the 20 minutes required before testing the next pair of stimuli. It effectively halved the time required for testing.

18.2.4   What is the best test to use for Sensory Evaluation II?

This question is not as easy to answer. Presumably the test should reproduce ordinary eating conditions as closely as possible. One obvious point is that a 2-AFC or 3-AFC should not be used, because during the ordinary eating experience, people do not have their attention called to concentrate on a particular attribute. So tests where consumers are allowed to choose which attributes they are to attend to, if any, are appropriate. The common duo–trio and triangle tests do not specify an attribute. If the consumer happens to attend to the attributes where differences occur, then they will notice the difference. If they do not attend to such attributes, they will not notice the difference. In an experiment in our laboratory, where consumers were performing difference tests between ice cream samples that varied in texture, the consumers did not notice any difference. But when the experimenter simply hinted at texture, nearly all consumers reported a difference. It would seem that texture is not an attribute that judges attend to, probably because as ice cream melts at room temperature, the texture keeps changing. So it is quite possible that consumers do not notice perceptible differences, because they do not attend to the appropriate attributes.

The triangle and duo–trio tests do not require judges to attend to a particular attribute, so they would be suitable for Sensory Evaluation II. However they are not very powerful. The same–different test, with a signal detection analysis (O'Mahony and Rousseau, 2002) is more powerful, although not as powerful as the 2-AFC or 3-AFC tests. However, its superior power to the triangle and duo–trio, make it an excellent candidate for Sensory Evaluation II. Besides its increased power and thus increased efficiency, it is also a more realistic test. In a real-life eating situation we might ask ourselves whether two foods are the same or different. We do not present ourselves with three foods and try to decide which two are the same and which one is different.

There is a problem with the same–different test, however. If data from several consumers are to be combined, the differences in the consumers' tau criteria add variance to the situation and this reduces d'. However, merely presenting the two products to be tested briefly beforehand, gives the consumers an idea of the scale of the difference to be detected, and allows them to adjust their tau criteria and so reduce the differences in their tau values. This reduction in the variance of tau increases d' (Rousseau et al., 1999). A mere single presentation is legitimate and realistic in this case. It would be quite usual for a consumer to taste two foods first and then taste them again before making up his or her mind. Similarly, re-tasting during a test, when desired, was also found to increase d' (Rousseau and O'Mahony, 2000).

After accepting the suitability of the same–different test, we might want to ask ourselves about the testing environment. A real-life environment might be difficult to reproduce. However, we try to get as close to reality as possible. A same different test with simultaneous presentation of the products will be the most sensitive. As the time between tasting the products increases, performance on the test becomes worse (Cubero et al., 1995; Avancini et al., 1999). This is true for inexperienced consumers. Judges who are very familiar with the products may not have their performance deteriorate as much, as the time interval between tasting the two products increases. This is because they are not comparing the two stimuli with each other; they are comparing the stimuli with familiar memory traces that are entrenched in their brain. Their cognitive strategy is different. The time between tasting then becomes less relevant, because the stimuli are not being compared with each other. This suggests that regular consumers, with their different cognitive strategies, need to be treated differently from occasional consumers.

These issues are important. It might be argued that if we were comparing two cereals with each other, these cereals may only be realistically compared at breakfast time on successive days, 24 hours apart. Presumably, to be realistic with a same–different test, it would need a 24 hour time interval between tasting the two stimuli. Yet, for regular users of the cereal with their different cognitive strategies, that is likely to render them more sensitive to differences, a shorter time between tasting the two stimuli might be more appropriate. These aspects of testing have hardly been researched. At the moment, to confront these problems, all we can do is use common sense and informed guesswork.

Some researchers base their Sensory Evaluation II decisions on Sensory Evaluation I testing. In other words, they used trained panellists to make decisions regarding consumers. Should trained panellists not be able to distinguish between two products, the reasonable assumption is made that less sensitive consumers will also not be able to do so. On the whole, this assumption is valid, although sometimes regular or long-term consumers can be more sensitive to differences than trained panellists. In this case, the long-term consumers should have been recruited as Sensory Evaluation I panellists. Yet, if the trained panellists can distinguish between two foods, the assumption is often made that untrained consumers might also be able to do this. This is a safe and conservative assumption and prevents foods that have altered sensory attributes from being sent to market. Yet, by following this strategy, some useful, cost-saving reductions might be missed. It often occurs that when trained panellists can tell the difference, untrained consumers cannot. So in this case, separate Sensory Evaluation II difference testing with consumers would seem a suitable addition to difference testing by trained panellists. Because the d' of trained panellists will be bigger than the d' of the consumers, it is possible to set up a scheme for measuring both and then use this to predict the consumers' d' values from the panellists' d' values (O'Mahony and Rousseau, 2002; Ishii *et al.*, 2007). This technique has the potential to save a lot of time and money by avoiding the need to have to keep testing consumers.

18.3   Preference tests

18.3.1   Paired preference with distinguishable products
Preference tests come under the general heading of consumer preference and acceptance testing. The logical way to present a preference test is as a paired preference test. Also, it is sensible when asking a preference question, to ensure that the consumer can distinguish between the two products. Logically, if the consumer cannot distinguish between the two products, he/she cannot have a realistic preference. The experimenter runs grave risks if he/she presents two confusable stimuli to the consumer and asks for a preference. The response of the consumer is more likely to be a result of response bias than of a real preference.

18.3.2   Testing for significance and the 'no preference' option
With difference tests, each judge performs several difference tests. If the consumer can discriminate between the two stimuli, the responses will be consistent; if not they will be inconsistent. Thus, significance is determined by repetition. If the person cannot tell the difference, his/her neutral (don't know, no difference) response is determined by repetition. In fact, a neutral 'don't know' or 'no difference' response is not allowed; consumers are forced to choose. Besides stabilising the beta criterion, forcing the consumer to choose solves the problem of consumers who may think that they cannot tell the difference when they can. People are not always good at judging how well they can discriminate. One judge in our laboratory protested that he was only guessing, but performed 20 2-AFC tests without error. The judge in question is now convinced that he could discriminate.

With a preference test, is not feasible to determine significance by repetition. This is because it is not feasible to keep repeating a preference test, if the products can be discriminated. Once your consumer has told you that he prefers the brown ice cream with the chocolate flavour to the pink one with strawberry flavour, it would be rather pointless to ask him the same question over and over again: he can simply repeat his answer, without ever tasting the ice creams. Therefore, in this case, we cannot use the strategy we use with difference tests. We cannot determine the neutral (no preference) response by repetition.

So, how do we determine the neutral (no preference) response? We simply give the consumer the option of responding with 'no preference'. We now have three response options, two indicating preference (prefer A, prefer B) and one indicating no preference. There is a further argument for the 'no preference' option. Without it, it can be difficult to interpret the data. If half the consumers were to report a preference for product 'A' and half report a preference for product 'B', no clear conclusion could be drawn if there were not a 'no preference' option. It would be possible that all the consumers had had no preference and they had chosen the products randomly. Yet, it would also be possible that half the consumers actually had preferred product 'A', while the other half had preferred product 'B'. The action to be taken by a food company would be very different in each case. The 'no preference' option avoids this problem.

Yet, there is still opposition to the 'no preference' option, because with three categories it is no longer possible to use a simple binomial statistical analysis. Because many researchers wish to use the simple binomial analysis, which they use for their difference tests, they have invented various ways of getting rid of any 'no preference' responses.

These have been reviewed but not necessarily recommended, in various texts (Odesky, 1967; Lawless and Heymann, 1998; Resurreccion, 1998; Marchisano et al., 2003; Angulo and O'Mahony, 2005). One recommendation is to simply ignore any 'no preference' responses. Another is to split them equally between the two preference options or split them proportionately between the preference options according to the respective preference frequencies. A more worrying option is to assign the 'no preference' responses to one of the two preference options, by simply tossing a coin (Falk et al., 1975). Any of these procedures might be relatively harmless in terms of the final conclusions drawn from the testing if there were only a negligible amount of 'no preference' responses. Yet, this is not always the case (Alfaro-Rodriguez et al., 2005; Angulo and O'Mahony, 2005). These recommendations all refer to ways of discarding or altering data to try to fit what are multinomial responses to a binomial statistical analysis. Surely it would be preferable to use a multinomial analysis for multinomial data.

There are several analyses available. A signal detection/Thurstonian analysis (Green and Swets, 1966; Alfaro-Rodriguez et al., 2005) is one possibility. Yet, if the experimenter did not wish to present the data in terms of d', there are various other analyses based on approaches such as maximum likelihood or Bayesian statistics (Bradley and Terry, 1952; Rao and Kupper, 1967; Draper et al., 1969). However, a simple chi-squared analysis is appropriate and has been investigated (Marchisano et al., 2003).

18.3.3   The chi-squared analysis and 'placebo' samples

The chi-squared analysis is not without its difficulties. There was a problem with deciding the expected frequencies. These are traditionally generated using a null hypothesis. Yet, to test observed frequencies against expected frequencies generated with a null hypothesis would be to test whether the data differed from a case where the responses were equally distributed over the three response options (prefer A, prefer B, no preference). Such an equal distribution would represent a situation where a third of the consumers had no preference, a third preferred 'A' and a third preferred 'B'. It might also represent a situation where consumers simply ignored the products, closed their eyes and randomly marked one of the three categories on the response sheets. Neither of these is equivalent to the case where consumers tasted the products and decided that they had no preference. Logically, the expected frequencies for the case where the consumers had no preference would be 100% 'no preference' responses and zero 'preference' responses. Yet, we have to realise that we are dealing with people and people do not always respond logically. It is very likely that they would have some response bias; the demand characteristics of the test might induce some preference responses from judges who had, in fact, no preference.

Ennis reports that several years ago, he mailed two cigarettes to a large number of consumers' homes for assessment on a variety of attributes (D.M. Ennis, personal communication). Consumers were also asked for their preferences and 40% reported preference for one of the cigarettes, 20% reported that they had no preference and 40% reported preference for the other cigarette. Yet, one cigarette had been taken from the initial part and the other from the final part of the same production run. They were essentially the same cigarette. Therefore, any preference expressed for one or other of the cigarettes would have been due to factors other than their sensory characteristics, such as response bias elicited by the demand characteristics of the test. Such a pattern of preference responses could thus be treated as representing the expected frequencies in the 'no preference' condition. Accordingly, any pattern of responses for products of different flavours could be treated as the corresponding observed frequencies, to be tested for significance against these 'no preference' (40–20–40) frequencies. The approach is analogous to using a placebo in medical research.

From Ennis's data, it would seem that 40–20–40 would be the appropriate 'expected frequency' distribution or 'placebo' for the case where consumers exhibited no preference. Yet, Marchisano et al. (2003) found that the frequencies varied with the products being tested, the types of responses allowed and the types of consumers tested. So it would seem that the expected frequencies would need to be determined for each specific testing situation. Accordingly, when they tested preference between two products with different flavours, using the 'no preference' option, they used these data as the observed frequencies for their chi-squared computation. For the expected frequencies, they took a separate sample of consumers from the same population and tested two putatively identical products from the same product niche, treating them in exactly the same manner as the two different flavoured products that they had tested. In fact, these response frequencies could be used as the expected frequencies for a whole range of preference tests, for products in that product niche. This approach has been used by a manufacturer of snack foods (Foley *et al.*, 2003).

This approach of Marchisano *et al.* (2003) made the assumption that the consumers who tasted the identical products, sampled from the same population, would provide an adequate 'placebo' test for the test sample of consumers who tested different products. The assumption was that the 'placebo' sample of consumers matched the test sample. This might be true for very large samples. Yet, it would be more logical to ensure a proper matching by using the same consumers for both the 'placebo' sample and the test sample. In other words, each consumer would test a pair of identical samples to obtain the expected frequencies for chi-squared and also test a pair of different samples to obtain the observed frequencies.

This approach is feasible, yet it raises some questions regarding methodology. Does it make any difference whether the identical placebo pair is presented first or whether it is presented second? It also opens up a simple additional approach to the statistical analysis. Instead of comparing the test frequencies with the placebo frequencies, it would be simpler merely to examine the test frequencies for those consumers, who had exhibited no response bias (reported no preference) in the placebo condition. Then, only preferences of unbiased consumers would be considered. Yet, the question remains how representative such consumers might be. Also, the sample size might be small. At the time of writing, these methodological issues are under investigation. The early results (Alfaro-Rodriguez *et al.*, 2007) indicate that it makes no difference whether the identical stimulus 'placebo test' is presented first or second. Yet, more research is needed before any definitive statements can be made.

Finally, there is the question of how to analyse the data. The result of the test is simply the proportion of consumers who prefer each product (prefer X, prefer Y) and the proportion who have no preference. This is what the test is designed to establish and what should be used for the appropriate business decisions. Tests of significance (chi-squared) indicate whether the data (observed frequencies) represent a real (significant) preference or 'no preference' trend, based on judgements of the sensory characteristics of the foods. The alternative (expected frequencies) is that consumers are acting as though the products were identical and thus have no preference and accordingly choose preference options because of response bias.

A measure of the overall strength of preference could be made in terms of d'. This would give a measure of how strongly the consumers tended to lean in the direction of preferring product X or preferring product Y. This can be a useful measure when comparing results between studies. Sometimes it is not obvious from an inspection of the various response frequencies whether one study shows a stronger preference for a particular product than another. An overall measure of preference such as d' can be useful here. Yet, it is important to stress that a d' analysis should not replace the basic analysis of the various frequencies of response (prefer X, prefer Y, no preference). To do this could lead to ambiguities. For example, it is perfectly possible to get a d' value of zero when everybody had a 'no preference' response or when preferences were divided equally between products 'X' and 'Y'. So although an overall measure of preference can be useful, it is not sufficient.

18.3.4   Ranking

The paired preference test is not the only type of preference measure. An alternative, when there are more than two products to compare, is simple ranking. A consumer may be presented with two foods and asked which one is preferred (or if there is no preference). With gentle coaxing, they might be persuaded to give a reason for their preference, although it is very important not to 'put words into his/her mouth'. The consumer may then be presented with a third product and asked where they would place that. Maybe they would put it in first place, last place or in the middle. Quite often, at this point, the consumer will spontaneously give a reason for their choice. A fourth or fifth product can be introduced in the same way. From a sample of consumers, a mean ranking is easy to compute. Because mean ranks have meaning only within the context of a particular experiment, the spacing between the ranks could be represented by R-index values instead (Pipatsattayanuwong et al., 2001). Thus, in this way, the rank order and the spacing between the ranks could be obtained, which is exactly what is obtained from using the mean values from a hedonic scaling procedure. The difference is that the spacing between the ranks is obtained from consumers' behaviour, while the spacing from a scaling procedure is obtained from the consumers' not very skilled numerical estimations.

18.4   How do we interact with the consumers?

If consumers are coming to some central location, the question becomes one of how to elicit their responses. One method is to sit the consumer down with the appropriate products and a response sheet. The consumer may be left in a quiet place so as to avoid distraction. Every group of researchers has its own set of techniques. In the author's laboratory we favour face-to-face testing. This comes from our background in psychophysical measurement. The consumer is made comfortable and good rapport is established with the experimenter. Whether the test is a difference test or a preference test, care is taken to make sure that the consumer knows exactly what is required. Care is taken, especially with a preference test, to make sure that the consumer understands that the experimenter is completely neutral; he has no particular interest in either product and does not care what the consumer's opinion might be. With a difference test, care is taken to ensure that the consumer does not feel that it is a test of their skill; they are merely helping out with a research project.

When the consumer is quite sure of the task, the testing begins. It may be a difference test or preference test from which numerical data are collected.

However, the consumer is encouraged to 'think aloud' so that qualitative data can be collected simultaneously. Comments are encouraged. But great care is taken to avoid formal questions that demand answers. The goal, here, is to avoid any suggestion effects or forced responses. Consumers can have a preference but be unable to describe why. They are good at saying what they like but are not so good at explaining why. Therefore, any formal questioning regarding why one product was chosen over another could result in ad hoc responses, simply made up for the occasion. They would carry little insight into the decision processes used during purchasing behaviour. Any questions have to be posed in such a way that if the consumer cannot express a reason for their choice, they need not respond. However, with good rapport, the consumer very often comes out with a whole string of spontaneous comments. Useful qualitative data are obtained while avoiding suggestion effects.

Preference or difference tests that involve the consumer having to respond to a whole set of attribute questions are worrying. The consumer will respond to them even if they refer to attributes that the consumer has never even noticed or does not feel are important. A lot of a relevant or even misleading data can be obtained in this way. Even consumer descriptions must be treated with great care. It must be borne in mind that the descriptive language used by consumers is the descriptive language of an untrained judge.

It should be remembered that suggestion is a very strong affect. Consumers will happily describe their reactions to and give ratings for attributes that are obviously not present in the products. The present author has even used suggestion to persuade television viewers that he could transmit smells over television. Some of the viewers thought they could smell new-mown hay. Some of these even had attacks of hay fever. Suggestion should not be underestimated. Because of this, the experimenter must be well trained so that nothing in his behaviour, language or even body language inadvertently suggests that the consumer should give a particular response. This is an occupation for well-trained and experienced experimenters.

Yet, testing at a central location would seem logically not as realistic as 'take home' testing. This is usually conducted in a serial monadic manner. A consumer is given a product to take home and use for a period of time. At the end of the period the consumer gives it some sort of rating. (It is often at this point that a battery of highly suggestive attribute questions are given.) On returning the first product, the consumer is given a second product to take home and assess and after this, a third product, and so on. It might be asked whether the consumer may have forgotten the exact details of the first product by the time they are testing the third or fourth product. An alternative procedure would be to give the consumer all the products at once and allow them to cross-compare them in their own time, until they have made a decision. Again a simple ranking might be sufficient. Whether this is a better approach is a question yet to be researched.

The disadvantage of 'take-home' tests is lack of control. Therefore it is important to make sure that the consumer is fully briefed and understands any special procedures that need to be followed. For example, some products cannot be tested on successive days; an interval of a day or so might be necessary between testing. Yet, the technique is important because the results of long-term use do not always agree with those of instant judgements at some central location.

Finally, the author is enjoying a particular testing situation, in his laboratory, at the present time. Difference and preference tests are being conducted on a set of products by a team that consists of an experienced sensory researcher, with assistance from the product developer along with his marketing person. This means that in their back-up role, the developer and marketer fully understand the nature of the test situation, witness the reactions of the consumers and help with the data analysis; this is far more convincing than any written report.

18.5   Sources of further information and advice

Recommended further reading on this topic is referenced in the chapter. Some of these references can be used for introductions to some of the topics discussed above.

For more discussion on the differences between consumer preference and acceptance testing, Sensory Evaluation I and Sensory Evaluation II, try:
O'Mahony, M. (1995) *Food Technology*, **29**, 72–82.

For a painless introduction to Thurstonian modelling and response bias, try:
Lee, H.S. & O'Mahony, M. (2004) *Food Science and Biotechnology*, **13**, 841–847.
or
O'Mahony, M., Masuoka, S. & Ishii, R. (1994) *Journal of Sensory Studies*, **9**, 247–272.

For a little more on Thurstonian modelling, try:
O'Mahony, M. & Rousseau, B. (2002) *Food Quality and Preference*, **14**, 157–164.
or
Ennis, D.M. (1993) *Journal of Sensory Studies*, **8**, 353–370.

For a recent look at preference testing, try:
Alfaro-Rodriguez, H., Angulo, O. & O'Mahony, M. (2007) *Food Quality and Preference*, **18**, 353–361.

For an example of preference by ranking, try:
Pipatsattayanuwong, S., Lee, H-S., Lau, S. & O'Mahony, M. (2001). *Journal of Sensory Studies*, **16**, 517–536.

Most introductory statistics texts will indicate methods for statistical analysis by binomial statistics or chi-squared. Concentrate on the books that simply describe how to analyse the data. Binomial tests do not usually require software. Chi-squared software appears in most introductory statistical packages.

For d' computations, the beta binomial and computations associated with Thurstonian modeling and d', the IFPrograms software package obtainable from short courses given by the Institute for Perception, Richmond, Virginia, USA, is the only comprehensive package available at the time of writing. Their website is www.ifpress.com

18.6   References

ALFARO-RODRIGUEZ, H., O'MAHONY, M. & ANGULO, O. (2005). Paired preference tests: d' values from Mexican consumers with various response options. *Journal of Sensory Studies*, **20**, 275–281.

ALFARO-RODRIGUEZ, H., O'MAHONY, M. & ANGULO, O. (2007). Be your own placebo: a double paired preference test approach for establishing expected frequencies. *Journal of Sensory Studies*, **18**, 190–195.

AMMONS, R. B. (1947). Acquisition of motor skill: I. Quantitative analysis and theoretical formulation. *Psychological Review*, **54**, 263–281.

ANGULO, O. & O'MAHONY, M. (2005). The paired preference test and the 'No Preference' option: was Odesky correct? *Food Quality and Preference*, **16**, 425–434.

ASTM (1968). *Manual on Sensory Testing Methods STP 434.* American Society for Testing and Materials, Philadelphia.

AVANCINI DE ALMEIDA, T. C., CUBERO E. & O'MAHONY, M. (1999). Same-different discrimination tests with interstimulus delays up to one day. *Journal of Sensory Studies*, **14**, 1–18.

BI, J. & ENNIS, D. M. (1998). A Thurstonian variant of the beta-binomial model for replicated difference tests. *Journal of Sensory Studies*, **13**, 461–466.

BI, J. & ENNIS, D. M. (1999a). The power of sensory discrimination methods used in replicated difference and preference tests. *Journal of Sensory Studies*, **14**, 289–302.

BI, J. & ENNIS, D. M. (1999b). Beta-binomial tables for replicated difference and preference tests. *Journal of Sensory Studies*, **14**, 347–368.

BI, J., TEMPLETON-JANIK, L., ENNIS, J. M. & ENNIS, D. M. (2000). Replicated difference and preference tests: how to account for inter-trial variation. *Food Quality and Preference*, **11**, 269–273.

BRADLEY, R. A. & TERRY, M. E. (1952). Rank analysis of incomplete block designs I. The method of paired comparisons. *Biometrika*, **39**, 324–345.

BROCKHOFF, P. B. (2003). The statistical power of replications in difference tests. *Food Quality and Preference*, **14**, 405–417.

BROCKHOFF, P. B. & SCHLICH, P. (1998). Handling replications in discrimination tests. *Food Quality and Preference*, **9**, 303–312.

BROWN, J. (1974). Recognition assessed by rating and ranking. *British Journal of Psychology*, **65**, 13–22.

CUBERO, E., AVANCINI DE ALMEIDA, T. C. & O'MAHONY, M. (1995). Cognitive aspects of difference testing: memory and interstimulus delay. *Journal of Sensory Studies*, **10**, 307–324.

DACREMONT, C., SAUVAGEOT, F. & DUYEN, T. H. (2000). Effect of assessor's expertise level on efficiency of warm-up for triangle tests. *Journal of Sensory Studies*, **15**, 151–162.

DESSIRIER, J-M. & O'MAHONY, M. (1999) Comparison of d' values for the 2-AFC (paired comparison) and 3-AFC discrimination methods, sequential sensitivity analysis and power. *Food Quality and Preference*, **10**, 51–58.

DRAPER, N. R., HUNTER, W. G. & TIERNEY, D. E. (1969). Which product is better? *Technometrics*, **11**, 309–320.

ELLIOT, P. B. (1964). 'Tables of *d′*', in Swets, J. A., *Signal Detection and Recognition by Human Observers*, New York, Wiley, 651–684.

ENNIS, D. M. (1990). Relative power of difference testing methods in sensory evaluation. *Food Technology*, **44**, 114, 116, 117.

ENNIS, D. M. (1993). The power of sensory discrimination methods. *Journal of Sensory Studies*, **8**, 353–370.

ENNIS, D. M. & BI, J. (1998). The beta-binomial model: accounting for inter-trial variation in replicated difference and preference tests. *Journal of Sensory Studies*, **13**, 389–412.

FALK, S. N., HENRICKSON, R. L. & MORRISON, R. D. (1975). Effect of boning beef carcasses prior to chilling on meat tenderness. *Journal of Food Science*, **40**, 1075–1079.

FOLEY, M., WILLIAMS, A., BADE, J., LANCASTER, B., POPPER, R. & CARR, B. T. (2003). Effect of preference-question format with and without sample differences. Abstracts of The 5th Pangborn Sensory Science Symposium, July 2003, Boston, Massachusetts, Abstract 043.

GREEN, D. M. & SWETS, J. A. (1966). *Signal Detection Theory and Psychophysics*. John Wiley, New York.

HARRIES, J. M. & SMITH, G. L. (1982). The two-factor triangle test. *Journal of Food Technology*, **17**, 153–162.

HERON, W. T. (1928). The warming-up effect in learning nonsense syllables. *Journal of Genetic Psychology*, **35**, 219–228.

ISHII, R., VIÉ, A. & O'MAHONY, M. (1992). Sensory difference testing: ranking *R*-indices are greater than rating *R*-indices. *Journal of Sensory Studies*, **7**, 57–61.

ISHII, R., KAWAGUCHI, H., O'MAHONY, M. & ROUSSEAU, B. (2007). Relating consumer and trained panels' discriminative sensitivities using vanilla ice cream as a medium. *Food Quality and Preference*, **18**, 89–96.

KUNERT, J. (2001). On repeated difference testing. *Food Quality and Preference*, **12**, 385–391

KUNERT, J. & MEYNERS, M. (1999). On the triangle test with replications. *Food Quality and Preference*, **10**, 477–482.

LAU, S., O'MAHONY, M. & ROUSSEAU, B. (2004). Are three-sample tasks less sensitive than two-sample tasks? Memory effects in the testing of taste discrimination. *Perception and Psychophysics*, **66**, 464–474.

LAWLESS, H. T. & HEYMANN, H. (1998). *Sensory Evaluation of Food Principles and Practices*. Chapman and Hall, New York.

LEE, H-S. & O'MAHONY, M. (2004). Sensory difference testing: Thurstonian models. *Food Science and Biotechnology*, **13**, 841–847.

MACMILLAN, N. A. & CREELMAN, C. D. (1991). *Detection Theory: A User's Guide*. Cambridge University Press, New York.

MARCHISANO, C., LIM, J., CHAO, H. S., SUH, D. S., JEON, S. Y., KIM, K. O. & O'MAHONY, M. (2003). Consumers report preferences when they should not: A cross-cultural study. *Journal of Sensory Studies*, **18**, 487–516.

MATA-GARCIA, M., ANGULO, O. & O'MAHONY, M. (2007). On warm-up. *Journal of Sensory Studies*, in press.

ODESKY, S. H. (1967). Handling the neutral vote in paired comparison product testing. *Journal of Marketing Research*, **4**, 199–201.

O'MAHONY, M. (1992). Understanding discrimination tests: a user-friendly treatment of response bias, rating and ranking *R*-index tests and their relationship to signal detection. *Journal of Sensory Studies*, **7**, 1–47.

O'MAHONY, M. (1995a). Sensory measurement in food science: fitting methods to goals. *Food Technology*, **29**, 72–82.

O'MAHONY, M. (1995b). Who told you the triangle test was simple? *Food Quality and Preference*, **6**, 227–238.

O'MAHONY, M. & GOLDSTEIN, L. (1987). Tasting successive salt and water stimuli: the roles of adaptation, variability in physical signal strength, learning, supra- and subadapting signal detectability. *Perception*, **12**, 425–426.

O'MAHONY, M. & ODBERT, N. (1985). A comparison of sensory difference testing procedures: sequential sensitivity analysis and aspects of taste adaptation. *Journal of Food Science*, **50**, 1055–1058.

O'MAHONY, M. & ROUSSEAU, B. (2002). Discrimination testing: a few ideas, old and new. *Food Quality and Preference*, **14**, 157–164.

O'MAHONY, M., GARSKE, S. & KLAPMAN, K. (1980). Rating and ranking procedures for short-cut signal detection multiple difference tests. *Journal of Food Science*, **45**, 392–393.

O'MAHONY, M., THIEME, U. & GOLDSTEIN, L. R. (1988). The warm-up effect as a measure of increasing the discriminability of sensory difference tests. *Journal of Food Science*, **53**, 1848–1850.

O'MAHONY, M., MASUOKA, S. & ISHII, R. (1994). A theoretical note on difference tests: models, paradoxes and cognitive strategies. *Journal of Sensory Studies*, **9**, 247–272.

PERYAM, D. R. (1958). Sensory difference tests. *Food Technology*, **12**, 231–236.

PFAFFMANN, C. (1954). Variables affecting difference tests. In *Food Acceptance Testing Methodology, A Symposium* (D. R. Peryam, F. J. Pilgrim and M. S. Peterson, eds.), pp. 4–20, National Academy of Sciences/National Research Council, Washington, DC.

PIPATSATTAYANUWONG, S., LEE, H-S., LAU, S. & O'MAHONY, M. (2001). Hedonic *R*-index measurement of temperature preferences for drinking black coffee. *Journal of Sensory Studies*, **16**, 517–536.

RAO, P. V. & KUPPER, L. L. (1967). Ties in paired-comparison experiments: a generalization of the Bradley-Terry model. *Journal of American Statistical Association*, **62**, 194–204.

RESURRECCION, A. V. A. (1998). *Consumer Sensory Testing for Product Development*, Aspen Publ. Inc., Gaithersburg, MD.

ROUSSEAU, B. & O'MAHONY, M. (1997). Sensory difference tests: Thurstonian and SSA predictions for vanilla flavored yogurts. *Journal of Sensory Studies*, **12**, 127–146.

ROUSSEAU, B. & O'MAHONY, M. (2000). Investigation of the effect of within-trial retasting and comparison of the dual–pair, same–different and triangle paradigms. *Food Quality and Preference*, **11**, 457–464.

ROUSSEAU, B., MEYER, A. & O'MAHONY, M. (1998). Power and sensitivity of the same–different test: comparison with triangle and duo–trio methods. *Journal of Sensory Studies*, **13**, 149–173.

ROUSSEAU, B., ROGEAUX, M. & O'MAHONY, M. (1999). Mustard discrimination by same–different and triangle tests: aspects of irritation and τ criteria. *Food Quality and Preference*, **10**, 173–184.

TEDJA, S., NONAKA, R., ENNIS, D. M. & O'MAHONY, M. (1994). Triadic discrimination testing: refinement of Thurstonian and sequential sensitivity analysis approaches. *Chemical Senses*, **19**, 279–301.

THIEME, U. & O'MAHONY, M. (1990). Modifications to sensory difference test protocols: the warmed up paired comparison, the single standard duo–trio and the A–Not A test modified for response bias. *Journal of Sensory Studies*, **5**, 159–176.

THUNE, L. E. (1950). The effect of different types of preliminary activities on subsequent learning of paired associate material. *Journal of Experimental Psychology*, **40**, 423–438.

VIÉ, A. & O'MAHONY, M. (1989). Triangular difference testing: refinements to sequential sensitivity analysis for predictions for individual triads. *Journal of Sensory Studies*, **4**, 87–103.

19

Thurstonian probabilistic approaches to new food product development

J. F. Delwiche, The Ohio State University, USA

19.1 Introduction

This chapter aims to give the reader insight into probabilistic approaches to new product development. It begins with an explanation of probabilistic, or Thurstonian, modeling of perception, without which it is impossible to understand the utility of these approaches. The remainder of the chapter is devoted to specific models currently utilized in certain sectors for new product development. These models are described, and their applications, advantages, and disadvantages are discussed.

19.1.1 Probabilistic vs deterministic views of the world

As Gertrude Stein once said, 'A rose is a rose is a rose.' This sense of constancy is how we typically view the world. We expect that if we look at a rose ten times, it will look the same intensity of red every time. If we sniff the rose ten times, we expect it to smell as 'sweet' each time. And if we stroke the rose with our fingers ten times, we expect it to feel as smooth each time. This is the deterministic view of the world. It assumes that if the stimulus remains constant, then so will the perceived sensations that arise from it. Such a view is typically used in new product development and in many of the models utilized by product developers. Thurstonian models do not assume this same constancy of perception. These probabilistic models assume that even when the physical stimulus remains constant, the perceived intensity of it does not. They assume that every time a physical stimulus is experienced, the perceived intensity will vary slightly. Sometimes the rose will smell a little 'sweeter,' sometimes a little less.

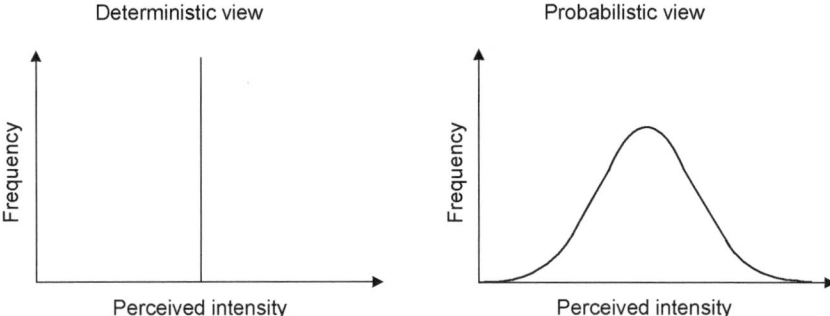

Fig. 19.1   Deterministic and probabilistic models differ in how they assume stimulus intensity will be perceived with repeated sampling of a stimulus.

These models assume that most of the time a physical stimulus will have a particular perceived intensity, and the more you deviate from this mean perceived intensity, the less often these instances of lower or higher perceived intensities will occur. It is assumed that if frequency of occurrence is plotted against perceived intensity, it will follow a normal distribution as in Fig. 19.1 – which is why Thurstonian models are known as probabilistic models. These models make no attempt to explain the source of the variation, but simply assume it exists. Variation in perceived intensity could arise from minor changes in sniffing, changes in attention, extraneous stimuli such as noises, variation in neural noise of the system, or any combination of these and other factors.

This seemingly small difference between these two approaches in the assumption of how a physical stimulus is perceived with repeated sampling has large implications for product development models. This can be illustrated by considering a simple case, such as comparison of two similar cheese crackers, A and B. Perhaps we are comparing our product (A) with a competitor's (B), or investigating the impact of a processing change, or are considering altering suppliers for a high-impact ingredient. Let's assume that cracker B is higher in overall intensity than cracker A. With a deterministic view, if a judge tastes each cracker type, he will always perceive cracker B intensity as stronger than cracker A, as represented in Fig. 19.2.

With the probabilistic view, each stimulus is represented as a separate intensity-frequency distribution. If cracker B was obviously stronger than cracker A, then the perceived intensity of the products can be represented as two non-overlapping curves, as in Fig. 19.3. When products are completely distinguishable, i.e. the curves do not overlap, then traditional Thurstonian modeling breaks down. Traditional Thurstonian modeling is effective only with confusable stimuli, which is typically represented by two overlapping curves, as in Fig. 19.4. The greater the intensity similarities, the greater the overlap between the curves will be.

As can be anticipated from the overlap of the intensity distributions, with a probabilistic view it is not expected that that cracker B will be perceived as stronger every time a judge tastes each cracker type (see Fig. 19.5). In fact, the

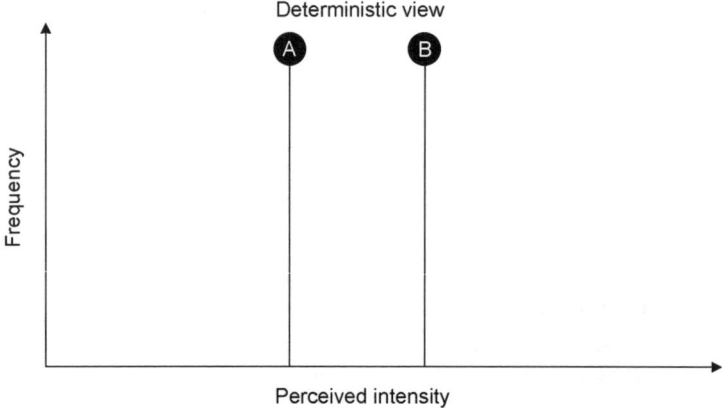

Fig. 19.2   With a deterministic view, product B is always perceived as stronger than product A.

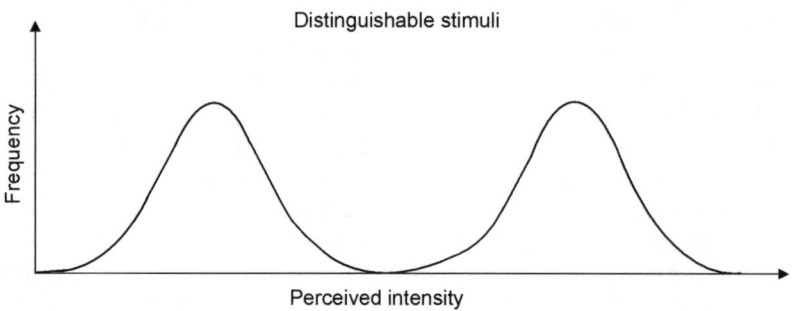

Fig. 19.3   If Product B is clearly different from Product A, the product intensities can be represented as two non-overlapping distributions.

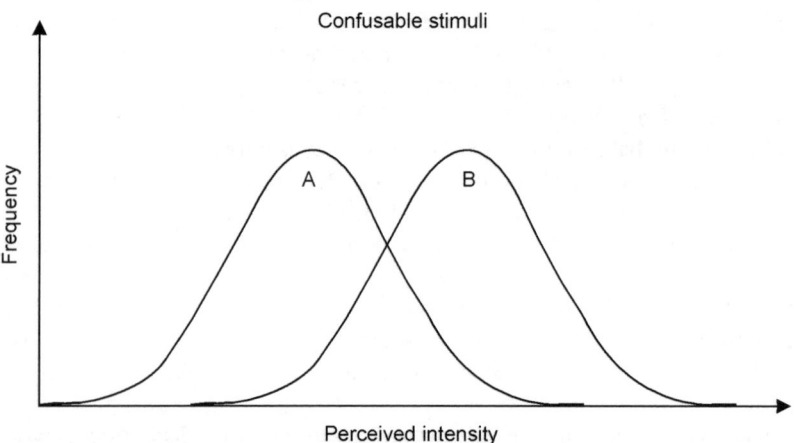

Fig. 19.4   If Products A and B are similar enough in intensity to be confusable, the product intensities can be represented by two overlapping curves.

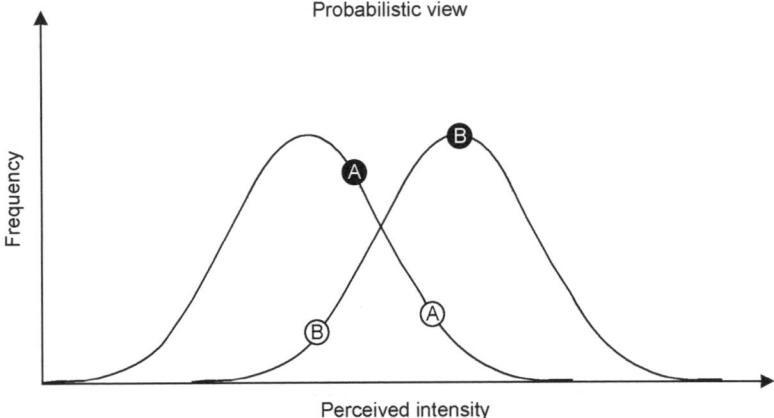

Fig. 19.5   With a probabilistic view, product B is usually perceived as stronger than product A (represented by black circles), but there are times when product A will be perceived as more intense than product B (represented by white circles).

probabilistic model predicts that there are occasions where cracker A will be perceived as stronger than cracker B. The Thurstonian model assumes that each time cracker A is tasted, the intensity perceived by the judge will be somewhere along the intensity range of the cracker A distribution, with the momentary perception of the product being represented by a circle on the intensity curve. Most of the time, the perceived intensity will be near the mean (as represented by the black circle on the cracker A distribution), but there are less frequent occasions where the perceived intensity will be higher (represented by the white circle on the cracker A distribution) or lower than this mean value (not shown). If the momentary intensity perceptions of crackers A and B are both near the mean of their respective intensity distributions, then cracker B will be perceived as stronger than cracker A (as represented by the black circles in Fig. 19.5). The white circles in Fig. 19.5 represent an occasion where the momentary intensity of cracker A is higher than the mean and the momentary intensity of cracker B is lower than the mean – an occasion where cracker A would be perceived as more intense than cracker B.

Anyone who has ever asked judges to assess products knows that they rarely give exactly the same rating to the same product. The deterministic view attributes these fluctuations in assessments entirely to a judge's responses to a constant physical signal; the probabilistic view attributes these fluctuations to both variations in the input the judge receives and variation in the judge's responses, as well as a myriad of other sources that add variation to the perception of the stimulus. Probabilistic models, while more complicated, are capable of explaining results that deterministic models cannot, such as the paradox of discriminatory non-discriminators (Byer, 1953; Frijters, 1981; Geelhoed *et al.*, 1994; Stillman, 1993). Another advantage of Thurstonian models is that these models can estimate the perceptual distance between products whether products are assessed by rating procedures or discrimination procedures.

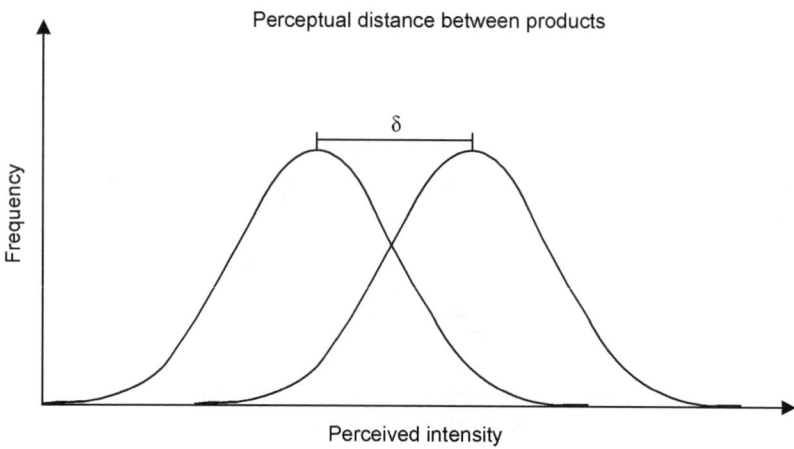

Perceptual distance between products

Fig. 19.6 The distance between the means measured in standard deviations is known as d' or δ.

19.1.2 Delta and d' – measuring the difference between products

When judges are asked to assess products, the actual population means (μ) of these products are unknown and are estimated from the judges' responses. Turning back to the cheese crackers, if the overall intensity of cracker A and cracker B are assessed by ratings, deterministic models can estimate the distance between products by determining the mean ratings of each cracker and taking the difference. This difference is measured in the same units as the rating scale. If the crackers are compared by a discrimination test, deterministic models can estimate the distance between products only by the number of correct responses. This makes comparison across methods extremely difficult.

Probabilistic models estimate the distance between products in a somewhat different manner. Thurstonian modeling was initially adopted in sensory evaluation in the assessment of discrimination procedure data. Specifically, it was used to quantify the distance between products in terms of delta (δ) or d'. Delta (or d') is the distance between the means of two product intensity distributions in terms of standard deviation (Fig. 19.6). The distance between the population means (μ_1 and μ_2) is δ, while the distance between sampling means ($\bar{X}_1$ and $\bar{X}_2$) is d'. Probabilistic models can also estimate δ from ratings data, making comparison of product differences determined with various methods more plausible. Many of the probabilistic methods that may be helpful in new product development use δ to quantify product differences.

19.2 Probabilistic models

With this background in mind, we can now discuss Thurstonian and probabilistic models that can be utilized for new product development. This chapter will discuss three such models: (1) Thurstonian ideal point modeling of just-about-right data, (2) probabilistic multidimensional scaling, and (3) landscape segmentation analysis. A fourth such model, the beta-binomial, was discussed in Chapter 18.

19.2.1   Thurstonian ideal point modeling

Thurstonian ideal point (TIP) modeling is one of the simplest Thurstonian models used in product development. It is an alternative way to analyze data obtained from just (about) right or 'JAR' scales (Lawless and Heymann, 1998), which are often used to help guide product developers in fine-tuning their products. As mentioned in Chapter 17, there is a multitude of ways JAR scale data can be analyzed (for additional discussion, see the Pangborn Sensory Science Symposium 2005 Workshop summary (Anon, 2005)). Since it is based upon JAR data, it is subject to all the same constraints of such data (for additional discussion, see Lawless and Heymann, 1998). As with all analyses, the insights provided to the development of new products is dependent upon the quality of the data collected. There is no analysis that can make up for data collected from the wrong subjects, poorly understood or missing attributes, or an insufficient sample size. The only fairly unique requirement of TIP modeling is that the data be categorical; thus to minimize data distortion, it is recommended that the scale used to collect ratings be categorical.

TIP modeling allows one to compare the JAR ratings of multiple products to a theoretical ideal product. Whereas most models are deterministic and conceptualize an ideal product as always receiving a rating of 'just right,' TIP modeling conceptualizes the ideal product as a probabilistic distribution centered on the 'just right' value. TIP modeling compares the probabilistic distribution of each product against the probabilistic distribution of the ideal (which is estimated from the subject responses to the actual products). Figure 19.7 illustrates a probabilistic distribution of a theoretical product with the ideal intensity and a product with a mean at 'too much;' both distributions have the same variance. Chi-square analysis comparing these distributions reveals that the two distributions shown below are significantly different (alpha = 0.001) if the proportions are observed for as few as ten cases.

TIP modeling is also very effective at demonstrating that a product could have a mean 'just right' rating, and but not be ideal because of 'heavy tails' (as in Fig. 19.8). Comparison of the ideal distribution to that with heavy tails with chi-square analysis reveals a significant difference between the two distributions, although in this instance the sample size must be at least 65 to obtain an alpha level of 0.001.

At the time of writing, the only practical way to do this sort of modeling is by the use of IFPrograms[TM] software, and the following details about output are specific to this program. It is likely that any future software programs that perform TIP modeling will include the same information. When IFPrograms[TM] evaluates JAR scale data with TIP modeling, multiple products are compared

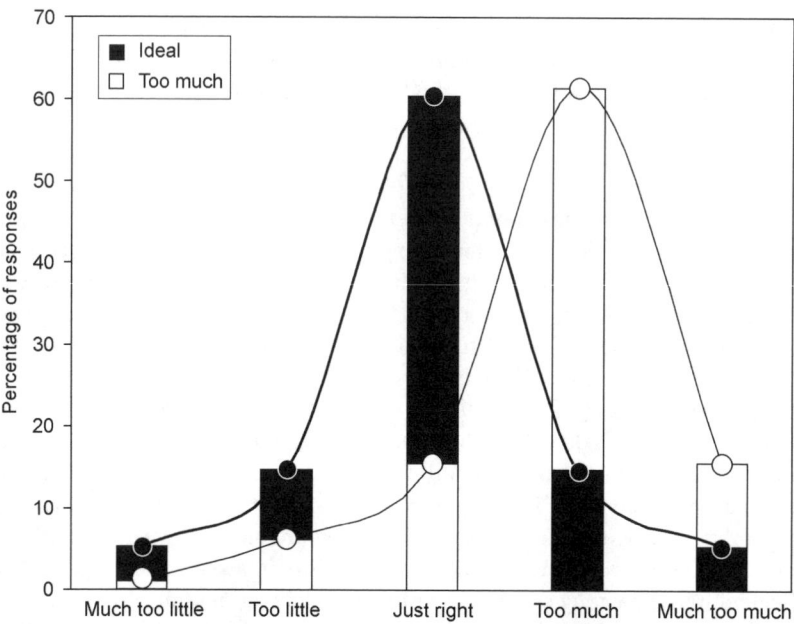

Fig. 19.7   Probabilistic distributions of a theoretical product at the ideal intensity and a product 'too high' in intensity with the same variance.

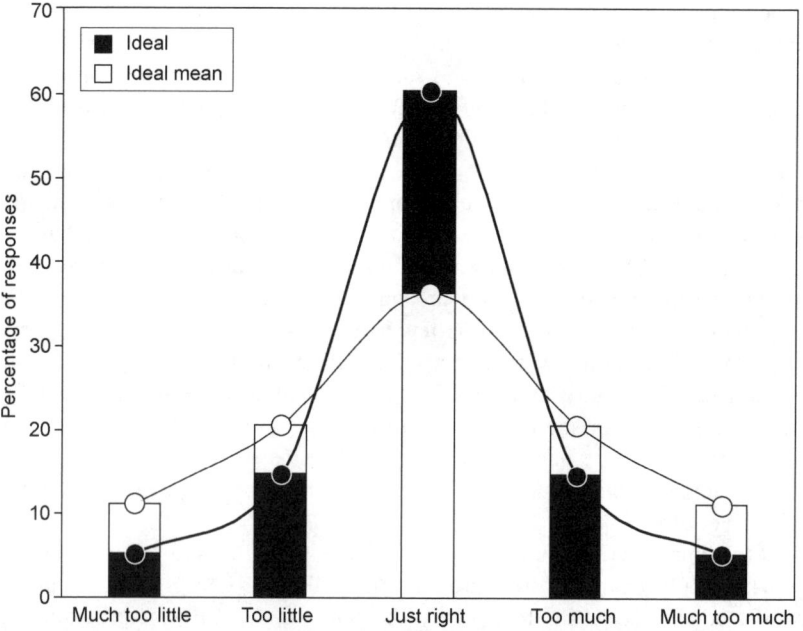

Fig. 19.8   Probabilistic distributions of a theoretical product at the ideal intensity and a product with the same mean intensity but greater variance.

with a theoretical product at the ideal (just right) level. The analysis compares the probabilistic distribution of each product against the probabilistic distribution of a theoretical product at the ideal level (as estimated from the responses to the actual products using the method of maximum likelihood). Attributes are considered one at a time, with the ideal distributions being estimated for each attribute separately. The shapes of the distributions, both the estimated ideal and the product distributions, are given as the proportion of responses in each category (symmetry of the curve around a just right level is assumed) for each distribution. The ideal distribution mean is set at zero (just right) and the means of the probabilistic distributions of the measured products are given relative to this point in terms of d'. Thus, if a measured product has a mean intensity that is one standard deviation above the estimated ideal level, its mean would be 1.0. Similarly, if a product's measured intensity is one and a half standard deviations below the ideal level, then its mean would be -1.5. In addition, the variance–covariance matrix associated with these means is given. TIP modeling also estimates the relative scale boundaries, determining the d' value that marks the boundary between categories (i.e. 'much too little' and 'too little').

This conversion of ratings into d' has several advantages. One of these advantages is that d' values can be calculated from both ratings data as well as from discrimination tests, potentially making it easier to compare results across various methods. Another advantage is that the conversion into d' values transforms the ratings into true scale values (Warnock et al., 2006). While rating scales are generally assumed to have equal interval spacing as in the top of Fig. 19.9, respondents often use the scales as though they were unequally spaced, as represented in the bottom of Fig. 19.9. Here, the end categories of the scale as well as the ideal category are being used less often than the intermediate categories. Such behavior is not uncommon in JAR assessments. As a result, the numbers associated with ratings are not evenly spaced, an assumption of parametric statistics (O'Mahony, 1986). In contrast, the d' values estimated from these ratings are evenly spaced and thus are more appropriate for analysis with parametric statistics, such as ANOVA.

Clearly, TIP modeling generates a great deal of output. What remains to be discussed is how it can be used in the development of new products. One of the

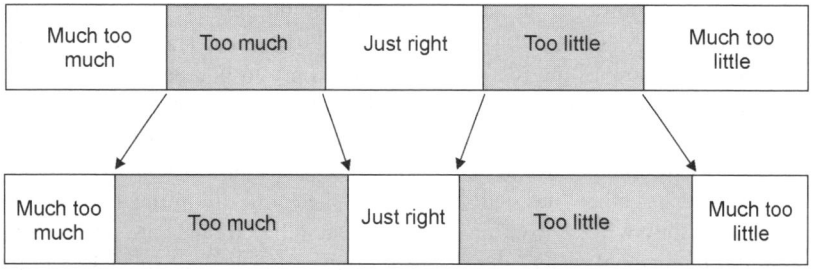

Fig. 19.9   Representation of how a judge may distort categories when making JAR assessments.

first things that can be done is to simply check how close the product mean levels are to the ideal level for all attributes. Because all means are d' values, a quick scan across them indicates which products are closest to the ideal and which will require the most adjustment for each attribute assessed. Next, the shape of the distributions can be compared to the ideal, either visually or by analyzing with chi-square (O'Mahony, 1986). This enables the detection of products that have a mean near the ideal level, but perhaps too much variance. Such variance can be due to variance in the product itself or be an indication of market segments that should be considered separately. If desired, examination of the category boundaries on the d' scale allows the researcher to examine the properties of the rating numbers, i.e. to see if categories were distorted by the judges. It is also possible to examine the correlations between the product means and either instrumental and/or descriptive analysis data. All this information can then be used in the optimization of a new product.

TIP analysis is used most effectively in combination with penalty analysis (discussed in Chapter 17). While TIP analysis excels at informing the researcher at how much the attributes deviate from the ideal, penalty analysis allows the product developer to prioritize the importance of the attributes to consumer acceptance of the products. It is possible for a large deviation from the ideal of a particular attribute to have a minimal impact on consumer acceptance (e.g. thickness of a chocolate bar), while for another attribute a small deviation may have a tremendous impact on acceptance (e.g. bitterness of a chocolate bar). Once the attributes have been prioritized by penalty analysis, TIP modeling gives the developer a fair way, via the sensitivity measure of d' instead of mixed physical measures such as millimeters and grams/liter, to compare how far each attribute is from the ideal. This approach allows the developer to see not only what is most important to product acceptance, but also how much physical variation can occur before the sensory profile is impacted. This has the added advantage of indicating what aspect will require the tightest regulation during production to ensure optimum quality is achieved.

19.2.2   Probabilistic MDS

Multidimensional scaling (MDS) has been used for many years by sensory researchers (for additional discussion, see Bieber and Smith, 1986; Lawless and Heymann, 1998; MacFie and Thomson, 1984; Popper and Heymann, 1996). It can be used to model how (dis)similar products are to one another or even to model hedonic data. From either direct or derived measures of dissimilarity, the assessed objects can be placed into multidimensional space (the number of dimensions is ultimately set by the researcher). The more similar two products are, the closer together they will be plotted; the more dissimilar they are, the further apart they will be from one another. Once the products are organized in this multidimensional space, the researcher can use a variety of techniques to reveal the attributes that influence the perception of product similarities (Lawless and Heymann, 1998). Multidimensional scaling is thus a way to model

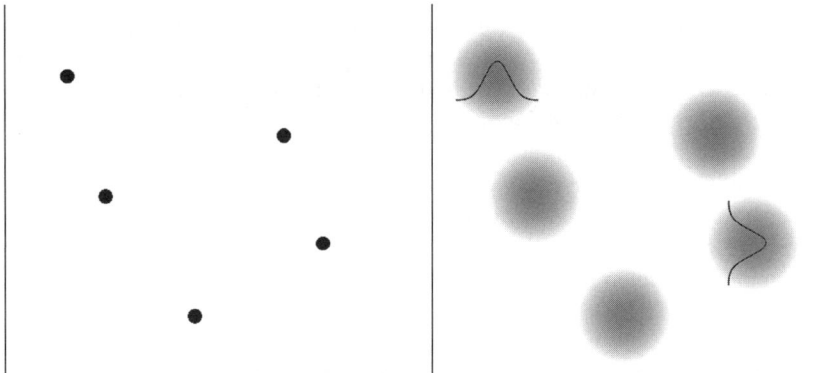

Fig. 19.10   Traditional deterministic MDS represents products as points (left) while probabilistic MDS represents products as probability distributions (right).

the relationship of complex products so that the underlying, or latent, structure of these relationships can be understood.

The traditional MDS models are deterministic, and represent each product as a single point in multidimensional space (Fig. 19.10, left). In contrast, probabilistic MDS represents each product as a distribution around a point of highest density; the density decreases as you travel away from this central point – which is basically a multidimensional representation of the probabilistic distributions illustrated in the previous section (Fig. 19.10, right). If the multidimensional space is two-dimensional, it can be envisioned as a circle or ellipse (depending upon whether or not the product variability is the same along each dimension). If the multidimensional space is three-dimensional, it can be envisioned as a sphere or egg-shaped solid. The advantage of representing each product as a distribution rather than a single point is the probabilistic MDS model can then disentangle product variance from the distance between products whereas deterministic MDS cannot and instead confounds the two (MacKay, 2005). For data sets with low variance, which may be the case with some instrumental measures of products, the difference between deterministic and probabilistic models is likely to be small. However, ratings data, even with highly trained individuals, generally show high variability. The structure of this data variance impacts how well deterministic MDS can recover the latent structure of the product relationships.

For modeling purposes, data variance structures can be classified via two features (MacKay, 2005). The first feature is whether the model estimates the same variance for each product or if it allows different products to have different variances. In other words, are all the products represented by probabilistic distributions of the same size in the multidimensional space? The second feature is whether the model estimates the same variance across all dimensions for a particular object (isotropic) or it permits different variances for the different dimensions (anisotropic). In other words, are the products represented by circles/spheres or by ellipses/egg-shaped solids? Deterministic MDS is most accurate at revealing the latent structure of the product relationships in instances where the variance is the same for all products and isotropic (MacKay, 2005). In all other instances, deterministic MDS tends to distort the configuration of the latent structure owing to the confounding of distance and variance. As a result, solutions require additional dimensions in order to minimize stress in the model. Since probabilistic MDS is able to differentiate between distance and variance, configurations arrived at using it are not similarly distorted, and as a result tend to require fewer dimensions to satisfactorily account for the variance than its deterministic counterpart. Because of the variance inherent in sensory evaluation data, probabilistic MDS is typically superior to deterministic MDS for modeling such data. Ironically, although the mathematics underlying probabilistic MDS are more complex than those underlying deterministic MDS, the actual estimated configurations are often simpler with probabilistic MDS and succeed in representing the products in lower-dimensional space (MacKay, 2005). Models of consumer acceptance tend to lose their utility when the number of dimensions needed to explain the latent structure rises above two or three dimensions. This means that modeling techniques, such as probabilistic MDS, that reduce the number of dimensions to model data variance will give the product developer an easier-to-use working model for estimating the impact of product attributes on consumer acceptance.

One of the most readily accessible programs for conducting probabilistic MDS is PROSCAL, which can be downloaded for free (http://www.proscal.com). This program can conduct probabilistic MDS on (dis)similarity ratings (either directly measured or derived from attribute intensity ratings) or, perhaps more importantly for the development of new products, it can use an unfolding model on consumer liking ratings (Coombs, 1976). If one inverses the traditional 9-point hedonic scale (Lawless and Heymann, 1998) such that better-liked products are given lower values and less-liked products higher values, then these values can be conceptualized as the distance from an ideal product and can be used to establish a multivariate space. If desired, consumers can be segmented, either by independent factors (sex, purchase history, socio-economic factors, etc.) or based upon the hedonic assessments themselves (e.g. dividing subjects into segments based upon their most liked product), and an ideal for each segment can be included in the unfolding model. This allows the researcher to learn how close the existing products are to each group's ideal product.

When the researcher has independently derived (dis)similarities and liking ratings, it is possible to analyze both sets of data simultaneously to determine the extent to which these datasets share a common space. Quite often, the resultant solution of common space analysis is more informative than models derived from either data set alone (MacKay, 2005). Of further benefit to product developers, from such common space analysis it is possible to estimate the percentage of consumers who will choose a product as their first choice, a concept known as 'perceptual shares' (MacKay, 2005). The perceptual share of each product can be determined for the entire market or each segment, depending upon the desires of the researcher. If the researcher has opted to consider segments, it is also possible to derive a measure of each segment's brand loyalty. Perceptual shares can also be used to predict the impact of proposed new products and to select strategies for optimal new product entry (for specific examples, see MacKay, 2005).

The discussion above has indicated several benefits to the use of probabilistic MDS, but has not addressed any of the difficulties with this approach. As mentioned earlier, probabilistic MDS is more mathematically complex than its deterministic counterpart, but since such calculations are no longer carried out by hand, this problem is relatively minor. Another drawback of probabilistic MDS is that it is not performed by most statistical software packages currently on the market. However, the availability of a free downloadable program minimizes this difficulty. The largest drawback of probabilistic MDS is that it requires the researcher to make far more decisions on a number of variables than does deterministic MDS. The researcher must indicate whether or not to allow variance to be set the same for all products or allowed to vary, whether to use an isotropic or anisotropic model, whether or not two data sets have enough in common to perform a common space analysis, etc. Once the researcher learns how to make these decisions (which unfortunately for the reader is beyond the scope of this chapter), probabilistic MDS is a powerful tool for the development of new products.

19.2.3   Landscape segmentation analysis

In recent years, interest has grown in probabilistic models based upon landscape segmentation analysis (LSA). This analysis is a subtype of probabilistic multi-dimensional scaling, but due to the current attention it is receiving, it is worth discussing separately. While LSA can be performed upon (dis)similarity ratings, it is most typically used with liking ratings. As mentioned in the previous section, these liking ratings are conceptualized as representing the distance for the rated product from the ideal. Both the products and the ideal are conceptualized as distributions rather than single point entities. From the liking ratings alone, through the process of unfolding, a map is created of both products and individual ideals in which each consumer is represented by his or her own distribution (Ennis, 2001). This focus on individual ideals is one of the distinguishing features of LSA. By plotting the individual ideals with the products, not only is it possible to subsequently determine the factors that are influencing liking, it also accounts for differences in individual scale usage (Ennis and Anderson, 2003). The individual ideal means are then used to create a density contour map, upon which is placed a scatter plot of the individual ideal means and the product means (see Fig. 19.11).

This scatter plot/contour map is extremely informative, allowing the researcher to see where products fall in the multivariate space relative to the individual ideal points and whether or not any market segments exist (Ennis and Anderson, 2003). If there are multiple segments, the researcher can examine how the current products are placed relative to these ideals and make strategic

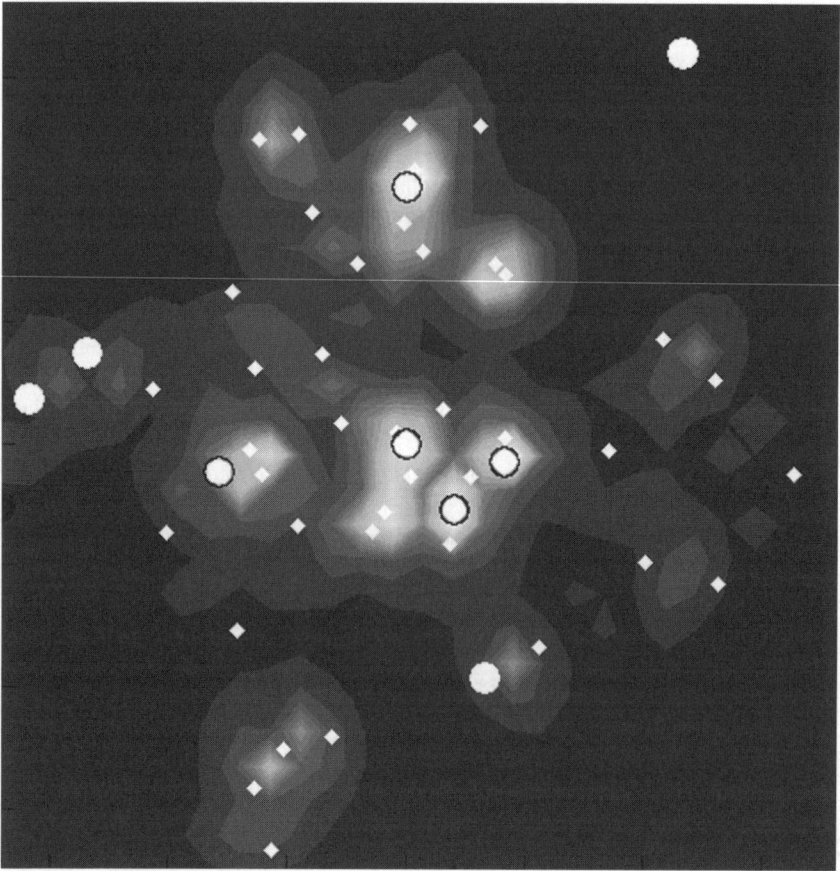

Fig. 19.11 Landscape segmentation analysis. White circles represent product means; white diamonds represent individual ideal means. The higher the density of individual ideal means, the lighter the contour map is.

decisions as to how to modify existing products or introduce new products. Perhaps the researcher wants to create a product that is no one's favorite but not objectionable to anyone. Alternatively, the researcher may wish to develop a product for a gap in the market and create one to be the favorite of a particular segment, or instead follow any of a number of other strategies. In addition, once the map is generated, other data about the products (descriptive analysis ratings, consumer assessments of traits, or instrumental data) can be added as vectors (Ennis, 2001; Ennis and Anderson, 2003). Such information is crucial in determining the dimensions that influence liking ratings and subsequently aiding the researcher in deciding how to alter product attributes to create the desired product.

Currently, only one software program performs this analysis without requiring extensive programming by the user: IFPrograms[TM]. While this greatly limits the availability of the analysis for many, it is possible to create similar maps with other programs capable of conducting probabilistic MDS. The IFPrograms™ version of LSA requires minimal understanding of the underlying modeling assumptions and does not require the researcher to make decisions about the model used as this is already set – it can be argued whether this is an advantage or disadvantage of the program, and this is largely dependent on the skill set of the researcher. It should be mentioned that a key distinguishing feature of LSA, the use of individual ideals, is not a practice endorsed by all statisticians and there are those who advise against it (MacKay, 2005). Finally, LSA requires that the data be collected on a categorical scale and that data be collected from a minimum of 200 subjects, a number that many researchers find prohibitive. Despite these limitations, many developers find the insights gleaned from this type of analysis invaluable in determining what type of product to create and what market niche to fill.

19.3   Future trends

It is difficult to predict the future of probabilistic models. While they are often superior in modeling data sets, they are generally more difficult to use than their deterministic counterparts. Quality does not always win out over availability – when is the last time you saw a beta video cassette or a movie on laser disc? Nor has it been proven that the advantages of these models outweigh the costs, which are largely in learning how to use such models. However, once one learns how to use such models, it becomes quite unsatisfactory to return to the use of their deterministic counterparts. These new probabilistic models embrace the variability in inherent in consumer responses instead of ignoring it, masking it, or trying to train it away. While it remains to be seen how widely the enthusiasm for these models will spread in the development of new products, those developers who dedicate themselves to mastering these sophisticated tools will undoubtedly be pleased by the insights they ascertain when using them. These models show great promise for new product developers seeking an edge over their competitors.

19.4   References

ANON (2005). Workshop summary: Data analysis workshop: getting the most out of just-about-right data. *Food Quality and Preference*, **15**(7–8), 891–899.

BIEBER, S.L. and SMITH, D.V. (1986). Multivariate analysis of sensory data: a comparison of methods. *Chemical Senses*, **11**, 19–47.

BYER, A.J.A., D. (1953). A comparison of the triangular and two taste-test methods. *Food Technology*, **7**, 185–187.

COOMBS, C.H. (1976). *A Theory of Data*. Mathesis Press, Ann Arbor, MI.

ENNIS, D.M. (2001). Drivers of Liking® for Multiple Segments. IFPress, Richmond, VA, **4**(1), 2–3.

ENNIS, D.M. and ANDERSON, J.L. (2003). Identifying Latent Segments. IFPress, Richmond, VA, **6**(1), 2–3.

FRIJTERS, J.E. (1981). An olfactory investigation of the compatibility of oddity instructions with the design of a 3-AFC signal detection task. *Acta Psychologica*, **49**(1), 1–16.

GEELHOED, E.N., MACRAE, A.W. and ENNIS, D.M. (1994). Preference gives more consistent judgments than oddity only if the task can be modeled as forced choice. *Perception & Psychophysics*, **55**(4), 473–477.

LAWLESS, H.T. and HEYMANN, H. (1998). *Sensory Evaluation of Food: Principles and Practices*. Chapman & Hall, New York.

MACFIE, H.J.J. and THOMSON, D.M.H. (1984). Multidimensional scaling. In *Sensory Analysis of Foods*, J.R.R. Piggot, ed., pp. 351–375. Elsevier, London.

MACKAY, D.B. (2005). Probabilistic scaling analyses of sensory profile, instrumental and hedonic data. *Journal of Chemometrics*, **19**, 180–190.

O'MAHONY, M. (1986). *Sensory Evaluation of Food*. Marcel Dekker, Inc., New York.

POPPER, R. and HEYMANN, H. (1996). Analyzing differences among products and panelists by multidimensional scaling. In *Multivariate Analysis of Data in Sensory Science*, T. Næs and E. Risvik, eds., pp. 159–184. Elsevier, Amsterdam.

STILLMAN, J.A. (1993). Response selection, sensitivity, and taste-test performance. *Perception & Psychophysics*, **54**(2), 190–194.

WARNOCK, A.R., SHUMAKER, A.N. and DELWICHE, J.F. (2006). Consideration of Thurstonian scaling of ratings data. *Food Quality and Preference*, **17**, 556–561.

20

Using auctions to estimate prices and value of food products

Y. Lohéac, ESC Bretagne Brest, France and S. Issanchou,
UMR FLAVIC INRA-ENESAD, France

20.1 Introduction

Perception of food quality is influenced by many factors, which can be categorised in three main sources: the product, the consumer and the context. As underlined by Oude Ophuis and van Trijp (1995), perceived quality is formed on characteristics that may have actually been experienced, or are believed to be associated with the product. The sensory characteristics are the most important experienced attributes that are closely related to the physical product. Many credence quality attributes such as healthfulness, naturalness, and environmental friendliness are formed on the basis of extrinsic characteristics such as the brand, label, and health claim. Indeed, these attributes induce expectations concerning the quality of the product and can even influence the perception of specific sensory characteristics and the global hedonic value of the product (see Chapter 10 and Anderson, 1973).

A large number of experiments have been conducted to evaluate the relative importance of intrinsic and extrinsic characteristics on consumers' hedonic perception and models developed by psychologists have been used in the food domain to interpret the empirical data and show how expectations induced by an external information and product judgement related to its sensory characteristics are combined to form a global hedonic perception (e.g. Cardello, 1994; Deliza and MacFie, 1996; Schifferstein, 2001). In these experiments, hedonic ratings are collected for the same consumers and the same products under different information conditions. Firstly, consumers indicate how much they like each product in a blind condition, i.e. they taste the product without any external information or with only generic information such as 'brut non-vintage Champagnes' (Lange *et al.*, 2002). Secondly, consumers indicate how much they expect to like the products on the basis of external information only (e.g. on the presentation of the champagne bottles). Finally they indicate how much they like each test product in a full information condition, i.e. they taste the product with external information. However, declaring you like a product very much does not mean that you are ready to buy it whatever its price even if it is for a special occasion. In other words, a consumer will not buy a product even if they like it very much if its market price is higher than the maximum price they are ready to pay for it. So, more and more researchers in the sensory domain use not only hedonic measurements but also purchase intent measurements, using methods incorporating tasting into procedure such as conjoint analysis or contingent valuation technique (e.g. Vickers, 1993; Solheim and Lawless, 1996; Moskowitz *et al.*, 1997; Malundo *et al.*, 2001; Bower *et al.*, 2003).

These approaches permit the estimation of the importance of specific attributes for purchase intent based on consumers' responses to combinations of different levels of these attributes. In these experiments, price is one of the test attributes. In the previously quoted studies, each product is presented with a stated price and consumers are asked if they would buy it (yes/no) and to rate their purchase intent on a scale. Another approach, referred as choice-based conjoint measurement, consists of presenting several scenarios (a scenario is composed of several products with different characteristics and prices) and asking consumers to choose for each scenario the most preferred product (see, for example, Lusk and Schroeder, 2004). Collected data can be analysed at the group level or at the individual level (Moskowitz and Silcher, 2006). However, this approach does not directly give individual reservation price (i.e. the maximum price a consumer is ready to pay), even if consumers are also asked to indicate how much they would be willing to spend on each test product (see, for example, Bower *et al.*, 2003).

In almost all cases no real purchase is required of the participants, so there are no financial consequences for them and these procedures do not require real economic commitments from the respondents. These approaches are thus categorised as hypothetical methods. It has been demonstrated that such hypothetical methods may suffer from a hypothetical bias (e.g. Neill *et al.*, 1994; Lusk and Hudson, 2004; Völckner, 2005). Indeed, in these studies willingness to pay elicited in a hypothetical context is higher than that elicited in a non-hypothetical context. In other words, consumers are more willing to declare to be ready to spend a certain amount of money for a given product than they are to really spend this amount of money. Economic commitment can be introduced in conjoint measurements: for example in a choice-based conjoint measurement, at the end of the experiment one scenario is randomly drawn and consumers are asked to buy the product they chose in this scenario (Lusk and Hudson, 2004). However, as pointed out previously, conjoint measurement does not enable us to obtain individual reservation prices. Auction mechanisms are an efficient way to place people in a non-hypothetical context, the real market environment, where they are compelled to reveal the value they put on different products (Combris *et al.*, 2001).

In this chapter, we will firstly present the principles of two methods: the Vickrey (Vickrey, 1961) and *n*th-price auctions, and the BDM procedure (Becker *et al.*, 1964). Then we will give examples of applications of auctions and we will illustrate the utility of using auctions to get information about the value attributed by consumers to different product characteristics and to better understand consumers' food choices. Finally we will discuss the advantages and limits of auctions and the possible way to get further to validate and/or improve the methods.

20.2   Estimate value of food product with auctions

20.2.1   General principles

Auctions interest two kinds of person: auctioneers and researchers. The former wants to sell goods at the highest prices that buyers will pay. Many kinds of goods can be the object of an auction and there exist various types of auctions depending on the product, culture and habits. Researchers interested in auctions study their parameters (game theory, equilibrium, efficiency, etc.) and use them to elicit the maximum (reservation) price that a buyer (consumer) would pay for a good. This price is the border between purchase and no purchase. The consumer does not buy a good for a price higher than his or her reservation price but buys a good at any price equal or lower than it and thus makes a surplus (difference between reservation price and sale price). There is an important difference between the aims of auctioneers and researchers. The former want to know a winner and are not interested in other bids, and researchers want to know the reservation price of each bidder and are secondarily interested by who wins the auction.

There are many types of auctions that can be used to elicit consumer willingness to pay for a product. Some are high voice auctions, as English (ascending-bid) or Dutch (descending-bid) auctions, and others are sealed-bid auctions, as Vickrey auctions (Vickrey, 1961)[1] and BDM (Becker–DeGroot–Marschack) method (Becker *et al.*, 1964), where biddings are made by paper or computer. These last two methods are the more commonly used in experimental economics. They theoretically lead bidders to truthfully reveal their value for the good in auction. This is their dominant strategy. Whatever the method used, when they answer the question 'what is the maximum price you would pay for this product?', bidders do not have an interest in announcing a price higher than they really want to pay for a good because it commits them to pay this price, and they do not have interest in announcing a price lower than that they want to pay because they risk not having the goods. Their interest is in announcing a price for which they cannot be unsatisfied and cannot lose anything because they agree to definitely buy the goods at a price equal or lower than their reservation price.

---

1. Lucking-Reiley (2000) gave some historical and interesting descriptions of Vickrey auctions.

Although it is possible to conduct hypothetical auctions, in our view the valuation of real goods (and real information) can occur only with an incentive system where participants really buy the goods to engage them in a real market environment including supply, demand, prices and exchanges. This is a major advantage of auctions. This environment includes a feedback for Vickrey auctions, i.e. all participants are publicly informed about the distribution of bids (prices) for this auction and about the winner and the price he or she pays.

In this chapter we describe auctions with only one round (valuation) for a given product (same product in same information condition) and where the valuation is made through willingness to pay without initial endowment. We choose this environment because it seems to be the nearest with a real purchase outside the laboratory. However, it is possible to use other protocols but it does not change the principles of auctions as described here for willingness to pay. The resulting prices could be somewhat different and there are still some debates about the impact of the different protocols in the experimental economics community. As stability of prices in auctions is discussed in experimental economics, some papers include multiple rounds for the same product in the same information condition in their experiments (e.g. Fox *et al.*, 2002; Jaeger *et al.*, 2004). Multiple rounds for same product/condition increase the duration of the experimental session. It certainly explains why many authors include a unique round by product/condition but multiply products and conditions (e.g. Lange *et al.*, 2002; Noussair *et al.*, 2004b; Stefani *et al.*, 2006). A possibility in all types of auctions is to endow the participants with a piece of a product and to invite them to bid for an alternative product. For example, participants were endowed with a typical pork sandwich and could buy an irradiated (to control *Trichinella*) pork sandwich using a Vickrey auction (Fox *et al.*, 2002). The initial endowment is also the basis of the willingness to accept process. The principle is that participants receive a product and say how much they want to exchange it. For example, Jaeger *et al.* (2004) endowed participants with a chocolate chip cookie which is clearly labelled to be without genetically modified (GM) ingredients; participants indicated how much they want (their willingness to accept) to exchange their GM-free cookie for a GM cookie. The difference between willingness to pay and willingness to accept is an important debate in experimental economics (see, for example, Plott and Zeiler, 2005). Last, it is possible to measure value of food product without auction, i.e. with choice experiment. With this method, participants are initially endowed with a product and could exchange it for various quantities of an alternative good (see for example Marrette *et al.*, 2006; Boutrolle *et al.*, 2006).

20.2.2   Implementation

An experiment could be organised to evaluate a food product. During this study, consumers are faced with real products (which can be observed and in some cases even be tasted) and a variety of information (nutritional, origin, production process, health, introduction of new technologies, etc.). The researchers are interested in consumers' reactions expressed through their willingness to pay for each product in each condition. Thus for each condition-product of an experiment, participants are invited to answer the following question 'what is the maximum price you would pay for this product?' The given price must be personal and could be greatly different from the market price.

At the beginning of each experimental session, the elicitation method and the associated incentive system are explained to the participants. It is also possible to introduce some training auctions at the beginning of the experiment. Indeed, hypothetical or real auctions with a side product could facilitate the comprehension of auction process (see, for example, Noussair *et al.*, 2004b). The speaker must underline to participants that they are in a real selling/buying process. If they announce a maximum price higher than the sale price, they must really buy the proposed product at this sale price. But the advantage is that they cannot buy a product at a price higher than they want. Of course, some adaptation must be introduced depending on the product and on information that is delivered. For instance, if the product does not exist, another way must be found; some options are presented in Section 20.5.

When there are many products, conditions or rounds in the same experimental session, the problem with repeated auctions that induce real purchases is the appearance of demand reduction or 'wealth diminution' effect for participants. To avoid this problem, the solution is to consider independently each product, condition or round by introducing an 'isolation effect' (Kahneman and Tversky, 1979). The principle is straightforward: rather than really play for each product, condition or round, only one of them is randomly drawn at the end of the experimental session, individually or for the whole group of participants. Thus, the wealth diminution effect does not take place and participants are in the same incentive condition for all products, conditions or rounds.

For each method and incentive system, we introduce an example of willingness to pay elicitation with five (famous) people in an experimental session for a marvellous red apple.

20.2.3   Vickrey and *n*th-price auctions

Vickrey auctions (second-price auctions; Vickrey, 1961) and other *n*th-price auctions are sealed-bid auctions (Lucking-Reiley, 2000) in which the winner does not pay the bid price but the next lower price than his or her bid. In a *n*th-price auction, all participants submit simultaneously written (on paper or on computer) bids without knowing the bids of the other participants. The auctioneer (or researcher) records the bids and reports them in an ordered manner for public information. Thus participants know all the bids existing for this auction. *N*th-price auctions are an expression of market environment with feedback on demand and selling price. As in all auctions, there is an interaction and competition between participants because only the highest bids win the auction. Participants bid against them. In spite of this competition, the dominant strategy for the bidders is to reveal their actual reservation price.

In the original Vickrey auction, only one product is put in the auction and the highest bid wins it. The winner buys the product but pays the second price, i.e. the second highest bid of the auction, and not at his or her bid price.

Example

The auctioneer (or researcher) wants to buy (wants to know the appreciation for) one marvellous red apple. Five participants (for who the researcher knows that they like apples) bid for this product: Rex (€1), Mike (€0.50), Lynette (€0.75), Susan (€0.80), and Mary-Alice (€0.30). The descending order of these bids is (in euro): 1 (Rex), 0.80 (Susan), 0.75 (Lynette), 0.50 (Mike), 0.30 (Mary-Alice). The winner of the Vickrey auction is Rex and he pays for this apple at the second highest bid: €0.80 (Susan's bid). Susan, Lynette, Mike and Mary-Alice do not buy any apple.

This method has been widely applied for valuation of food products, e.g. for champagne (Lange *et al.*, 2002; Combris *et al.*, 2006b), red wine (Lecocq *et al.*, 2005), beef steaks (Lusk *et al.*, 2004), orange juice and chocolate bars (Noussair *et al.*, 2004a), apples, potatoes and bread (Rozan *et al.*, 2004), pork sandwiches (Fox *et al.*, 2002), etc.

Because there is only one winner in Vickrey auction, there is no incentive for the participants who have a low reservation price for the product; they never can win the auction and thus could be inclined to under-bid compared with their actual reservation price. Conversely, those who absolutely want the product could be inclined to over-bid.

To avoid the lack of incentive when there is only one winner, the principle of Vickrey auction could be extended to more than one winner (with one product by winner). Nth-price auction is the generalisation of Vickrey auction. Uniform-price auction is one possibility. This is just a variation of the second price auction. The auctioneer (experimenter) determines the number $(n - 1)$ of winners before the auction. The $n - 1$ highest bidders win the auction, buy each one unit of the good and pay the nth price. For instance, if there are two winners, this is a third-price auction: the winners are the two highest bidders and they pay one unit of the product at the third price.

There is no limit to this uniform-price auction, except the number of bidders and the number of products in auction. If the number of winners is too high, there are too many products in auction, the sale price risks to be low and it is not certain that bidders will be likely to give their real reservation price, and in particular the high-value bidders are not inclined to give a really high bid because they are sure to win.

Example

Consider now that there are two marvellous red apples in auction, thus there are two winners who receive one apple each and pays it at the third highest price. Thus, Rex and Susan both receive a red apple and pay €0.75 for it (Lynette's bid). Lynette, Mike and Mary-Alice do not buy an apple.

We identify some authors who used this method for food product valuation: a 3rd-price auction (Platter *et al.*, 2005), a 4th-price auction (Umberger *et al.*,

2002) and a 5th-price auction (Hoffman *et al.*, 1993; Jaeger *et al.*, 2004). The argument is that it enables more participants to join in 'while still being relatively simple to explain and administer' (Jaeger *et al.*, 2004).

As in 2nd-price auctions, in the nth-price auction the number of winners is known with certainty. Thus, the experimenter can be faced with over-bidding and under-bidding for those who absolutely want to win and for those who absolutely do not want to win. To avoid these biases, the experimenter can introduce uncertainty regarding the number of winners by using the random nth-price auction proposed by Shogren *et al.* (2001). These authors pointed out disadvantages associated with previous methods (fixed nth-price auction and BDM method, presented in Section 20.2.4), in particular those insincere bidders who bid far below or above their actual reservation price. They proposed a new mechanism designed to engage otherwise disengaged insincere bidders and showed that bidding a real private value is a dominant strategy in a random nth-price auction.

The experimenter explains at the beginning of the session that the number of winners for the auction is unknown. After the bidding part of the session, a number is randomly drawn (by experimenter, computer or anyone else) between 1 and the number of participants minus one $(P - 1)$ to determine the number of winners $(n - 1)$ in the auction. The $(n - 1)$ winners with the highest bids buy one unit of the product and pay for it at the nth-price, i.e. the bid just below the lowest bid of the winners.

Example
In this situation where there are five consumers $(P = 5)$, there are between one and four $(P - 1)$ goods in the auction. After bidding, a number $(n - 1)$ is drawn between $(1, 2, 3, 4)$ which determines the number of winners whose pay the nth-price. If the number 3 is drawn, the three higher bidders (Rex, Susan and Lynette) win one apple each and pay €0.50 for it (Mike's bid, the fourth). Mike and Mary-Alice do not buy anything.

The random nth-price auction method was used in recent studies for food valuation: on beef steaks (Lusk *et al.*, 2004), and spelt (Stefani *et al.*, 2006). As the authors underlined, this auction combines the best elements of two classic demand-revealing mechanisms: the Vickrey auction and the BDM mechanism (which is presented in Section 20.2.4).

In nth-price auctions, the experimenter can also have his or her own reservation price for the test product, i.e. a minimum acceptable bid price (the same for all variants presented during the session). Participants are informed that no selling would occur if the highest bid happened to be lower than this price. Of course, this price is not disclosed before the end of the session (for instance, Combris *et al.*, 2006b; Platter *et al.*, 2005).

20.2.4   BDM method (Becker–DeGroot–Marschack)
The BDM method (Becker *et al.*, 1964) is not, strictly speaking, an auction, since there is no competition between participants in a market environment

(number of goods available is not limited). Each participant could be a winner/buyer and it is possible to run an experimental session with only one participant, which is not possible in other auctions. Rather than comparing participants' prices between them to determine the exchange price, as in nth-price auctions, the BDM method proposes comparing the reservation prices of all participants to a random price drawn in a predefined price distribution. It is possible to illustrate this as a comparison between a buying price and a selling price.

The selling prices come from a price distribution determined by the experimenter. The simplest distribution is a uniform one with regular steps. For instance, the distribution could be made up by ten counters from 0.1 to 1 with 0.1 steps (in euros). However, a more realistic distribution must reflect the distribution of prices on the market of the product studied. This distribution could be constructed from a database on prices and market shares.

After the valuation part of the experimental session, the experimenter proceeds to the selling price draw. There are two possibilities at this level. First, only one selling price is collectively randomly drawn for the whole session. Second, each participant randomly draws an individual selling price. Whatever is the method used, the principle of the incentive system is the following: (1) if the randomly drawing (selling) price is higher than the reservation (buying) price, there is no exchange, i.e. the participant does not buy any product; and (2) if the selling price is equal to or lower than the buying price, the participant buys the product at the selling price. To draw only one selling price for a session presents the advantage of being quick. However, if the drawing price is the highest of the distribution, it is possible that nobody could buy the product (no winners), and if the drawing price is the lowest, it is possible that everybody buys the product at a very low price, which could lead the experimenter to go bankrupt (even if he or she does not care about financial loss). The individual draw method has the advantage that it diversifies the risk and increases individuals' commitment.

Example

To illustrate this method, we use the previous example where five consumers give a bid for one marvellous red apple: Rex (€1), Mike (€0.50), Lynette (€0.75), Susan (€0.80) and Mary-Alice (€0.30). The preliminary step before the experiment is to determine the price distribution (selling price) of the goods on which consumers bid. Here we can retain a uniform distribution from €0.05 to €1.50 with steps of €0.05.

At this level, there are two possibilities:

1. Only one sale price is randomly drawn for the whole session by one participant: €0.65 by example. In this case, Rex, Susan and Lynette who have a reservation price higher than the selling price win a red apple and pay €0.65. Mike and Mary-Alice who have a reservation price lower than the selling price, do not buy this product.

2. One sale price is randomly drawn by each participant, thus:
   - Rex draws the selling price €0.55 (lower than his reservation price: €1.00), he buys one apple at this price (he saves €0.45, 1.00 − 0.55);
   - Mike draws the selling price €1.30 (higher than his reservation price: €0.50), he does not buy apple;
   - Lynette draws the selling price €0.05 (lower than her reservation price: €0.75), she buys one apple at this price (she saves €0.70, 0.75 − 0.05);
   - Susan draws the selling price €0.80 (equal to her reservation price: €0.80), she buys one apple at her reservation price;
   - Mary-Alice draws the selling price €0.35 (higher than her reservation price: €0.30), she does not buy the apple.

As with the Vickrey auction, the BDM method has been widely used to value food products, e.g. for cookies (Lusk and Fox, 2003; Noussair et al., 2004a,b), beef steaks (Lusk et al., 2004), apples, potatoes and bread (Rozan et al., 2004), orange juice (Combris et al., 2006a) and bread (Issanchou et al., 2006).

20.2.5   Similarities, differences and efficiency of methods

The methods presented previously are incentive-compatible and any participant could buy a product at a price lower than the maximum price he or she is willing to pay for it. They are near to a market process where buyers compare their reservation price to the market price of a product. They also are theoretically equivalent and enable us to elicit a willingness to pay of consumers for a good that may be considered genuine.

Even if these methods are theoretically equivalent, they can give different empirical results. Many studies have compared incentive valuation methods in terms of efficiency to elicit realistic willingness to pay for a product. Several authors (Shogren et al., 2001; Lusk, 2003b; Lusk et al., 2004; Combris and Ruffieux, 2005) present the advantages and disadvantages of each method very effectively. However, there is no evidence that any one method is more efficient than the others and experiments underlined the mixed results, both with food product and with other goods. For instance, Noussair et al. (2004c) found that the Vickrey auction was more effective than the BDM procedure: it induces less bias and less dispersion in the training phase in which items with induced values were proposed in the auction. Rutström (1998) observed the same result. Rozan et al. (2004) found that bids from BDM were higher than from 2nd-price auction. Lusk et al. (2004) underlined that 2nd price, random nth price and BDM gave equivalent results in initial bidding round, but by final bidding rounds 2nd-price was higher than BDM which was higher than random nth-price. Shogren et al. (2001) also observed this inversion of order (2nd price > random nth price, then 2nd price < random nth price) between two distinct conditions of a same game. Knetsch et al. (2001) found that 2nd-price auctions gave higher bids than a 9th-price auction. Combris et al. (2001) observed only very little differences between BDM method and Vickrey auction for valuation of champagne.

Völckner (2005) did not found consistent significant difference between these two methods.

In a fixed nth-price auction the number of participants in an experimental session may affect bids. Using a 4th-price auction, Umberger and Feuz (2004) found that the average bid price increased as the panel size increased from 6 to 11 then slightly decreased from 11 to 12. On this point of view, BDM method presents a real advantage because there is no impact of the number of participants in an experimental session.

A problem could arise when a participant does not want the product in auction for any reason (for example because the participant has recently bought a similar product and does not want to buy another one at the time of the experiment). The consequence is that they under-bid and this bid does not indicate any reservation price. To avoid this problem, it is possible to introduce the preliminary question, 'do you want to buy this product?' (the negative answer is coded as a null price). Thus the auction is constructed in two steps:

1. Do you want to buy this product? Yes/No.
2. If you want to buy this product, what is the maximum price you would pay for it? (We discuss this further in Section 20.3.)

To our knowledge, this question has not been not introduced in many experiments. Lusk and Fox (2003) and Combris et al. (2006a) introduced this question in their experiments using the BDM method. Another option is to tell participants that they could submit a bid of zero if they do not want to buy the product (see, for example, Umberger and Feuz, 2004, who used a 4th-price auction). Participants who give zero bids for all products or never want to buy the products are qualified as 'non-buyers'; they are not in the market and their data can be excluded for the estimation of the mean willingness to pay. Participants who buy at least one of the test products (or give at least one non-zero bid) are considered as 'active buyers' (Rozan et al., 2004). This procedure can be used without any problem with BDM because the possible answers of a given participant has no influence on the behaviour of the other participants. In contrast, it can be dramatic in nth-price auction where the number of bidders must be relatively large. This procedure was used in an experiment where a BDM procedure and a 2nd-price auction were compared (Rozan et al., 2004); the authors observed more non-buyers with the 2nd price than with BDM. They suggest that this difference is due to the fact that participants with low reservation prices have a lower expected value for winning the auction with 2nd price than with BDM. Therefore 'they are more likely to forego participation in the auction'. Thus, in a 2nd-price auction, the non-buyers would not only correspond to participants not interested by the product but also to participants with low reservation prices.

The debate concerning the efficiency of the presented methods is still open. However, as pointed out by Combris and Ruffieux (2005), the nth-price auctions can induce a competitive behaviour between participants, leading to over-bids. It seems that random nth-price auction mixes the advantage of Vickrey auction (active market environment with feedback) and of the BDM method (each participant is engaged because each has a chance of winning). However, the explanation of this random nth-price auction does not appear to be intuitive and straightforward for participants. In contrast, the BDM procedure is easily understood by consumers. Moreover the method can be used with a low number of participants without any impact on the results, and can even be implemented with only one person. Thus BDM can be used not only in laboratory settings but also in field settings (Lusk *et al.*, 2001; Wertenbroch and Skiera, 2002; Lusk and Fox, 2003). However, in a multiple participant session, BDM and a random nth-price auction are more difficult to manage from a practical point of view in terms of products that must be available for sale as there is uncertainty regarding the number of winners. Finally, if the same consumers are invited to several successive sessions, one cannot exclude, with a nth-price auction, a possible collusion between consumers to give low prices.

20.3   Using information from auctions to understand food choices

Experimental auctions have been particularly used by agricultural economists to determine the price premium that consumers would be ready to pay for safer foods from a microbiological point of view (e.g. Hayes *et al.*, 1995; Buzby *et al.*, 1998; Shogren *et al.*, 1999), or from a chemical point of view with a case concerning heavy metals in apples, potatoes and bread (Rozan *et al.*, 2004) or pesticides in apples (Roosen *et al.*, 1998). Such data tells us if the extra cost induced by the modification of the production process can be compensated by a higher market price. Other experiments have been conducted to determine the acceptance of novel packaging (e.g. a vacuum-skin packaging for beef steak, Hoffman *et al.*, 1993), novel technologies such as irradiation (e.g. Fox *et al.*, 1998, 2002; Shogren *et al.*, 1999), and genetic modification (see Lusk *et al.*, 2005, for a meta-analysis of experimental auctions on genetically modified foods). In such cases it is particularly important to examine the number of zero bids (or the number of participants who refuse to buy this new product, if in a first step, consumers are asked if they want to buy the product as presented previously in Section 20.2.5) among the active buyers (i.e. among those who do not give always zero bids or who accept to buy the traditional product). Indeed, this number would give an indication about the percentage of boycotters.

Auctions have been also used to evaluate how some sensory characteristics were valued by consumers. In this domain, a quite large number of studies concerns meat. These experiments only provide information about a sensory characteristics such as tenderness (e.g. Feldkamp *et al.*, 2005; Alfnes and Rickertsen, 2006), or include a taste test (Lusk *et al.*, 2001), and/or combine both (Lusk *et al.*, 2001). If sensory studies demonstrate the importance of tenderness on meat liking, these valuation studies determine the premium price attributed to a given increase (experienced or not) of tenderness. Moreover

Alfnes and Rickertsen (2006) demonstrated the interest of categorising the beef into three tenderness classes as they observed that the total value of beef increased by 8% when consumers had information about tenderness categories.

More recently auctions have been used to compare the value attributed to intrinsic characteristics such as the sensory properties, and to external characteristics such as the brand (for champagne by Lange *et al.*, 2002; for wine by Lecocq *et al.*, 2005); the origin (Stefani *et al.*, 2006), and the legal definition of the product (Combris *et al.*, 2006b), or the information related to the composition of the product such as the fibre content in bread and the associated health-related information (Issanchou *et al.*, 2006). Such data are useful not only for product development but also for public policy on health as they provide understanding of consumers' reactions and in particular how they account for sensory pleasure and health in their food choices.

In experimental economics, as in sensory analysis, a key point is to conduct the experiment with participants who consume the type of product to be studied. In experimental economics, it is also important to recruit regular buyers of the test product. Moreover one concern in experimental economics is to recruit consumers of all socio-economic classes. To achieve this, participants can be sampled in representative districts of all socio-economic classes of the area where the study is conducted (Lange *et al.*, 1999). Such a method effectively permits the recruitment of consumers who would not answer an advertisement in a newspaper. Moreover, it is also possible to introduce other questionnaires about attitudes, habits, knowledge, etc. on food, nutrition, health, etc. These parameters permit the experimenter to define the panel precisely, to control heterogeneity between participants, to identify homogeneous groups and behaviour profiles (people with the same characteristics) and to compare them.

The published data do not allow the most influential sociodemographic characteristics to be identified: first because not all authors tested the same set of variables, and second because results differ between studies. However, it is very likely that the effect of individual characteristics is product dependent. For instance, Lange *et al.* (2002) found that older consumers (>40-years-old) gave higher prices than young consumers, and that women gave more importance than men to the brand in comparison to their hedonic appreciation when valuing the champagnes in full information condition (tasting while being informed about the brand). Lecocq *et al.* (2005) found that women had a lower willingness to pay for red wine but did not find any age effect, and found that consumers with a higher household income had a lower willingness to pay. Menkhaus *et al.* (1992) found that the difference in willingness to pay for beef with a vacuum-skin packaging in comparison to a standard packaging decreased with the number of people in the household; surprisingly they found that this difference is lower for participants with a full-time employment in comparison with part-time and unemployed; they also observed a greater difference in willingness to pay for the two types of packaging for participants with a larger income; they thus conclude that 'a reasonable target market for beef in vacuum-skin package is consumers with higher incomes'. Lusk *et al.* (2001) observed that younger consumers and women were willing to pay more for a guaranteed tender steak than other consumers but did not observe any impact of income. In this study, the sociodemographic characteristics did not have the same impact on willingness to pay and on preferences. Indeed age and education have a significant positive impact on the probability that a consumer prefers tender steak. There is also a number of cases where no significant impact of gender, age, education, and incomes was observed.

Consumer reactions (variations in willingness to pay) to introduction of information on food, notably on health, could be linked with sociodemographics and also with attitudes, habits, knowledge, etc. to identify a typology of groups with similar behaviours. For example, Roosen *et al.* (1998) noticed that the parents of small children had a higher willingness to pay for apples without neuro-active pesticides. Such a result permits one to evaluate the efficiency of policy formation. Shogren *et al.* (1999) found that a positive attitude towards irradiation prior to the experiment is related to a higher willingness to pay for irradiated chicken. They also observed that reading a US Department of Agriculture leaflet led participants to increase their willingness to pay for this new product. This reveals that information could help to increase the acceptance of a novel technology. In conclusion, it is clear that the data concerning the influencing sociodemographic characteristics permit a better understanding of consumers' behaviour and thus can be useful for a public decision maker for defining and targeting prevention policy or for a private decision maker for his/her marketing policy.

There also exists an emergent literature which includes experimental games (risk aversion, time preference, altruism, etc.) and/or psychological scales (impulsivity, sensation seeking and other personality traits, etc.) in valuation experiments to explore the impact of variables which are not directly observable (e.g. Combris *et al.*, 2006a). Some recent experimental papers underline differences between participants which could be linked with experimentally revealed characteristics, for example: drug users are more risk seekers (Blondel *et al.*, 2007), and self-employed workers are also more risk seekers (Colombier *et al.*, 2007).

20.4   Auctions and other methods: advantages, disadvantages and complementarities

Several authors have collected hedonic scores and reservations prices on the same set of products, either with a between-subject design (Lange *et al.*, 2002), or with a within-subject design (Noussair *et al.*, 2004a, Völckner, 2005; Combris *et al.*, 2006b; Stefani *et al.*, 2006). On average the two methods led to the same ranking of products. Moreover, it was observed that the relative importance of sensory properties and of labelling on the global evaluation of the product was identical with both methods (Lange *et al.*, 2002; Stefani *et al.*, 2006). So one can wonder what the contribution of auction methods is, or, in other words if it is worth using auctions and what auctions tell us that hedonic measurements do not tell. This is all the more important since using an auction procedure needs more explanation and training and requires longer sessions than hedonic measurements.

First, it is clear that the way a hedonic scale is used may differ among individuals. Thus, a given score does not necessarily correspond to the same value for different consumers. In contrast, reservation prices are expressed in a monetary unit which has a common value for all participants, even if its utility differs among individuals depending, in particular, on their income. Second, when collecting hedonic scores, the experimenter cannot know the hedonic score below which a consumer will not buy a product whatever its price. However, a reservation price equal to 0 definitely means that the consumer does not want to buy the product. Having this information is particularly important when studying novel technologies which are not necessarily accepted by all consumers (Noussair et al., 2004a). Hedonic measurements and auctions can be complementary. Apart from cases of boycotting, collecting hedonic scores and reservations prices with the same consumers provides the hedonic limit between which a product will not be bought. Using this approach on orange juices, Combris et al. (2006a) found that below a specified hedonic level, there is no purchase of product and observed that this level could be relatively low.

Besides the fact that auctions enable supplementary information to be col-lected, unlike compared to hedonic measurements, a detailed examination of the results collected in the studies where both approaches were used enabled the authors to reveal differences in consumers' behaviour when responding to these two questions. Owing to the higher involvement of consumers in a non-hypothetical situation, one may conjecture that auctions will reveal product differences more effectively than hedonic measurements. However, in a blind condition, i.e. the most relevant condition to study the impact of sensory properties on preferences, it has been observed than individual coefficients of variation were higher and presented a higher heterogeneity for hedonic scores than for prices (Lange et al., 2002).

It also appears, in these studies where both methods were compared, that when giving a reservation price, consumers do not only indicate their own preferences, they are also influenced by their knowledge or beliefs about the market price of the product. So, reservation prices do not only reflect a personal value attributed to the characteristics of the product but also a common market value. This can explain why when the champagne bottles were presented either alone or in combination with tasting, Lange et al. (2002) observed better product discrimination with the Vickrey method than with hedonic measurement. This can also explain why Issanchou et al. (2006) observed a higher ranking in reservation price than in hedonic rating for a baguette which looked like a premium rather than a standard product. This result illustrates that hedonic measurement and auctions refer to different systems of values attributed to the product characteristics. This is why Lange et al. (2002) observed that the individual characteristics that influence hedonic ratings are not the same as those influencing reservation prices. For example, it was found that women are more influenced than men by the extrinsic information when giving a hedonic score to champagnes in the full information condition (i.e. with tasting in presence of the bottle) but this gender difference disappears for reservation prices.

In conclusion, it is clear that hedonic measurements and the auction procedures do not assess the same constructs (Noussair *et al.*, 2004a). Thus the choice of the method depends on the question. In order to get information about the impact of the intrinsic properties of a product, hedonic measurement in a blind condition seems the most relevant method. As soon as external information is introduced, it is clear that the question of interest is 'Is there an added value of its label and how it interacts with the value attributed to the sensory properties?' In such a situation, a non-hypothetical approach such as the auction procedures described in this chapter is certainly the most relevant. Indeed, these methods permit one to know the monetary value attributed to a given product while taking into account the economical constraint faced by the consumers. However, there are some limitations in the auction procedures which will be presented in the next section.

20.5 Limits of experimental auctions and future trends

First, it is necessary to have the products and thus it is not possible to test a product presenting a combination of characteristics that do not exist yet but could potentially be developed. To test products that do not exist on the market, it is necessary to use a hypothetical procedure. To avoid the hypothetical bias, one option is to use 'cheap talk' (Lusk, 2003a). In this context, 'cheap talk' consists of explaining to participants the hypothetical bias before asking them how much they are willing to pay for a given new product (see Lusk, 2003a, for a cheap talk script). Such 'cheap talk' was used by Lusk (2003a) in a survey on genetically modified rice where a contingent valuation technique mechanism (a double-bounded dichotomous choice question) was used. Lusk found that cheap talk effectively reduces willingness to pay for consumers without declared knowledge about genetically modified foods but not for knowledgeable consumers. As pointed out by Lusk, further research is needed to have a better idea of the conditions in which cheap talk is effective at reducing hypothetical bias.

Second, most of the experiments described in the literature were conducted in a laboratory setting. This context permits good control of the conditions such as product presentation (e.g. lightning, and temperature, a key point in case of tasting) and also permits specific information to be selected and thus studied. However, a laboratory setting may limit the external validity of the results. In other words, it is possible that the absolute prices obtained in a laboratory setting differ from what people would pay in a real buying condition.

Moving the auction from the laboratory to a field setting presents several advantages (Lusk and Fox, 2003): first, it allows the population of interest to be targeted better; second, it puts participants in a context where typical purchasing decisions are made. Lusk and Fox have compared willingness to pay values obtained, for three new cookies with a BDM procedure, in a laboratory and in a store. Results showed that in-store participants tend to be significantly more likely to give zero values than in-lab participants. Authors indicated that this result could be due to the fact that in the store more participants are cash-constrained or to the fact that there are a large number of substitute products in the store. The presence of substitute products is indeed a major difference between a laboratory and a retail context. When analysing data of 'engaged bidders' (i.e. those giving at least a non-zero bid for one of the cookies), Lusk and Fox found that the mean willingness to pay was significantly higher in the store than in the laboratory. They explained their results in the context of commitment cost theory. In a store, once consumers have chosen to buy a specific product, they are more 'impatient' about purchasing it; they have more information about prices of substitutes and thus are more certain about products' relative value; and, finally, they do not expect to gather more information about product's value in the future. These results need to be confirmed for different products with different unitary costs and different buying frequencies. However, they are in accordance with the fact that willingness to pay values obtained in a laboratory for different bottles of champagne was lower than the corresponding market prices.

Even if the field experiments reported in the literature have some advantages compared with the laboratory experiments in terms of participating population, and in terms of context (place, presence of substitutes), they still have some major differences with a real buying situation. First, even in the case of multiple round auctions for a given product, all the bids are submitted within a short period of time. So, high bids can be obtained because the cost of doing so on a single occasion is low and/or may result from curiosity, appeal for novelty which could vanish with repeated purchase and consumption. This result reinforced the opinion that the price premium obtained in experimental auctions for new safer products certainly exceeds the extra cost that people would actually pay on the long term in a real retail market. Second, in experimental auctions conducted either in a laboratory setting or in a field setting, participants' attention is focused on specific product characteristics (Combris and Ruffieux, 2005). In a real purchase situation, there is a lot of available information and consumers probably pay less attention to a specific information than in a situation (whatever the place) where they are explicitly asked to participate in an experiment and where specific information is highlighted. Thus it has been shown that in a laboratory setting bids for products containing genetically modified organisms were similar after tasting them without any information than when they were presented in their original packaging, i.e. as they can be seen in a shop; but bids decreased once the labels were presented on a large screen and participants were invited to read them (Noussair et al., 2002). Finally, as soon as an experimenter is present, participants could give a socially desirable answer. Such a bias is more likely to happen in a hypothetical experiment than in a non-hypothetical experiment (Alfnes and Rickertsen, 2006). However, when bids are collected on a single occasion such a bias may not be completely excluded for some products and in particular for products with ethical dimensions, and/or products from consumers' own region/country.

A large number of experiments have demonstrated the importance of using non-hypothetical settings for collecting data about the monetary value attributed by consumers to a product with given characteristics. It has also been shown that consumers are able to give different values to close variants of a given food product. Nevertheless, it is clear that more research is needed to study the external validity of experimental auctions conducted in a laboratory setting. It is necessary to compare results collected in laboratory auctions and actual purchase data. To our knowledge, there is only one report where demand curves contracted from bids have been compared with demand curves from actual purchases, made via door-to-door sales, and where no significant differences were found (Brookshire et al., 1987). So, there is a need to perform such comparisons between data collected in a real purchase environment over a quite long period of time and without focusing consumers' attention on specific product characteristics with data from auctions collected in different situations. In particular, there is a need for comparison on a single occasion versus several occasions; with versus without focusing consumers' attention towards specific product characteristics; with versus without the option to buy or not to buy the products; and with a similar or different number of variants presented at the same time. Performing these comparisons on different types of products will give a better idea of the conditions in which data from laboratory auctions have external validity.

Further research must also be devoted to a more systematic study of the individual characteristics influencing willingness to pay in laboratory settings as well as in field settings.

20.6   Sources of further information and advice

Food product valuation through auctions is a research field in which sensory evaluation is more and more combined with experimental economics and marketing tools. A book written by Jason Lusk and Jason Shogren entitled *Experimental Auctions: Methods and Applications for Economic and Marketing Research*, published by Cambridge University Press (2007) is a very good source of information and advice to choose a method and conduct food valuations. The Charles Holt webpage is a good gateway to the world of experimental economics (http://www.people.virginia.edu/~cah2k/). Many economic journals are interested by experimental economics and auctions, for instance: *Experimental Economics, Journal of Economic Behavior & Organisation, Journal of Economic Psychology, American Journal of Agricultural Economics*. In the field of sensory science, *Food Quality and Preference* regularly published papers presenting experiments using experimental auctions.

20.7   References

ALFNES, F. and RICKERTSEN, K. (2006), 'Experimental methods for the elicitation of product value in food marketing research'. In *Primary Industries Facing Global Markets: The Supply Chains and Markets for Norwegian Food* (Ed, Asche, F.) Universitetsforlaget, Chapter 11, 268–291.

ANDERSON, R. E. (1973), Consumer dissatisfaction: the effect of disconfirmed expectancy on perceived product performance, *Journal of Marketing Research*, **10**, 38–44.

BECKER, G. M., DEGROOT, M. H. and MARSCHAK, J. (1964), Measuring utility by a single-response sequential method, *Behavioral Science,* **9**, 226–232.

BLONDEL, S., LOHÉAC, Y. and RINAUDO, S. (2007), Rationality and drug use: An experimental approach, *Journal of Health Economics*, **26**(3), 643–658.

BOUTROLLE, I., DELARUE, J., ARRANZ, D. and COMBRIS, P. (2006), How to elicit consumers' preference for a low added-value food product using WTP and WTA measurements: development of an incentive-compatible choice experiment, *Appetite*, **47**, 259.

BOWER, J. A., SAADAT, M. A. and WHITTEN, C. (2003), Effect of liking, information and consumer characteristics on purchase intention and willingness to pay more for a fat spread with a proven health benefit, *Food Quality and Preference*, **14**, 65–74.

BROOKSHIRE, D. S., COURSEY, D. L. and SCHULZE, W. D. (1987), The external validity of experimental economics techniques: analysis of demand behavior, *Economic Inquiry*, **25**, 239–250.

BUZBY, J. C., FOX, J. A., READY, R. C. and CRUTCHFIELD, S. A. (1998), Measuring consumer benefits of food safety risk reductions, *Journal of Agricultural and Applied Economics*, **30**, 69–82.

CARDELLO, A. V. (1994), In *Measurement of Food Preferences* (Eds, MacFie, H. J. H. and Thomson, D. M. H.), Blackie Academic & Professional, London, pp. 253–297.

COLOMBIER, N., DENANT-BOEMONT, L., LOHÉAC, Y. and MASCLET, D. (2007), Risk aversion: an experiment with self-employed workers and salaried workers, *Applied Economics Letters*, forthcoming.

COMBRIS, P. and RUFFIEUX, B. (2005), La révélation expérimentale des préférences des consommateurs, *INRA Sciences Sociales*, 4.

COMBRIS, P., LANGE, C. and ISSANCHOU, S. (2001), Assessing the effect of information on the reservation price for champagne: second price compared to BDM auction with unspecified price bounds, *8th Oenometrics*, 21–22 May, Napa Valley.

COMBRIS, P., ISSANCHOU, S. and LOHÉAC, Y. (2006a), Food decision, information and personality, *IAREP-SABE Conference*, 5–8 July, Paris, France.

COMBRIS, P., LANGE, C. and ISSANCHOU, S. (2006b), Assessing the effect of information on the reservation price for champagne: what are consumers actually paying for?, *Journal of Wine Economics*, **1**, 75–88.

DELIZA, R. and MACFIE, H. H. (1996), The generation of sensory expectation by external cues and its effect on sensory perception and hedonic ratings: a review, *Journal of Sensory Studies*, **11**, 103–128.

FELDKAMP, T. J., SCHROEDER, T. C. and LUSK, J. L. (2005), Determining consumer valuation of differentiated beef steak quality attributes, *Journal of Muscle Foods*, **16**, 1–15.

FOX, J. A., SHOGREN, J. F., HAYES, D. J. and KLIEBENSTEIN, J. B. (1998), CVM-X: calibrating contingent values with experimental auction markets, *American Journal of Agricultural Economics*, **80**, 455–465.

FOX, J. A., HAYES, D. J. and SHOGREN, J. F. (2002), Consumer preferences for food irradiation:

how favorable and unfavorable descriptions affect preferences for irradiated pork in experimental auctions, *Journal of Risk and Uncertainty*, **24**, 75–95.

HAYES, D. J., SHOGREN, J. F., SHIN, S. Y. and KLIEBENSTEIN, J. B. (1995), Valuing food safety in experimental auction markets, *American Journal of Agricultural Economics*, **77**, 40–53.

HOFFMAN, E., DALE, J. M., DIPANKAR, C. and RAY, A. (1993), Using laboratory experimental auctions in marketing research: a case study of new packaging for fresh beef, *Marketing Science*, **12**, 318–338.

ISSANCHOU, S., GINON, E., LOHÉAC, Y. and MARTIN, C. (2006), Effect of fibre information on consumers' willingness to pay for French baguettes, *A Sense of Diversity. Second European Conference on Sensory Consumer Science of Foods and Beverages*, 26–29 September, The Hague, The Netherlands.

JAEGER, S. R., LUSK, J. L., HOUSE, L. O., VALLI, C., MOORE, M., MORROW, B. and TRAILL, W. B. (2004), The use of non-hypothetical experimental markets for measuring the acceptance of genetically modified foods, *Food Quality and Preference*, **15**, 701–714.

KAHNEMAN, D. and TVERSKY, A. (1979), Prospect theory: an analysis of decisions under risk, *Econometrica*, **47**, 263–291.

KNETSCH, J. L., TANG, F.-F. and THALER, R. H. (2001), The endowment effect and repeated market trials: is the Vickrey auction demand revealing?, *Experimental Economics*, **4**, 257–269.

LANGE, C., ROUSSEAU, F. and ISSANCHOU, S. (1999), Expectation, liking and purchase behaviour under economical constraint, *Food Quality and Preference*, **10**, 31–39.

LANGE, C., MARTIN, C., CHABANET, C., COMBRIS, P. and ISSANCHOU, S. (2002), Impact of the information provided to consumers on their willingness to pay for champagne: comparison with hedonic scores, *Food Quality and Preference*, **13**, 597–608.

LECOCQ, S., MAGNAC, T., PICHERY, M.-C. and VISSER, M. (2005), The impact of information on wine auction prices: results of an experiment, *Annales d'Economie et de Statistiques*, **77**, 37–57.

LUCKING-REILEY, D. (2000), Vickrey auctions in practice: from nineteenth century philately to twenty-first century E-commerce, *Journal of Economic Perspectives*, **14**, 183–192.

LUSK, J. L. (2003a), Effects of cheap talk on consumer willingness-to-pay for golden rice, *American Journal of Agricultural Economics*, **85**, 840–856.

LUSK, J. L. (2003b), Using experimental auctions for marketing applications: a discussion, *Journal of Agricultural and Applied Economics*, **35**, 349–360.

LUSK, J. L. and FOX, J. A. (2003), Value elicitation in retail and laboratory environments, *Economics Letters*, **79**, 27–34.

LUSK, J. L. and HUDSON, D. (2004), Willingness-to-pay estimates and their relevance to agribusiness decision making, *Review of Agricultural Economics*, **26**, 152–169.

LUSK, J. L. and SCHROEDER, T. C. (2004), Are choice experiments incentive compatible? A test with quality differentiated beef steaks, *American Journal of Agricultural Economics*, **86**, 467–482.

LUSK, J. L., FOX, J. A., SCHROEDER, T. C., MINTERT, J. and KOOHMARAIE, M. (2001), In-store valuation of steak tenderness, *American Journal of Agricultural Economics*, **83**, 539–550.

LUSK, J. L., FELDKAMP, T. and SCHROEDER, T. C. (2004), Experimental auction procedure: Impact on valuation of quality differentiated goods, *American Journal of Agricultural Economics*, **86**, 389–405.

LUSK, J. L., JAMAL, M., KURLANDER, L., ROUCAN, M. and TAULMAN, L. (2005), A meta-analysis of genetically modified food valuation studies, *Journal of Agricultural and Resource Economics*, **30**, 28–44.

MALUNDO, T. M. M., SHEWFELT, R. L., WARE, G. O. and BALDWIN, E. A. (2001), An alternative method for relating consumer and descriptive data used to identify critical flavor properties of mango (*Mangifera indica* L.), *Journal of Sensory Studies*, **16**, 199–214.

MARRETTE, S., ROOSEN, J., BLANCHEMANCHE, S. and VERGER, P. (2006), Health information and the choice of fish species: an experiment measuring the impact of risk and benefit information, Working paper 06-WP 421. Center for Agricultural and Rural Development, Iowa State University, Ames, Iowa 50011-1070.

MENKHAUS, D. J., BORDEN, G. W., WHIPPLE, G. D., HOFFMAN, E. and FIELD, R. A. (1992), An empirical application of laboratory experimental auctions in marketing research, *Journal of Agricultural and Resource Economics*, **17**, 44–55.

MOSKOWITZ, H. R. and SILCHER, M. (2006), The applications of conjoint analysis and their possible uses in Sensometrics, *Food Quality and Preference*, **17**, 145–165.

MOSKOWITZ, H. R., KRIEGER, B. and BARASH, J. (1997), The impacts of product acceptability, brand value and price on responses of foodservice professionals to bulk turkey, *Journal of Food Quality*, **20**, 533–546.

NEILL, H. R., CUMMINGS, R. G., GANDERTON, P. T., HARRISON, G. W. and McGUCKIN, T. (1994), Hypothetical surveys, provision rules, and real economic commitments, *Land Economics*, **70**, 145–154.

NOUSSAIR, C., ROBIN, S. and RUFFIEUX, B. (2002), Do consumers not care about biotech foods or do they just not read the labels?, *Economics Letters*, **75**, 47–53.

NOUSSAIR, C., ROBIN, S. and RUFFIEUX, B. (2004a), A comparison of hedonic rating and demand-revealing auctions, *Food Quality and Preference*, **15**, 393–402.

NOUSSAIR, C., ROBIN, S. and RUFFIEUX, B. (2004b), Do consumers really refuse to buy genetically modified food?, *The Economic Journal*, **114**, 102–120.

NOUSSAIR, C., ROBIN, S. and RUFFIEUX, B. (2004c), Revealing consumers' willingness-to-pay: a comparison of the BDM mechanism and the Vickrey auction, *Journal of Economic Psychology*, **25**, 725–741.

OUDE OPHUIS, P. A. M. and VAN TRIJP, H. C. M. (1995), Perceived quality: a market driven and consumer oriented approach, *Food Quality and Preference*, **6**, 177–182.

PLATTER, W. J., TATUM, J. D., BELK, K. E., KOONTZ, S. R., CHAPMAN, P. L. and SMITH, G. C. (2005), Effects of marbling and shear force on consumers' willingness to pay for beef strip loin steaks, *Journal of Animal Science*, **83**, 890–899.

PLOTT, C. R. and ZEILER, K. (2005), The willingness to pay–willingness to accept gap, the 'endowment effect,' subject misconceptions, and experimental procedures for eliciting valuations, *American Economic Review*, **95**, 530–545.

ROOSEN, J., FOX, J. A., HENNESSY, D. A. and SCHREIBER, A. (1998), Consumers' valuation of insecticide use restrictions: an application to apples, *Journal of Agricultural and Resource Economics*, **23**, 367–384.

ROZAN, A., STENGER, A. and WILLINGER, M. (2004), Willingness-to-pay for food safety: an experimental investigation of quality certification on bidding behaviour, *European Review of Agricultural Economics*, **31**, 409–425.

RUTSTRÖM, E. E. (1998), Home-grown values and incentive compatible auction design, *International Journal of Game Theory*, **27**, 427–441.

SCHIFFERSTEIN, H. N. J. (2001), In *Food, People and Society: a European Perspective of Consumers' Food Choices* (Eds, Frewer, L., Risvik, E. and Schifferstein, H.), Springer Verlag, Berlin, pp. 73–96.

SHOGREN, J. F., FOX, J. A., HAYES, D. J. and ROOSEN, J. (1999), Observed choices for food safety in retail, survey, and auction markets, *American Journal of Agricultural Economics*, **81**, 1192–1199.

SHOGREN, J. F., MARGOLIS, M., KOO, C. and LIST, J. A. (2001), A random *n*th-price auction, *Journal of Economic Behavior & Organization*, **46**, 409–421.

SOLHEIM, R. and LAWLESS, H. T. (1996), Consumer purchase probability affected by attitude towards low-fat foods, liking, private body consciousness and information on fat and price, *Food Quality and Preference*, **7**, 137–143.

STEFANI, G., ROMANO, D. and CAVICCHI, A. (2006), Consumer expectations, liking and willingness to pay for specialty foods: do sensory characteristics tell the whole story?, *Food Quality and Preference*, **17**, 53–62.

UMBERGER, W. J. and FEUZ, D. M. (2004), The usefulness of experimental auctions in determining consumers' willingness-to-pay for quality-differentiated products, *Review of Agricultural Economics*, **26**, 170–185.

UMBERGER, W. J., FEUZ, D. M., CALKINS, C. R. and KILLINGER-MANN, K. (2002), US consumer preference and willingness-to-pay for domestic corn-fed beef versus international grass-fed beef measured through an experimental action, *Agribusiness*, **18**, 491–504.

VICKERS, Z. M. (1993), Incorporating tasting into a conjoint analysis of taste, health claim, price and brand for purchasing strawberry yogurt, *Journal of Sensory Studies*, **8**, 341–352.

VICKREY, W. (1961), Counterspeculation, auctions, and competitive sealed tenders, *Journal of Finance*, **16**, 8–37.

VÖLCKNER, F. (2005), Biases in measuring consumers' willingness to pay, *Research Papers on Marketing and Retailing. No. 25*, University of Hambourg.

WERTENBROCH, K. and SKIERA, B. (2002), Measuring consumers' willingness to pay at the point of purchase, *Journal of Marketing Research*, **39**, 228–241.

21

The use of partial least squares methods in new food product development

M. Martens, Matforsk, Norway, M. Tenenhaus, HEC School of
Management, France, V. Esposito Vinzi, ESSEC Business School,
France and H. Martens, Matforsk, Norway

21.1 Introduction

Successful food product development requires a clear vision and understanding
of the interaction between consumers and product in a context and time perspective. Common to the consumer-led and producer-led food product development is the enormous amount of information available from marketing surveys,
consumer studies, sensory analysis, product/process facts, i.e. consumer,
sensory, physicochemical/instrumental data. The success stems from the ability
to link the various types of data. We can easily fall into two extremes of either
drowning in data or oversimplifying the reality. Multivariate data analysis offers
tools to reveal underlying relationships in sets of various data. Food product
development is a multivariate challenge partly because of the complexity of the
biological materials, and partly because human perception mostly is based on
many stimuli.

'Partial least squares' (PLS) represents a certain modelling principle that
describes multivariate observations in terms of their patterns of co-variation. Two
closely related methodologies based on the PLS principle will be presented here –
PLS regression (PLSR) and PLS path modelling. Behind both of these PLS
developments, one motive has been inductive (explorative) – discovery of
predictive relationships in data tables – and another has been deductive (confirmative) – testing theories about causality. Throughout the past decades, they both have
proven helpful in making studies of complex real-world systems feasible and
statistically valid, delivering results upon which strategic decisions can be taken.

The present chapter will first give a short historical and motivational overview of PLS methods followed by a user-friendly theoretical introduction. A short layperson's guide to the PLS methods will then correspond to examples in practice and reflections upon future trends.

21.2   PLS methods

21.2.1   Short history

In order to make strategic and innovative decisions, we need relevant, reliable, timely and understandable information. The best information about the real world would come from the real world itself. However, the world is under indirect observation (Jöreskog and Wold, 1982). We seldom have direct access to the real thing: neither with our senses nor with our instrumental measuring tools. This may create methodological limitations, statistical uncertainties and cognitive interpretation problems. Realising this, an approach to multivariate data analysis is called soft modelling which works in a latent space opposite to hard modelling working on the manifest variables requiring hard, *a priori* assumptions. PLS methods and their 'mother' principal component analysis (PCA) belong to the former approach, while many central methods from traditional statistics belong to the latter, such as analysis of variance (ANOVA), ordinary least squares regression (OLSR) and maximum likelihood-structural equation modelling (ML-SEM).

Within the PLS community two different cultures have arisen from the original work by Herman Wold (Wold, 1982), namely one direction developed within chemistry/chemometrics (Wold *et al.*, 1983a; Martens and Næs, 1991) and another within econometrics (Lohmöller, 1989; Tenenhaus *et al.*, 2005a), often referred to as the PLS regression models and PLS path modelling, respectively. Furthermore, the development and use of PLS methods within sensometrics are thoroughly documented (e.g. Bech *et al.*, 2000; Martens and Martens, 1986; Næs and Risvik, 1996). The recent effort to combine PLSR and PLS path modelling is, among others, driven by the need to link technological and marketing data in food product development.

21.2.2   PLS from a user perspective

What is more important for the final sensory quality of a product – the raw materials or the processing method – and how can they be optimised? What are the relationships between product descriptors and consumer liking? What is more important for the consumer – a product's intrinsic quality or its branding? Can food choice be predicted? To what extent can consumer responses be explained by the consumers' background, e.g. age, gender, attitude and habits? Which consumer segments prefer which products, and how can they be reached?

Answering these questions is a realistic goal in new product development of food products. Typically we have a low number of product versions (in data-analytical terminology 'samples' or 'objects'), on which we collect preference

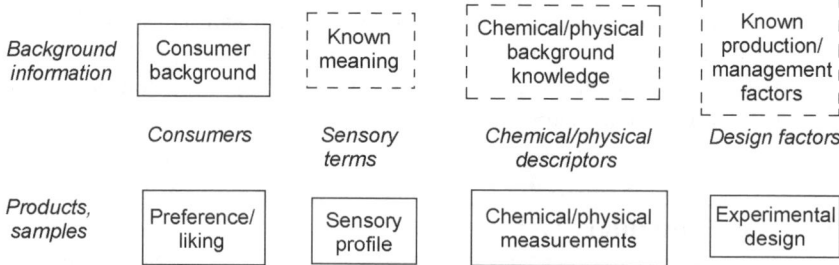

Fig. 21.1 Typical tables of input data. A given set of products or samples may be described by consumer responses, sensory profile, instrumental measurements and experimental design or production facts. Background information usually exists on the individual consumers, and sometimes also for the other sets of variables (dotted lines).

and/or liking data from many consumers, as well as sensory profile data, chemical/physical and/or instrumental data. In addition we usually have data about the product ingredients, process facts, origins and/or design factors as exemplified by the lower set of data tables in Fig. 21.1. Together, these data tables create a so-called 'multi-block' modelling situation: several different sets of descriptors for the same set of samples. Such data sets, which are the main focus of this chapter, are easily studied by PLS regression as well as by PLS path modelling.

This multi-block situation can be extended in different ways: if it is natural to split the samples into classes (*a priori* known or empirically discovered) to be modelled separately, we have what is called a 'multi-class' situation. Furthermore, we may have additional background information as illustrated by the upper set of data tables in Fig. 21.1, e.g. demographic facts, attitudes and/or habits about the consumers (solid lines) or about the other types of product descriptors (dotted lines). Together with other potential types of background knowledge, this creates a so-called 'multi-matrix' situation. Some more recent PLS modifications for analysing these more complicated situations will also be outlined.

In general, a user of data analytical methods wants to find the most important relationships between various types of data, be it at a predictive or a causal level. Here are five good reasons for choosing PLS methods:

1. PLS data modelling provides both overview and detailed insight within and between otherwise overwhelming data tables. The nature of the PLS modelling allows more versatile and informative insight into the data than most other modelling techniques.
2. PLS data modelling simplifies cross-disciplinary communication, since it can model many types of data simultaneously.
3. PLS treats natural co-variation between different variables as a stabilising advantage, not a 'collinearity problem'. It works well even in cases with far more variables than samples. This is very important in product development situations where the number of new products to be tested is often six to ten, while it is desirable to measure them with a number of types of variable.

4. PLS is suited for empirical model validation, to guard against over-optimistic interpretation.
5. PLS is tolerant to reasonable amounts of missing values (dependent on software).

The main critiques sometimes levelled against PLS methods may be assessed as follows:

- 'Multivariate methods show nothing more than univariate methods can do'. The summary graphics used so extensively in the PLS cultures gives the user much better overview than, for example, stacks of print-outs from lots of separate univariate analyses, and it gives user-friendly detection of significant information in large data tables.
- 'By using multivariate data analysis we can get whatever results we want'. Just as a good text editor program does not make everybody a good writer, PLS software does not make everybody a good researcher. But the validation methods used in PLS are usually simple to use and simple to understand.
- 'Linear models are not appropriate for "non-linear" consumer data'. PLS methods can handle curved relationships. Beyond that, non-linear modelling usually gives marginal benefit and makes the data analysis much more complicated.

It is hoped that the PLS theoretical overview (Sections 21.2.3 and 21.2.4), shortened into a layperson's guide (Section 21.3) and thereafter used in examples (Section 21.4), will illustrate this.

21.2.3   PLS regression (PLSR)

Example and terminology
Let us consider a hypothetical example: we want to model the observed appetite of children (y) from a set of p different body size measurements X, consisting of the measured length of left leg (x_1), right leg (x_2), left arm (x_3), right arm (x_4), ..., shoe size (x_p). From the X- and y-data of, say, $n = 100$ children, we want to determine ('estimate') the model $y \approx f(X)$. If the modelling process is successful, then the correlation r between the observed and predicted appetite is high, and this enables us to predict appetite y from body-measures X in many more children.

In the following, lower-case italics (e.g. p or r) symbolise scalars (individual numbers or data-points), lower-case bold-faced italic letters (e.g. x or y) symbolise that they are vectors (columns of data); upper-case bold-faced italic letters (e.g. X or Y) represent matrices (two-way tables of data). The symbol $\approx$ means 'is approximately equal to'.

The arrow scheme of Fig. 21.2a illustrates a two-block regression situation: One set of p variables $X = [x_k, k = 1, 2, 3, ..., p]$ is going to 'explain' (or predict) another set of q variables $Y = [y_j, j = 1, 2, ..., q]$ via a mathematical relationship: $Y \approx f(X)$. In the figure there are only two Y-variables and three X-variables, but in most cases many more variables can (and should) be included. In our example, the X-variables could be the children's body size measurements

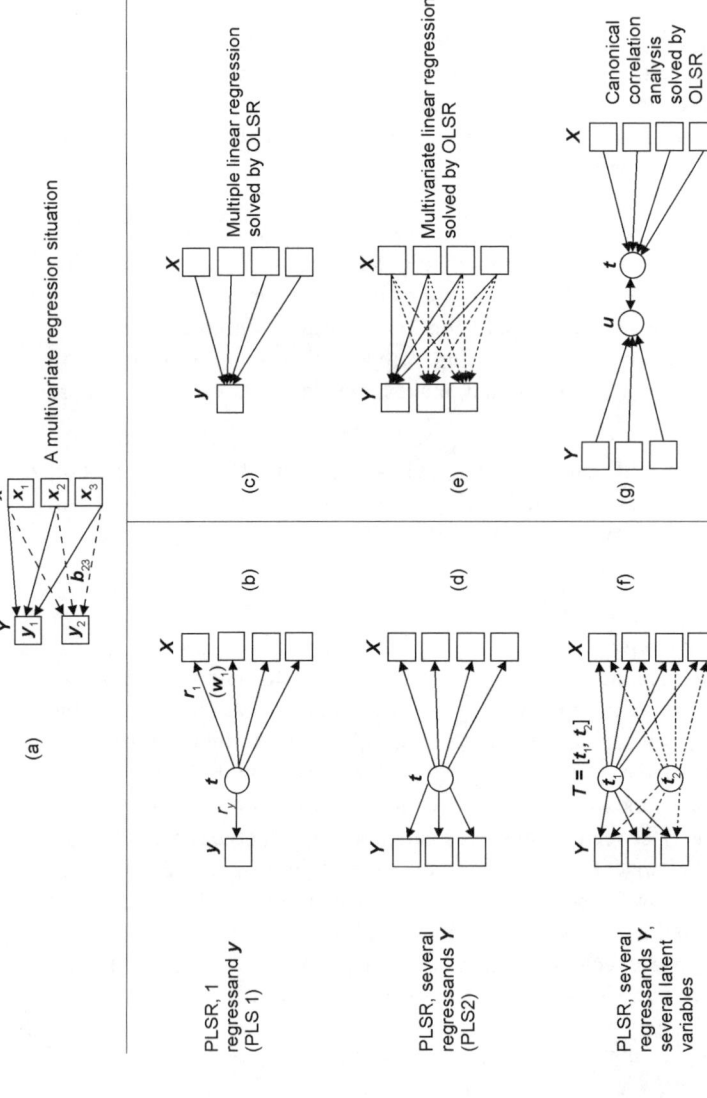

Fig. 21.2 PLSR vs OLSR. (a) Several input variables $X = [x_1, x_2, x_3, \dots]$ are to be described by several input variables $Y = [y_1, y_2, \dots]$. (b) PLS regression for a single Y-variable via a single latent variable t which is a weighted (w) sum of the X-variables. r = correlation coefficients between t and the input variables. (c) OLS regression for a single Y-variable from X. The arrows collide if the X-variables are strongly intercorrelated. (d) PLS regression for several Y-variables, via a single latent variable t from X. (e) OLS regression for several Y-variables from X. (f) PLS regression for several Y-variables via two latent variables t_1 and t_2 from X. (g) Canonical correlation analysis between several Y-variables and several X-variables via latent variable t from X and u from Y.

and the Y-variables could include various characteristics of the children beyond appetite (y_1), e.g. vocabulary size (y_2), reading speed (y_3), ..., liking of apples (y_q). Although other alternatives exist (e.g. artificial neural nets), there are in most cases very good reasons to choose a simple *linear* model type: $Y \approx XB$. The regression coefficients $B = [b_{11}, ..., b_{p1}, b_{12}, ..., b_{p2}, ..., b_{1q}, ..., b_{pq}]$ linking all the X-variables to all the Y-variables, are unknown and have to be found based on the X- and Y-data from n samples (e.g. the $n = 100$ children). An important statistical question is how to estimate so many regression coefficients *reliably*, given the available data from only n samples. A more general question is how to find and get an overview of *all* the informative data-patterns within and between these blocks of descriptors.

How PLSR differs from traditional statistical regression
The rest of Fig. 21.2 explains why PLSR works so well for analysing such multivariate regression situations. Three versions of PLSR (left side) are compared with alternative tools from traditional statistics (right side) based on ordinary least squares (OLS). In these illustrations, the traditional semiotic symbols from path modelling are used. Squares represent manifest variables (observed, measured or known attributes). Circles represent latent variables (variables not directly observed, but derived as some linear combination of a set of manifest variables). Arrows point from variables that are either known or estimated to variables that we want to learn how to predict.

The arrow schemes in Fig. 21.2b and c illustrate two different solutions to the simplest multivariate linear regression (MLR) situation, in which one 'regressand' or 'endogenous variable' y is to be described reliably as a function of a set of p 'regressor' or 'exogenous' variables $X = [x_1, x_2, ..., x_p]$: $y \approx Xb$.

The traditional statistical methods based on assumptions of full-rank estimate the coefficient values $b_1, b_2, ..., b_p$ *independently of each other*. These estimates become imprecise when the p X-variables are correlated and thus do not carry independent information about y. This full-rank 'formative' modelling assumption may work well if indeed each of the X-variables carries unique information. But alas, in our little example, children's left and right arms and legs, etc., tend to grow at more or less the same relative pace. Hence, many of our present X-variables tend to increase *together* as children grow. Such a network of inter-correlations should in fact be expected in most real-world experimental data sets, be it in marketing surveys or chemical quality measurements. The world is generally under indirect observation, so each individual underlying source of variability can seldom be observed directly and uniquely.

Figure 21.2b and c illustrate how differently PLSR and OLSR handle intercorrelations between X-variables. In the traditional 'formative' method OLS regression ('OLSR', Fig. 21.2c), natural co-variation between different X-variables creates a 'collinearity problem', namely a collision between the contributions of different X-variables to y, making it impossible to estimate each and every regression coefficient $b_1, b_2, ..., b_p$ independently. Moreover, the X-variables are then assumed to be noise-free. In contrast, the PLSR (Fig. 21.2b) is a 'reflective' method, where the collinearity between the X-variables is sum-marised into a *latent variable t*, which is a weighted average (linear com-bination) of *all* the X-variables. This latent X-variable or 'PLS component' (PC) t is defined so that t represents the most dominant common pattern of co-variation in the X-variables, while rejecting much of the noise in X.

In our example, the dominant pattern in X corresponds to the physical maturity of each child. This latent variable t is then used for modelling y. Just how the latent variable t was constructed from the input variables in X is defined by the so-called 'weights' or 'loading weights' as illustrated for the first X-variable in Fig. 21.2b by (w_1); how these weights are found will be described later. On the other hand, how well this latent variable correlates to each of the input variables is represented by the arrows from t to Y and X, and are in PLSR called 'correlation loadings' as illustrated by 'r' for y. Usually, the weights w and the correlation loadings r are rather similar.

The use of latent variables eliminates the collinearity problem: with increas-ing intercorrelation between the variables in X, the traditional OLSR solution becomes less and less stable and thus more and more sensitive to noise in the data, while the PLSR solution just becomes more and more stable. The arrow from t to y represents the Y-loading and shows how well y is modelled by t in this data set, in terms of the correlation (r, a number between -1 and $+1$) between t and y. Similarly the X-variables themselves can be reflected by t.

The arrow schemes in Fig. 21.2d and e illustrate PLSR and OLSR in the 'multivariate regression' situation, where a set of regressand variables $Y = [y_1, y_2, ..., y_q]$ are to be related to the set of X-variables, $Y \approx XB$. In the above example, X again represents the body size descriptors and Y represents the children's liking of apples, vocabulary size etc. Both the PLSR solution (Fig. 21.2d) and the traditional OLSR solution (Fig. 21.2e) remain the same. The OLSR works quite well if the X-variables vary independently, but creates massive collinearity collisions when the X-variables show strong co-variation. On the other hand, the PLSR simply uses a latent variable from X to reflect the patterns of co-variation, in both X and Y.

The PLSR for one regressand y (Fig. 21.2b) was originally called 'PLS1', to distinguish it from the multi-Y version of PLSR (Fig. 21.2d), which was then termed 'PLS2'. Today it is realised that PLS1 is just a special case of PLS2.

In most real-world data sets there is more than one clear pattern of co-variation. For the present example on children, in addition to the general *size* component that correlates positively to all the body measures of the children, an *obesity* component might be expected, correlating positively to body weight but negatively to the length measurements. In PLSR this is simply handled by increasing the number of latent variables (PLS components, 'PCs'): $T = [t_1, t_2,... t]$. These 'super-variables' t_1, t_2, ... from X are completely uncorrelated with each other in the available sample set. The optimal number of such A is determined from the data, e.g. by so-called cross-validation.

Figure 21.2f illustrates this for the case with $A = 2$ PLS components. In both cases the Y-variables may or may not be intercorrelated among themselves. But while these Y-intercorrelations are left unused in the multivariate OLSR, the PLSR uses it to stabilise the solution further. As an extreme contrast, the traditional OLS method of canonical correlation analysis (CCA, Fig. 21.2g) not only requires each of the X-variables to carry independent information, but also requires the same for each of the Y-variables. So it works only when neither the X-variables nor the Y-variables are too correlated.

Model estimation

The OLSR and PLSR estimates of the regression coefficients are identical if the number of PLSR components A is chosen to be equal to the number of X-variables (p). Unfortunately, this full-rank solution $B_{A=p}$ is often bad, because too many independent parameters have been estimated from data of too few samples, and this may lead to so-called variance inflation.

For many data sets the optimal A is often just 2 or 3, and two- or three-dimensional bi-linear models are easy to inspect graphically, as the subsequent examples will show. If the optimal $A > 3$, it can be easier to look at the so-called 'reduced-rank' linear summary model $Y \approx XB_A$. The PLSR regression coefficient B_A represents the regression coefficients B in the general model $Y = XB$ (Fig. 21.2a). It is obtained by a combination of the bi-linear PLSR parameters, as described, for example, in Martens and Martens (2001). But compared with the OLS estimate of B (Fig. 21.2e), the reduced-rank PLSR estimate is stabilised, because the minor, noisy variation patterns in X (those beyond the optimal rank A) are simply ignored.

Figure 21.3 illustrates how the main parameters of each component are estimated in the one-block PCA and the two-block PLSR methods PLS1 and PLS2. In all three cases the latent variable or PC score vector t is a linear combination of the X-variables, and is used reflectively for modelling the X-variables themselves.

The methods differ in how the weights for the consecutive PCs from X are defined. While the PCA only attempts to approximate X itself, PLSR attempts to model both X and Y. In PCA (Fig. 21.3a) the weights ensure that each consecutive PC vector t describes as much as possible of the variation among all the X-variables – this means that in PCA t displays maximum covariance with itself. In PLS1 (Fig. 21.3b) t instead has maximum covariance with y, while in PLS2 (Fig. 21.3c) t has maximum covariance with a linear combination of all the Y-variables, u.

One of the two main benefits of the PLSR is that even in situations where the X-variables are highly intercorrelated, the modelling of Y from X is simple – just find the optimal number of PCs, A, and use this in the prediction model. Hence the method is reasonably safe to use in practice. The second, even more important benefit is the graphical interpretability provided by the various PLSR model parameters, the scores $t_1, t_2, \ldots$, the correlations between these scores and the original X- and Y-variables, the reduced-rank regression coefficients B_A, etc. Hence the method provides both overview and detailed insight into the actual data set at hand, not only into a presumed 'population' which the data set is intended to represent.

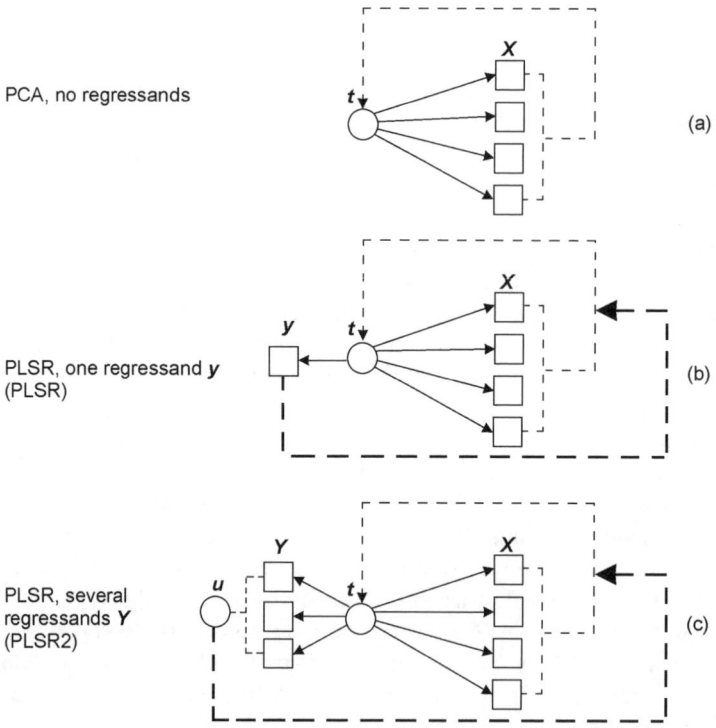

PCA, no regressands

PLSR, one regressand **y**
(PLSR)

PLSR, several
regressands **Y**
(PLSR2)

(a)

(b)

(c)

Fig. 21.3 Estimation of weights $w_1, w_2, w_3, w_4, \ldots$ that define how latent variable t is obtained as a weighted sum of the X-variables. (a) Principal component analysis (PCA): t describes maximum covariation within X. (b) PLSR with one Y-variable ('PLS1'): t describes maximum covariation with y. (c) PLSR with several Y-variables ('PLS2'): t describes maximum covariation with u, which is the most relevant weighted sum of the Y-variables.

Note that PLSR is asymmetrical, in the sense that it is geared towards predicting Y from X. The choice of what is X and what is Y is pragmatic and does not have to be related to causal considerations.

Special cases and extensions
When external design information is available, this can be used as either X or Y in PLSR. To distinguish between them and to emphasise the fact that both use conventional two-block PLSR, Martens and Martens (2001) gave them the abbreviations APLSR and DPLSR, respectively. Both involve known design data (see Fig. 21.1), but the thought models behind them are rather different. If we use known design variables as X and a set of response variables as Y, then the PLSR allows us to study the effect of the design variables on the response variables, like an enhanced multivariate ANOVA, revealing main effects, interaction effects, etc. (Martens and Martens, 1986). The graphical function-ality of PLSR provides overview, while the re-sampling (cross-validation/jack-knifing) allows significance testing. Thus APLSR means 'Anova-like PLSR'.

The opposite abbreviation, DPLSR means 'discriminant PLSR', and concerns the use of known design variables as Y (e.g. indicators of which known class each sample in the training set belongs to) and a set of response variables as X. The PLSR then helps us combine the response variables in such a way that we can discriminate between different classes of samples, and use this to classify new samples. This use of PLSR was originally called PLS discriminant analysis (abbreviated 'PLS-DA', Wold *et al.*, 1983b), and represents a user-friendly alternative to the traditional OLS-based full-rank linear discriminant analysis (LDA). Further improvements to the PLS-DA have been suggested for the multi-class case (Barker and Rayens, 2003; Nocairi *et al.*, 2005).

In the above examples the regression is performed over rows in Fig. 21.1, to find the relationships between blocks of variables describing the same samples. Conversely, the regression may be performed over columns between two data tables. For consumer data (Fig. 21.1), we may find, for example, how their patterns of preference/liking (Y) are related to their patterns of background descriptors (X) or vice versa.

A number of extensions of the PLSR have been developed over the years. For instance, if X is a three-way data table (e.g. 10 samples × 15 sensory descriptors × 12 sensory panellists), the so-called 'N-PLS' developed by Bro (1996) may be used instead of using the panel average in conventional PLSR. Moreover, if background information Z about the Y- or X-variables exist, so that the three blocks of data form an 'L-shaped' data structure, then various forms of 'L-PLSR' (Martens *et al.*, 2005a,b; Section 21.4.3) or 'Domino-PLS' (Martens, 2005) may be used. So-called PLS typological (or clusterwise) regression (Esposito Vinzi *et al.*, 2004) tolerates clusters of samples (products or consumers) with more coherent models, compared with a single, global PLSR model applied to the whole collection of samples.

Multi-block situations arise when three or more types of variables are available for the same set of samples, e.g. preference/liking data, sensory profile data, chemical/physical measurements and experimental design information for $n = 10$ products (Fig. 21.1). Special multi-block PLSR extensions also exist (see, e.g., Tenenhaus and Hanafi, 2007), which can give more detailed graphical insight. However, the many blocks may instead pragmatically be merged into two blocks and analysed by the standard two-block PLSR in the strongly 'data-driven' tradition of chemometrics. On the other hand, if there is a need to test a more detailed causal theory, the method of PLS path modelling is available from the more 'theory-driven' traditions of marketing and econometrics.

21.2.4   PLS path modelling

PLS path modelling can be seen as a multi-block generalisation of the one-block PCA and the two-block PLSR. First, the simple PLSR will be recast as a two-block path model. To accommodate the fact that the two PLS methods stem from two different scientific cultures, the display of the predictive direction of the arrow scheme has been turned around: while Fig. 21.2f represents the one-

(a)

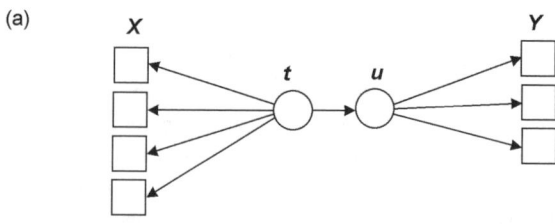

(b)

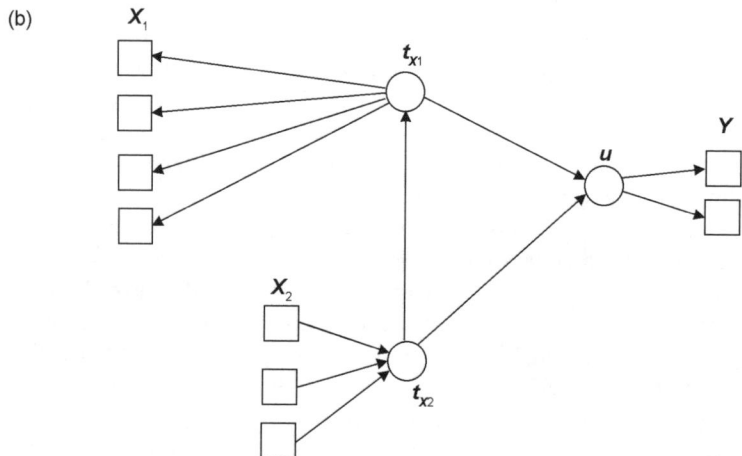

Fig. 21.4   Examples of PLS path models. (a) PLS regression (Fig. 21.2d) visualised as conventional path model. Two blocks X and Y modelled via their first latent variables t and u, in a structural equation where Y is predicted by X, and hence u is predicted by t. (b) Three blocks X_1, X_2 and Y modelled via their first latent variables t_{X1}, t_{X2} and u, in a structural equation where block X_1 affects X_2 and Y, while block X_2 affects only Y and not X_1. In this example, X_1 and Y are modelled reflectively ('Mode A') while X_2 is modelled formatively ('Mode B').

component PLSR in the right-to-left direction of the statistical regression convention ('Y is a function of X'), Fig. 21.4a shows the same method in the left-to-right direction of the Path Modelling convention ('X leads to Y').

A more extended PLS path model is now outlined, to illustrate how causal assumption details can be built into the PLS model. Figure 21.4b shows an example of a three-block PLS path model in which two blocks of X-variables (X_1 and X_2) are assumed to affect a third block of variables, Y. Each block is usually represented by one latent variable – the exogenous regressor blocks X_1, X_2, ... are represented by t_{X1}, t_{X2} and the endogenous regressand block Y by u. The PLS path modelling first models each single block at a time in terms of its latent variable(s) in order to find the relevant weights assigned to each observed variable in the construction of the corresponding latent variable. Then it updates the solutions by taking into account the assumed causal links between the blocks, and finally reiterates this procedure until convergence.

Two particular deviations from the two-block PLSR are illustrated in the PLS path model (Fig. 21.4).

Reflective vs formative latent variables
Block X_1 in the example is defined to be approximated by a latent variable t_{X1} of the same type as in PCA and PLSR ('Mode A'). In contrast, block X_2 is defined like the X-variables (regressors) in, for example, full-rank OLS regression (Fig. 21.2c): each variable in block X_2 is expected to provide unique information in the way the block's latent variable t_{X2} is formed ('Mode B'). This may be useful if the input variables are observed independently.

In our example, block X_1 could represent various body measures of children, and the latent variable t_{X1} could represent the physical maturity of the children. In contrast, the variables in block X_2 could be the children's habitual consumption of fruits and vegetables, their parents' income, their gender, etc., and t_{X2} could thus represent the children's age-independent predisposition for liking apples. The two variables in block Y could represent two different product types, say, red apples and green apples; this 'liking'-block is here approximated by its first latent variable u in a reflective model.

Inner structure model
Since there are more than two blocks of variables, their latent variables can be related to each other in several different ways. The structural equation system behind the path model in Fig. 21.4b shows that both blocks X_1 and X_2 are expected to affect block Y. In addition, block X_1 is expected to affect X_2.

So let us assume that Y represents children's liking of different products, X_1 represents the physical measures and hence the physical maturity of the same children and X_2 represents some sociodemographic background descriptors and hence their predisposition for liking the given products. The model in Fig. 21.4b would then imply that we expect both the summary of physical measures X_1 and the combination of some sociodemographic background X_2 to affect children's liking of apples Y, via their latent variables. But we also expect the sociodemographic background descriptor block X_2 to affect the physical measures block X_1 (e.g. due to nutritional habits or genetics in the family), irrespective of their liking of these apples. Moreover, we assume that each sociodemographic background descriptors to provide independent information about the latent predisposition variable t_{X2}. Finally, we expect the latent variables t_{X1} and t_{X2} to provide information Y. Later in this chapter we illustrate how the sign and the quantitative size of each of the relationships (arrows) in this arrow scheme can be estimated and displayed by using a PLS path modelling program. PLS path modelling concerns analysis of multi-block data in light of current knowledge, and may be used either deductively, for statistical testing of various causal hypotheses, in analogy to the more restrictive technique of LISREL (Jöreskog and Wold, 1982), and/or inductively in a prediction-relevance oriented discovery process based on graphical inspection of, for example, the different block scores, in analogy to other multi-block bi-linear methods.

The causal assumptions on which the user subjectively chooses to build his or her path model should be assessed critically. If the underlying causality theory is weak, and the purpose of the data modelling is primarily to predict a set of Y-variables from one or more blocks of X-variables, then a conventional two-block PLSR might be easier, owing to its lower need for theoretical assumptions.

21.3   Layperson's guide to PLS methods

21.3.1   General

Based on the more theoretical overview of PLS methods above, we pick out some important issues of relevance to the users. A more detailed 'Layman's guide to multivariate data analysis' can be found in Martens *et al.* (1983), later expanded (Martens and Martens, 2001, p. 51).

In general, there are six steps in a research and development (R&D) project in order to reach from question via a data model to answer:

I: Defining purpose/objective/question to be answered.
II and III: Experimental planning and experimental work.
IV: Preprocessing and quality control of data.
V: Data analysis, including choice of model and validation.
VI: Conclusion, including graphical presentation.

In product development the overall purpose is often to predict and understand consumer behaviour, preferences, expectations, attitudes, wants and needs in relation to product characteristics (i.e. step I). Experimental design, etc. (i.e. steps II–IV) are of course extremely important for getting good data, but that will not be outlined here. In the following we shall rather concentrate on steps V and VI above of relevance to product development.

21.3.2   Typical input data tables

The various types of data illustrated in Fig. 21.1 can be translated into a more data analytical terminology as in Fig. 21.5 to be used in the following sections.

- Product samples = samples $(1, 2, \ldots, n)$. The products must be relevant to the purpose, i.e. target population, and span interesting product variation. Fewer than ten product samples are common. But this is not recommended, because some degree of repetition between the samples is desired so that the reproducibility of the conclusions can be assessed. A 'class' here means a set of samples or objects.
- Variables $(1, 2, \ldots, p)$. Let the same n samples be evaluated by, e.g., 100 consumers $(p = 100)$ giving their liking response for n samples (Y). A 'block' is here used to mean a set of variables. In the two-block situation, we may have analytical sensory description of the same samples (X). In the multi-block situation we may also have chemical and instrumental measurements (X_2) and experimental design information about the product (X_3).

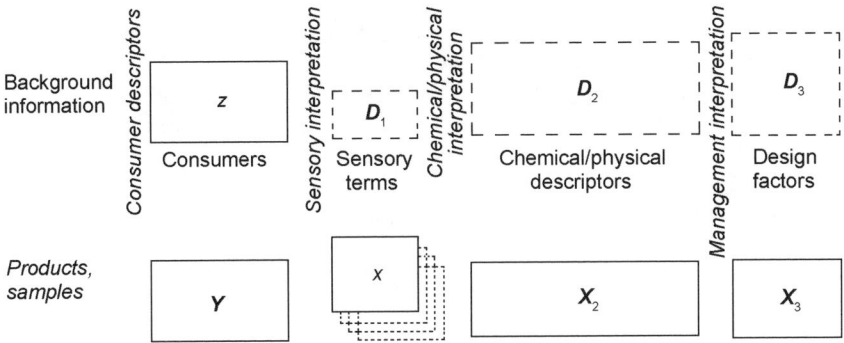

Fig. 21.5 Multi-matrix modelling situations. A chosen set of products or samples are described by several types of variables, $Y, X, X_2, X_3, \ldots$, forming a multi-block data structure. Some of the data tables (blocks of variables) may be N-way, e.g. the sensory responses for individual assessors are outlined behind the sensory block X. Background information usually exists for the individual consumers (Z), so that Z, Y and X form an 'L-shaped' data structure. Other background data tables $D_1, D_2, D_3, \ldots$ (dotted lines) extend this to a more general multi-matrix data structure.

- Data tables. The samples constitute rows and the variables are columns in a two-way ($n \times p$) data table (also called a matrix).

In a general 'multi-matrix' situation, we may, in addition to Y and X, have information about the consumer's background through questionnaires (Z = e.g. demographic variables, attitudes) as well as additional information and background knowledge about the sensory (D_1) and/or instrumental measurements (D_2). The sensory variables in D_1 may be groupings of the sensory terms into odour, taste or texture (and, e.g., further classes in a flavour wheel). Likewise the chemical/physical variables in D_2 may be interpreted and classified into nutritional or health aspects. A third type of information concerning product design, may be interpretations by the management in a company, thus examples of variables in D_3 could be cost or environmental factors. All together we can put the data into a multi-matrix scheme (Fig. 21.5) for linking the various types of data.

21.3.3   Choosing a suitable data model
Referring to Fig. 21.5 we can, in broad terms, get an overview of connecting the purpose to a data analytical model.

One data block (X or Y or Z, etc, or a combination of several blocks into one)
- Purpose: find the main patterns of systematic variation in input data, e.g. for visual quality control of any input data.
- Data analysis: principal component analysis (PCA).

Two data blocks (X and Y)
- Purpose: find the main systematic variation patterns from one data table X that have relevance also for another data table Y. This allows us to interpret

the structures within and between X and Y, and the obtained model may be used for predicting Y from X in future samples.

- Data analysis: PLS regression of Y on X (PLSR). Which variables to put in X and which to put in Y is up to the user. It is usually a good idea to try a couple of combinations in order to gain full insight (quite easy in dedicated PLSR software). The special cases of using known design or classification variables as X (APLSR) or as Y (DPLSR) are very useful, for analysis of design-effects and for discrimination/classification, respectively (see, e.g., Martens and Martens, 2001; Barker and Rayens, 2003).

Multi-way regression (N-way X and Y)
- Purpose: find the systematic relationships within a set of samples, where several modes of variables in X have relevance to Y. Figure 21.5 illustrates a case where three-way sensory block X (samples × terms × assessors) can be related to conventional two-way Y.
- Data analysis: Lengard and Kermit (2006) analysed such data by three-way PLSR. Dijksterhuis *et al.* (2005) combined conventional PLSR with generalised Procrustes analysis of three-way X.

Three data blocks (X, Y and Z)
- Purpose: find the main systematic variation patterns in both X and Z that are of relevance to Y.
- Data analysis: endo-L-PLSR (e.g. Martens *et al.*, 2005a,b), exo-L-PLSR (Sæbø *et al.*, 2007) or *N*-way L-PLS (Plaehn and Lundahl, 2006).

Multi-block PLSR (Y, X₁, X₂, X₃, ...)
- Purpose: find the main systematic variation patterns between multiple blocks of data measured on the same set of product samples.
- Data analysis: multi-block PLSR or PLS path modelling (see, e.g., Tenenhaus *et al.*, 2005a; Esposito Vinzi *et al.*, 2007).

Multi-matrix PLSR (Y, X₁, X₂, X₃, ..., and Z, D₁, D₂, D₃, ...)
- Purpose: multi-blocks in two dimensions, i.e. find the main systematic variation patterns among several data matrices.
- Data analysis: domino PLSR (see, e.g., Martens, 2005)

21.3.4   Typical output results
Results from a PLS analysis in an applied situation are often graphically expressed in the following terms, to be used in Section 21.4.

- *Prediction error vs number of PLSR components*: show the model's ability to predict Y from X via PLSR components. The optimal number of PLSR components t_1, t_2, ... for a data set should be taken as the lowest number of components (e.g. 2 or 3) that makes the cross-validated prediction error in Y from X low. These components are considered 'significant'.

- *Scores*: show the main relationships modelled between the product samples, by plotting the first few significant latent variables t_1, t_2, ... (circles in Fig. 21.2f) pair-wise against each other.
- *Correlation loadings*: show the main relationships modelled between the variables, by plotting the correlations r (arrows in Fig. 21.2b) between the first few significant latent variables t_1, t_2, ... and each of the input variables x_1, x_2, x_3, ..., y_1, y_2, ...
- *Regression coefficients*: show the full multivariate model, summarised over all the significant PLSR components, as reduced-rank regression coefficients B_A in the linear $Y \approx XB_A$. Use the perturbations in B_A in the cross-validation to assess the uncertainty in B_A ('jack-knifing') and look for X–Y relations that are statistical significant.

21.4   Examples of PLS methods in practice

In the frame of the layperson's guide above, we shall now show examples of data models for two data blocks, three data blocks and multi-block data. Each example refers to Figs 21.1 and 21.5.

21.4.1   Sensory-chemical relationships: PLSR

The first example concerns a classical use of two-block PLSR of relevance to product development: relating Y = sensory data to X = chemical and experimental design data (cf. Fig. 21.1).

Objectives

Is there any relationship between variations in sensory and chemical properties of a certain type of product? To what extent can chemical measurements be used to predict sensory quality? What is the effect of varying ingredients or process parameters on the sensory quality?

Data description

Frøst *et al.* (2001) investigated the effects of fat contents and various processing factors (homogenisation and addition of thickener, whitener and cream aroma) on milk. Sixteen milk products were produced in a factorial design and assessed with respect to 15 sensory properties (whiteness, yellowness, creamy smell and flavour, total fattiness, etc.).

Results

As shown in Frøst *et al.* (2001) each of the 15 sensory properties (panel averages) was found to be significant in univariate ANOVA. However, this information does not tell us anything about the underlying patterns of co-variation between the sensory terms. Therefore the 15 sensory properties (Y) were related to the fat content (0.1, 1.3 and 3.5% fat milk) and the processing factors (X) by PLSR. Cross-validation showed that the first two PLS components t_1 and t_2 from X had predictive validity for Y. Their correlation loadings, showing how t_1 and t_2 correlate to X and Y, are shown in Fig. 21.6a. It reveals one main variation pattern linking, for example, increasing sensory yellowness to increasing fat content, from 0.1% via 1.3% to 3.5% fat. The other main variation correlation pattern associates three of the design factors, thickener, whitener and homogenisation, to, for example, sensory whiteness, although more weakly. To illustrate how such correlation loading plots are to be understood, the score plot of the individual samples (t_1 vs t_2) is shown twice: in Fig. 21.6b the fat percentage is outlined, while in Fig. 21.6c the three design factors thickener, whitener and homogenisation are outlined (high vs low levels). Moreover, back in the correlation loading plot, a cluster of sensory descriptors (creamy-flavour, thickness-visual, total fattiness, creaminess, thickness, residual mouth-fill and glass coating) is anti-correlated with another cluster (boiled milk-

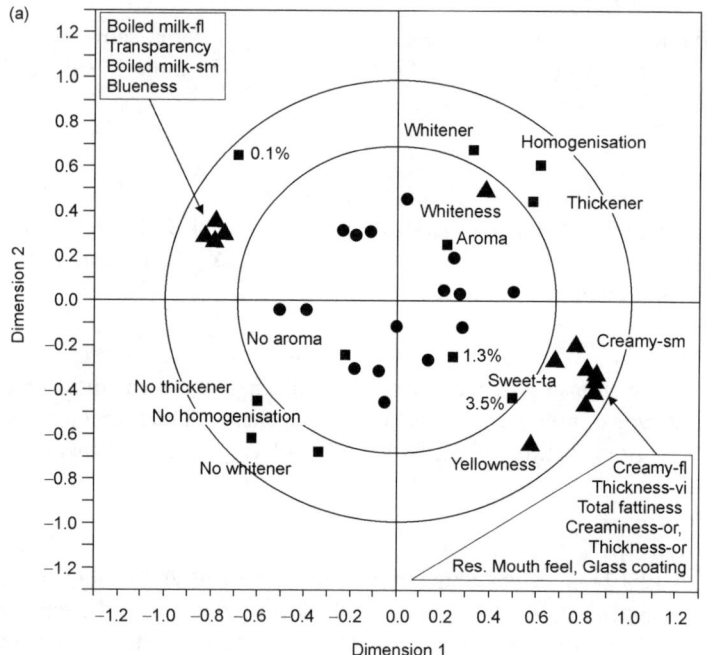

Fig. 21.6   (a) APLSR correlation loadings for the two first dimensions showing differences among the 16 products; see Frøst *et al.* (2001, p. 332). Reprinted with copyright permission from Elsevier, 2006. ▲ Sensory descriptors, ■ design factors and ● products. For clarity, product names are not shown. The inner and outer circles represent 50 and 100% explained variance, respectively. (b) Score plot t_1 vs t_2. Identical fat levels are connected and indicated as (0 = 0.1%, 1 = 1.3% and 3 = 3.5%) and the fat level direction from the correlation loadings is outlined. (c) Same score plot t_1 vs t_2, with identical production factor levels connected and named with respect to presence or absence: thickener (T/t), whitener (W/w) and homogenisation (H/h). Arrows show overall whitener and thickener level direction and homogenisation level direction based on the loadings.

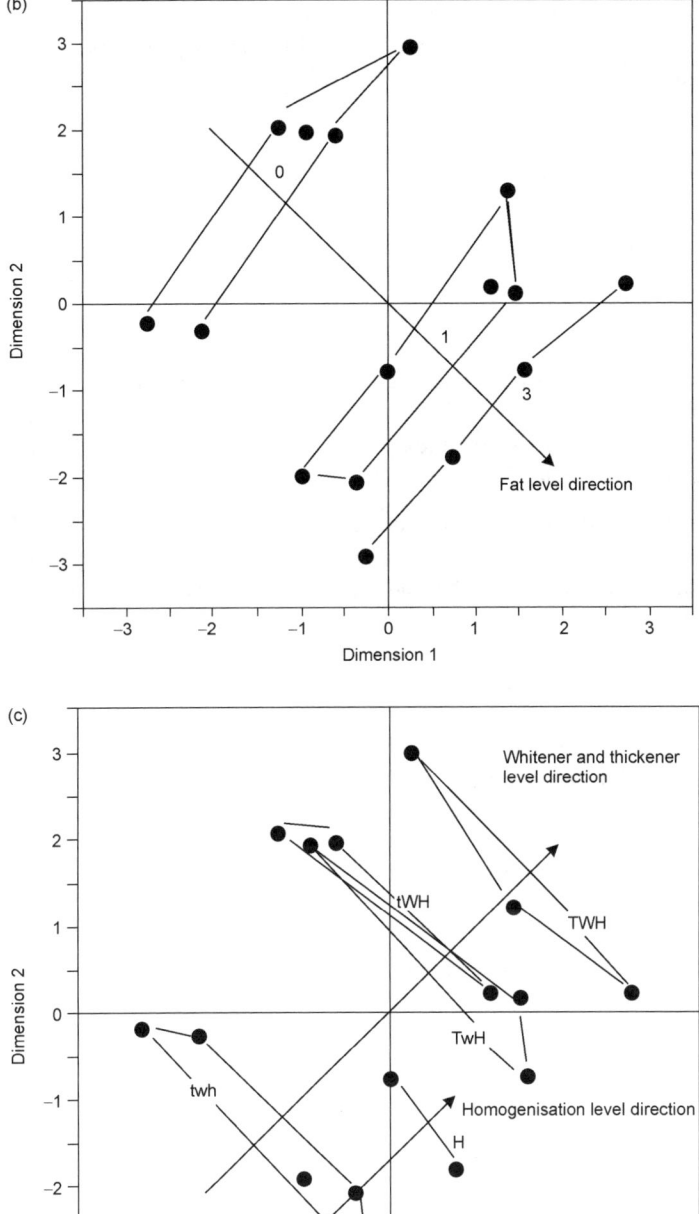

Fig. 21.6   Continued

flavour, transparency, boiled milk-smell and blueness); the direction of this contrast shows that it is primarily associated to the fat percentage, but to some degree also to the addition of thickener, etc.

Discussion
So, what did we get out of the data by applying multivariate APLSR beyond that from univariate ANOVA?

- The PLSR results (Fig. 21.6) revealed that sets of sensory properties co-varied.
- Different sets of sensory properties were differently affected by the fat content and process factors.
- The experiment showed larger sensory differences between 0.1% and 1.3% fat than between 1.3% and 3.5%. So the fat does not affect the properties of milk in a linear fashion, but the linear PLSR picks up such non-linearities and reveals them graphically.

The present results may explain the success of a number of new low-fat milk products (0.5–0.7% fat) since 2000: they mimic the sensory quality of milk with higher fat content, owing to the non-linear sensory response to fat content!

21.4.2   Consumer-sensory relationships: PLS preference mapping (PrefMap)

The next example illustrates the consumer-driven question: to what extent will an analytical sensory description of a product relate to consumer preferences? Relating to Fig. 21.5, we here refer to Y = different consumer segments' preferences for a set of products and X = sensory descriptive profile for the same products (panel average).

Objective
What are the relationships between consumer liking and sensory description of a product?

Data description
The data come from a preference mapping project on hard and semi-hard cheeses (Westad *et al.*, 2004). For sensory profiling 15 different varieties of cheeses were selected. Based on sensory evaluation by a trained panel and use of PCA, six of the cheese samples (here named 1, 2, ..., 6) were taken to a consumer test. In total 177 consumers evaluated liking of the six samples. The consumers were segmented into 12 more or less distinct groups (with technical names S1, SCORES1, CL50S0, ..., CL?FC, ..., SCORES3) by various data analytical methods (not shown here). The average product preferences in the segments and the mean preference of all 177 consumers (PREFMEAN) were defined as Y-variables and subjected to external preference mapping, i.e. regressed on 22 sensory descriptive terms as X-variables. A PLSR was then

employed on the *X* and *Y* data. For 'bi-plot' visualisation, dummy variables for the six products were correlated to the PLS components and included in the correlation loading plot.

Results

Results from the PLSR are shown in Fig. 21.7 in terms of correlation loadings (i.e. how the input variables correlate to t_1 and to t_2). The data were found to span two main components. The first, horizontal component associates high PREFMEAN with high acid flavour and odour, low bitter and sharp flavours, etc. Consumer segments CL50S1 and SCORE?FC respond like PREFMEAN. However, in the vertical direction consumer segments SCORES1 and SL50S0 respond in the opposite direction to segments SCORES3 and S3 with regard to products 4 vs 5 and 6; their preference disagreement concerns stickiness, fattiness, juiciness and solubility vs graininess and hardness. Product 6 is seen to be particularly bitter and sharp with low acid flavour. Product 5 has particularly high solubility, fattiness and low graininess, hardness and elasticity, while product 4 has almost the opposite characteristics.

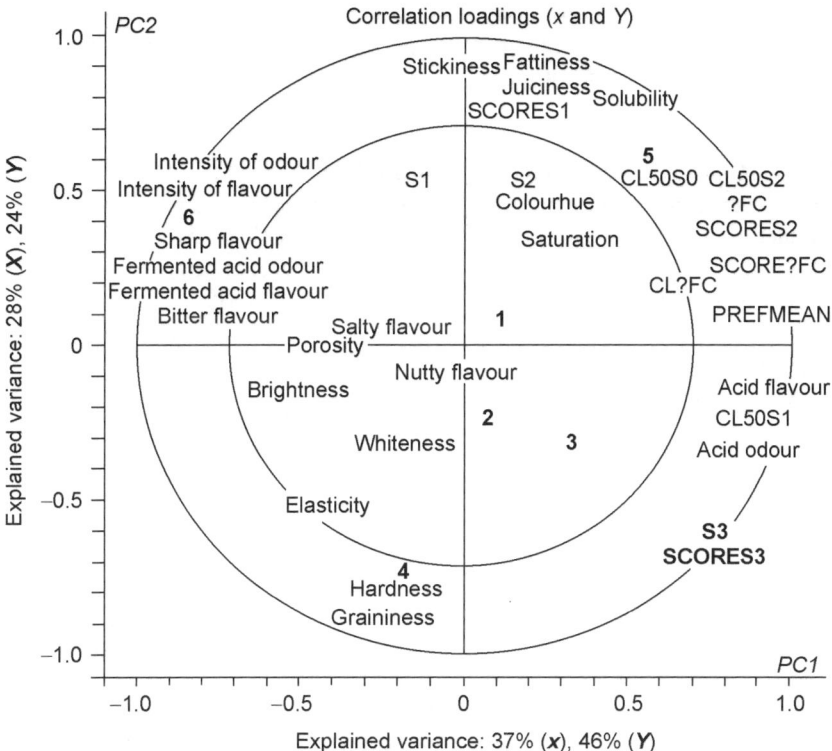

Fig. 21.7 Correlation loadings for components 1 and 2 for external preference mapping with consumer segments as response variables in PLS regression. See Westad *et al.* (2004, p. 685). Reprinted with copyright permission from Elsevier, 2006.

Discussion
The example illustrates a phenomenon often obtained, namely that flavour and texture often form two independent patterns of co-variation. Moreover, it stresses the importance of looking for segments among consumers, and not only using their overall mean (here: PREFMEAN). The different strategies for revealing these segments by, for example, fuzzy clustering and by analysing the consumer data table transposed are described by Westad *et al.* (2004). In a huge questionnaire study of consumer attitudes to health and pleasure, the survey data were segmented and results presented in *one* effect plot as the combined significance level and sign of the regression coefficients from the optimal PLSR model (Martens *et al.*, 2005a,b). Other clustering techniques (Sahmer *et al.*, 2006) and PLS methodologies (Tenenhaus *et al.*, 2005b) are developed under the PrefMap concept.

PrefMap has proven to be a valuable tool in product development as shown in Chapter 23. Nevertheless, there are (at least) two meanings of 'preference mapping': one that refers to certain statistical tools for analysing one block internal data (e.g. PCA) or external two block data (e.g. PLSR as used here). The other meaning is reflecting the idea of understanding the drivers of consumer preferences and then stands for a broader interpretation. That again points to the need for linking two or more data sets together, leading towards multi-block and multi-matrix PrefMap situations.

21.4.3   Consumer liking-background-product: L-PLSR

The next example aims at validating consumers' liking by simultaneously taking product properties and consumer background information into account. This is an example of Y = consumer liking, X = product characteristics and Z = consumer background in Fig. 21.5, forming a L-shaped PLSR.

Objective
What are the relationships between consumer liking and both consumer background characteristics and product descriptors?

Data description
In a beer study (Mejlholm and Martens, 2006) one purpose was to stimulate the development of new beer types on the Danish market. Ten beer samples covered four types of beer: lager, strong lager, ale and wheat beer, representing both new and established beers on this market. The beers were evaluated by a trained sensory panel (nine panellists) using the nine sensory descriptors colour, body, bitter, carbonation, alcohol, fruity, floral, spicy and grainy/roasted. Consumers answered a questionnaire about their background (e.g. gender, age, habits and attitudes towards beer) as well as tasting the same ten beers in a hedonic test. L-PLSR was used to reveal relationships between consumer liking (Y), product descriptors (X = sensory, chemical and design data) and consumer background information (Z).

Results

Results could then be presented in one L-PLSR correlation loadings plot (Fig. 21.8). PC1 indicated a difference in liking between men and women related to types of beer (ale vs lager) described in sensory terms (spicy vs carbonation). PC2 spanned an age variation from young consumers preferring wheat beer to elderly, liking strong beer with high percentage alcohol tasting bitter. Overall the established beers were given higher liking scores than the new types of beer. The dots in the figure represent individual consumers, indicating different consumer segments. The knowledge achieved from this study is to be used in product development and marketing of beer to the target consumer group.

Discussion

As in the previous example in Section 21.4.2, the present example also shows significant differences ($p < 0.001$) between the products for all the sensory attributes, using univariate ANOVA. By use of various PLS methods, it was possible to study co-variation patterns among the samples. First, two different two-block PLSR models were studied (*Y* vs *X* and *Y* vs *Z*). Then a simultaneous

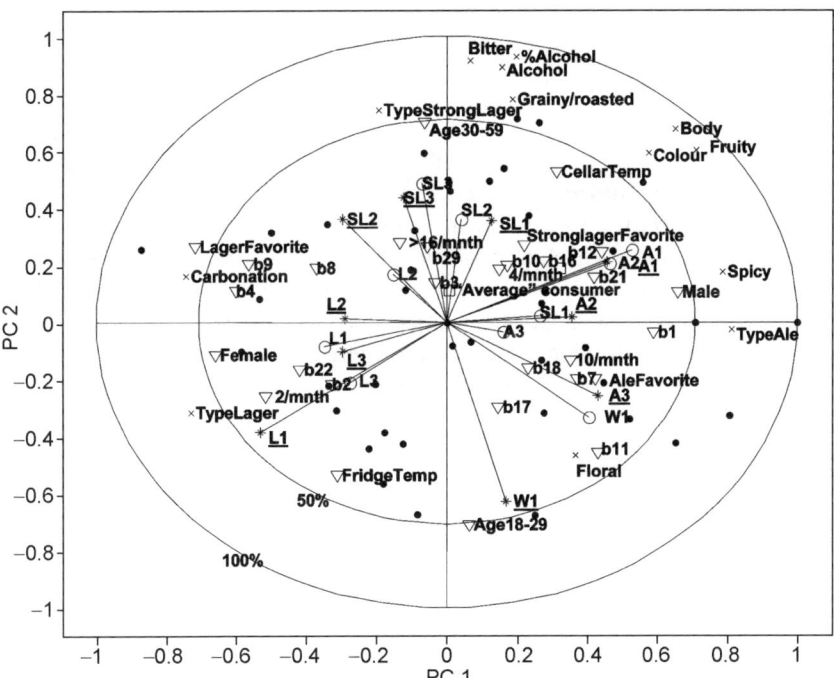

Fig. 21.8 L-PLSR correlation loadings for PC1 vs PC2: * = beers, by product descriptors *X*, o = beers, by liking *Y*'s correlation to consumer background *Z*, × = product descriptors *X*, ∇ = consumer background *Z*, ● = individual consumers' liking *Y*. b1–b29 are background variables of minor interest. The inner and outer ellipses represent 50% and 100% explained variance respectively. See Mejlholm and Martens (2006, p. 114). Reprinted with copyright permission from Elsevier, 2006.

three-block L-PLSR between X, Y and Z was performed (Fig. 21.8). This revealed which product properties X were preferred by which consumer types Z, via their links to the consumer response data Y. Other examples show how L-PLSR can be used in consumer attitudes studies (Kubberød et al., 2006).

21.4.4   Sensory-global quality: PLS path modelling
In the last example we are going to use PLS path modelling on wine data in a multi-block setting (Fig. 21.5), where $X_1, \ldots, X_4$ represent four different sensory blocks and y represents the overall product quality. The data have been collected by C. Asselin and R. Morlat and fully described in Escofier and Pagès (1988). These data have already been analysed by PLS and generalised Procrustes analysis (GPA) in Tenenhaus and Esposito Vinzi (2005) and by PLS multi-block analysis in Tenenhaus and Hanafi (2007).

Objective
In this new analysis, the objective is to relate the wine testing results to the global quality of the wine, taking into account the chronological aspect of the experiment.

Data description
A set of 21 red wines with Bourgueil, Chinon and Saumur origins are described by 27 sensory descriptive terms plus an assessment of global quality. These 28 variables were grouped into blocks (Figure 21.9a) and organised according to their chronological order: X_1 (smell at rest) $\rightarrow$ X_2 (view) $\rightarrow$ X_3 (smell after shaking) $\rightarrow$ X_4 (tasting) $\rightarrow$ y (global quality). In the structural equation of the chosen path model each block is assumed to affect all subsequent blocks via their first latent variable. The double circles symbolise that the latent variables ($\bigcirc$) are displayed without their associated measured variables ($\square$); for completeness even the single variable y is represented by a latent variable.

The individual sensory blocks are represented in Fig. 21.9b, code as:

- X_1 = *Smell at rest*: Rest1 = smell intensity at rest, Rest2 = aromatic quality at rest, Rest3 = fruity note at rest, Rest4 = floral note at rest, Rest5 = spicy note at rest.
- X_2 = *View*: View1 = visual intensity, View2 = shading (from orange to purple), View3 = surface impression.
- X_3 = *Smell after shaking*: Shaking1 = smell intensity, Shaking2 = smell quality, Shaking3 = fruity note, Shaking4 = floral note, Shaking5 = spicy note, Shaking6 = vegetable note, Shaking7 = phenolic note, Shaking8 = aromatic intensity in mouth, Shaking9 = aromatic persistence in mouth, Shaking10 = aromatic quality in mouth.
- X_4 = *Tasting*: Tasting1 = intensity of attack, Tasting2 = acidity, Tasting3 = astringency, Tasting4 = alcohol, Tasting5 = balance (acidity, astringency, alcohol), Tasting6 = mellowness, Tasting7 = bitterness, Tasting8 = ending intensity in mouth, Tasting9 = harmony.

Results

The path coefficients of the chosen model shown in Fig. 21.9b have been estimated with PLS-Graph. The reflective Mode A was chosen for the 'outer' relationships to the observed or manifest variables (MVs). For estimating the inner path models linking the latent variables (LVs) to each other (Fig. 21.9a),

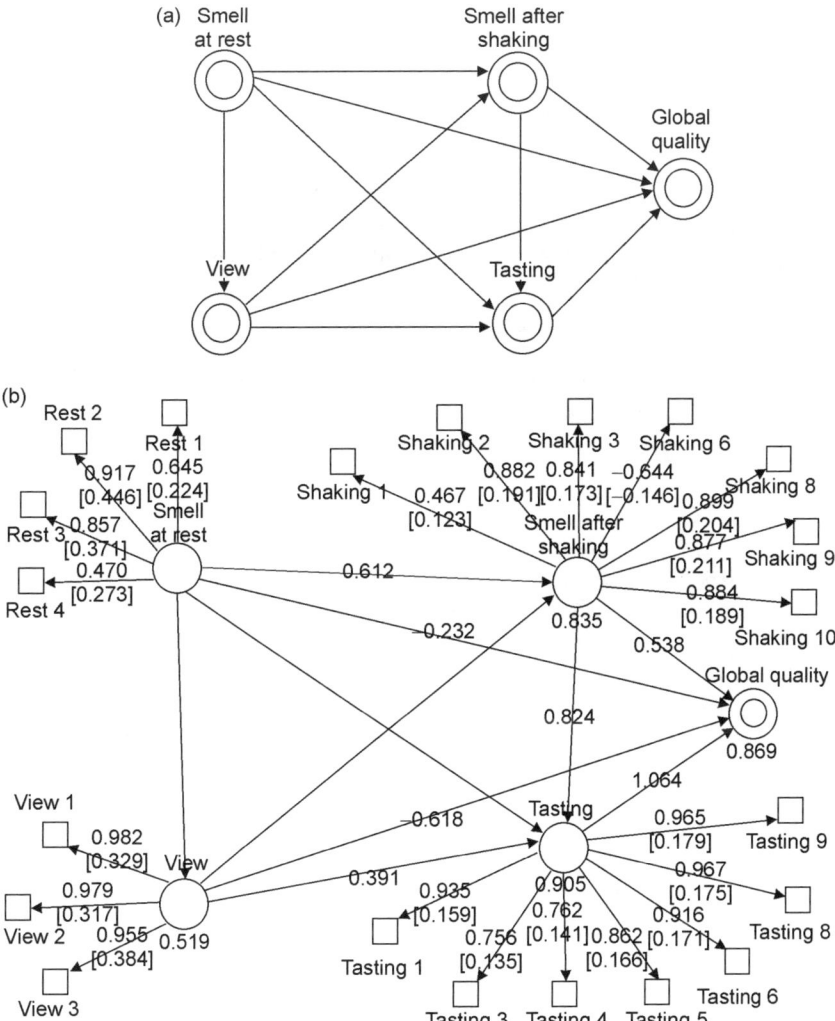

Fig. 21.9 PLS path modelling of wine quality assessments. (a) A causal model of how global quality (1 variable) is related to the consecutive steps of the tasting experiment (smell at rest → view → smell after shaking → tasting). (b) The PLS path modelling results, using the so-called 'Centroid' scheme and OLS regression for estimating the inner structural equations. The parameters were validated by bootstrap, using the 'construct level changing option' of the re-sampling procedure of PLS-Graph. (c) PLS regression coefficient estimates of inner structural equations, with approximate 95% confidence intervals computed by leave-one-out jack-knife.

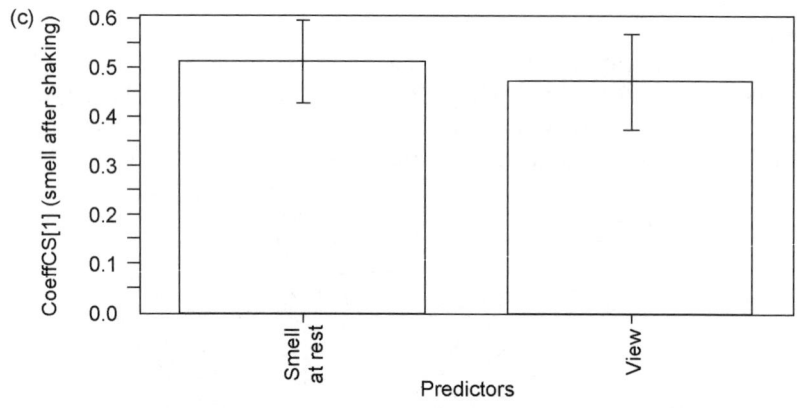

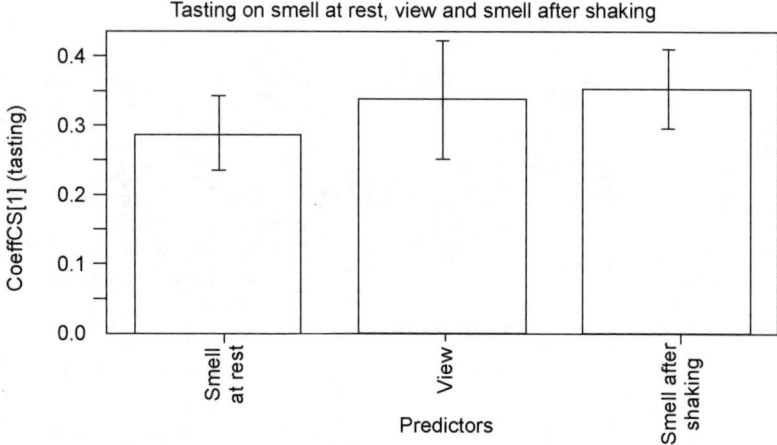

Tasting on smell at rest, view and smell after shaking

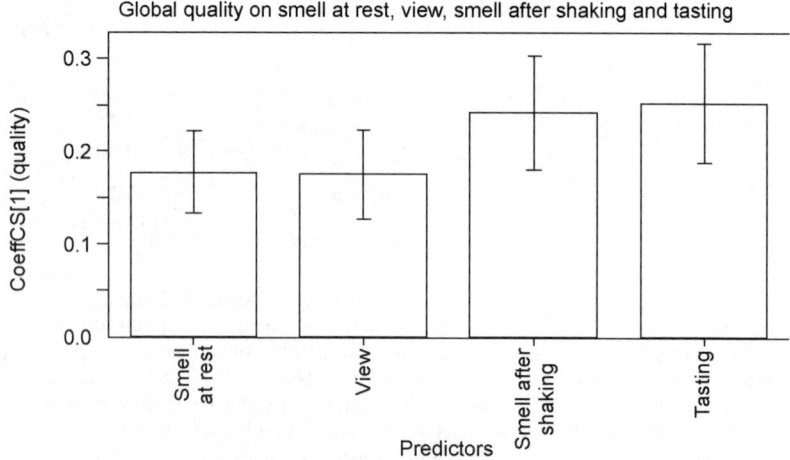

Global quality on smell at rest, view, smell after shaking and tasting

Fig. 21.9   Continued

the conventional full-rank OLS regression was used initially, according to the tradition in PLS path modelling.

Keeping only the MVs with significant weights, we get the results shown in Fig. 21.9b. For each MV two numbers are shown: the correlations between the MVs and their LV are given above the arrows connecting the MVs to their LV. The weights used for the construction of the LV estimate as a linear combination of its MV are shown in brackets below the arrow. The numbers given on the arrows connecting the LVs are the regression coefficients for each structural equation. The degree of fit (R^2) is given below the endogenous LVs.

Discussion

The weights of the manifest variables in Fig. 21.9b make sense: they are all positive except for Shaking6 = Vegetable note, a wine characteristic that is very negatively correlated with global quality (-0.82). All the other retained characteristics are positively correlated with wine quality.

All the regression coefficients in the inner relation model should also ideally be positive. Figure 21.9b shows that is not the case. The reason is that the latent variables from the four sensory blocks are highly intercorrelated, as shown in Table 21.1. This generates severe multi-collinearity problems (Fig. 21.9a) when the full-rank OLS regressions of conventional PLS path modelling was used for estimating the inner relationships. Therefore the OLSR was replaced by reduced-rank PLSR for the inner relationships. Three different PLSR regression models were estimated, using the endogenous LVs Smell after shaking, Tasting or Global quality as *Y*-variable and the LVs of the preceding blocks as *X*-variables, respectively.

The regression coefficients of these PLS regression models are shown in Fig. 21.9c. The PLSR for Smell after shaking yielded almost the same coefficients for Smell at rest and View. The PLSR for Tasting now gave positive coefficients to Smell at rest, View and Smell after shaking. The PLS regression for Global quality gave positive weights to all explanatory LVs. Global quality seems more related to Tasting and Smell after shaking than to Smell at rest and View.

Figure 21.9c shows the regression coefficients with approximately 95% confidence for the structural equations of Fig. 21.9a. In order to obtain positive

Table 21.1 Correlations between the latent variables

	Smell at rest	View	Smell after shaking	Tasting	Global quality
Smell at rest	1.000				
View	0.720	1.000			
Smell after shaking	0.877	0.810	1.000		
Tasting	0.749	0.874	0.916	1.000	
Global quality	0.606	0.597	0.826	0.862	1.000

regression coefficients in these inner relationships, it was necessary to use only one PLS component for each model. With conventional OLSR solution of the structural equations related to the endogenous variables Smell after shaking, Tasting and Global quality, the fitted R^2 values of 0.835, 0.905 and 0.869 were obtained. The corresponding fitted R^2 from PLSR with one component were equal to 0.830, 0.835 and 0.630, respectively. The latter results are probably more realistic, both in terms of signs, confidence intervals and R^2.

Hence, in this example we have illustrated how PLS path modelling may be used for assessing details in multi-block situations. Moreover, we have demonstrated the successful combination of techniques from PLS path modelling and PLS Regression.

21.5   Future trends

21.5.1   Afraid of sensory and consumer data?
Can we trust sensory and consumer data? Are they not too noisy, showing very little systematic variation?

Today two controversies within sensory and consumer studies are often revealed. The first concerns fear of sensory and consumer data in general. This is clearly expressed in traditional product development where, on the one hand, the product developer (often a technologist or natural scientist) trusts only so-called 'precise instrumental measurements' as 'objective' knowledge about the product. On the other hand, the marketing people in a company (often economists or psychologists) reject the idea of objectivity and product focus, instead focusing on consumer lifestyle and subjectivity as sources for 'true' knowledge necessary for economical success. However, during the past decade increased knowledge about information processing in the brain has become available. In this context, multivariate data analysis with graphics may be considered as an extended help for the scientist's brain to extract relevant systematic patterns, thus guard against wishful thinking (getting 'whatever results we want'; see Section 21.2.2). Thus, we expect the 'subjectivity–objectivity' dichotomy to diminish in the future, as far as new product development is concerned. Or at least, although it may seem provocative to say, that using multivariate data analysis turns 'subjective' data into more 'objective', interpretable and statistically valid overviews and conclusions for strategic decision making, thus reducing the risk of failure in the market for new products.

The second controversy concerns reliability of qualitative versus quantitative methods; i.e. interview techniques versus various scaling techniques, at the extremes. From a data analytical point of view there is no clear distinction. We expect the near future to show that the need for a strict qualitative–quantitative distinction will vanish. The reason is that even interviews create results that need to be summarised, interpreted and critically assessed by someone. Transcribing the qualitative results into data tables (e.g. indicators for different words, word-combinations or frequencies thereof) and analysing them along with other available data may reduce the investigator's subjective influence on the conclusions. Even in such cases the PLS methods can be useful.

Trusting data from sensory and consumer analyses may be strengthened by an increased understanding and use of PLS methods in order to get valid results in future NPD.

21.5.2   Is consumer-led production better?

Roughly speaking, in the mid-1980s, companies led the product development, while consumers just had to accept what they got. If anything, gastronomy, with the one-man chef creating new dishes, has taken this role today. Nevertheless, present and future trends reverse this into a consumer-driven product development. But is this better? Would it not be more realistic to talk about consumer- and producer-led developments with *extensive feedback mechanisms*?

New product development (NPD) is nowadays a hot field within management where competence about consumers as well as product technology is combined to ensure innovativeness of the company in a global market. Furthermore, the traditional quality criteria terminology with focus on 'customer satisfaction' is today extended to include customer perceived qualities in a wider context (e.g. European consumer satisfaction index (ECSI)). An example of how PLS path modelling can be used to describe causes and consequences of customer satisfaction in the frame of ECSI, is shown by Tenenhaus *et al.* (2005a).

Questions concerning feedback mechanisms and search for causality lead us to the discussion of directions of arrows in the PLS methods.

21.5.3   Merging the PLS cultures

New product development requires efficient methods for revealing how new opportunities in products can meet new opportunities in the market. The success in such a complex process depends on the researchers' ability to generate new insight, as well as their creativity and critical sense. Data modelling is only one element in this process, but an important one.

The PLS principle allows efficient and versatile data modelling. Therefore PLS methods are finding increasing use in new food product development. But for historical reasons they have existed in two distinct scientific cultures. Sensory science has been a common ground ('sensometrics'). Elsewhere the data-driven PLS regression is used in the natural sciences, while slightly more theory-driven PLS path modelling is used in the social sciences. Today's many PLS software packages belong to either one culture or the other. The combination of PLS causal and explorative methods can be further justified as they share the common objective of prediction.

However, the two-block PLSR may be understood theoretically as a special case of PLS path modelling ('Mode A' in both X and Y, using several components, with deflation on X only). And in practice several PLSR models can be quickly defined and tested, using interactive PLSR software. Both explorative and confirmative thinking are important. The newer multi-block PLSR methods are almost indistinguishable from general PLS path modelling. Thus, the PLS cultures are now merging, and so, probably, will PLS software systems. This development is driven in part by the need to link technological and marketing data in food product development – but it also reflects the general trend towards integration between hard and soft sciences.

The PLS principle of finding the main 'underlying harmonies' within and between different data tables is just one approach to data analysis. With proper validation and graphics, it can be used effectively even by non-statisticians – provided they get a good PLS software system and put some time and effort into learning it (see below). This is important since the number of data sets requiring proper data analysis is far higher than available number of willing and able statisticians in the R&D market.

21.6   Sources of further information and advice

Realising that there has been an explosion in PLS activities worldwide, the present authors limit themselves to give a few sources of information concerning relevant software packages and addresses to the authors' respective affiliations. Furthermore, readers may find it useful to look deeper into the journals and books with several publications on PLS methods and applications in the list of references below.

21.6.1   PLS software packages
In the present chapter mainly The Unscrambler and PLS-Graph have been used. More information about these and other programs can be found at respective websites.

Packages developed mainly in the field of chemometrics/sensometrics:

The Unscrambler^{TM}: www.camo.no
SIMCA: www.umetrics.com
PLS Toolbox: www.eigenvector.com

Packages developed mainly in the field of econometrics/marketing:

PLS-graph: www.plsgraph.com
SmartPLS: www.smartpls.de
SPAD-PLS (developed under the ESIS (European Satisfaction Index System) project of the European Commission and now commercialised by the French SPAD Test&Go company in Paris.) www.spadsoft.com

21.6.2   Contact addresses
The present authors have initiated, arranged and/or participated at PLS conferences since 1982. The first PLS international symposium was arranged

in France in 1999, the fifth one (PLS'07) will be in Norway in September 2007 (see www.PLS07.org) and the next one is planned to be held in China in 2009.

Norwegian Food Research Institute (Matforsk)
Osloveien 1
N-1430 Aas, Norway
www.matforsk.no

University of Copenhagen
Department of Food Science
Rolighedsvej 30
DK-1958 Frederiksberg C, Denmark
www.ifv.kvl.dk

HEC School of Management (GRECHEC)
Rue de la Liberation
78351 Jouy-en-Josas, France
www.hec.fr

ESSEC Business School
Department of Information and Decision Systems
Avenue Bernard Hirsch B.P. 50105
95021 Cergy-Pontoise
France
www.essec.edu

University of Naples 'Federico II'
Department of Mathematics and Statistics
Via Cintia, 26 – Complesso Monte S. Angelo
80126 Naples, Italy
www.dms.unina.it

21.7 References

BARKER M and RAYENS W S (2003), 'A partial least squares paradigm for discrimination', *Journal of Chemometrics*, **17** 166–173.

BECH A C, JUHL H J, HANSEN M, MARTENS M and ANDERSEN L (2000), 'Quality of peas modelled by a structural equation system', *Food Quality and Preference*, **11** 275–281.

BRO R (1996), 'Multiway calibration. Multilinear PLS', *Journal of Chemometrics*, **10** 47–61.

DIJKSTERHUIS G, MARTENS H and MARTENS M (2005), 'Combined Procrustes analysis and PLSR for internal and external mapping of data from multiple sources', *Computational Statistics and Data Analysis,* **48** 47–62.

ESCOFIER B and PAGÈS J (1988), *Analyses factorielles simples et multiples,* Dunod, Paris.

ESPOSITO VINZI V, LAURO C and AMATO S (2004), 'PLS typological regression: algorithmic, classification and validation issues', in Vichi M, Monari P, Mignani S, Montanari

A (eds), *New Developments in Classification and Data Analysis*, Springer-Verlag, Heidelberg, 133–140.

ESPOSITO VINZI V, CHIN W W, HENSELER J and WANG H (2007), *Handbook of Partial Least Squares: Concepts, Methods and Applications*, Springer-Verlag, Heidelberg.

FRØST M B, DIJKSTERHUIS G and MARTENS M (2001), 'Sensory perception of fat in milk', *Food Quality and Preference*, **12** 327–336.

JÖRESKOG K and WOLD H (1982), *Systems Under Indirect Observation – Parts I and II*, North-Holland, Amsterdam.

KUBBERØD E, DINGSTAD G I, UELAND Ø and RISVIK E (2006), 'The effect of animality on disgust response at the prospect of meat preparation – an experimental approach from Norway', *Food Quality and Preference*, **17** 199–208.

LENGARD V and KERMIT M (2006), '3-way and 3-block regressions in consumer preference analysis', *Food Quality and Preference*, **17** 234–242.

LOHMÖLLER J B (1989), *Latent Variables Path Modeling with Partial Least Squares*, Physica-Verlag, Heidelberg.

MARTENS H (2005), 'Domino PLS: a framework for multi-directional path modelling', in Aluja T, Casanovas J, Vinzi V E, Morineau A and Tenenhaus M, *PLS and Related Methods, Proceedings of the PLS'05 International Symposium*, SPAD * Test & Go Group, Paris, 125–132.

MARTENS M and MARTENS H (1986), 'Partial least squares regression', in Piggott J R, *Statistical Procedures in Food Research*, London, Elsevier Applied Science Publishers, 293–359.

MARTENS H and MARTENS M (2001), *Multivariate Analysis of Quality. An Introduction*, Chichester, John Wiley and Sons.

MARTENS H and NÆS T (1991), *Multivariate Calibration*, 2nd edn, Chichester, John Wiley and Sons.

MARTENS H, WOLD S and MARTENS M (1983), 'A layman's guide to multivariate data analysis', in Martens H and Russwurm H Jr (eds), *Food Research and Data Analysis*, London, Elsevier Applied Science Publishers, 473–492.

MARTENS H, ANDERSSEN E, FLATBERG A, GIDSKEHAUG L H, HØY M, WESTAD F, THYBO A and MARTENS M (2005a), 'Regression of a data matrix on descriptors of both its rows and of its columns via latent variables: L-PLSR', *Computational Statistics and Data Analysis*, **48** 103–123.

MARTENS M, FRØST M B and MARTENS H (2005b), 'Consumer attitudes to health and pleasure – survey data studied by PLSR', in Aluja T, Casanovas J, Vinzi V E, Morineau A and Tenenhaus M, *PLS and Related Methods, Proceedings of the PLS'05 International Symposium*, SPAD * Test & Go Group, Paris, 431–437.

MEJLHOLM O and MARTENS M (2006), 'Beer identity in Denmark', *Food Quality and Preference*, **17** 108–115.

NOCAIRI H, QANNARI E M, VIGNEAU E and BERTRAND D (2005), 'Discrimination on latent components with respect to patterns. Application to multicollinear data', *Computational Statistics and Data Analysis*, **48** 139–147.

NÆS T and RISVIK E (1996), *Multivariate Analysis of Data in Sensory Science*, Amsterdam, Elsevier Science.

PLAEHN D and LUNDAHL D S (2006), 'An L-PLS preference cluster analysis on French consumer hedonics to fresh tomatoes', *Food Quality and Preference*, **17** 243–256.

SAHMER K, VIGNEAU E and QANNARI E F (2006), 'A cluster approach to analyze preference data: choice of the number of clusters', *Food Quality and Preference*, **17** 257–265.

SÆBØ S, MARTENS M and MARTENS H (2007), 'Three-block data modeling by endo- and exo-LPLS regression', in Esposito Vinzi V, Chin W W, Henseler J and Wang H, *Handbook of Partial Least Squares: Concepts, Methods and Applications*, Springer-Verlag, Heidelberg.

TENENHAUS M and ESPOSITO VINZI V (2005), 'PLS regression, PLS path modeling and generalised Procrustean analysis: a combined approach for multiblock analysis', *Journal of Chemometrics*, **19** 145–153.

TENENHAUS M and HANAFI M (2007), 'How to use PLS path modeling for analyzing multi-block data sets', in Esposito Vinzi V, Chin W W, Henseler J and Wang H, *Handbook of Partial Least Squares: Concepts, Methods and Applications*, Springer-Verlag, Heidelberg.

TENENHAUS M, ESPOSITO VINZI V, CHATELIN Y-M and LAURO C (2005a), 'PLS path modeling', *Computational Statistics and Data Analysis,* **48** 159–205.

TENENHAUS M, PAGÈS J, AMBROISINE L and GUINOT C (2005b), 'PLS methodology to study relationships between hedonic judgements and product characteristics', *Food Quality and Preference*, **16** 315–325.

WESTAD F, HERSLETH M and LEA P (2004), 'Strategies for consumer segmentation with applications on preference data', *Food Quality and Preference*, **15** 681–687.

WOLD H (1982), 'Soft modeling: The basic design and some extensions', in Jöreskog K and Wold H, *Systems Under Indirect Observation – Part II*, North-Holland, Amsterdam, 1–54.

WOLD S, MARTENS H and WOLD H (1983a), 'The multivariate calibration problem in chemistry solved by the PLS method', in Ruhe A and KÅgstrom B, *Proceedings of the Conference on Matrix Pencils*, March 1982, Lectures Notes in Mathematics, Springer-Verlag, Heidelberg, 286–293.

WOLD S, ALBANOC C, DUNN III W J, ESBENSEN K, HELLBERG S, JOHANSSON E and SJOSTROM M (1983b), 'Pattern recognition: finding and using regularities in multivariate data', in Martens H and Russwurm Jr H, *Food Research and Data Analysis*, Applied Science, Barking, 147–188.

22

Case study of consumer-oriented food product development: reduced-calorie foods

J. Bogue and D. Sorenson, University College Cork, Ireland

22.1 Introduction

Reduced-calorie foods represent very significant product development opportunities for food manufacturers as consumers select reduced-calorie or reduced-fat foods to limit their intake of calories or fat. However, the development of reduced-calorie foods that will gain consumer acceptance poses many technical and marketing challenges for new product development (NPD) personnel in terms of optimising the marketing (extrinsic) and sensory (intrinsic) attributes of these foods. These extrinsic and intrinsic attributes have a strong influence on consumer acceptance of such foods.

A reduced-calorie food refers to a food product with lower amounts of calories, through either calorie or fat reduction, in relation to regular products. Reduced-calorie foods are central to contemporary diets as consumers are concerned about what foods they eat and the links to dietary-related diseases such as heart disease and cancer. The significance of reduced-calorie foods is increasing as consumers seek healthy options across product categories. Furthermore, nutrition experts recommend a two-pronged approach to maintaining a healthy diet in terms of, firstly, lowering caloric intake and, secondly, consuming palatable low-fat options (*Retail News*, 2000). A central issue to exploiting product opportunities in the reduced-calorie foods category is whether consumers are willing to make trade-offs in terms of the sensory characteristics of such products, and the extent of these trade-offs, as food firms modify recipes to develop foods low in calories or fat (van Trijp and Steenkamp, 1998).

22.2 Consumer trends and healthy eating

As healthy eating has become a significant consumer food trend so has the increased interest in reduced-calorie foods. This healthy trend has resulted from factors such as: the increased consumer awareness of health issues; the identification of scientific links between diet and health; reports that have recommended healthier diets; and the increased cost of healthcare to governments (Rabobank, 1999). Furthermore, obesity is a major public health problem for many Western countries as well as countries such as China and Japan (British Nutrition Foundation, 1999). Obesity remains a priority issue for society and public health programmes. For example, the United States has the highest rate of obesity of any industrialised nation with more than half of all adults, and 25% of all children, overweight or obese, with between 15 and 20% of the population in Europe obese (Blundell, 2000; Schlosser, 2001). To encourage positive dietary behaviours, consumers are advised to use guides to aid in the planning of their daily food choices, such as the food pyramid devised by the United States Department of Agriculture (USDA) (Health Promotion Unit, 1998). They are advised to consume foods at the top of the food pyramid such as confectionery, cakes and high-fat snacks, sparingly. Food products such as bread, cereals and potatoes, at the base of the pyramid, should be consumed more frequently.

Consumers are also advised to reduce their daily caloric intake of fat derived from saturated fatty acids, and to reduce the daily caloric intake derived from total fat, associated with a reduced risk from cardiovascular heart disease (CHD), cancer, strokes and diabetes, and a reduced incidence of being overweight or obese (Hollingsworth, 1996; Richardson, 1996). For example, US consumers have been specifically advised to limit their daily caloric intake from fat to 30% or less, with no more than 10% of total daily calories derived from saturated fat (Best, 1991; Sheng et al., 1996). In addition, the UK Government's long-term aim is to reduce the average percentage of total daily calories derived from fat to no more than 35% (Golding et al., 2001). However, the barriers to the adoption of reduced-calorie and reduced-fat diets include: the reduction in the sensory properties of foods; increased product costs; the lack of family support; and an inability to judge the calorie or fat content of diets (Lloyd et al., 1995). Notwithstanding this, health and wellness represent two significant food trends driving NPD activities in the global food and beverages market. In particular, reduced-calorie foods and beverages have come to dominate the global healthy food and beverages market in terms of both volume and value sales. Global value sales of reduced-calorie foods and beverages were US$63.2bn in 2002, where reduced-calorie dairy products and reduced-calorie beverages accounted for 53% and 33% of global value sales respectively in 2002 (Leatherhead Food Research Association, 2004).

22.3   Reduced-calorie foods and beverages: marketing and technological challenges

Although the global reduced-calorie food and beverages market experienced high levels of NPD activity across categories since 1990, the failure rates for new reduced-calorie products are reportedly high (Leatherhead Food Research Association, 2004). In that context, a number of critical product design and marketing factors are believed to constrain the further development of the reduced-calorie food and beverages market. These include: an increased media and consumer interest in new dietary weight regimes; the choice of product descriptors used on reduced-calorie food and beverage labels; and the differing penetration levels of reduced-calorie products both within, and across, categories. In addition, poor consumer expectations, in terms of trade-offs between health benefits and sensory pleasure, have also affected the growth of the global reduced-calorie food and beverages market.

The labelling and correct use of descriptors in the reduced-calorie products market are problematic issues for marketers, particularly in view of consumers' expectations of reduced-calorie products. Research has highlighted the possibility of utilising certain product descriptors when marketing products to different segments. For example, previous research highlighted that female consumers preferred the label 'fat-free' while male consumers preferred the term 'healthy' (Bogue *et al.*, 1999). However, with increased consumer interest in reduced-calorie products there have also been increased opportunities for label abuse where mislabelling has been reported in the use of descriptors such as 'healthy', 'light', 'fat-free' or '90% fat-free' (O'Rourke, 1999). Research has revealed that consumers were often confused by, and consequently were mistrustful and suspicious of, many descriptors associated with reduced-calorie foods and beverages (Hamilton *et al.*, 2000; Patterson *et al.*, 2001). Therefore the marketing of health benefits associated with reduced-calorie products necessitates the communication of such benefits in a manner that consumers will regard as credible (Grunert, 2000; Hamilton *et al.*, 2000).

Consumer expectations remain one of the fundamental marketing constraints to increasing the growth of the reduced-calorie food and beverages market. Importantly, penetration of reduced-calorie products varies dramatically both within and across product categories reflecting consumers' preferences (Rittman, 2003). For example, consumers' perceptions of chocolate confectionery and baked goods as indulgent products have hindered the development of these reduced-calorie categories (Leatherhead Food Research Association, 2004). Furthermore, cross-cultural differences in consumer acceptance of reduced-calorie products are evident across Europe. Specifically, the penetration level of reduced-calorie products ranged from 9% and 10% for Germany and the United Kingdom to 3% and 5% for Italy and Spain respectively in 2002 (Leatherhead Food Research Association, 2004). In contrast, the United States represented the largest market for reduced-calorie foods and accounted for 70% (US$44.2bn) of global value sales in 2002.

Consumers' negative expectations of the sensory quality of reduced-calorie products represent the most fundamental NPD constraint to market growth of many reduced-calorie food and beverage categories. The sensory properties of fats make a diet rich and flavourful, and the fat content of different foods and beverages has a significant influence on flavour and odour release and textural structure, which needs to be taken into account for reduced-calorie products. Consequently, the removal of fat from a food or beverage, to achieve the desired caloric reduction, often necessitates changes to the product's formulation, which makes it technologically difficult to formulate products that are equivalent to the standard products in terms of flavour, texture and appearance (O'Donnell, 1993).

Therefore, the challenge that remains for food technologists, in the medium to long term, is to remove the fat from foods to gain the desired fat reduction while simultaneously replacing the fat with ingredients such as fat replacers that safely impart the flavour, texture and mouthfeel associated with fat but with fewer calories (Pszczola, 1996). However, from a technological perspective, research has suggested that, in the short to medium term, firms are more likely to realise higher levels of consumer satisfaction, and ultimately higher levels of new product success, where firms select products that are naturally low in fat (Bogue and Ritson, 2004). This view was based upon observations of consumers' sensory preferences for a range of reduced-fat products, where reduced-fat milks and reduced-fat yoghurts were more preferred than reduced-fat Cheddar-type cheeses. This was attributed to the deterioration in the sensory quality of hard cheese as a consequence of the extent of the fat removal necessary to produce a reduced-fat Cheddar-type hard cheese (Drake and Swanson, 1997). From a marketing perspective, it has also been argued that new reduced-calorie products should be marketed and branded differently from their regular calorie equivalents as brand extensions could be expected to raise consumers' expectations based on their knowledge of the regular product (Bogue *et al.*, 1999; Cardello, 1993). A pertinent question that should therefore be raised is: 'How can firms increase the new product success rates for reduced-calorie foods and beverages'?

22.4   New product development success factors

Although NPD represents an important strategic orientation for firms, the food innovation process is a complex, costly and risky process, where 90% of new products introduced to the market fail within the first year of launch (Traill and Grunert, 1997). Product development is a knowledge-intensive process where the generation of new ideas and concepts requires detailed knowledge of both products and consumers. Therefore, firms that effectively manage knowledge throughout the NPD process create more evident values in a firm's offering in order to effectively meet consumers' needs. However, Jensen and Harmsen (2001) noted that few organisations had implemented a range of factors that

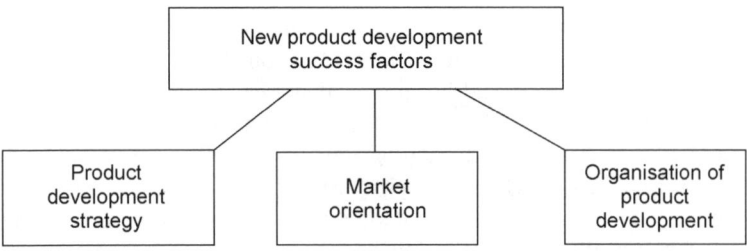

Fig. 22.1   New product success factors. (Source: Harmsen, 1994.)

could improve NPD success and organisational performance (see Fig. 22.1). In particular, researchers argue that the degree of market orientation in the food sector is low and that well-organised market information systems are scarce in food marketing systems (Bogue, 2001; Grunert *et al.*, 1996).

In order to manage consumer knowledge effectively throughout the NPD process, numerous researchers suggest a market-oriented approach to innovation, as market-oriented firms are considered proficient at the generation and dissemination of information (Kleinschmidt and Cooper, 1995; Kohli and Jaworski, 1990). Harmsen (1994: 294) defined market-oriented NPD as: 'development based on an understanding of the needs and wants of consumers as well as an integration of this understanding into products'. In market-oriented organisations, consumers are viewed as co-designers in the NPD process since they can make an effective contribution to new food product design, and the integration of the consumer with the NPD process can best be achieved at the pre-development stages of concept ideation, concept screening and optimisation (Bogue, 2001; Cooper, 1993). The incorporation of consumers' value-creation at the pre-development or concept stages of the NPD process make organisations better able to adapt to changes in consumers' needs, reduce uncertainty in NPD, and can ultimately lead to higher quality and consumer satisfaction. However, it is not sufficient to solely generate information on consumer needs. The information also has to be disseminated to NPD team members and incorporated into the decision-making processes on product design (Zhang and Doll, 2001).

So how does a multifunctional team work towards the design of an optimal product for target consumers? Concept optimisation research, which focuses on the early stages of the NPD process, leads to a more systematic and multifunctional approach to product development (Earle, 1997). There exist a number of contemporary research techniques, which can utilise both technical and marketing information, that promote closer integration between the marketing and technical functions, for the process of food product optimisation. These include: focus groups, quality functional deployment (QFD), sensory analysis and conjoint analysis. However, the uptake of formal concept optimisation research methodologies remains low or is applied in an ad-hoc fashion, which is considered a significant contributor to low success rates in product development worldwide (Nijssen and Frambach, 2000; Wind and Mahajan, 1997).

In Section 22.5 an NPD case study is presented on the development of a range of innovative reduced-calorie 'on-the-go' beverages. The research presented in this case study explores the concept of managing consumer knowledge at the early stages of the NPD process, through applying it to the development of a range of innovative reduced-calorie 'on-the-go' beverages. In terms of new product and process improvement, the approach adopted in this case study provides a framework by which firms can integrate 'voice of the consumer' information more effectively, and manage consumer knowledge more efficiently, at the early stage of the NPD process. This market-oriented approach to NPD can assist firms identify, screen and optimise new product concepts, in this case reduced-calorie 'on-the-go' beverage concepts, at relatively low cost during the pre-development stages of the NPD process.

22.5   New product development case study: reduced-calorie on-the-go beverages

Successful new product launches require an in-depth understanding of markets and trends in order to anticipate changing consumer needs and preferences. For example, the 'on-the-go' convenience food and beverages market has experienced high levels of NPD activity in recent years in line with changing consumer food and lifestyle trends. As Eurofood (2001) explained: 'the traditional three meals a day becomes a rushed five as consumers are planning their meals around their lives, and not their lives around meal time'.

There is a growing trend towards new foods and beverages that combine both health and convenience, which is attributed to: an increase in both the number of food service outlets and variety of meal solutions available (Datamonitor, 2005); sociodemographic changes, the increasing prevalence of individualistic lifestyles, and consumers' perceived time constraints (Mintel, 2004); the broadening of distribution channels away from specialist dietetic outlets and into the mainstream food and beverages market (Leatherhead Food Research Association, 2003); rising consumer interest in diet and health (Boyle and Emerton, 2002); an acceleration in the decline of traditional fixed mealtimes (Groves, 2002); and the increasing use of single-serve packs and individual portion formats (Eurofood, 2001). Consequently, many retailers and food manufacturers are engaged in increased levels of NPD activity in the reduced-calorie 'on-the-go' category in response to consumers' dietary and lifestyle changes.

22.5.1   New product development and on-the-go products

Product development activities in this dynamic product category can be viewed as a continuum ranging from the repositioning and adaptation of existing product offerings to the development of new products, particularly in relation to portable packaging, portion size and the nutritional quality of 'on-the-go' foods and beverages, in order to exploit market opportunities in the reduced-calorie 'on-the-go' food and beverages category. Consequently, NPD activity in the global reduced-calorie 'on-the-go' food and beverage category has focused primarily on: new flavour line extensions; new package design and formats; new positioning strategies; and differentiation in terms of functional benefits (Boyle and Emerton, 2002).

Manufacturers of weight loss foods and beverages such as the Slim-Fast Company, and manufacturers of meal replacement foods and beverages such as Abbot Laboratories, repositioned their existing products to broaden their respective brand's appeal to the healthy lifestyle sector (Roberts, 2003). In contrast, other food and beverage manufacturers have been more proactive in terms of their NPD activities in the reduced-calorie 'on-the-go' food and beverages category. For example, some manufacturers have been engaged in incremental innovations such as line extensions through the adaptation of existing products such as Kellogg's 'Special K' bar, while other firms have pursued the development of more innovative nutrient-dense 'on-the-go' products such as Sanitarium's 'Up and Go' and Otsuka's 'Energen' beverage ranges (Leatherhead Food Research Association, 2004). In particular, the reduced-calorie 'on-the-go' food and beverage category represents the fastest growing segment within the overall reduced-calorie food and beverages market, as consumers seek ways to cope in a time-challenging and demanding environment (Leatherhead Food Research Association, 2003). However, the reduced-calorie 'on-the-go' food and beverage category, although well established in the United States and Japan, remains underdeveloped within European markets.

22.5.2   Reduced-calorie on-the-go foods and beverages: marketing and technological challenges

Though NPD opportunities exist for innovative foods and beverages positioned on a health (e.g. reduced-calorie) or health and convenience (e.g. reduced-calorie 'on-the-go') platform, the development of these products from both a technical and marketing perspective presents considerable challenges to NPD practitioners. From a purely technological perspective, reduced-calorie 'on-the-go' beverages must deliver two critical product design success factors to gain consumer acceptance and increase consumer repurchase probability: satiety and sensory pleasure.

The successful adoption of reduced-calorie 'on-the-go' beverages by consumers clearly necessitates a dramatic change in consumer behaviour, in addition to overcoming perceptual and attitudinal barriers towards replacing '*solid foods*' with '*liquid beverages*' for 'on-the-go' consumption. These issues therefore present a considerable challenge to marketers in terms of developing effective positioning, communication and pricing strategies (extrinsic attributes) for reduced-calorie 'on-the-go' beverages, as well as identifying the optimal intrinsic product design attributes for high levels of consumer acceptance. Consequently, the overall objective of the work presented in this NPD case study was the identification of innovative reduced-calorie 'on-the-go' beverage concepts with high levels of consumer satisfaction, targeted at different market segments.

This case study is presented in three distinct sections, which reflect both the three key pre-development stages in the NPD process, and the different research methods used at each stage of this case study: concept ideation through focus groups; concept screening through sensory analysis; and finally concept refinement and optimisation through conjoint analysis.

22.5.3   Concept ideation in new product development: focus group results

The focus group methodology has emerged as an extremely significant exploratory design tool that can facilitate the integration of 'voice of the consumer' information at the very early stages of the NPD process. In that context, focus groups are particularly appropriate for concept ideation and concept screening at the early stages of the NPD process as focus groups enable product developers gain direct contact with potential users or purchasers of products. In particular, focus groups are considered especially valuable to product developers where consumers' needs are poorly understood, and where group discussions provide concentrated, well-defined and pre-filtered data (Fitzpatrick, 1997).

The objective from this exploratory aspect of the NPD process was to gather qualitative information from consumers that would provide product design information. Forty-six respondents were recruited by means of convenience sampling to participate in six focus groups held in three different centres in Ireland (see Table 22.1). An experienced moderator designed a focus group guide in line with best practice, and conducted the focus groups, which were audiotape recorded and lasted approximately 1 hour and 20 minutes. All focus group participants were rewarded with a small payment of €40 for their time and effort. The qualitative data was transcribed from the audiotape recordings. The focus group transcriptions were then coded using the computer package N6™ (QSR International, 2002). The N6™ software package facilitated the process of identifying, coding and retrieving information for further analysis.

The focus group discussions revealed that some participants did not see a role for reduced-calorie products in their present diet, but suggested that these products may have a role to play in their diet at some stage in the future: 'If I started to get obese, I would probably cut down and consume reduced-calorie products' (FG2). Product descriptors have a key function to play in the marketing of reduced-calorie foods, particularly in terms of consumer purchase behaviour, and ultimately, acceptance of reduced-calorie products, and a number of researchers reported on the diverse product descriptors preferred by those who consumed reduced-calorie food products (Bogue *et al.*, 1999; Geraldi, 1992). Overall, focus group discussants were confused with descriptors such as 'light' or 'healthy'. In this study, the acceptance of product descriptors varied between groups, which again highlighted the different market segments. For example, Focus Group 4 respondents did not like the descriptor 'diet': 'I think "diet" is a bit severe' (FG4).

Table 22.1   Sociodemographic profile across focus groups (FG)

Profile	FG1	FG2	FG3	FG4	FG5	FG6
Group size	8	8	8	7	8	7
Gender						
Male	2	0	1	2	4	4
Female	6	8	7	5	4	3
Age group (years)	18–34	18–24	25–44	35–54	25–34	18–34
Marital status						
Single	4	7	2	0	4	7
Married	4	1	5	6	2	0
Separated/divorced	0	0	1	0	0	0
Cohabiting	0	0	0	1	2	0
Education level						
Primary level	0	0	0	0	0	0
Intermediate/junior certificate	0	0	1	0	0	0
Leaving certificate	0	2	4	4	2	2
Vocational	0	0	0	0	2	0
Third level	8	6	3	3	4	5
Social class	B, C1	C1, C2	C1, C2	B, C1	C2, D	B, C1
Location	Cork	Cork	Limerick	Dublin	Dublin	Limerick

The introduction of product prompts during focus group discussions can help understand consumers' perceptions of competitive products and also packaging design issues with regard to the marketing of reduced-calorie products. Participants reported that the packaging design of reduced-calorie products was not as colourful as regular products, which projected an inferior image of the product to the focus group discussants: 'Why reduce the [package] colour on the outside? You are implying that it [a reduced-calorie product] will taste horrible' (FG5).

The taste of foods had an important influence on food choice for the majority of participants, and the perception that reduced-calorie foods were more 'artificial' was evident from the focus group discussions. The range of reduced-calorie foods available on the supermarket shelves did not impress participants across all focus groups. For certain participants, reducing the quantity of regular products consumed, i.e. the full-fat version, was a preferable option than the consumption of reduced-calorie alternatives: 'I think reduced-calorie foods taste so artificial. I would rather eat a small amount of the regular foods than eat reduced-calorie foods' (FG4). However, some discussants held more positive views on reduced-calorie foods, which highlighted the different segments in the market: 'I think reduced-calorie foods are so close to the originals, I don't know why people don't eat them [reduced-calorie foods]' (FG6).

Focus group participants were then introduced, by means of product prompts and information on a flipchart, to reduced-calorie 'on-the-go' beverage concepts, which reflected the range of products that could potentially be developed in a market-oriented fashion. The focus group discussions revealed that respondents held mixed attitudes towards missed meals and eating 'on-the-go'. Older focus group participants appeared less likely to miss meals and were most negative towards the concept of eating on-the-go. In contrast, a number of younger participants across focus groups reported missed meals during weekdays, and breakfast in particular: 'I usually concentrate on just getting into work and if I have time when I get in, I'll grab something to eat' (FG6). Significantly, consumers expected 'on-the-go' beverages consumed at breakfast time to be low in fat and calories, to be high in fibre, and to be satiating. The focus group discussions revealed that the key extrinsic and intrinsic product design attributes that influenced consumers' purchase intention towards new reduced-calorie 'on-the-go' beverages included: price, functional health benefits, flavour, the reduced-calorie carrier or base product, package size and packaging format.

The reduced-calorie carriers or base products (rice milk, oat bran and carrot, fruit soup and vegetable smoothie) investigated in this study were chosen based upon their satisfactory micronutrient and macronutrient profiles, and superior textural and satiety attributes. However, from discussions with focus group participants it appeared that reduced-calorie cereal-based and fruit and milk-based carriers were deemed most appropriate for consumption in the morning time. In particular, non-carbonated beverages that contained fruit pieces were perceived to be more natural and more satiating than smooth-style beverages. In contrast, participants were less accepting of reduced-calorie 'fruit soup' and 'vegetable smoothie' 'on-the-go' beverage concepts for consumption in the morning time. Instead, focus group discussants associated the reduced-calorie 'fruit soup' and 'vegetable smoothie' 'on-the-go' beverage concepts with consumption at lunchtime.

The final two components of this reduced-calorie case study entailed a quantitative analysis of consumers' preferences for key intrinsic and extrinsic attributes of reduced-calorie 'on-the-go' beverage concepts, from both a sensory and marketing perspective. The sensory and marketing studies in this case study were conducted separately over two sessions with 300 consumers.

22.5.4   Concept screening in new product development: sensory analysis results

Sensory analysis is an important tool for the development of new food products with increasing recognition of the role sensory perception plays in food choice, and has come closer into the domain of marketing as product developers become more aware of its role in food choice. In recent times, sensory science has developed to provide detailed information on consumer acceptance of foods and has become a significant marketing technique (Bogue et al., 1999). Sensory analysis provides marketers with an understanding of product quality, direction for product quality, evaluations of new product concepts, and profiles of competitors' products from a consumer perspective (Bogue et al., 1999; Moskowitz, 1991; Shukla, 1994).

The sensory analysis component of this reduced-calorie case study constituted the second data set. Samples of the four potential reduced-calorie carriers (rice milk, oat bran and carrot, fruit soup and vegetable smoothie) were developed by the technical partners involved in this project, and the fat content was standardised (1% fat) across all four samples. Sensory analysis research, incorporating both Descriptive Sensory Analysis and consumer preference tests, was then undertaken to quantitatively investigate consumers' sensory preferences for potential carriers for a range of reduced-calorie 'on-the-go' beverage concepts. The sensory analysis research was carried out based upon the importance of the carrier attributes to consumers, in terms of choosing between alternative reduced-calorie beverage concepts.

Descriptive sensory analysis
Descriptive sensory analysis can be used to describe the sensory properties of a product and quantify their intensities (Resurreccion, 1998). An experienced sensory panel of 12 assessors took part in the descriptive sensory analysis at a sensory research facility. The development of a full descriptive vocabulary, to describe the sensory characteristics of the reduced-calorie carriers, was carried out over two group discussions and the panel of assessors evaluated the reduced-calorie carriers for flavour, afterflavour, appearance, odour and mouthfeel. Fifteen descriptors were developed, and defined, to characterise the reduced-calorie carriers: six for flavour, four for odour, two for appearance, two for mouthfeel and one for afterflavour.

Following this, the sensory panel quantitatively evaluated the four reduced-calorie carrier samples using the descriptive terms generated, in line with best practice (Bogue *et al.*, 1999). Each sample was assigned a random three-digit code and the order of presentation was balanced, within sensory sessions, to avoid first order and carry-over effects (MacFie *et al.*, 1989). Some 50 ml of each sample was presented for analysis, in a wine glass covered with a clock glass, to retain odour, at ambient temperature (21 °C). The panel members were presented with deionised water to cleanse their palates between samples. The samples were scored for odour, flavour, afterflavour, appearance and mouthfeel using unstructured 100 mm line scales, anchored at the ends with extremes for each descriptive term. The unstructured scale offered the advantage of not having a fixed set of categories (Land and Shepherd, 1984). The intensity of each descriptive term was recorded for each of the samples and the data were collected using Compusense five v3.0 (Compusense, 2001).

Discrimination between reduced-calorie carriers, using the descriptive vocabulary, was tested using one-way analysis of variance (ANOVA) (O'Mahoney, 1986), with the statistical package SPSS v11 (SPSS, 2003). This established if there were significant differences between the samples and, if so, which attributes were responsible for the differences. ANOVA carried out on the mean panel scores of the 15 descriptors for the reduced-calorie carriers showed that 10 out of the 15 descriptors significantly discriminated ($p \leq 0.05$) between the four reduced-calorie carrier samples (see Table 22.2).

Table 22.2   Results of ANOVA on descriptive analysis of four reduced-calorie carriers averaged across assessors and carriers

Descriptor	Sample Vegetable smoothie	Fruit soup	Rice milk	Oat bran and carrot	p value
Appearance – watery	36.31	45.90	42.51	39.02	0.063
Appearance – creamy	21.92	61.81	66.84	23.13	0.000*
Odour – fruity	34.01	65.31	34.21	29.13	0.024*
Odour – strength	49.81	34.21	35.61	42.60	0.063
Odour – creamy	26.31	49.94	55.38	20.12	0.010*
Odour – sweet	22.38	29.61	32.27	24.36	0.071
Flavour – sweet	22.51	41.29	38.21	26.36	0.044*
Flavour – strength	64.11	28.34	34.91	41.29	0.009*
Flavour – acidic	11.10	28.21	27.34	18.42	0.051
Flavour – creamy	17.92	47.92	46.09	23.91	0.000*
Flavour – greasy/fatty	15.56	35.31	28.46	9.52	0.007*
Flavour – tainted	49.34	13.94	11.67	34.86	0.000*
Afterflavour – lingering sweetness	32.76	42.98	47.37	40.51	0.056
Mouthfeel – smooth	46.21	67.60	70.10	52.31	0.013*
Mouthfeel – mouthcoating	10.27	32.51	38.37	22.31	0.011*

* Significant at the $p = 0.05$ level

Consumer preference tests

The next stage in the sensory work involved gathering consumer preference scores on the four reduced-calorie carriers. Subjective tests with untrained consumers can be used to investigate different sensory preferences and this, in conjunction with descriptive sensory analysis, can help identify those sensory attributes that drive consumers' preferences (Bogue *et al.*, 1999; Stone and Sidel, 1995). A sample of 300 consumers was recruited to take part in a preference test for a range of new 'on-the-go' beverage concepts. The beverages were not identified to consumers as reduced-calorie beverages.

Consumers blind-tasted the different reduced-calorie carriers in isolated cubicles. Each sample was labelled with a different randomly generated three-digit code, and the order of tasting was balanced to account for first order and carry-over effects (MacFie *et al.*, 1989). Some 50 ml of each sample were given to participants, and the reduced-calorie carriers were sampled at ambient temperature (21 °C). Consumers were presented with each sample separately, and were asked to give preference ratings for their like, or dislike, for the sensory characteristics of each sample, on a 9-point Likert scale. Consumers were provided with water and were encouraged to take a short break and to cleanse their palate between samples.

Consumers' preference scores were analysed using ANOVA to see if significant differences existed between reduced-calorie carrier samples, in terms of consumers' preference scores. A principal component analysis (PCA) of the consumer preference data was then conducted using Guideline v7.5 (CAMO, 1998)

Table 22.3　Respondent sociodemographic profile

Sociodemographic variable	Category	Sample (N)	Sample (%)
Gender	Male	115	39.0
	Female	179	61.0
Age group (years)	18–24	78	26.0
	25–34	114	38.0
	35–44	60	20.0
	45–54	30	10.0
	55–65	18	6.0
Marital status	Single	163	54.0
	Married	104	35.0
	Cohabiting	22	7.0
	Separated/divorced	6	2.0
	Widowed	5	2.0
Educational status	Completed primary level only	18	6.0
	Completed secondary level only	78	26.0
	Completed third level	204	68.0
Employment status	Employed full time	189	63.0
	Employed part time	36	12.0
	Student	48	16.0
	Retired	12	4.0
	Unemployed	15	5.0

to give an internal preference map (Greenhoff and MacFie, 1994). MacFie and Thomson (1994: 381) stated that preference mapping offered: 'a way of superimposing either a preference vector or an ideal point, for each consumer involved in the study, on a multi-dimensional representation of properties or products'. The sociodemographic profile of the 300 consumers who took part in the preference tests of the four reduced-calorie carrier samples is shown in Table 22.3.

The results of the internal preference mapping for the reduced-calorie carrier samples are shown in Fig. 22.2. The first two PCs accounted for 75% of the explained variation. In Fig. 22.2 the preference of each consumer is represented by a black diamond. The direction and distance of a consumer from the midpoint of the graph illustrate the intensity and clarity of that consumer's preference for the range of reduced-calorie carriers tested. When principal components are graphed, the relationships between samples rated can be seen and the position of the samples on the graph, in relation to each other, is determined by the consumer loadings. To interpret the analysis, both scores and loadings must be viewed together. The position of samples relative to one another in the two-dimensional space (PC1 versus PC2) is determined by the way individual consumers express their preferences for each sample on the 9-point Likert scale provided. Those samples close to one another are of near equal importance, in terms of preference, for consumers.

From Fig. 22.2 it can be seen that consumer preference was towards the reduced-calorie fruit soup FSOUP and rice milk RMILK carriers. The most

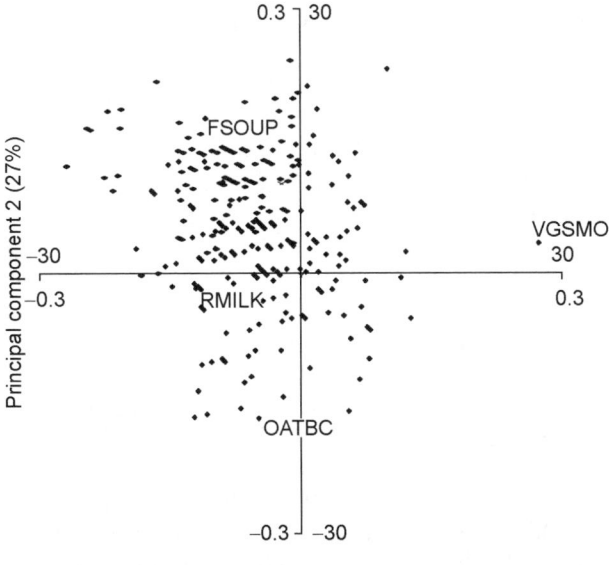

Fig. 22.2   Internal preference mapping of consumer preference scores for four reduced-calorie carriers.

preferred reduced-calorie carrier was FSOUP (mean score 6.31 out of 9), which was characterised as having a 'creamy' appearance, a 'fruity' odour, a 'sweet', 'creamy' and 'greasy/fatty' flavour, and a 'smooth' mouthfeel (see Table 22.2). Consumers also had a preference for the reduced-calorie carrier RMILK (mean score 5.87 out of 9). RMILK was characterised by a 'creamy' appearance, a 'creamy' odour and flavour, a 'lingering sweetness' afterflavour and a 'smooth' mouthfeel (see Table 22.2). The reduced-calorie oat bran and carrot carrier OATBC did gain a degree of acceptance among consumers, although its mean preference score was low (mean score 5.32 out of 9). However, the reduced-calorie vegetable smoothie carrier VGSMO, which was characterised by odour 'strength', flavour 'strength' and flavour 'tainted', was least preferred by consumers in this study (mean score 4.56 out of 9) (see Table 22.2).

The sensory analysis aspect of this work helps the product developer identify which reduced-calorie product is most preferred by consumers and also the key attributes that drive consumer acceptance of such a product. The next step in the design process is to use the information generated to identify product concepts and get consumers to rate those concepts. The sensory analysis information can be used in combination with the focus group results to help design the conjoint-based consumer study. The conjoint analysis technique is used to predict the key extrinsic and intrinsic product design attributes that influence consumers' preferences for innovative reduced-calorie 'on-the-go' beverages, as well as the trade-offs consumers make between alternative reduced-calorie 'on-the-go' beverages.

22.5.5   New product concept refinement and optimisation through conjoint analysis

Conjoint analysis is a multivariate technique that models purchase decision-making processes through an analysis of consumer trade-offs among hypothetical multi-attribute products. The conjoint analysis technique views a product as a combination of a set of attribute levels, where varying these attribute levels, according to a statistically determined design, facilitates the estimation of utility values, which determine consumers' total utility or overall judgement of a product (Green and Srinivasan, 1978). In that sense, conjoint analysis mimics real choice situations where respondents are required to simultaneously consider many dimensions of alternatives. Information generated through conjoint analysis provides consumer-driven information to R&D personnel regarding the nature of consumers' preferences between alternative product attribute levels that can aid the new product design process (Green and Krieger, 1991; Hair *et al.*, 1998).

The conjoint analysis component of this reduced-calorie case study constituted the third and final data set. The product attributes, and associated attribute levels, used in this study were derived from a combination of the focus group and sensory analysis results, and also from discussions with the technical R&D personnel involved in this multidisciplinary NPD research project (see Table 22.4). The Orthogonal Design Procedure in SPSS then generated 22 hypothetical reduced-calorie 'on-the-go' beverages, four of which were holdout beverage profiles, based upon the selected attributes and associated attribute levels. The four holdout beverage profiles would be rated by consumers, but not used in the estimation of utility values, to determine how consistently the conjoint models could predict consumers' preferences for hypothetical reduced-calorie on-the-go beverage concepts that would not be evaluated by consumers (Hair *et al.*, 1998). A conjoint-based survey was designed that presented 22 hypothetical reduced-calorie 'on-the-go' beverages for consumer evaluation using a 9-point Likert scale. Additional questions that related to consumers' eating habits, lifestyles and their sociodemographic information were also included in the conjoint survey.

Three hundred consumers who participated in the sensory analysis study were invited to complete the conjoint-based survey (see Table 22.3). The questionnaires were then analysed using SPSS v11 (SPSS, 2003). The individual level conjoint analysis procedure in SPSS determined the importance of each attribute and attribute level to consumers' preferences for new reduced-calorie 'on-the-go' beverages. The conjoint analysis revealed that the price, health benefit, flavour and the carrier attributes were most important in terms of choosing between alternative reduced-calorie 'on-the-go' beverage concepts. Pearson's R (0.998) and Kendall's tau (0.974) values were high and indicated strong agreement between the averaged product ratings and the predicted utilities from the conjoint analysis model.

Agglomerative hierarchical cluster analysis was employed initially to determine the desired number of clusters. This preliminary segmentation process

Table 22.4 Attributes and attribute levels used in the conjoint-based survey

Attribute	Level
Flavour	Orange
	Banana
	Strawberry
Carrier	Reduced-calorie rice milk
	Reduced-calorie oat bran and carrot
	Reduced-calorie fruit soup
Health benefit	Contains probiotic cultures and selected nutrients to boost the immune system
	Contains probiotic cultures and selected nutrients to aid the digestive system
	None
Pack size	250 ml
	330 ml
	500 ml
Packaging	Plastic bottle
	Pouch
	Carton
Price	€1.50
	€2.50
	€4.00

suggested that three to five clusters existed for reduced-calorie 'on-the-go' beverages. A five-cluster solution was finally chosen to reflect the variation in consumers' preferences that might exist in the marketplace. *K*-Means cluster analysis was then used to segment respondents into five clusters of consumers with similar preferences for reduced-calorie 'on-the-go' beverages based on attribute utility patterns (see Table 22.5). In Table 22.5 the highest utility values are in bold and the lowest utility values are in italics. The sociodemographic profile of each cluster, which helped further distinguish between segments, is also presented in Table 22.5. This market segmentation process enables NPD personnel to target reduced-calorie 'on-the-go' beverages at consumers with similar preferences.

Reduced-calorie on-the-go cluster profiles
The sociodemographic profile of Cluster 1 was biased towards well-educated (78%), young adults (43.9%) and females (60.7%) (see Table 22.5). Cluster 1 contained 41 consumers and this segment was most influenced by the flavour attribute when choosing between alternative reduced-calorie 'on-the-go' beverage concepts. This segment most preferred orange (1.32) and strawberry (1.03) flavoured reduced-calorie 'on-the-go' beverages, and least liked banana flavoured reduced-calorie 'on-the-go' beverages (−2.35). The price attribute was extremely important to this cluster, and these consumers preferred low-priced (€1.50) (0.68) and medium-priced (€2.50) (0.14) reduced-calorie 'on-the-go' beverages. The health benefit attribute was also important to this cluster in

Table 22.5   Averaged utility values and sociodemographic profiles across clusters

Attribute	Factor/attribute level	Cluster 1	Cluster 2	Cluster 3	Cluster 4	Cluster 5
Flavour	Orange	**1.32**	0.08	0.21	**0.57**	**0.35**
	Banana	*−2.35*	*−0.27*	*−0.62*	*−0.31*	−0.14
	Strawberry	1.03	**0.19**	**0.42**	−0.26	*−0.21*
Carrier	Reduced-calorie rice milk	**0.19**	0.05	0.92	−0.08	0.55
	Reduced-calorie oat bran and carrot	*−0.25*	*−0.16*	*−2.34*	*−0.10*	**1.12**
	Reduced-calorie fruit soup	0.06	**0.10**	**1.42**	**0.18**	*−1.68*
Health benefit	Contains probiotic cultures and selected nutrients to boost the immune system	**0.40**	**0.20**	0.10	1.16	**0.79**
	Contains probiotic cultures and selected nutrients to aid the digestive system	0.11	0.18	**0.34**	**1.23**	0.23
	None	*−0.51*	*−0.38*	*−0.43*	*−2.39*	*−1.02*
Pack size	250 ml	*−0.03*	*−0.14*	*−0.22*	*−0.02*	0.03
	330 ml	**0.05**	0.00	*−0.21*	*−0.28*	**0.25**
	500 ml	−0.02	**0.13**	**0.44**	**0.30**	*−0.28*
Packaging	Plastic bottle	**0.16**	0.02	**0.29**	**0.07**	*−0.31*
	Pouch	*−0.19*	**0.04**	*−0.28*	0.06	**0.44**
	Carton	0.03	*−0.06*	−0.01	*−0.13*	−0.13
Price	€1.50	**0.68**	**0.71**	**0.69**	**1.18**	**1.05**
	€2.50	0.14	0.13	−0.03	0.37	0.12
	€4.00	*−0.82*	*−0.84*	*−0.66*	*−1.55*	*−1.16*
Cluster size		41	114	42	67	36

Sociodemographic variable	Factor	Cluster 1 (%)	Cluster 2 (%)	Cluster 3 (%)	Cluster 4 (%)	Cluster 5 (%)
Gender	Male	39.3	48.2	42.9	35.4	21.9
	Female	60.7	51.8	57.1	64.6	78.1
Age group (years)	18–24	43.9	31.6	26.2	18.5	3.1
	25–34	31.7	33.3	35.7	38.5	62.5
	35–44	12.2	21.1	23.8	26.2	15.6
	45–54	7.3	9.6	11.9	12.3	6.3
	55–65	4.9	4.4	2.4	4.6	12.5
Marital status	Single	68.4	53.8	52.4	43.1	50.0
	Married	24.4	24.7	49.5	46.2	37.5
	Separated/divorced	4.8	0.9	2.4	1.5	3.1
	Cohabiting	2.4	7.9	4.8	9.2	9.4
	Widowed	–	12.8	–	–	–
Educational status	Completed primary level only	2.4	14.0	–	–	–
	Completed secondary level only	19.5	21.1	47.6	32.3	9.4
	Completed third level	78.0	64.9	52.4	67.7	90.6
Employment status	Employed full time	68.3	56.8	66.7	63.1	68.8
	Employed part time	12.2	8.8	16.7	12.3	15.6
	Student	14.6	21.9	11.9	12.3	3.1
	Retired	–	10.8	–	3.1	–
	Unemployed	4.9	1.7	4.8	7.7	12.5

terms of its preferences. Cluster 1 most preferred reduced-calorie 'on-the-go' beverages that contained probiotic cultures and selected nutrients to either 'boost the immune system' (0.40) or 'aid the digestive system' (0.11). This segment also preferred reduced-calorie rice milk (0.19) and fruit soup (0.06) carriers to a reduced-calorie oat bran and carrot carrier (−0.25) (see Table 22.5).

Cluster 2, which contained 114 respondents, considered the price attribute most significant in terms of choosing between alternative reduced-calorie 'on-the-go' beverage concepts. The health benefit and flavour attributes were also significant to this segment. Cluster 2 most preferred strawberry flavoured (0.19) reduced-calorie 'on-the-go' beverage concepts that contained probiotic cultures and selected nutrients to 'boost the immune system' (0.20). This cluster contained a near equal percentage of male and female consumers, who were single (53.8%) and in full-time employment (see Table 22.5).

Cluster 3 was biased towards females, who represented approximately 57% of that cluster. The carrier was most important for Cluster 3, and this segment preferred reduced-calorie fruit soup (1.42) and reduced-calorie rice milk (0.92) carriers in that order. Price was the second most important attribute to Cluster 3 and these consumers expressed the greatest preference for low-priced (€1.50) (0.69) reduced-calorie 'on-the-go' beverage concepts. This segment also gave a relatively high utility value for pack size across clusters (see Table 22.5).

Cluster 4 scored the highest utility value for the health benefit attribute across clusters, and therefore, could be considered health driven in terms of its purchase preferences (see Table 22.5). Specifically, Cluster 4 preferred reduced-calorie 'on-the-go' beverages that contained probiotic cultures and selected nutrients to either 'aid the digestive system' (1.23) or 'boost the immune system' (1.16). Price was the second most important attribute to Cluster 4, and this segment expressed the greatest preference for low priced (€1.50) (1.18) reduced-calorie 'on-the-go' beverage concepts. Interestingly, Cluster 4, the health-driven segment, was biased towards females and adults aged 35 years and older.

The carrier attribute was most important to Cluster 5, and this segment most preferred reduced-calorie oat bran and carrot-based (1.12) 'on-the-go' beverages. Price was the second most important attribute to Cluster 5, and this segment expressed the greatest preference for low priced (€1.50) (1.05) reduced-calorie 'on-the-go' beverage concepts. The health benefit attribute was also important to this segment of consumers, and similar to Clusters 1 and 2, most preferred reduced-calorie 'on-the-go' beverages that contained probiotic cultures and selected nutrients to 'boost the immune system' (0.79). The sociodemographic membership of Cluster 5 was biased towards females (78.1%), and adults aged between 25 and 34 years (62.5%). This cluster also contained the highest percentage of respondents across clusters that completed third level education (see Table 22.5).

Conjoint model predictions for reduced-calorie on-the-go beverages
The next step in the process is to design and evaluate alternative new product concepts using the group level simulation analysis procedure in SPSS. The

group level simulation analysis procedure in SPSS requires the generation of hypothetical product concepts that are not evaluated in the original survey. These hypothetical new product concepts can represent new market (competitor) entrants, alternative marketing strategies or, in this case, new product offerings that a firm may wish to commercialise. In that sense, the group level simulation analysis technique represents a powerful tool which can assist product development personnel predict consumers' preferences for new hypothetical product concepts at the early or concept stages of the NPD process.

In this study the hypothetical reduced-calorie 'on-the-go' beverage concepts were generated following rigorous analysis of the qualitative, sensory and cluster analysis data, and from discussions with the technical partners involved in this project. In particular, interpreting the cluster analysis results for the purpose of designing the simulation analysis research must be approached carefully. For example, the group level simulation analysis procedure in SPSS could be used to identify reduced-calorie 'on-the-go' beverages specifically targeted at each segment identified in this study. This strategy is most appropriate when consumers' preferences differ markedly across clusters, and in competitive markets where a firm needs to segment selectively in order to gain a superior competitive advantage in the marketplace. However, the group level simulation analysis technique was used in this case study to identify a limited number of reduced-calorie 'on-the-go' beverages that would appeal to a number of consumer segments. This strategy is most appropriate in emerging markets or where consumers' preferences are relatively similar across clusters. In fact, it appeared from Table 22.5 that Clusters 1 and 4 exhibited relatively similar preferences for orange flavoured fruit soup-based reduced-calorie 'on-the-go' beverages. Similarly, Clusters 2 and 3 appeared to exhibit relatively similar preferences for strawberry flavoured fruit soup-based reduced-calorie 'on-the-go' beverages. In contrast, Cluster 5 exhibited different preferences for reduced-calorie 'on-the-go' beverages, and was therefore excluded from the group level simulation analysis.

Overall, a Kendall's tau value of 1 for the four holdouts was obtained which suggested perfect agreement between the holdout ratings and the model predictions. It was therefore possible to analyse consumers' preferences for alternative reduced-calorie 'on-the-go' beverage concepts using choice simulators, both maximum and probability (BTL and Logit) modelling, across clusters. These models were used to estimate preference scores associated with each hypothetical reduced-calorie 'on-the-go' beverage concept included in the simulation analyses. Although the maximum utility model assumes respondents chose only beverage concepts with the highest predicted utility scores, the probability models assume respondents rarely make decisions using such precise notions of utility (Hair et al., 1998). Importantly, the group level simulation analysis across clusters provided for a more market-oriented approach to NPD whereby the preferences of each segment were taken into account when optimising the product design formulation for reduced-calorie 'on-the-go' beverages.

Seven hypothetical reduced-calorie 'on-the-go' beverage concepts (RCBEV 1–RCBEV 7) were generated for the group level simulation analysis across clusters (see Table 22.6). RCBEV 4 and RCBEV 5 were chosen for inclusion in the group level simulation analysis as they represented hypothetical reduced-calorie 'on-the-go' beverage concepts, which according to the cluster analysis results (Table 22.5) would yield high predicted preference scores for Clusters 1 and 4 respectively. Similarly, RCBEV 1 was included in the group level simulation analysis as this beverage concept was expected to yield high predicted preference scores for Clusters 2 and 3. However, new product concepts that combine the optimal product design attributes may not represent commercially feasible new products. This simplistic approach to new product design neglects the multi-faceted nature of consumer food choice, where the interplay between market-related factors such as price, and product-related factors such as sensory perception and user benefit, ultimately influence consumers' cognitive food choice motives. Therefore, four further hypothetical reduced-calorie 'on-the-go' beverage concepts (RCBEV 2, RCBEV 3, RCBEV 6 and RCBEV 7) were included in the simulation analysis, in order to identify which consumer segments would be expected to make trade-offs between key market and product-related attributes, when evaluating alternative reduced-calorie 'on-the-go' beverages.

Cluster 1
The conjoint models predicted that Cluster 1 would most prefer the reduced-calorie 'on-the-go' beverage RCBEV 4 (mean score 7.5 out of 9). This beverage was described as an orange flavoured reduced-calorie rice milk-based 'on the-go' beverage. RCBEV 4 contained added probiotic cultures and selected ingredients to boost the immune system, and retailed at €1.50 per 500 ml plastic bottle. It was evident from Table 22.5 that this segment of consumers was also receptive towards fruit soup-based carriers. Therefore, the conjoint models revealed that Cluster 1 would also give a high predicted preference score for RCBEV 5 (mean score 7.2 out of 9), which was a fruit soup-based variant of RCBEV 4 (see Table 22.6). Although RCBEV 4 and RCBEV 5 were expected to receive high predicted preference scores for Cluster 1, these reduced-calorie 'on-the-go' beverage concepts were not considered commercially feasible, owing to their very low (€1.50 per 500 ml) retail price.

In that context, the conjoint models predicted that Cluster 1 would not make trade-offs between the health benefit and price attributes when evaluating alternative reduced-calorie 'on-the-go' beverage concepts. For example, Cluster 1 would be expected to be more receptive towards RCBEV 7 than RCBEV 6 according to the predicted preference scores and probability (BTL and Logit) models. Both hypothetical reduced-calorie 'on-the-go' beverages (RCBEV 6 and RCBEV 7) were identical to RCBEV 5 in respect of the carrier, flavour, pack size and packaging attribute levels, but differed in terms of the health benefit and price attribute levels. Specifically, RCBEV 7 (mean score 6.8 out of 9) was a medium-priced (€2.50 per 500 ml) variant of RCBEV 5, while RCBEV

Table 22.6 Results of the group level simulation analysis for reduced-calorie on-the-go beverage concepts across clusters

Attributes/preference scores	RCBEV 1	RCBEV 2	RCBEV 3	RCBEV 4	RCBEV 5	RCBEV 6	RCBEV 7
Flavour	Strawberry	Strawberry	Strawberry	Orange	Orange	Orange	Orange
Reduced-calorie carrier	Fruit soup	Fruit soup	Fruit soup	Rice milk	Fruit soup	Fruit soup	Fruit soup
Health benefit	Boost the immune system	None	Boost the immune system	Boost the immune system	Boost the immune system	None	Boost the immune system
Pack size	500 ml	500 ml	500 ml	500 ml	500 ml	500 ml	500 ml
Packaging	Plastic bottle	Plastic bottle	Plastic bottle	Plastic bottle	Plastic bottle	Plastic bottle	Plastic bottle
Price	€1.50	€1.50	€2.50	€1.50	€1.50	€1.50	€2.50
Cluster 1 (pref. score)	6.2 out of 9	5.5 out of 9	5.9 out of 9	7.5 out of 9	7.2 out of 9	6.5 out of 9	6.8 out of 9
Cluster 2 (pref. score)	7.6 out of 9	7.1 out of 9	6.9 out of 9	6.4 out of 9	6.6 out of 9	6.0 out of 9	5.8 out of 9
Cluster 3 (pref. score)	7.8 out of 9	7.1 out of 9	7.3 out of 9	6.4 out of 9	6.9 out of 9	6.3 out of 9	6.6 out of 9
Cluster 4 (pref. score)	6.4 out of 9	5.7 out of 9	6.0 out of 9	7.2 out of 9	7.8 out of 9	6.4 out of 9	7.4 out of 9

6 was a non-functional variant of RCBEV 5 retailing at €1.50 per 500 ml (mean score 6.5 out of 9) (see Table 22.6).

Cluster 2

The conjoint models predicted that Cluster 2, the largest segment identified in this study, would most prefer the reduced-calorie 'on-the-go' beverage RCBEV 1 (mean score 7.6 out of 9) (see Table 22.6). This beverage was described as a strawberry flavoured reduced-calorie fruit soup-based 'on the-go' beverage. RCBEV 1 contained added probiotic cultures and selected ingredients to boost the immune system, and retailed at €1.50 per 500 ml plastic bottle. However, this reduced-calorie 'on-the-go' beverage concept was not considered commercially feasible for a functional drink, owing to its low (€1.50 per 500 ml) retail price. An increase in price from €1.50 per 500 ml (RCBEV 1) to €2.50 per 500 ml (RCBEV 3) was also expected to yield a relatively high predicted preference score (mean score 6.9 out of 9) for Cluster 2. Importantly, the group level simulation analysis revealed that Cluster 2 would make trade-offs between the health benefit and price attributes. Specifically, if a competitor's beverage RCBEV 2, a non-functional variant of RCBEV 1, were to be launched on the market in the future then the conjoint models predicted that Cluster 2 would be expected to choose RCBEV 2 (mean score 7.1 out of 9) over RCBEV 3 (see Table 22.6).

Cluster 3

It was evident from Table 22.5 that Clusters 2 and 3 exhibited similar preferences for reduced-calorie 'on-the-go' beverages. Not surprisingly, it was expected that Cluster 3 would most prefer the reduced-calorie 'on-the-go' beverage RCBEV 1 (mean score 7.8 out of 9) (see Table 22.6). However, unlike Cluster 2, this segment would not be expected to make trade-offs between the health benefit and price attributes. Therefore, the conjoint models predicted that Cluster 3 would be most receptive towards RCBEV 3 (mean score 7.3 out of 9), a medium-priced variant (€2.50 per 500 ml) of RCBEV 1, and less receptive towards RCBEV 2 (mean score 7.1 out of 9), which was a non-functional variant of RCBEV 1 (see Table 22.6).

Cluster 4

Finally, the conjoint models predicted that Cluster 4 would most prefer the reduced-calorie 'on-the-go' beverage RCBEV 5 (mean score 7.8 out of 9). This beverage was described as an orange flavoured reduced-calorie fruit soup-based 'on the-go' beverage. RCBEV 5 contained added probiotic cultures and selected ingredients to boost the immune system, and retailed at €1.50 per 500 ml plastic bottle. Again, this reduced-calorie 'on-the-go' beverage concept was not considered commercially feasible, owing to its low (€1.50 per 500 ml) retail price. Importantly, Cluster 4, the health-driven segment, would not be expected to make trade-offs between the health benefit and price attributes when evaluating alternative reduced-calorie 'on-the-go' beverage concepts. Therefore, it was predicted that these consumers were more likely to purchase the medium-priced beverage RCBEV 7 (mean score 7.4 out of 9) and less likely to purchase the non-functional beverage RCBEV 6 (mean score 6.4 out of 9), which retailed at €1.50 per 500 ml plastic bottle (see Table 22.6).

The final outcome of the conjoint analysis process is the identification of two commercially feasible reduced-calorie 'on-the-go' beverage concepts. RCBEV 3 and RCBEV 7 are strawberry and orange flavoured fruit soup-based reduced-calorie 'on-the-go' beverages respectively, which contain probiotic cultures and selected ingredients to boost the immune system, and retail at €2.50 per 500 ml plastic bottle (see Table 22.6). These product design specifications can now be used by the multidisciplinary NPD team to guide the technical development of consumer-led reduced-calorie 'on-the-go' beverages, targeted at a number of market segments, with relatively high predicted levels of consumer acceptance.

22.5.6   Case study conclusion

The consumer-led approach to NPD presented in this case study made it possible to identify the optimal combination of product design attributes, for reduced-calorie 'on-the-go' beverage concepts, targeted at a number of potential market segments using information generated from qualitative and quantitative research. This approach to NPD also provides guidance to firms in terms of suitable communication, positioning and pricing strategies for reduced-calorie 'on-the-go' beverages.

22.6   Summary

New food product development is a multidisciplinary knowledge intensive process, which necessitates the generation, dissemination and management of knowledge across all functions involved in the development of new foods and beverages. In particular, the early stages of the NPD process represent extremely critical stages for managing knowledge of both internal technological capabilities and external measures of consumers' needs. The results of this case study highlight the value of concept optimisation research methodologies to managing knowledge in the early or concept development stage of the NPD process. Gathering consumers' views during the early stages of the new product design process through focus groups, sensory analysis and conjoint analysis helps identify potential strategic marketing opportunities for innovative product concepts. It also provides a systematic framework for managing the gathering of consumer knowledge during the new food product development process.

The market for reduced-calorie food and beverages is expected to grow as consumers seek improved reduced-calorie versions of regular products. The current trend towards healthy eating offers new product opportunities for firms particularly in view of the failure of many reduced-calorie products, across categories, to meet consumers' expectations. The reduced-calorie food and beverage market offers new product opportunities for those firms that understand consumers' preferences and choice motives, in terms of those market and sensory factors that influence preferences, and can target products effectively at specific market segments. In particular, a key conclusion arising from this NPD case study is the nature of reduced-calorie food and beverage products where firms need to understand the extrinsic and intrinsic product design attributes that influence consumers' preferences. Firms that adopt a market-oriented approach to NPD, through the use of advanced concept optimisation research techniques, will benefit from a deeper understanding of consumers' value systems. This in turn can assist firms identify key market segments and more accurately make strategic marketing decisions for reduced-calorie foods and beverages. Concept optimisation research techniques promote a multidisciplinary approach to NPD, which can help firms to manage knowledge more effectively and efficiently between functional disciplines involved in the NPD process.

22.7   Sources of further information and advice

BRODY A L and LORD J B (2000). *New Food Products for a Changing Marketplace.* Lancaster, PA: Technomic Publishing Company.

COOPER R G (1993). *Winning at New Products*, 2nd Edition. Reading, MA: Addison-Wesley Publishing Company.

DESHPANDÉ R (1999). *Developing a Market Orientation.* Thousand Oaks, CA: Sage Publications.

FULLER G W (1994). *New Food Product Development: From Concept to Marketplace.* Boca Raton, FL: CRC Press.

JONES T (1997). *New Product Development: An Introduction to a Multifunctional Process.* Butterworth-Heinemann.

JONGEN W M F and MEULENBERG M T G (2005). *Innovation in Agri-food systems: Product Quality and Consumer Acceptance.* Wageningen, The Netherlands: Wageningen Academic Publishers.

MOSKOWITZ H R, SEBASTIANO P and SILCHER M (2005). *Concept Research in Food Product Design and Development.* Ames, IO: Blackwell Publishers.

Product Development and Management Association – *Visions Magazine*: http://www.pdma.org/visions

Journal of Product Innovation Management: http://www.pdma.org/journal/

MAPP – Aarhus School of Business: http://www.asb.dk/research/centresteams/centres/mapp.aspx

UCC – Department of Food Business and Development: http://www.ucc.ie/academic/foodecon/

22.8   References

BEST D (1991), 'Getting more for less: the challenges of fat substitution', *Prepared Foods*, **160**(6), 72–73, 75, 77.

BLUNDELL J E (2000), 'Health and calcium of the dairy product', in *Proceedings of Santé et Nouveax Produits Laitiers* (*Health and New Dairy Products*), 6–9 March, Europel, Annecy, France.

BOGUE J (2001), 'New product development and the Irish food sector: a qualitative study of activities and processes', *The Irish Journal of Management incorporating IBAR*, **22**(1), 171–191.

BOGUE J and RITSON C (2004), 'Understanding consumers' perceptions of product quality for lighter dairy products through the integration of marketing and sensory information', *Acta Scandinavia*, **1**(2), 67–77.

BOGUE J, DELAHUNTY C, HENRY M and MURRAY J (1999), 'Market-oriented methodologies to optimise consumer acceptability of Cheddar-type cheeses', *British Food Journal*, **101**(4), 201–316.

BOYLE C and EMERTON V (2002), *Food and Drinks Through the Lifecycle*. Leatherhead, Surrey: Leatherhead International.

BRITISH NUTRITION FOUNDATION (1999), *Obesity*, The Report of the British Nutrition Foundation Task Force. Oxford: Blackwell Science.

CAMO (1998), *The Guideline Methodology*, Guideline+ v.7.5 1996-1999, Oslo, Norway: CAMO ASA.

CARDELLO A V (1993), 'What do consumers expect from low-cal, low-fat, lite foods?', *Cereal Foods World*, **38**(2), 96–99.

COMPUSENSE (2001), *Compusense five v3.0*. Ontario, Canada: Compusense Inc.

COOPER R G (1993), *Winning at New Products*, 2nd edn, Reading, MA: Addison-Wesley.

DATAMONITOR (2005), 'Consumer beat: is low-carb Kaput?', *Restaurants and Institutions*, **115**(2), 14.

DRAKE M and SWANSON B (1997), 'Silver bullets and low fat technologies', *Prepared Foods*, **166**(7), 37–38, 40, 45, 46.

EARLE M D (1997), 'Changes in the food product development process', *Trends in Food Science and Technology*, **8**(1), 19–24.

EUROFOOD (2001), 'Food on-the-go increasing as habits change', *Eurofood*, 10 May.

FITZPATRICK L (1997), 'Qualitative concept testing tells us what we don't know', *Marketing News*, **31**(12), 26.

GERALDI R C (1992), 'The European marketplace: is the lite revolution going global?', *Manufacturing Confectioner*, **71**(4), 53–57.

GOLDING C, CADE J, LAWTON C and GREENWOOD D (2001), 'Comparison of low and high fat consumers in the UK Women's Cohort Study', in *Society for the Study of Ingestive Behaviour, International Conference Proceedings*, 26–30 June, Philadelphia, United States.

GREEN P E and KRIEGER A M (1991), 'Product design strategies for target-market positioning', *Journal of Product Innovation Management*, **55**(1), 20–31.

GREEN P E and SRINIVASAN V (1978), 'Conjoint analysis in consumer research: issues and outlook', *Journal of Consumer Research*, **5**(2), 103–123.

GREENHOFF K and MACFIE H J H (1994), 'Preference mapping in practice', in MacFie H J H and Thomson D M H, *Measurement of Food Preferences*. London: Blackie Academic and Professional.

GROVES A M (2002), *Food Consumption 2002*. Watford: IGD Business Publication.

GRUNERT K G (2000), 'Three issues in consumer quality perception and acceptance of dairy products', in *Proceedings of Santé et Nouveax Produits Laitiers (Health and New Dairy Products)*, 6–9 March, Europel, Annecy, France.

GRUNERT K G, HARTVIG LARSEN H, MADSEN T K and BAADSGAARD A (1996), *Market Orientation in Food and Agriculture*. Boston, MA: Kluwer Academic.

HAIR J F, ANDERSON R E, TATHAM R L and BLACK W C (1998), *Multivariate Data Analysis*, 5th edn. Englewood Cliffs, NJ: Prentice-Hall.

HAMILTON J, KNOX B, HILL D and PARR H (2000), 'Reduced fat products: consumer perceptions and preferences', *British Food Journal*, **102**(7), 494–506.

HARMSEN H (1994), 'Tendencies in product development in Danish food companies – report of a qualitative analysis', MAPP Working Paper No. 17, Project No. 2, February.

HEALTH PROMOTION UNIT (1998), *A Guide to Healthy Food Choices*, Dublin: Health Promotion Unit, Department of Health and Children.

HOLLINGSWORTH P (1996), 'The leaning of the American diet', *Food Technology*, **50**(4), 86, 88, 90.

JENSEN B and HARMSEN H (2001), 'Implementation of success factors in new product development – the missing links?', *European Journal of Innovation Management*, **4**(1), 37–52.

KLEINSCHMIDT E J and COOPER R G (1995), 'The relative importance of new product success determinants – perception versus reality', *R & D Management*, **25**(3), 281–298.

KOHLI A K and JAWORSKI B J (1990), 'Market orientation: the construct, research propositions, and managerial implications', *Journal of Marketing*, **54**(2), 1–18.

LAND D G and SHEPHERD R (1984), Scaling and ranking methods, in Piggott J R, *Sensory Analysis of Foods*. London: Elsevier Applied Science.

LEATHERHEAD FOOD RESEARCH ASSOCIATION (2003), *Drinks On The Go – International Trends and Developments*. Leatherhead, Surrey: Leatherhead International.

LEATHERHEAD FOOD RESEARCH ASSOCIATION (2004), *Low and Light Food and Drinks. International Trends and Developments in Weight Control*. Leatherhead, Surrey: Leatherhead International.

LLOYD H M, PAISLEY C M and MELA D J (1995), 'Barriers to the adoption of reduced-fat diets in a UK population', *Journal of the American Dietetic Association*, **95**(3), 316–322.

MACFIE H J H and THOMSON D M H (EDS) (1994), *Measurement of Food Preferences*. London: Blackie Academic & Professional.

MACFIE H J, BRATCHELL N, GREENHOFF K and VALLIS I V (1989), 'Designs to balance the effect of order of presentation and first-order carry-over effects in hall tests', *Journal of Sensory Studies*, **4**(2), 129–148.

MINTEL (2004), *Healthy Eating – Ireland*. London: Mintel International Group Ltd.

MOSKOWITZ H R (1991), 'Optimizing consumer product acceptance and perceived sensory quality', in Graf E and Saguy I S, *Food Product Development from Concept to the Marketplace*. New York: Van Nostrand Reinhold.

NIJSSEN E J and FRAMBACH R T (2000), 'Determinants of the adoption of new product development tools by industrial firms', *Industrial Marketing Management*, **29**(2), 129–131.

O'DONNELL C D (1993), 'Reduced-fat foods: still the no. 1 formulation challenge', *Prepared Foods*, **162**(13), 19–21, 24–25, 28–31, 34–35.

O'MAHONEY M (1986), *Sensory Evaluation of Food*. New York: Marcel Drekker.

O'ROURKE R (1999), *European Food Law*. Bembridge, UK: Palladian Law Publishing.

PATTERSON R E, SATIA J A, KRISTAL A R, NEUHOUSER M L and DREWNOWSKI A (2001), 'Is there a consumer backlash against the diet and health message?', *Journal of the American Dietetic Association*, **101**(1), 37–41.

PSZCZOLA D (1996), 'Low-fat dairy products: getting a new lease on life', *Food*

Technology, **50**(9), 32.

QSR INTERNATIONAL (2002), *N6 (Non-numerical Unstructured Data Indexing Searching & Theorizing) Qualitative Data Analysis Program*. Melbourne, Australia: QSR International Pty Ltd. Version 6.0.

RABOBANK (1999), *The Fight for Stomach Share*. Special Report, Rabobank International, Food and Agribusiness Research, September. Utrecht, Netherlands: Rabobank.

RESURRECCION A V A (1998), *Consumer Sensory Testing for Product Development*. Gaithersburg, MD: Aspen.

Retail News (2000), 'The dairy best', *Retail News*, January, 51–53.

RICHARDSON D P (1996), 'European food industry perspectives on healthy eating', *World of Ingredients*, October, 39–40, 42, 44–45.

RITTMAN A (2003), 'Retro desserts: from old to new again', *Prepared Foods*, **172**(10), 47–56.

ROBERTS W A (2003), 'Bar none', *Prepared Foods*, **172**(4), 18–20.

SCHLOSSER E (2001), *Fast Food Nation*. London: Penguin.

SHENG M, BROCHETTI D, DUNCAN S E and LAWRENCE R A (1996), 'Hedonic ratings of reduced-fat food products: factors affecting ratings from female and male college-age students', *Journal of Nutrition in Recipe and Menu Development*, **2**(2), 31–40.

SHUKLA T P (1994), 'Sensory research in food development and marketing', *Cereal Foods World*, **39**(11/12), 876.

SPSS (2003), *Statistical Package for Social Sciences v11*. Chicago, IL: SPSS Ins.

STONE H and SIDEL J L (1995), 'Strategic applications for sensory evaluation in a global market', *Food Technology*, **49**(2), 80, 85–86, 88–89.

TRAILL B and GRUNERT K (EDS) (1997), *Product and Process Innovation in the Food Industry*. London: Blackie Academic and Professional.

VAN TRIJP J C M and STEENKAMP J E B M (1998), 'Consumer-oriented new product development: principles and practice', in Jongen W M F and Meulenberg M T G, *Innovation of Food Production Systems*. Wageningen: Wageningen Pers.

WIND J and MAHAJAN V (1997), 'Issues and opportunities in new product development: an introduction to the special issue', *Journal of Marketing Research*, **34**(1), 1–12.

ZHANG Q and DOLL W J (2001), 'The fuzzy front end and success of new product development', *European Journal of Innovation Management*, **4**(2), 95–112.

23

Preference mapping and food product development

H. MacFie, Hal MacFie Training Services, UK

23.1 Introduction

Preference mapping is a key tool in portfolio management as it enables marketing personnel to view and understand the marketplace and see where the competitors are moving to and where to position their own products. When linked to descriptive sensory data it also enables new product development (NPD) staff in R&D to understand how to manipulate the sensory properties of products to move to desired positions in the marketplace and both marketing and R&D can understand the key sensory attributes that respondents are using to make assessments about liking. It is important to realise that it is a source of reformulations or hybrids of current products. It is not a source of totally novel products; for those one should apply some of the nine tests outlined by van Kleef *et al.* (2005).

This chapter outlines the main techniques currently in use, with their pros and cons, and indicates how to design and collect suitable data for preference mapping. The application of various decision-making tools to enable optimal products to be identified and their sensory properties determined are illustrated. Finally some current research areas are described with an assessment of their importance to the practitioner.

Preference mapping really began with the paper by Carroll in 1972. He outlined the ideas of internal and external preference mapping, thus confusing generations of workers. To anyone in R&D used to the idea of an internal panel and an external panel it is clear that with external preference mapping you would be using the external panel to get the map and vice versa for internal. Absolutely incorrect! Carroll was a mathematician, not a sensory scientist, and internal refers to the fact that you obtain the map of the products internally from the consumer liking data and external refers to the fact that the map is based on data collected outside of the consumer data.

The first food-based preference mapping paper to my knowledge was on meat (Horsfield and Taylor, 1976). With the wider availability of the internal preference map program MDPREF and the very timely text on multidimensional scaling by Schiffman *et al.* (1981), papers started to appear. In the references I have tried to collect all the preference mapping references from the *Journal of Sensory Studies*, *Food Quality and Preference* and *Journal of Food Science* to provide a useful resource for readers to follow up on methods or particular food categories. This list is not exhaustive. However, the real engine for the dissemination of preference mapping concepts was provided by the sensory and food-based market researchers in the USA and Europe who soon realised that being able to connect up liking and descriptive data and present the so-called sensory drivers and marketplace maps gave them a perceived competitive edge over the competition. Alternative black box methods started to appear and there still remain a large number of hybrid preference mapping techniques that have never been published.

A second reason for the success of preference mapping is that it also satisfies two widespread, but possibly erroneous, assumptions that many workers and new product developers make when working with consumers. These are that the general consumer finds it difficult to articulate and score many perceptual attributes of foods and that the so-called experts that traditionally make quality assessments of foods are not representative of the general consumer population. If you make these assumptions it naturally follows that the optimum solution is to just ask consumers a minimum set of affective questions and recruit a non-technical consumer with excellent sensory facilities who is trained to recognise and articulate and score sensory attributes of foods. The two data sets can then be related by some statistical linkage method.

The growth of preference mapping has thus been engineered on the basis of conducting simultaneous central location tests and sensory descriptive analysis on the same products. Along the way there have been a number of developments and understandings that have increased the accuracy, precision and reduced bias of the data we collect and we will discuss these first before elaborating on the statistical linkage methods.

23.2   Conducting central location trials

Central location trials (CLTs) are sequential monadic tests conducted in a location such as a church hall, market research facility, etc. The respondents are either pre-recruited or come in directly off the street and assess a number of the products in question. They usually complete a thorough demographic and consumption questionnaire at the end of the test. The foremost advantage of the CLT over the home use test (HUT) is that there is total control over the preparation and presentation of the samples. This can be essential if some kind of comparative claim is going to be made against other products. It is also quick; with enough recruiters and temporary booths it is possible to complete the study in one day. The key advantage over monadic is that the consumer is able to express a number of opinions that enable us to work out which sensory attributes appear to be positively or negatively correlated with their liking and to establish a product map that defines a common space on which we can project liking directions or ideal points.

CLTs are conducted as marketplace screens, for competitive benchmarking and to compare the performance of a number of prototypes with the current product. If the analysis is going to involve preference mapping of any sort then a practical minimum number of products to be assessed is 6. However, most practitioners would recommend 8 as a realistic minimum and try for 12 or more. The reasons for this will become clear when we discuss the methodology. These numbers often disturb product developers, who are used to assessing 3 or 4 products. However, much of the cost of a respondent goes into the recruitment and transport arrangements to get them to the facility. Once they are there, it is relatively cheap to keep them there for a while, and in fact this means they will be more likely to turn up since the reward will be larger, so experimenters try to have respondents sample a larger number of products than one would normally undertake in a sensory test. Naturally a reasonable recovery time between assessments is maintained.

The process of selecting which products to include in the test is relatively easy for a control versus prototype test using a statistically based fractional design. However, the method of selecting samples from the marketplace is usually more tortuous, as marketing and R&D will have some very firm views on which products must be included or excluded. The presence of a sensory panel that can perform sensory descriptive analysis (SDA) is invaluable here. An iron rule for consumer trials should be that it is generally not worth putting two samples in a CLT that are indistinguishable to the sensory panel. Another rule to ensure success in recovering a reasonable map is not to restrict the samples to those that are generally likely to be market leaders as this will give too small a range of variation. The aim should be to select a set that will span the space of the marketplace. A typical scenario for product selection is to collect all the products one can, including any prototypes, and bench test them to select a subset of say 20–30 to go forward to SDA. On completion of the SDA, perform a principal components analysis and select a set of 12 or 15 samples that span the space, circle them and send the map over to marketing and R&D for their views. When the list of samples come back that absolutely must be included in the trial, make a second iteration, attempting to avoid products with identical sensory properties going into the trial. If there are problems or political battles, then it is relevant to make the point that it will be possible to project samples for which there is an SDA profile into the map to see where they lie, even if they were not submitted to the CLT.

Having selected the samples, the next question will be whether to conduct a complete trial, in which all respondents assess all samples, or conduct an incomplete trial. Preference map practitioners will always push for a complete trial, as many cluster analysis and internal preference mapping algorithms require a complete set, and there is little doubt that this must be the optimum solution. If an incomplete design is necessary then a rough rule for preference mapping trials is that each respondent should assess at least 75% of the total set (Hedderley and Wakeling, 1995). In addition a balanced incomplete block design is used so that every product is used equally often across the respondents.

With the samples selected, we can turn our attention to selecting the respondents. Marketing can usually provide profiles of typical users and quotas are set up for age, gender, location and usage level. In some instances previous trials have characterised psychographic profiles and a key set of recruitment questions can be used to select quotas of the different profile groups.

We then come to the thorny question of how many respondents to recruit. One approach is to use previously obtained estimates of residual standard deviations to calculate the number of respondents that will give you the precision you require. Hough et al. (2006) have provided a very useful guideline for this. They examined a number of CLTs across many products and countries and concluded that a typical residual standard deviation for these trials was 0.23 of the scale length. So if one was working with the 9-point hedonic scale there would be a scale length of 8 (9 − 1) and the residual standard deviation (RSD) would be 0.23×8 (1.84). So using this value and power tables, if we set a Type I error of 5% (5% chance of finding a significant difference between two samples when none exists) and Type II error of 10% (10% chance of concluding there is no difference between two samples when one exists), we can work out the number of respondents needed in a trial to be able to detect a pre-specified difference.

Table 23.1 indicates that for a pre-specified difference of 10% of the scale length, i.e. 0.8 in terms of the 9-point scale a number of 112 respondents would be required for the Type I and II settings above. This is remarkably close to the minimum number used by many practitioners of 120. However this number is not going to be sufficient for a trial in which you may be wishing to analyse certain subgroups or are expecting to detect and study naturally occurring segments in the population. A rough guideline if you are thinking of analysing groups separately is to recruit at least 50 people per group and ideally more. It is important to realise that some individuals have great trouble in giving numbers to their sensory experiences and their ratings appear to be completely uncorrelated with anyone else's. There is also the frustrating group of non-discriminators that give the same score to every sample. So it is necessary to over-recruit so that we have at least 40–50 respondents in each important subgroup.

Another problem with the 50 respondents per segment rule is that segments do not typically appear to be the same size. It often happens that there is one large group and a few smaller segments. As will be seen later, it is becoming common practice to identify the clusters and then work out products that can satisfy as many segments as possible, rather than analyse each segment and develop a specific product. In these cases it is not necessary to characterise each

Table 23.1 Number of consumers needed for an acceptability test (from Hough *et al.*, 2006)

RMSL[a]	α (%)[b]	d[c]	β (%)[d]		
			20	10	5
0.14	10	0.2	7	9	11
	5	0.2	8	11	14
	1	0.2	12	15	18
	10	0.1	25	34	43
	5	0.1	32	42	52
	1	0.1	47	59	71
	10	0.05	98	135	170
	5	0.05	124	166	205
	1	0.05	184	234	280
0.23	10	0.2	17	23	29
	5	0.2	22	29	35
	1	0.2	32	40	48
	10	0.1	66	91	115
	5	0.1	84	112	138
	1	0.1	124	158	189
	10	0.05	262	363	459
	5	0.05	333	445	551
	1	0.05	495	631	755
0.3	10	0.2	29	39	49
	5	0.2	36	48	59
	1	0.2	53	68	81
	10	0.1	112	155	196
	5	0.1	142	190	235
	1	0.1	211	269	322
	10	0.05	446	617	780
	5	0.05	566	757	936
	1	0.05	842	1072	1284

[a] RMSL = root mean square error divided by scale length.
[b] α % = probability of Type I error.
[c] d = difference in means that is sought in the experiment (scale 0–1).
[d] β % = probability of Type II error.

cluster too intently and a figure of 150–180 respondents allows for three segments. However some practitioners routinely identify six to nine clusters in the population and therefore sampling numbers of 300 plus are recommended. This seems a little high to the current author; a figure of between 150 and 200 should suffice for most markets.

The sampling plan for each respondent is the next task. Following MacFie *et al.* (1987), the Williams Latin squares (Williams, 1949) that balance the occurrence of each sample in each position, and the number of times each sample follows every other sample are now used routinely. If the incomplete design is being used, the designs proposed by Wakeling and MacFie (1995) are utilised. Ball (1997) has discussed how to keep the design as balanced as possible throughout the sampling process so that, in the case of an unexpected halt in the flow of consumers into the test facility, a relatively high degree of balance is maintained. For multi-session trials in which the respondent may be returning to the facility two, three or more times, the Design Express program (Design Express version 1.6, 2006, Qi Statistics Ltd, UK) gives easy access to designs that are as balanced as possible, even for incomplete designs. Note that, although these designs are universally used, there are those who maintain that a random-isation methodology would be preferable (Kunert, personal communication).

These Williams designs do not remove the carry-over effect of one sample on to another from the data. They just distribute any carry-over effects evenly among the samples. One effect that can be removed is the tendency of respondents in sequential monadic trials to score the first product more highly than the rest. The use of a warm-up or dummy sample is recommended here. This sample is presented to the respondents with the full questionnaire and they complete it normally, believing it to be part of the test. However the results are discarded as being part of the learning process. It is often thought that a warm-up sample is needed only on the first session of a multi-session test. However, Fig. 23.1 shows the mean liking score in each position across a balanced design for some beers across 3 days of testing. On the first day a warm-up sample was presented and the results discarded. However, no warm-up was presented on subsequent days. The

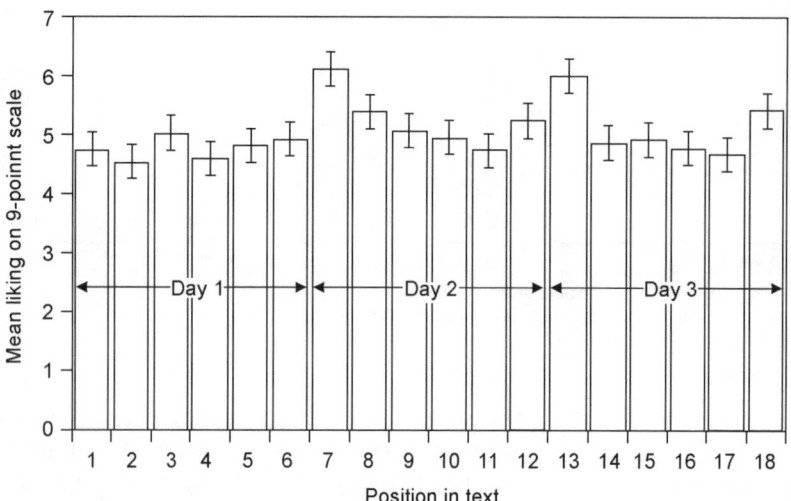

Fig. 23.1 Average liking scores of a CLT test of 18 beers for positions 1 to 18 across 3 days of testing. On the first day a warm-up sample was presented prior to the start of the test. Warm-up samples were not presented prior to testing on days 2 and 3. Evidence of a higher score given at the start and end of the sessions 2 and 3 is clear. The vertical scale is the mean liking on a 7-point hedonic scale.

effects on the data are clear, as is the usual occurrence of a higher score at the end of a session as respondents realise they have finished for the day or are about to collect their recompense. This could be minimised by informing respondents that they are going to score one more sample than is actually the case. The case for a warm-up sample is not so strong when the interest is only on aggregated results across all the respondents as positional effects can be removed as part of the analysis of variance estimation procedure. However in cases where each subject is going to be input to a cluster analysis or an internal preference map, or input as a separate y-variable in an external preference mapping procedure, these effects are difficult to remove by adjustment, and may significantly influence the scores of a respondent. In our view a warm-up sample is an essential component of successful preference mapping.

The next question is: how do we select a sample to be a warm-up? The author's advice on this stems from some unpublished data in which he presented 12 fragrances to 60 subjects using the same order of samples with no warm-up. This was then repeated using three other orders with different sets of 60 respondents. Analysis of the data showed that the order that had happened to contain the least liked sample first had given a different order of samples means than the other three orders. It was concluded that the halo effect of the first sample had lifted the scores of this sample. However, subjects were simply not willing to score every other sample one or two points higher so a re-evaluation of the scale took place with resulting distortion of the final sample means. The author's advice is therefore to try to use a sample that is in the middle of the space and should score in the middle of the liking range. The presence of sensory descriptive analysis data here is again invaluable as one can select a product in the middle of the space defined by the first two principal components.

One further aspect of sample design that needs to be considered is whether to replicate one or two samples in the sequential monadic test. This gives an estimate of respondent reliability and has been used by some practitioners as a way to screen out poorly performing respondents. It also gives extra degrees of freedom to enable a test of respondent by sample interaction in the aggregated analysis of variance. In the author's experience, clients are usually too eager to put an extra prototype in and therefore do not often include a replicate, but it can be very useful on occasion.

The next step before commissioning the CLT is the design of the question-naire. It is usual for experimenters to come down with a self-designed extensive list of questions they would like the respondents to answer. Here I take a fairly tough view and advise clients and researchers to ask a minimum set of questions, ideally just one: How much do you like this product? The rationale for this is that most respondents eat foods, consume beverages, shave, apply make-up, wash, shampoo, drive cars all the time and do not think deeply about the sensory variations in these experiences. By making them complete extensive question-naires on every product, the likelihood is that they will start to re-evaluate their opinions, often for reasons of so-called impression management (making you think they are someone they are not) and so the resulting scores do not relate to their daily appreciation and the resulting product will not be liked in the 'real world'.

The author is particularly opposed to the use of just-about-right (JAR) questions in these preference mapping studies. Earthy *et al.* (1997) reported how the addition of JARs to a questionnaire in a CLT of chocolate mousses varying in sweetness and ratio of dark to milk chocolate altered the overall liking of products. Popper *et al.* (2004) repeated the trial with extra conditions, including intensity of attributes and other liking questions as well as JARS, and showed that the presence of attribute liking question and the presence of JARS altered the ranking of the overall liking of products. In my view many respondents do not carry ideal points around in their head but this can be easily imposed on them by the presence of JAR questions. This is not to say that I am implacably opposed to JAR questions. The use of penalty analysis to help reformulation is increasing and readers should consult Chapter 17 by Lori Rothman for further details. However, I maintain that, for an accurate view of consumer liking that is essential for unbiased preference mapping, these JAR questions, and ideally all other attribute questions, should be asked as a separate exercise. Note that Earthy *et al.* (1997) indicated there to be no differences between asking the hedonic question on its own and asking the hedonic question first, so the latter is not a solution.

A very important issue to be considered before completing the preparation of the CLT trial is how the product is going to be presented to the respondent. Here normal use should be emulated as far as possible, without losing control of the standardisation of presentation that is a key characteristic of CLT trials. After all, if 99.9% of respondents only ever drink gin in the presence of tonic water, what is the point of getting them to assess neat gin? Similarly if you take sugar in your tea, never eat olive oil spread on its own, only ever eat sushi with a drop of soya sauce, then the presentation should be modified accordingly. Posri (2001) showed that modifying the presentation of tea to respondents' usual strength, milk and sugar dosage gave improved correlation with subsequent HUT results. However, the modification had to be standardised for each respondent. When respondents were permitted to prepare the tea themselves, the correlation was reduced, presumably because variation in preparation reduced the perceived differences.

A final issue to be dealt with when commissioning CLTs is who is going to visit on the first day of each series to ensure that the trial is being conducted correctly. This must be someone who understands the trial, is an experienced tester and has the authority to bring the trial to a halt if standards are not being observed. Skipping this operation is not an option unless you have complete trust in the team conducting the trial.

A brief summary of the recommendations is given in Table 23.2. If you observe all of these, you are well on the way to producing the kind of data that will produce reliable maps that can be used to predict liking of other products within the sensory space you have selected.

Table 23.2   Recommendations for conducting a successful CLT for a preference mapping exercise

---

- Minimum number of samples is 6, ideal is 8–12.
- Conduct a descriptive analysis on a larger set first.
- Do not select samples to go into the CLT that are sensorily indistinct from a sample that has already been selected.
- Span the space of the marketplace in selecting the samples.
- Conduct a complete trial so that each respondent assesses all samples.
- If not complete, each respondent should test at least 75% of the samples.
- Use target users.
- Select 120 respondents minimum or, if three or more segments are expected, add 50 for each additional segment expected.
- Use a Williams design to order the sample for balance across position of presentation and to balance any possible carry-over effects.
- Use a warm-up sample at the start of each session. This should be something in the middle of the range.
- Inform the respondents that they will be tasting one more product than they actually will be.
- Emulate the normal mode of use for sample presentation, e.g. tea with milk, sushi with soi sauce, but control so that the treatment of each sample is identical.
- Arrange to visit the CLT facility at the start-up on the first day.

23.3   Analyses

23.3.1   Preliminary analyses

Upon receiving the data from the market research company, or from the data preparation team if you have conducted it yourself, it is worth conducting some screening analyses to check what has happened. I find the 'pivot table' option in Excel absolutely invaluable here. The first question is whether the recruiting quotas have been achieved, and frequency tables such as that shown in Table 23.3 quickly tell the tale. If a category scale such as the 9-point hedonic scale has been used then it is usual to calculate the frequency table and the top three and bottom three box percentages as shown in Table 23.4. Note that in this table the two winning products are easily identified. However there are many products where nearly half the respondents expressing high degree of liking and the other

Table 23.3   Number of respondents: gender by consumption frequency

Consumption frequency	Male	Female	Grand total
1	136	112	248
2	168	240	408
3	616	672	1288
4	392	248	640
Grand total	**1312**	**1272**	**2584**

Table 23.4 Overall liking: number of respondents using each category of the 9-point scale

Product name	1	2	3	4	5	6	7	8	9	Grand total	Top three box (%)	Bottom three box (%)
A	7	13	36	52	81	69	43	15	7	**323**	20.1	17.3
B	19	17	34	66	56	64	46	15	6	**323**	20.7	21.7
C	12	19	36	56	77	58	47	10	8	**323**	20.1	20.7
D	27	40	49	58	59	42	30	14	4	**323**	14.9	35.9
E	5	2	7	27	68	84	80	36	14	**323**	40.2	4.3
F	4	6	13	46	61	78	75	33	7	**323**	35.6	7.1
G	10	7	20	39	73	74	64	32	4	**323**	31.0	11.5
H	4	16	39	50	68	68	43	28	7	**323**	24.1	18.3
Grand total	**88**	**120**	**234**	**394**	**543**	**537**	**428**	**183**	**57**	**2584**		

half expressing a high degree of disliking. It is clear that a simple overall liking ranking is not going to encapsulate the opinions of these two groups and a more multidimensional approach is called for. These preliminary tables can be sent to the client almost immediately. The winners and losers are identified and the necessity for further analyses is immediately clear.

23.3.2   Cluster analysis

Cluster analysis may be thought of as a preliminary to preference mapping, although I prefer to view it as the lowest level of preference mapping in terms of assumptions about how the population is behaving. With cluster analysis the only assumption is that not all people are displaying the same liking patterns, and with deterministic cluster analysis we hypothesise that each respondent lies in one of several distinct clusters or is considered an outlier. For deterministic analyses there are two algorithms that are provided by all the main statistical packages, hierarchical cluster analyses and non-hierarchical cluster analyses, and we will briefly outline their mode of operation.

The first stage of any cluster analysis must be to define an index of similarity or dissimilarity between respondents. Recall that for this operation we are going to restrict ourselves to the data matrix containing the liking ratings of N respondents on P products. We are going to transform this matrix to a P by P symmetric matrix of (dis)similarity and then use the clustering algorithm to segment the respondents into clusters as shown in Fig. 23.2. Although there are literally dozens of possible indices of (dis)similarity we will discuss only three, and their analogues for the case where ranking data have been collected.

The key question is whether you want to work with the raw data or adjust respondents for the differences in scale usage. The most commonly used raw index is Euclidean distance, which is a multivariate extension of the Pythagoras distance between two points. The ideas are shown in Fig. 23.3 and the distance

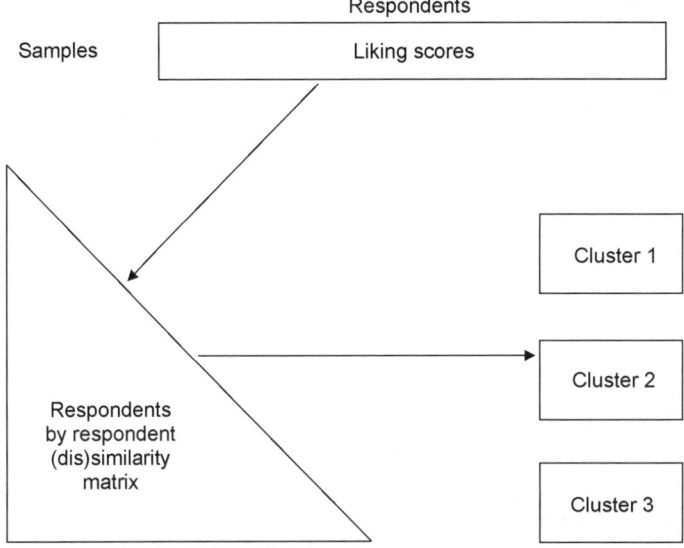

Fig. 23.2   Cluster analysis schemata: liking data are transformed to a similarity or dissimilarity matrix and a clustering algorithm is used to identify the number and memberships of clusters among the respondents.

Liking scores

Product	Mary	Diana	Jane	Sally
A	5	5	2	8
B	6	7	3	6
C	7	8	5	4
D	8	8	7	2

X_1 X_2 X_3 X_4

p products

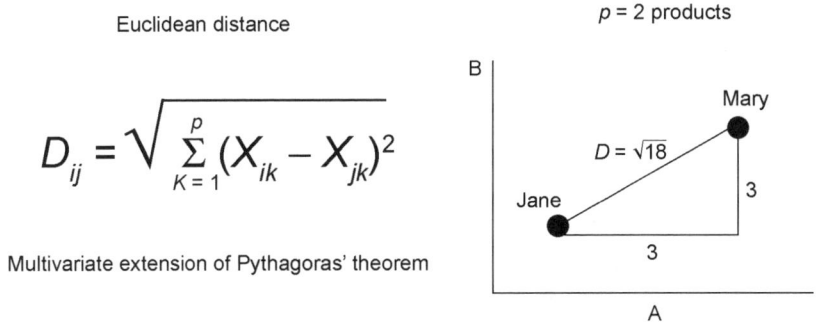

Euclidean distance

$$D_{ij} = \sqrt{\sum_{K=1}^{p}(X_{ik} - X_{jk})^2}$$

Multivariate extension of Pythagoras' theorem

$p = 2$ products

$D = \sqrt{18}$

Fig. 23.3   Calculation of Euclidean distance between two respondents liking scores across all samples. The exact calculation between two respondents for just two samples is shown. For more samples the distance is simply the sum of squared distance between liking scores of corresponding samples in the set.

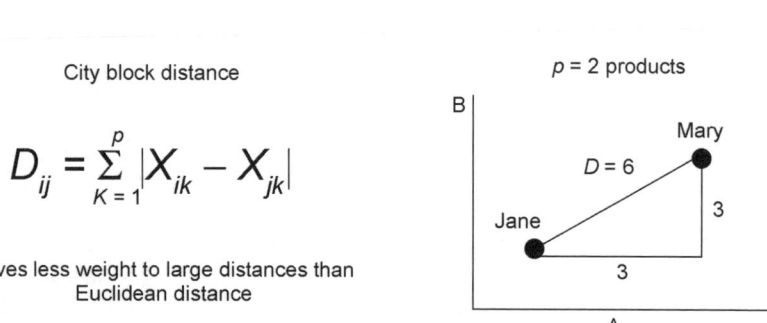

Liking scores

	X_1	X_2	X_3	X_4
Product	Mary	Diana	Jane	Sally
A	5	5	2	8
B	6	7	3	6
C	7	8	5	4
D	8	8	7	2

p products

City block distance

$$D_{ij} = \sum_{K=1}^{p} |X_{ik} - X_{jk}|$$

Gives less weight to large distances than
Euclidean distance

p = 2 products

$D = 6$

Fig. 23.4 Calculation of city block (Manhattan) distance between two respondents liking scores across all samples. The exact calculation between two respondents for just two samples is shown. For more samples the distance is simply the sum of the absolute distance between liking scores of corresponding samples in the set.

matrix between four respondents is also shown. A very useful alternative to the Euclidean distance is the city block distance, also termed the Manhattan distance, so named because it is not possible to walk through a block of buildings and so we have to walk around the perimeter. As it is simply the sum of the absolute distances between corresponding scores on a set of samples between two respondents, it is very easy to calculate. The calculations are shown in Fig. 23.4. The key advantage of this measure over the Euclidean distance is that in the latter a single large difference, when squared, can dominate the result. This is not going to be anywhere near as influential in city block.

Since respondents have not been calibrated in any way before performing the rating task, there is a compelling argument that the variation in range usage is nothing to do with perceived product differences and more to do with personality differences. Standardisation of scores by subtraction of a respondent mean and division by respondent standard deviation is an obvious method of adjustment. However, the calculation of a Pearson correlation coefficient between two respondents' scores is perhaps even more attractive as the measure has a known range and is intuitively understood by experimenters. It also does not require a transformation of the data and so results for cluster means are easily expressed in the raw data. The ideas are shown in Fig. 23.5 and in the resulting correlation matrix in Fig. 23.6 we see that the outlier, Sally, is in fact simply scoring all products inversely to the other respondents.

Having selected a suitable measure of (dis)similarity, we form the symmetric

Liking scores

Product	Mary X_1	Diana X_2	Jane X_3	Sally X_4
A	5	5	2	8
B	6	7	3	6
C	7	8	5	4
D	8	8	7	2

Correlation

$$D_{ij} = \mathrm{Corr}(X_i, X_j)$$

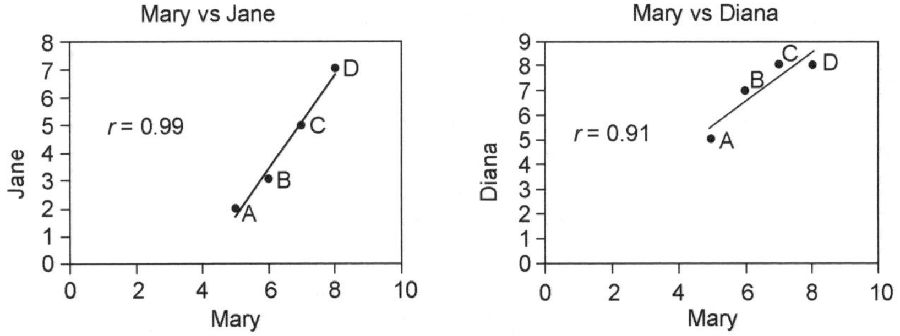

Fig. 23.5 Pearson correlation between two respondents' liking scores across all samples. The correlation and pairwise plot between two pairs of respondents is shown.

matrix that defines the interrelationships between the respondents and move on to select a clustering algorithm. We will discuss hierarchical cluster analysis (sometimes termed agglomerative hierarchical classification, AHC) first. All versions of this algorithm work by first considering each respondent as a separate cluster. Using a single linkage strategy, the two most similar respondents join to form a new cluster at the (dis)similarity level between them, the next two respondents join at the higher similarity level, respondents join clusters

Liking scores

Product	Mary X_1	Diana X_2	Jane X_3	Sally X_4	
A	5	5	2	8	
B	6	7	3	6	p products
C	7	8	5	4	
D	8	8	7	2	

	Correlation between vectors of values			
	Mary	Diana	Jane	Sally
Mary	0	0.91	0.99	−1
Diana	0.91	0	0.85	−0.91
Jane	0.99	0.85	0	−0.99
Sally	−1	−0.91	−0.99	0

Fig. 23.6 The four-respondent correlation matrix that would form the input to a cluster analysis. Respondent Sally clearly likes completely opposite products to the other three.

of other respondents at the dissimilarity level of their nearest neighbour in the cluster. The process by which respondents join each other, and then clusters join each other is accurately recorded in a dendrogram. The basic ideas are demonstrated in Fig. 23.7.

Early users of single linkage analysis found a tendency for the clusters to join up in chains, and so a system in which the clusters joined at the (dis)similarity level of their two furthest neighbours was devised. Although this technique, appropriately termed furthest neighbour analysis, certainly gave more compact clusters, it still depended very much on the relative positions of just two respondents and was found to give unstable solutions. A preferable, and widely adopted, algorithm joins two clusters at the average (dis)similarity level between all possible cross-cluster member pairings. This technique is termed average linkage or, sometimes, unweighted mean pair group. Another widely used algorithm is Ward's method, which attempts to find clusters that show

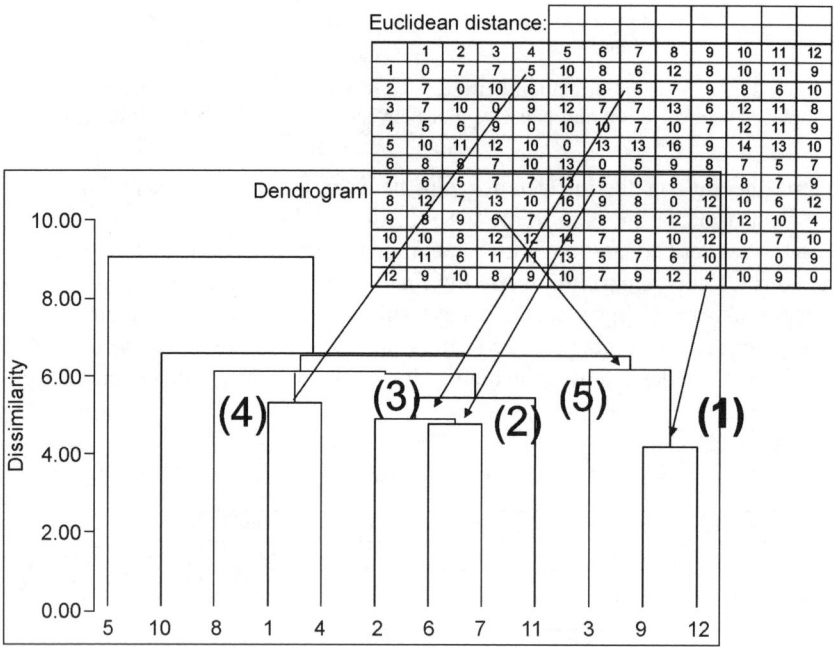

Fig. 23.7 Schematic indicating how a single linkage clustering algorithm applied to Euclidean distance matrix of 12 respondents would create the dendrogram illustrated. In the step (1) respondents 9 and 12 are identified as having the minimum Euclidean distance between each other. These two respondents are placed on the right of the dendrogram and vertical lines drawn up to be joined at the corresponding distance reading on the vertical scale. In step (2) respondents 6 and 7 are identified as being the next closest to each other and their vertical lines are joined at the respective distance. In step (3) respondent 2 is close to 7 and joins the already-formed cluster of respondents 6 and 7. The dendrogram thus faithfully traces the exact formation of clusters, cluster of clusters until all respondents have joined. Clearly respondent 5 is the outlier in this group.

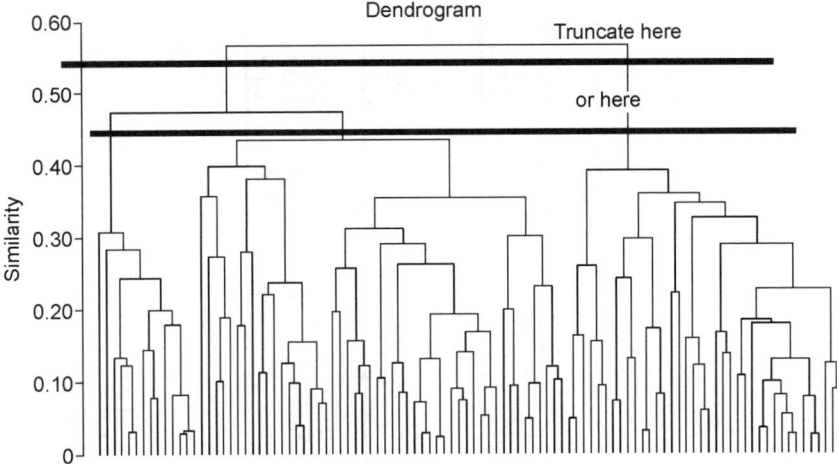

Fig. 23.8   Dendrogram of an average linkage hierarchal cluster analysis based on Euclidean distance of 102 respondents who have rated 10 varieties of apple. The bold horizontal lines represent two possible truncation strategies, leading to two or three clusters.

maximum ratio of between to within variance (similar to F ratio in an analysis of variance).

The number of clusters and their membership is decided by placing a horizontal line across the dendrogram as shown in Fig. 23.8. Most cluster analysis programs have a procedure to detect and identify the number of clusters. However, many users prefer to make that decision themselves. This can be an iterative process, inspecting the clusters at various similarity level cut-offs. In Fig. 23.8 we could take two or three clusters. One point to remember is that if demographic frequency analysis is going to be used by marketing personnel, these clusters should not be too small (i.e. greater than 50). However, sometimes we use cluster analysis as a data reduction process to reduce the number of respondents to a reasonable reduced set for preference mapping and in this case then much smaller clusters can be accommodated.

Once the number of clusters has been determined we can calculate the means and inspecting these enables us to understand what characterises the difference between the clusters, and perhaps give them descriptive labels. Very often they polarise on one or two of the samples. For example in Fig. 23.9 we show the liking means for the two cluster solution and we can see the 'green apple' liker cluster and 'red apple' cluster that are usually identified in dessert apple eating populations in the United Kingdom (Daillant-Spinnler, 1996). The demographic frequency tables enable us to characterise any unusual patterns. For example in Table 23.5 we see that there are more older people in the 'green apple' liker group.

The case study in Fig. 23.10 demonstrates how useful the demographic frequencies can be. The dendrogram revealed four clusters of likers for prototypes

Composition of the clusters:

Cluster	1	2
Within	4310.306	2309.580
Size	64	38

Cluster centroids:

Cluster	Gibson's Green	Johnson's Red	Golden Delicious	Granny Smith	Pink Lady	Fuji	Top Red	Braeburn	Royal Gala	Sun Gold
Cluster 1	4.875	4.697	6.739	6.794	6.219	4.836	3.459	5.928	3.808	6.989
Cluster 2	4.771	9.137	7.592	3.466	4.605	7.176	6.363	4.374	6.574	3.782

Central observations:

Cluster	Gibson's Green	Johnson's Red	Golden Delicious	Granny Smith	Pink Lady	Fuji	Top Red	Braeburn	Royal Gala	Sun Gold
Cluster 1 (L27)	5.800	5.000	5.700	5.100	3.300	5.100	6.200	5.700	1.700	6.600
Cluster 2 (L80)	3.300	9.900	7.900	2.300	2.100	6.900	7.600	1.100	7.000	2.700

These are the observations that appear to lie in the middle of the cluster space

Fig. 23.9 Typical cluster analysis output after truncation. The size of each cluster in terms of number of respondents and the corresponding within groups sums of squares are given at the top of the figure. The table of cluster centroids (means) show the average values for each product using all the respondents in a cluster. It appears that cluster 1 likes the Golden Delicious, Granny Smith, Pink Lady and Sun Gold varieties. Cluster 2 likes Johnson's Red very much and Golden Delicious. Golden Delicious is the best single variety to sell as it is scored well by both clusters. The table of most typical member of each cluster appears at the bottom of the table. This individual has the lowest average distance from every other member of the cluster. This table is often produced by cluster analysis programs but is not generally used in preference mapping applications.

Table 23.5 Demographic variable by cluster frequency tables for the two cluster solution of the data from Figs 23.8 and 23.9. Cluster 1 has more old members (Age 2). Cluster 1 has many more A–B working status members (Class 1). There is not much effect due to whether they have children or not

Count of row	Cluster		Grand total
	1	2	
Age			
1	26	20	**46**
2	38	18	**56**
Grand total	**64**	**38**	**102**
Class			
1	40	10	**50**
2	24	28	**52**
Grand total	**64**	**38**	**102**
Children			
1	36	25	**61**
2	28	13	**41**
Grand total	**64**	**38**	**102**

of a new personal product. Inspecting the means in Fig. 23.11 would indicate that the best prototype to invest in would be Sample E. However, the frequency table in Fig. 23.11 indicates that Cluster One contains primarily purchases of the competitor and Cluster Two contains primarily buyers of our current product. Since the new product is going to be an addition to the portfolio, we will invest in Sample F, liked by all the other clusters except our current purchase group.

In considering which algorithm to use, we believe nearest neighbour and furthest neighbour can be discarded as being too dependent on the relative positions of individual respondents. To decide between average linkage and Ward's is more difficult as both have their supporters. However, it is helpful to consider the role of two types of outliers in this data. There are respondents who give scores that are totally different from the rest of the population and who should therefore join the other clusters last in average linkage. We can term these 'obvious outliers'. Then there are respondents that do not belong to any particular cluster but lie somewhere between two clusters. We can term these respondents 'in-lying outliers'.

In practice with CLT liking data the author has found that average linkage coupled with the Pearson correlation coefficient gives reasonable clusters and is resistant to obvious outliers as these show up on the left hand side of the dendrogram. However it is known that in-lying outliers may influence the formation of clusters and it is for this reason that many users are moving towards probabilistic methods such as latent class analysis. These are discussed further in Section 23.4.

With regard to Ward's method, it must be said that this algorithm certainly gives very compact clusters. However, the author believes, and others have commented on this (Wakeling, personal communication), that obvious outliers,

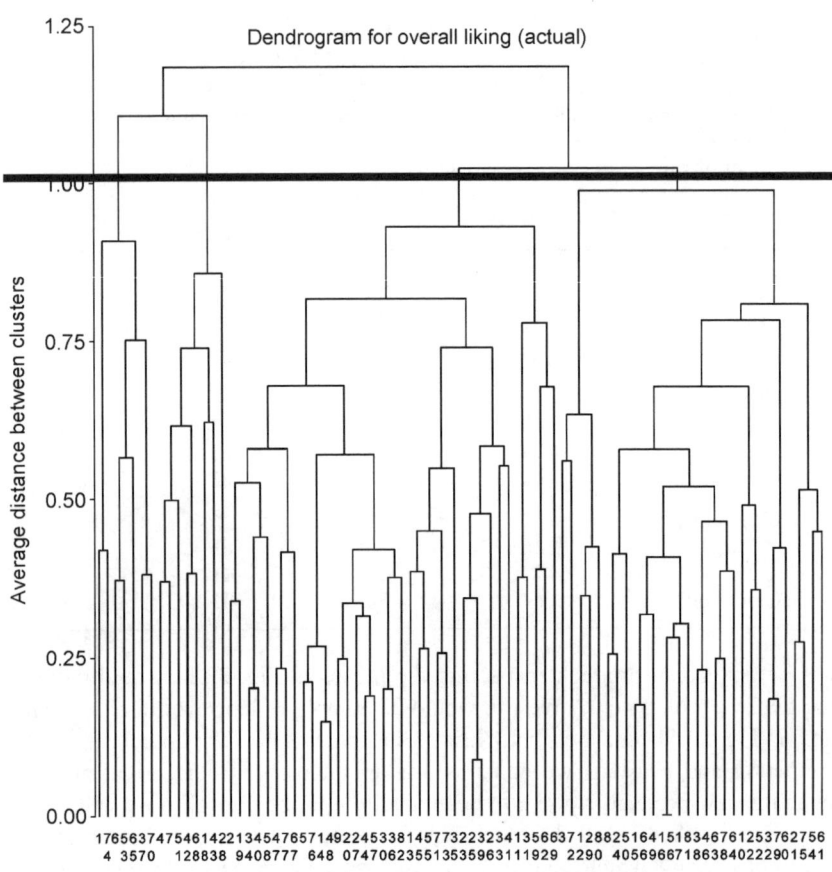

Fig. 23.10 Dendrogram of an average linkage cluster analysis applied to Euclidean distances of respondent liking of a personal product. Although two of the clusters are small, the decision was taken to truncate to establish four clusters.

as well as in-lying outliers, tend to get included in the clusters and can be influential. This makes the solutions unstable. The Ward's method also tends to seek clusters of roughly equal size and this is not necessarily true for food or beverage-based clusters. The author has often found one large cluster, with two or three lesser clusters. So my current recommendation for deterministic clustering is to use a Pearson correlation coefficient (or standardised data) with average linkage and use inspection to determine the level at which to truncate. However, Ward's method sometimes gives the most satisfactory clusters.

23.3.3 Internal preference mapping
In terms of assumptions, the basic tenet of internal preference mapping is that there is a shared perceptual space of a set of foods, beverages or personal products that are perceived to a larger or smaller extent by each respondent. However, the liking

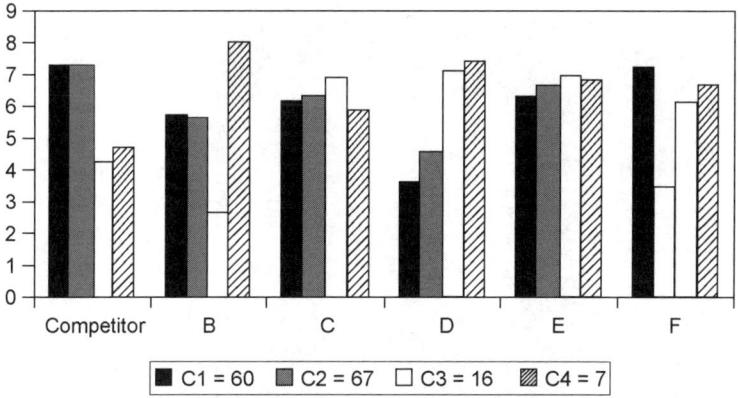

Table of brand by cluster					
Brand (Q7 usual brand)	Cluster				
Frequency	1	2	3	4	Total
Competitor	48	11	7	4	70
Own brand	12	56	9	3	80
Total	60	57	16	7	150

Fig. 23.11 The mean liking scores for the four clusters of Fig. 23.10 are given for the competitor and prototypes B to E. On the surface prototype E looks the best one to invest in. However, the new product is a roll out and the frequency table below indicates that members of Cluster 1 are primarily competitor buyers and Cluster 2 primarily include our current product buyers. Investment in product F offered the best possibility to take share from the competitor with minimum cannibalisation of current product.

of individuals may vary according to which sensory properties they like or do not like. With regard to deterministic internal preference mapping, it is also assumed that individuals focus on a single or highly correlated group of sensory attributes and rate according to how much or how little of those attributes are present. Figure 23.12 shows an idealised version. This is the so-called vector model and vectors are shown for two individuals. Individual 123 is rating in one direction and individual 37 is rating in a different direction. However, both are seen to be basing their ratings on a common underlying configuration of the samples. The other asterisks indicate the ends of the vectors for other individuals. As this is a principal component solution the length of the vector is proportional to the percentage of an individual's scores that are accounted for in the picture. In this idealised example we can see that most are well away from the origin and are there quite well explained. In most plots we will also plot the unit circle which represents the theoretical maximum length of a vector (all variation projects into the space).

The utility of the model is well demonstrated in this figure. It appears that there at least two separate groups of respondents, and we could consider launching two products into this marketplace. One is placed at around 10

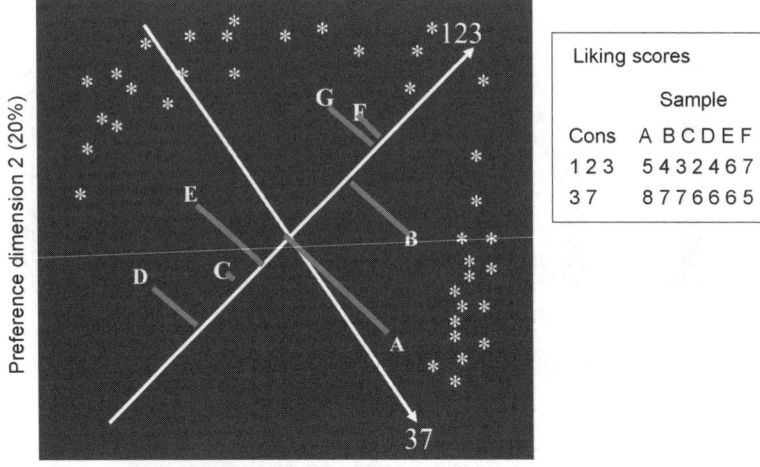

Fig. 23.12   Internal preference mapping stylised solution. The positions of the products in the internal preference mapping space indicate samples that are presumably close perceptually as the respondents score the likings of these products to be the same. However there is a large degree of difference of opinion between respondents. The vectors of increasing liking for respondent 123 and 37 are shown. The asterisks represent the arrow heads of the vectors of remaining respondents.

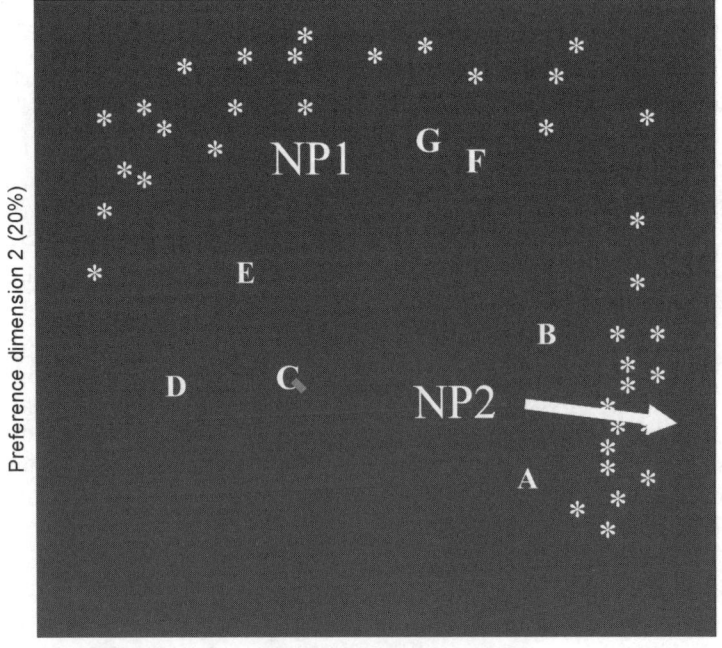

Fig. 23.13   The internal preference map of Fig. 23.12 showing the positions of two possible new products NP2 and NP2 that would compete successfully against the current products.

o'clock (NP1) and another at around 5 o'clock (NP2) as shown in Fig. 23.13.

We obtain the plots of the samples and the vectors of the respondents by applying principal components analysis (PCA) to the liking data as shown in Fig. 23.14. It is usual to apply PCA to the liking data using the correlation matrix, which is equivalent to removing the mean and dividing by the standard deviation of each individual respondent's scores. In Fig. 23.15 a more typical example of internal preference mapping of eight varieties of Gala apples from a CLT with 104 respondents is shown. Note that only 40.3% of the total variance is shown in the first two dimensions. Although this is a good recovery of the major sample differences, the percentage of variance is never going to be near 100% as there will always be a good deal of individual variance, as is shown by the large scatter of individual respondents. Product developer clients will not be happy with plots representing this proportion of variance. However suppose we cluster first, find three clusters (Fig. 23.16) and then do internal preference mapping on the clusters as shown in Fig. 23.17. Now 94% of the variance is explained and it is possible to

Sample	Consumer					Mean
	Fred	John	Kate	Pat	Sue	
A	9	6	2	6	2	5
B	6	9	9	2	6	6.4
C	2	6	9	6	9	6.4
D	6	2	2	9	6	5
E	6	6	6	6	6	6

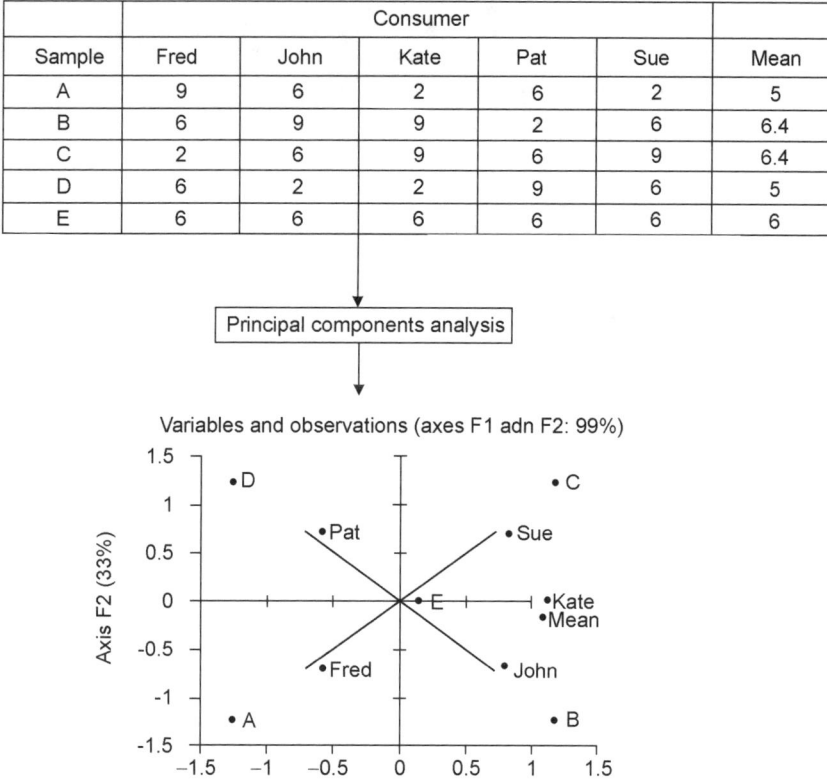

Fig. 23.14 Internal preference mapping is simply principal component analysis of the liking data. The input matrix and the output bi-plot from a principal component analysis of an idealised data set are shown.

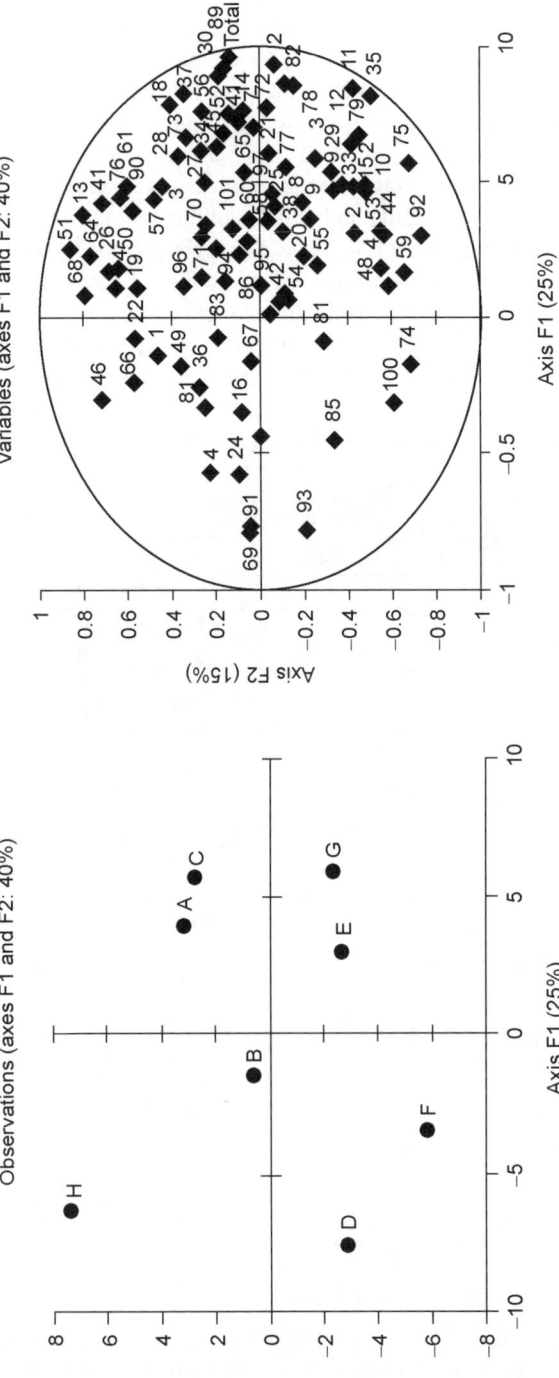

Fig. 23.15 Internal preference mapping of eight gala apples rated for liking by 104 respondents. The figure on the left shows the samples A to F plotted against the first two principal components. The figure on the right shows the 104 respondents. Each point is on the best fit vector of increasing liking for a respondent. The distance of each point from the origin is proportional to how much variation in a respondent's scores are projected into this two-dimensional space.

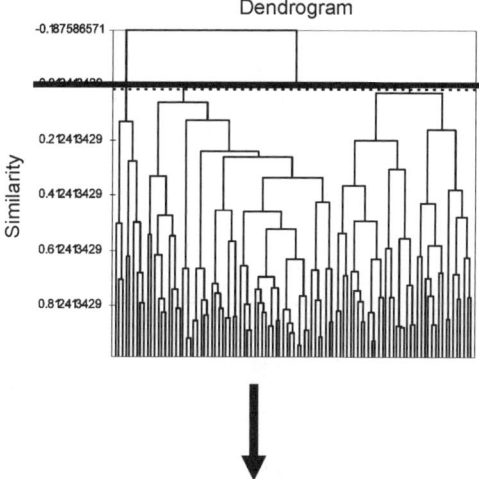

Class centroids:

Class	A	B	C	D	E	F	G	H
1	6.128	5.451	5.992	4.774	6.100	3.674	5.956	6.472
2	6.085	5.553	6.185	4.028	6.085	5.545	6.866	3.709
3	4.478	6.289	5.344	6.967	4.678	6.456	5.633	7.411

Fig. 23.16 Cluster analysis of the gala apple data. Two main clusters are indicated and the centroids are shown.

show both samples and clusters on the same bi-plot; clients will feel more comfortable with this solution. Effectively the individual variance has been compressed as part of the clustering step. It is easy to understand why this is done so often in practical applications of preference mapping.

To project the sensory variables on to these plots we simply calculate the correlation coefficient of the panel sensory means with the PC scores and then use these correlations as coordinates to plot the variables on a graph with range −1 to +1 on each axis as shown in Fig. 23.18. These points have the same properties as the points that represent respondents in the loading plots: variables that plot on to the unit circle are completely projected; variables that plot on the same line passing through the origin are completely correlated in the space that is shown. Similarly, in a bi-plot such as that shown in Fig. 23.18, we can infer what needs to be done to a product to move it around the space. For example, suppose we are currently producing product C in Fig. 23.18 and marketing asks us to move it to match competitor A. We can see that we need to increase 'toughness', decrease the capacity to 'break easily' and leave 'juiciness' and 'large bit' sensory scores the same (as the directions of these attributes from the origin are at right angles to the desired direction of change).

We can use the same rationale to project in the demographic variables such as gender, age, region of origin, etc. We simply calculate the means scores given to each sample by one level of the variable, such as females, and then correlate this into the plot using the same rationale as in Fig. 23.18.

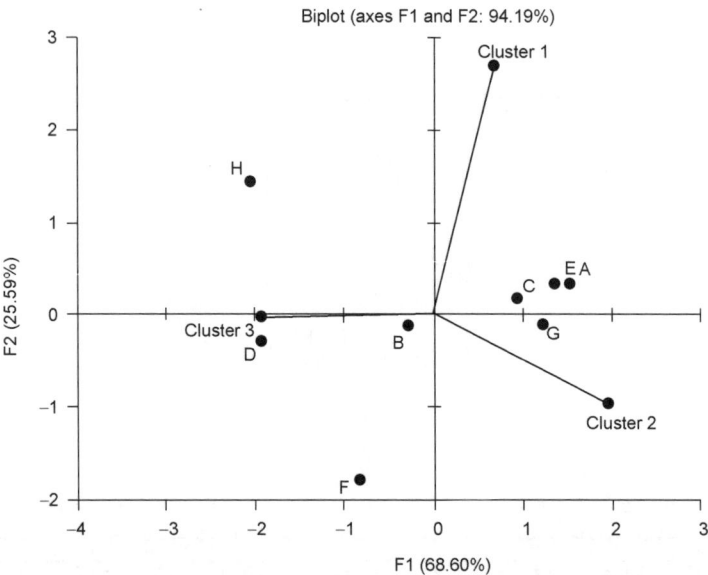

Fig. 23.17   Internal preference mapping of the cluster means (covariance matrix). Notice that the percentage of variation explained in this plot is over 94%, which will be much better received by the clients than the 40% value of Fig. 23.15. A similar configuration of samples is shown, although all the liked samples are compressed.

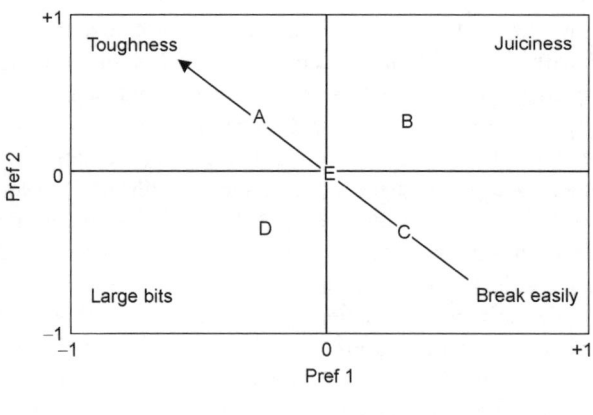

	Pref 1	Pref 2
Toughness	−0.7	0.7
Juiciness	0.7	0.7
Breaks easily	0.4	−0.7
Large bits	−0.7	−0.4

Fig. 23.18   Superimposing the sensory variables on to an internal preference map using correlations with the first two preference dimensions (PCs). The table on the left shows the correlations of four sensory attribute means with the first and second preference scores. The figure on the right indicates the position on rescaled plot with axis end-points ±1. The arrow indicates a direction of increasing toughness.

23.3.4   Identifying the sensory properties of a desired product – reverse regression

The internal preference map with sensory and demographic variables super-imposed gives a valuable view of the products from the marketplace that have been selected for trial and sent out to respondents for central location testing, and the sensory correlates and useful demographic variables. Using cluster means instead of individual respondents simplifies the bi-plot and increases the percentage variance explained. A likely outcome of marketing personnel viewing and understanding this plot is that they identify a position in the space that they would like to move a current product to. The question then arises as to what the sensory properties of this product would be. This problem is solved using reverse regression.

The reverse regression procedure is illustrated and explained using a CLT conducted on mozzarella cheese. These data were reported in Pagliarini *et al.* (1997). The results of the internal preference mapping are shown in Fig. 23.19. The loading plot on the right is a fine example of segmentation. Respondents who plot on the positive side of the first PC like the cow-based cheeses and dislike the (authentic) cheeses made from water-buffalo. Respondents who plot on the negative side of the first PC have exactly the opposite opinion. Suppose that we decide to enter this marketplace with two products that will score best in each segment. In Fig. 23.19 we have denoted two possible product positions P4 ($-25,0$) and P5 ($20,-5$). In practice, one might decide to make few more test products in each region but for the purpose of illustration we will restrict these to two.

The regression set-up is shown in Fig. 23.20. The set of sensory panel mean scores is shown on the right and forms the dependent matrix Y. Labelling the first two PCs of the preference mapping PCA as F1 and F2, we see that F1 and F2 are the independent variables in the regression. For each sensory variable we wish to set up a regression model to predict from F1 and F2, and then we can predict the sensory scores that would be obtained if we set (F1, F2) as P4 ($-25,0$) or P5 ($20,-5$). The results are shown together with the significance of the regression in Fig. 23.21. It is important not to make predictions where the regression is not significant, so in this case we do not give much credence to the predicted values for 'Elastico' as it is not significant.

23.3.5   External preference mapping

External preference mapping is level 3 of the hierarchy of preference mapping methods. Level 1 – cluster analysis – assumes only that the respondents fall into different segments. Level 2 – internal preference mapping – does not necessarily assume different groups but does assume a shared perceptual space. Level 3 – external preference mapping – assumes a shared perceptual space but additionally that we have an external definition of that perceptual space. This is usually the first few principal components of the sensory panel scores, but could also be the first few dimensions of a repertory grid study (Baxter *et al.*,

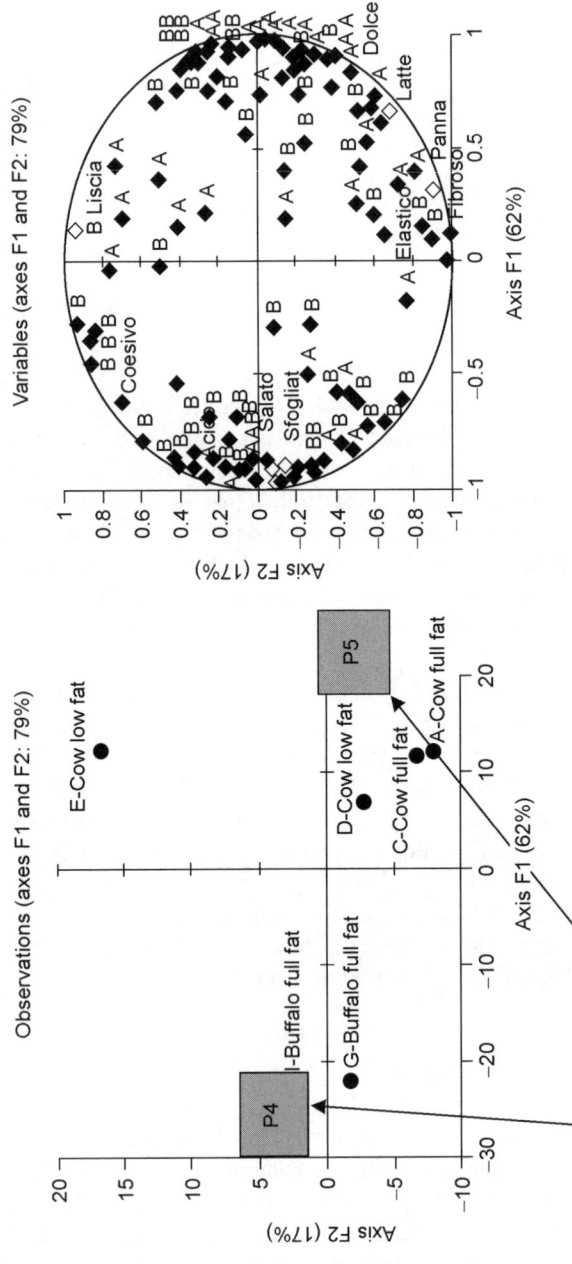

Fig. 23.19 Identifying possible positions that winning products might occupy in the internal preference space. This preference map of mozzarella cheeses indicates a two-segment population. Product P4 should give a higher liking score for that segment that likes buffalo-based cheese. Product P5 should give a higher liking score to the other segment that prefers cow-based cheese. Placing products as indicated would give a score of −25 for P4 on the first dimension and +20 for P5. On the second dimension P4 = 0 and P5 = −5. Note in the figure on the right that the vectors of 12 sensory attributes given by the Italian descriptive panel have been projected.

F1	F2	Acido	Coesivo	Dolce	Elastico	Fibroso	Yogurt	Liscia
12.573	-7.815 A - Cow full fat	2.87	3	5.27	4.97	7.47	3.4	5.63
12.368	-6.576 C - Cow full fat	3	3.2	4.87	5.7	7.13	2.83	6.27
7.055	-2.734 D - Cow low fat	2.9	3.43	5.13	4.23	6.73	3.27	5.73
12.375	16.806 E - Cow low fat	3.6	4	3.57	4.13	3.77	2.87	8.03
-21.650	-1.709 G - Buffalo full fat	4.83	4.13	3.27	4.63	6.03	5.07	6.03
-22.721	2.028 I - Buffalo full fat	5.1	3.77	2.97	4.73	5.87	5.1	6.47
-25.000	0.000 P4	5.09	3.99	3.04	4.63	5.96	5.25	6.18
20.000	-5.000 p5	2.46	3.09	5.43	5.02	7.06	2.62	6.04
	Average	3.72	3.59	4.18	4.73	6.17	3.76	6.36
	RegCoeffs F1	-0.054872	-0.016063	0.045757	0.004154	0.008217	-0.05958	0.007055
	RegCoeffs F2	0.031163	0.034753	-0.066797	-0.041866	-0.146301	-0.01049	0.091888

F1	F2	Salato	Latte	Panna	Sfogliat	Succoso
12.573	-7.815 A - Cow full fat	4.2	5.43	4.27	5.43	5.3
12.368	-6.576 C - Cow full fat	3.57	5	3.73	4.5	4.66
7.055	-2.734 D - Cow low fat	3.3	5.3	3.8	5.4	5.47
12.375	16.806 E - Cow low fat	3.6	3.73	2.3	4.57	4.7
-21.650	-1.709 G - Buffalo full fat	4.7	3.63	3.17	6.23	6.23
-22.721	2.028 I - Buffalo full fat	6	3.83	3	6.87	7.33
-20.000	0.000 P4	5.21	3.85	3.13	6.47	6.69
20.000	-5.000 p5	3.26	5.45	3.98	4.62	4.59
	Average	4.23	4.49	3.38	5.50	5.62
	RegCoeffs F1	-0.048894	0.032057	0.012207	-0.048297	-0.053638
	RegCoeffs F2	-0.001703	-0.063687	-0.072189	-0.016701	-0.009499

Fig. 23.20   The regression set up to predict the sensory properties of P4 and P5. F1 and F2, the internal preference mapping score variables are the independent variables and the sensory attribute scores are the dependent (Y) variables. We then regress each of the sensory variables in turn on the two independent variables and use the model to obtain the predicted values for P4 and P5 shown.

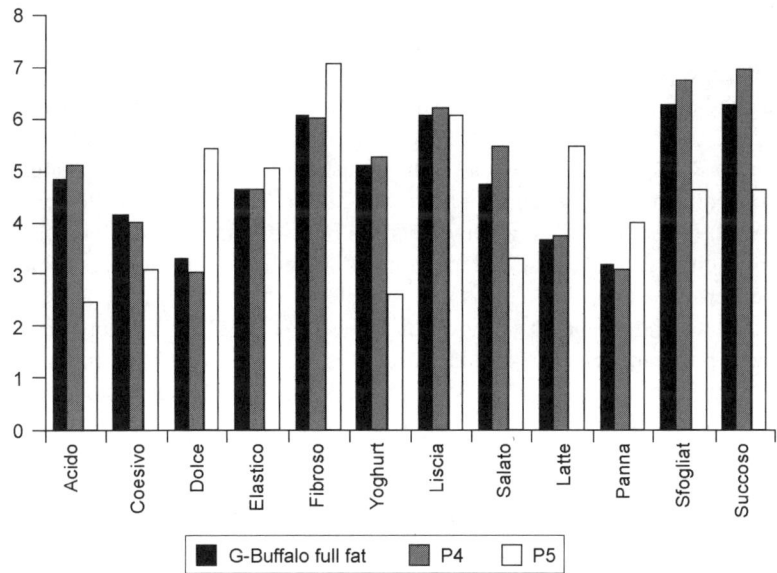

Fig. 23.21   Predicted sensory properties of P4 and P5 plotted against buffalo sample G. This chart shows that P4 (the new buffalo cheese) is predicted to score higher than buffalo on the descriptors *Acido*, *Salato*, *Sfogliat* and *Succoso* that were used by the Italian descriptive panel. P5 (the new cow fat cheese) is very different from the buffalo on these parameters.

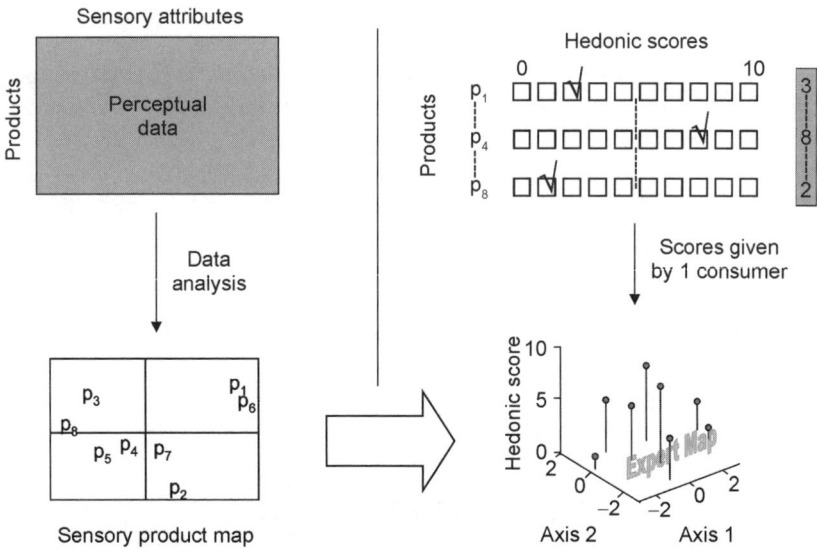

Fig. 23.22 External preference mapping schemata I. The perceptual data on the left is projected down into fewer principal components. These are placed at the base of the figure on the right. Each consumer's scores can then be plotted and a surface visualised. Thanks to J.M. Siefferman for permission to publish this graphic.

1998; Carbonell *et al.*, 2007), or the dimensions of multidimensional scaling trial of pairwise difference cores, or simply the perceptual scores from the consumer responses in the CLT.

In methodological terms, external preference mapping is most accurately described as PC response surface methodology (RSM) since effectively we conduct a RSM analysis of overall liking, cluster overall liking, or separately on each respondent's liking scores. Figures 23.22 and 23.23 illustrate the concepts very well; the sensory data, assumed here to be two dimensional for illustrative purposes only, form the base and the liking of an individual respondent is plotted in the vertical dimension. It is then possible to fit a wide range of possible surfaces to each respondent as shown on the left of Fig. 23.23. The top surface is an ideal point model. The middle surface is an anti-ideal point model. The bottom surface is a so-called saddle surface with an ideal point model in one sensory direction and a vector in the other.

There is thus no need for specialised software to do external preference mapping. Any multiple regression package will do. The most general form of the model for a two dimensional sensory space with components F1 and F2 is:

$$\text{Liking} = a + b1^* \text{ F1} + b2^* \text{ F2} + c1^*\text{F1}^2 + c2^*\text{F2}^2 + d^*\text{F1}^*\text{F2}$$

The full form of this model is termed the quadratic surface model. However, since F1 and F2 are almost always principal components, they are independent of each other and the sum of their cross-products is zero. We thus arrive at the elliptical ideal point model shown in Fig. 23.24 and written as:

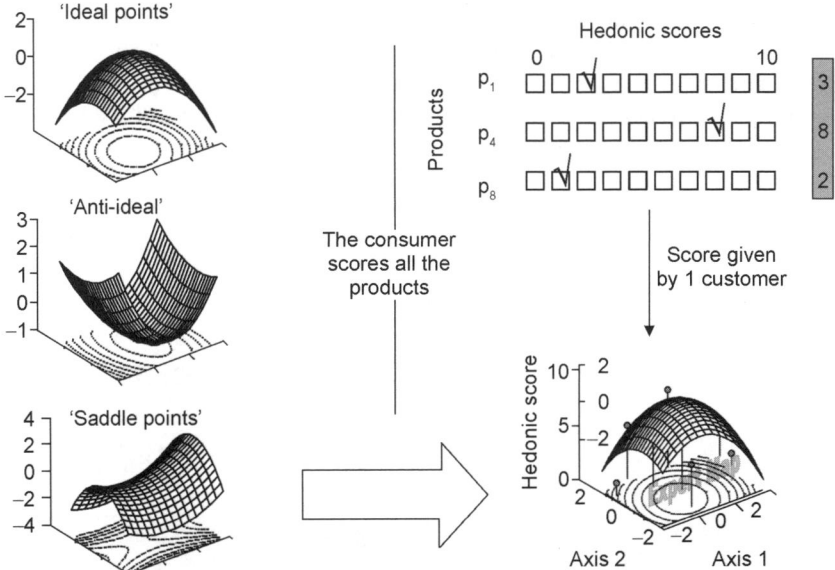

Fig. 23.23 External preference mapping schemata II. The three possible ideal point models that may be used to explain consumer hedonic scores in terms of sensory data are shown. Ideal point models represent consumers where their ideal point is within the current space of the products. Anti-ideal point models represent consumers with a clear idea of what they do not like. Saddle point models represent consumers who have an ideal point in one of the perceptual directions but have a vector model (ideal point outside the current space) in the other. Thanks to J.M. Siefferman for permission to publish this graphic.

$$\text{Liking} = a + b1* \text{F1} + b2* \text{F2} + c1*\text{F1}^2 + c2*\text{F2}^2$$

If $c1 = c2$ we obtain the circular ideal point model shown in Fig. 23.24. (Note that if $c1$ and $c2$ are positive then it is an ideal point model but if they are negative it is an anti-ideal point model.) If $c1$ and $c2$ are not significant we arrive at the vector model similar to that used in internal preference mapping. There is thus a simple hierarchy of models moving from vector to circular and elliptical ideal point models that can be tested in an analysis of variance framework.

The usual method to represent the models on the sensory space is shown in Fig. 23.25. Vector models are represented by an arrow in the direction of increasing liking, where the length of the arrow is proportional to the goodness of fit. Ideal point models are used to calculate the coordinates of the optima in the sensory space and the ideal point is plotted together with an indication of the model type. Thus no brackets indicates an ideal point model, $(-)$ indicates an anti-ideal point model and $(+-)$ indicates a saddle point.

23.3.6 Fitting clusters, individuals and contour plotting

After cluster analysis the fitting task is quite reasonable and it is usual to fit the overall liking mean variable as well. If segmentation is severe it is likely that this will not fit significantly. It also possible to project in the demographic variables on to the plot for interpretive purposes as was done for the internal preference maps.

One problem that will often be encountered is that a cluster will not be fitted significantly by any of the models. One possible reason for this is that the

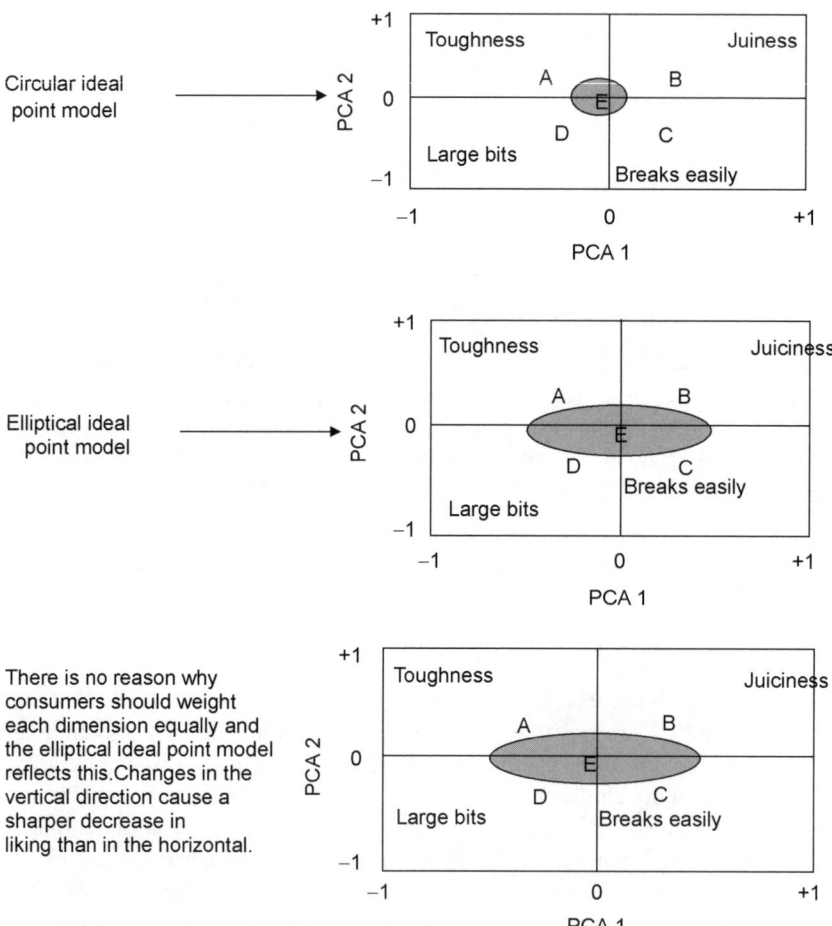

Circular ideal point model

Elliptical ideal point model

There is no reason why consumers should weight each dimension equally and the elliptical ideal point model reflects this.Changes in the vertical direction cause a sharper decrease in liking than in the horizontal.

The mathematical expression for a circular ideal point model
liking = $b0 + b1 (X1) + b2 (X2) + b3 (X1)^2 + b4 (X2)^2$
X1 = PCA 1
X2 = PCA2
$b0$ = intercept $b1$, $b2$, $b3$, $b4$ regression coefficients
The maxima or minima occurs when the first and second derivatives are zero.

Fig. 23.24　A circular ideal model looking down on to the sensory space from above. Shaded region is the ideal area for this consumer. As the product moves in any direction away from this area it is liked less. There is no reason why consumers should weight each dimension equally and the elliptical ideal point model reflects this. Changes in the vertical direction cause a sharper decrease in liking than in the horizontal.

Quadratic surface model

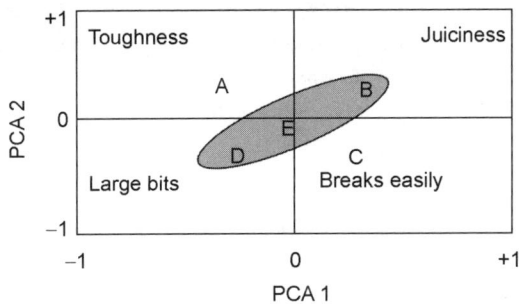

The general quadratic model allows for covariance between the X variables

The mathematical expression for a circular ideal point model
liking = $b0 + b1 (X1) + b2 (X2) + b3 (X1)^2 + b4 (X2)^2 + b5 (X1)(X2)$
X1 = PCA 1
X2 = PCA2
b0 = intercept $b1$, $b2$, b3, $b4$, $b5$ regression coefficients
The maxima or minima occurs when the first and second derivatives are zero.

Fig. 23.24   Continued

External preference map

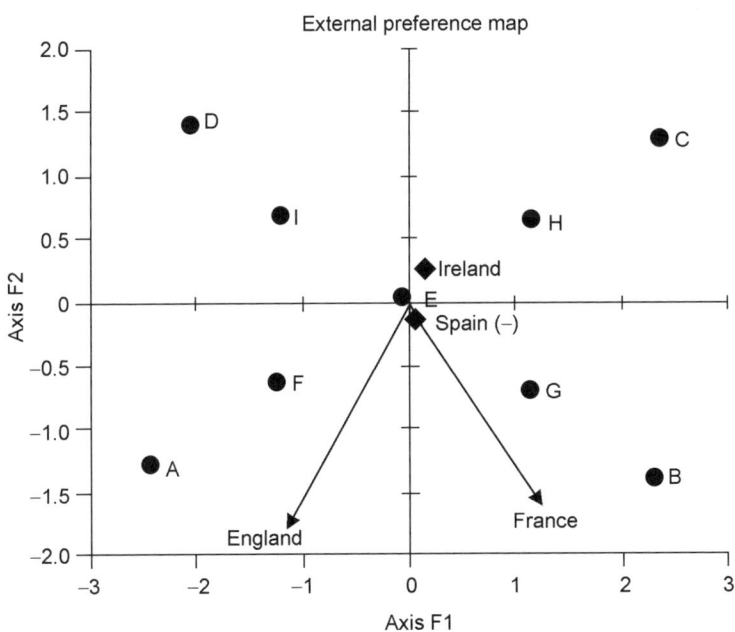

Fig. 23.25   Conventions for displaying the various preference mapping models on to the map. In this map of various countries CLT results on tasting prototypes of a confectionery product, France and England are vector models. Ireland is an ideal point model with the diamond marking the position of the ideal point. Spain is an anti-ideal point.

sensory space on which you are fitting may not be representative of the percep-tual space that is being used by this cluster. For example, suppose you have done the sensory PCA on data that includes appearance, texture, flavour and aroma data. Typically assessors are more comfortable with appearance and texture attributes and these can be highly correlated and tend to dominate the PCA. However, if the non-fitted cluster is sensitive only to differences in aroma and flavour, then repeating the PCA on this subset of attributes will produce a sensory space that is likely to lead to some well-fitting models. It is my current practice to produce spaces separately for appearance, texture and mouthfeel, and flavour aroma and aftertaste, and fit each of the clusters to each of these spaces. This gives maximum possibility to find good fits and makes a lot more interpretive sense. Another possible reason for a non-fitting space is that the clusters are concentrating on a few attributes and a re-weighted PCA is required. This is discussed further by Jaeger *et al.* (1998) who develop a technique of re-weighting based on the correlations of the sensory attributes with the internal preference map space.

Danzart (1998) introduced the idea of defining contours on the sensory space showing the number of consumers who are predicted on the basis of their initially fitted model to score greater than a fixed score, called a cut-off score. This can be very useful to identify regions of suitable prototypes and gives product developers more room in which to innovate, pare costs, etc. On the 9-point hedonic scale a suitable cut-off might be 6.5 to 7.5. Of course, not all consumers will have a significant model so this can mean that many respon-dents' scores are not used, even when using a liberal $p < 0.2$ significance criterion. Using the cluster models and multiplying by the number of people in a cluster is sometimes more convincing.

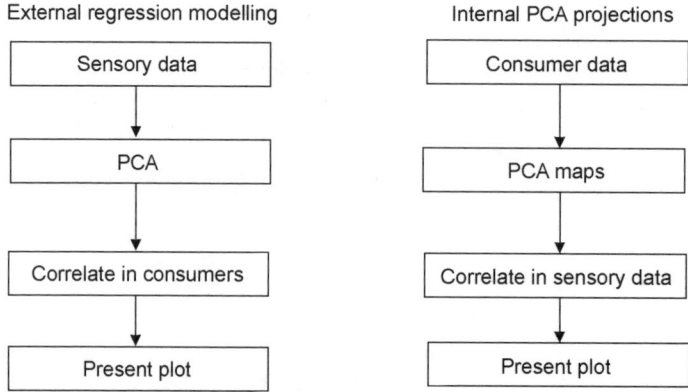

Fig. 23.26 Flow charts denoting the sequence of operations for internal and external preference mapping. The key difference is in the source of the map on which the clusters or respondents will be projected.

23.3.7   Internal versus external preference mapping

The basis of the two approaches are given in Fig. 23.26. The major advantage of Internal preference mapping is that the space is defined from the consumer responses. Many CEOs and marketing personnel would be deeply suspicious of a space defined by ten women who come in three days a week part time, which is exactly how the external space is often defined. We hope that the two spaces are identical because there are many technical advantages to using external. We have a more sophisticated hierarchy of models than the vector based model used by PCA on liking. We can build up contours of liking as discussed in the previous section. If a new product comes on to the market, we can have it profiled by the sensory panel and see where it fits into the sensory space and predict its liking overall or within cluster without doing a CLT. If the spaces are not identical, we should attempt to understand why and search for a re-weighting (PCA on the covariance matrix?) as discussed by Jaeger *et al.* (1998).

23.4   Recent developments in preference mapping

23.4.1   Probabilistic modelling

Probabilistic modelling examines the distribution of points in space and identifies areas where there are high concentrations of individuals in the space. These areas, sometimes called latent classes, are defined on the basis of distributional assumptions, and the probability that each individual belongs to each of the classes is calculated. This effectively reveals the in-lying outliers that lie within the space but are not close to any latent class. Although theoretically superior to hierarchical and non-hierarchical cluster analysis, the experience of the author with the Latent Gold software has not been convincing enough for the author to switch. Courcoux and Chavanne (2001) have discussed the approach in detail, including a version of external preference mapping. MacKay and Zinnes (1995) developed a version of probabilistic multidimensional scaling that is based on ideal point assumptions and recovers a map of products. The software is freely available at the time of writing (www.proscal.com). In a wine tasting example, Mackay (2005) showed that the ideal point recovery is superior to deterministic multidimensional scaling (this is recovery of a space based on distances between points and may not be the same as PCA-based internal preference mapping). The sensory data that are available are fitted separately using a multidimensional scaling solution based on distances between the products and then a common space fit of both the sensory and the consumer space is fitted. In an interesting development, Mackay (2006) demonstrated how to use the model with simulation methods to estimate how often a new product would be liked better against a set of products. It must be emphasised that this is a simulation of how well the product would do in a CLT when tested blind and is not an estimate of how well the branded product would do in a marketplace. Nevertheless this is a very useful development and should be adopted in all preference mapping application software. Ennis and Johnson (1994) have developed a more refined model that allows for individuals to have wandering ideal points based on earlier work by De Soete *et al.* (1986). A commercial package directly targeted at the preference mapping application is available with a very nice graphical interface (www.ifpress.com).

23.4.2   Partial least squares approaches

Partial least squares (PLS) is an algorithm that is applied to estimate a model based on a set of variables in a matrix X that will predict another set of variables based in set Y measured on the same samples. It is thus a kind of multivariate multiple regression package. An obvious way to do this would be to do a PCA on the X set first and then regress each Y separately on the most important PCs. This method, called principal components regression, is widely used but in theory suffers from the fact that the PCs are derived based on the variance in X and do not take any of the Y variation into account. In addition the covariance or correlation between the Y variables is not taken into account. Thus with 100 Y variables that are highly correlated one will have 100 models that are probably all very similar but this is not explicitly revealed in the analysis.

PLS extracts pairs of directions, one from the X set and one from the Y set that display maximum correlation or covariance and explain as much variation as possible in both sets. The correlations between the variables in both sets are thus acknowledged. Successive pairs are extracted that are independent of previous pairs (i.e. PLS1 direction in X is uncorrelated with PLS2 direction in X). The number of pairs extracted is determined by seeing how well the next pair contributes to the prediction of samples when they are left out of the analysis. With no improvement in prediction the process terminates. The set-up for a preference mapping analysis is given in Fig. 23.27 and the solution for the mozzarella cheese application discussed earlier is given in Fig. 23.28. This is almost identical to the internal preference mapping solution. Helgesen *et al.*

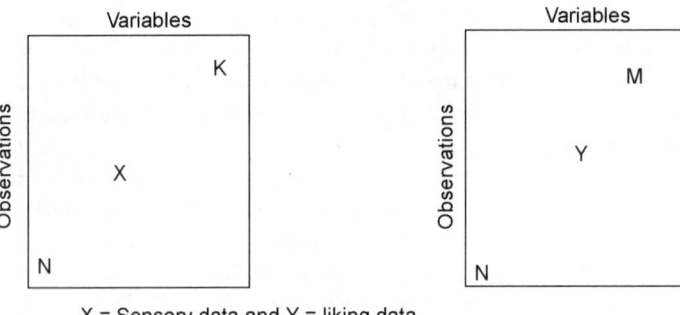

X = Sensory data and Y = liking data
X = principal components of sensory and Y = liking data
X = instrumental and Y = sensory data

Fig. 23.27   Possible PLS data set-ups. X is the data that is going to be used to predict Y. With external preference mapping the X data are the sensory panel mean data and Y is the matrix of individual respondent liking scores.

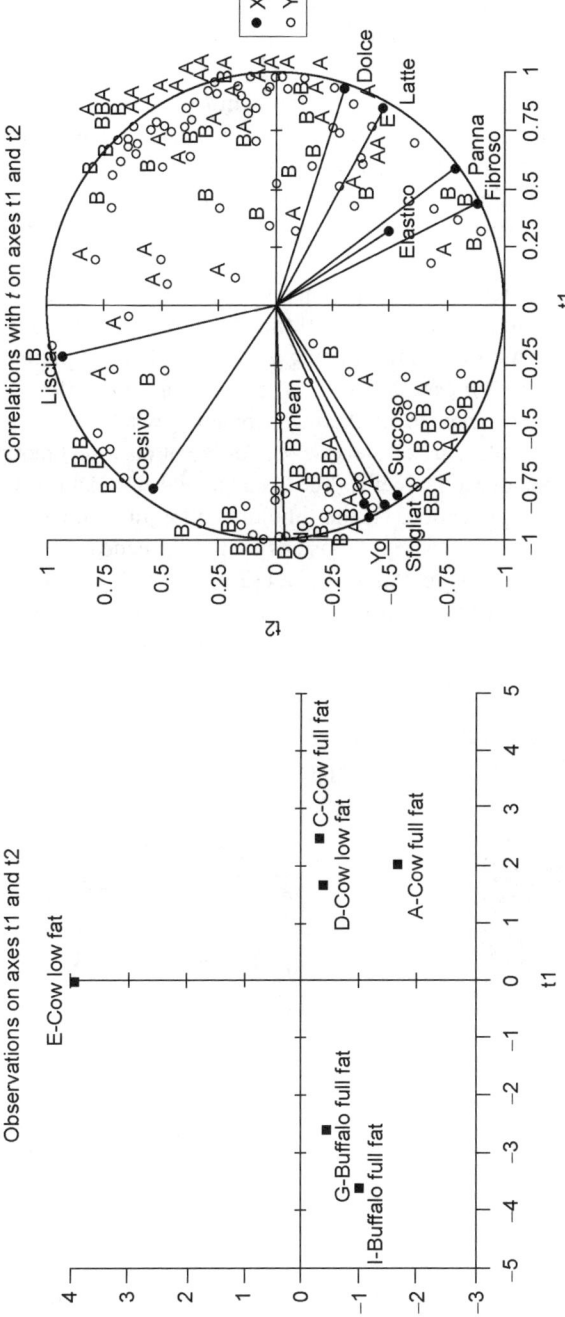

Fig. 23.28 The sample and loading plots of the first two PLS dimensions obtained by fitting the liking data Y to the sensory panel mean data X of the mozzarella data.

(1997) were among the first to apply PLS to a preference mapping application. The method is a linear model so it is effectively fitting a vector model of each person to the sensory data without the need to do a preliminary PCA of the sensory data. If squared terms for the sensory data are added into the X set, then the analysis effectively becomes a quadratic model and ideal points can be incorporated. In my experience this is not often done. PLS is very useful for reverse regression as it is possible to estimate a model for all the sensory attributes in one operation, and prediction is quick and easy to set up. It is not so easy to spot which attributes are not being fitted significantly and attention must be paid to this.

There are a number of developments to the basic PLS model that are potentially very useful. L-shaped PLS (Martens *et al.*, 2005) uses the set-up shown in Fig. 23.29 and is essentially using the consumer demographic variables as part of the fitting process as well as the sensory variables. This is an extremely attractive concept to marketing personnel, who are often frustrated that the segments extracted by cluster analysis or latent class techniques bear no relation to any demographic cells. See Plaehn and Lundahl (2005) for a preference mapping application. At the time of writing the output given by the technique is complex, and simpler visuals will be needed to convince the marketing and R&D personnel that this is a useful development. We need more experience with this technique to find a really convincing example where the fitting process produces a different solution from that obtained simply by fitting the sensory and superimposing the demographic cell mean attributes (e.g. female vector and male vector) on to the map.

A second extension of PLS is to use it to model the Y set when there are different blocks of data in the X space. For example in Fig. 23.30 from

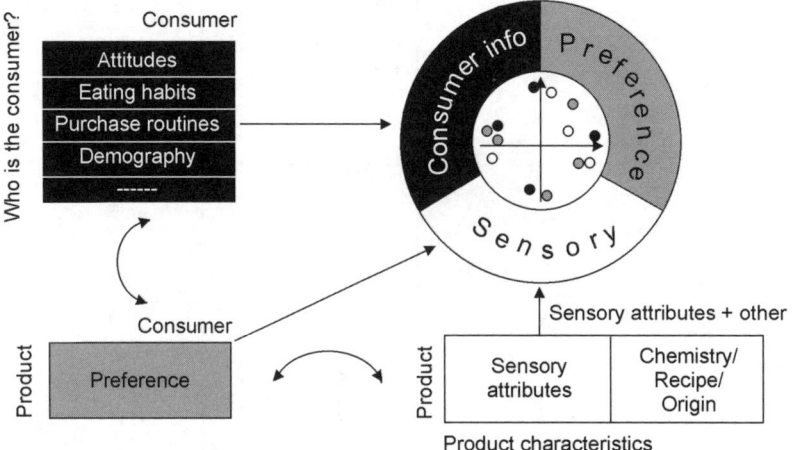

Fig. 23.29   The set-up for L-shaped PLS. In addition to modelling the relationships between liking and sensory or physical chemical, the possible relations between demographic and other personality variables with liking are explored. This figure was presented by Frank Westad of Matforsk at the 2004 Sensometrics conference.

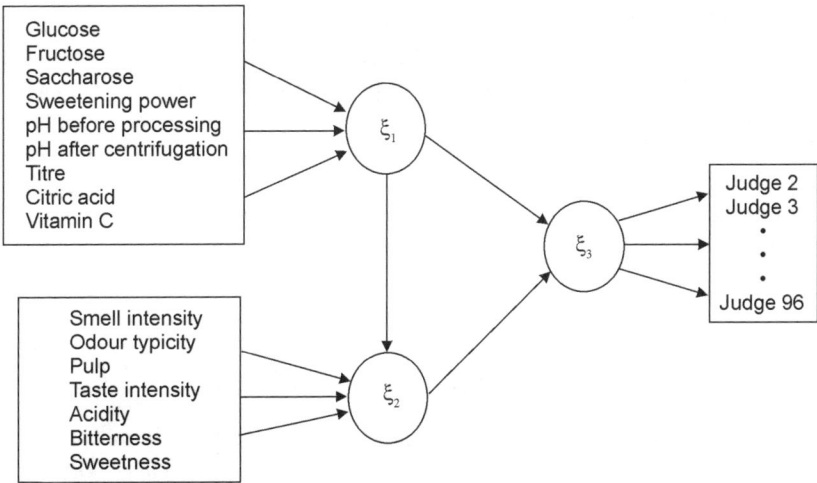

Fig. 23.30 A multi-block set up in which the X set is composed of two blocks of data. The latent variables ξ that are considered to be underlying the physical relations are connected as shown. This figure is reproduced from Tenenhaus *et al.* (2005).

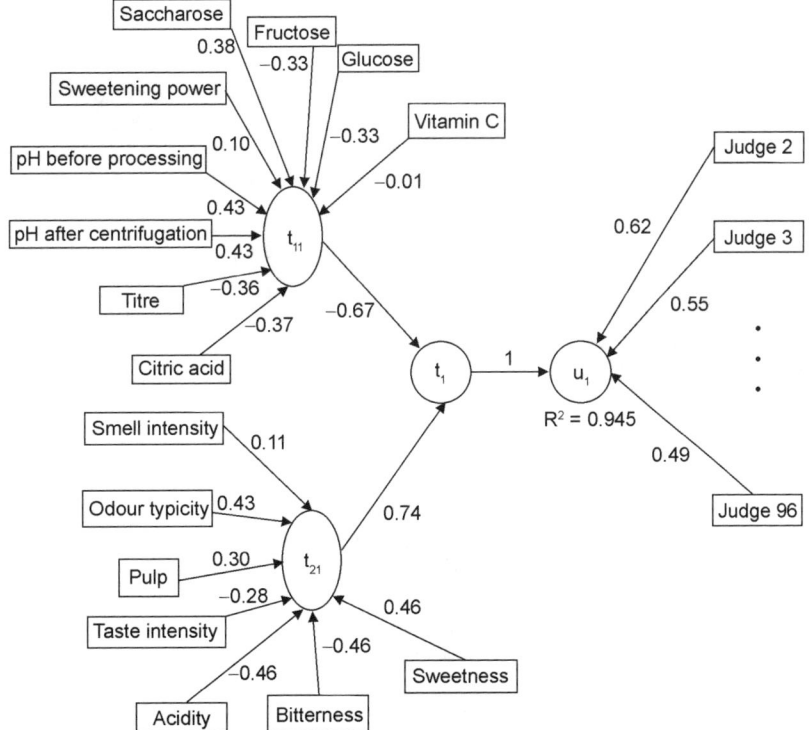

Fig. 23.31 An estimated hierarchical multi-block PLS model. Non-significant loadings are in italics. This figure is reproduced from Tenenhaus *et al.* (2005).

Tenenhaus *et al.* (2005) there is a sensory set and a physicochemical set in the X and we can consider models that might represent how these two components contribute. Although in theory we can apply the structural equation modeling framework developed by Jöreskog and Sörbom (1989), in practice we need over 200 samples for this and we rarely have so much data so multi-block PLS can be very useful here. Wold *et al.* (1996) proposed the extension to hierarchical multi-block models (see Fig. 23.31); the present author believes this to be a most important step forward as we can now start to consider how the senses interact and in what order they come together to influence liking

23.5   Sources of further information and advice

As mentioned in the text, I have attempted to collect as many publications on preference mapping as I can from *Food Quality and Preference*, *Journal of Sensory Studies* and other food journals. It is hoped that the reader will find an application close to their interest. There are sections in many sensory books including Lawless and Heymann (1998).

23.6   Acknowledgement

Thanks are due to Anne Hasted of QI Statistics for Figs 23.3 and 23.4.

23.7   References and further reading

ARDITTI S (1997), 'Preference mapping: a case study', *Food Quality and Preference*, **8**, 323–327.

BALL R D (1997), 'Incomplete designs for the minimization of order and carry-over effects in sensory analysis', *Food Quality and Preference*, **8**(2), 111–118.

BÁRCENAS P, PÉREZ DE SAN ROMÁN R, PÉREZ ELORTONDO F J and ALBISU M (2001), 'Consumer preference structures for traditional Spanish cheeses and their relationship with sensory properties', *Food Quality and Preference*, **12**, 269–279.

BAXTER I A, JACK F R and SCHRODER, M J A (1998), 'The use of repertory grid method to elicit perceptual data from primary school children', *Food Quality and Preference*, **9**, 73–80.

BOWER J A and SAADAT M A (1998), 'Consumer preference for retail fat spreads: an olive oil based product compared with market dominant brands', *Food Quality and Preference*, **9**, 367–376.

BRUECKNER B, SCHONHOF I, SCHROEDTER R and KORNELSON C (2007), 'Improved flavour acceptability of cherry tomatoes. Target group: children', *Food Quality and Preference*, **18**, 152–160.

CARBONELL L, IZQUIERDO L and CARBONELL I (2007), 'Sensory analysis of Spanish mandarin juices. Selection of attributes and panel performance', *Food Quality and Preference*, **18**, 329–341.

CARROLL J D (1972), 'Individual differences and multidimensional scaling', in R. N. Shepard, A. K. Romney and S. Nerlove (Eds.), *Multidimensional Scaling: Theory and Applications in the Behavioral Sciences* (Vol. 1, pp. 105–155). Seminar Press, New York.

CASPIA E L, COGGINS M W, YOON Y and WHITE C H (2006), 'The relationship between acceptability and descriptive sensory attributes in cheddar cheese', *Journal of Sensory Studies*, **21**, 112–127.

CHO H-Y, CHUNG S-J, KIM H-S and KIM K-O (2005), 'Effect of sensory characteristics and non-sensory factors on consumer liking of various canned tea products', *Journal of Food Science*, **70**(8), 532–538.

COURCOUX P and CHAVANNE P C (2001), 'Preference mapping using a latent class vector model', *Food Quality and Preference*, **12**, 369–372.

DAILLANT-SPINNLER B, MACFIE H, BEYTS P and HEDDERLEY D (1996), 'Relationships between perceived sensory properties and major preference directions of 12 varieties of apples from the Southern Hemisphere', *Food Quality and Preference*, **7**, 113–126.

DANZART M (1998), 'Statistiques descriptives, cartographie des préférences', in Coordinateur SSHA, *Evaluation sensorielle: Manuel méthodologique*, pp. 219–246; pp. 290–297, Technique et documentation. Lavoisier, Paris.

DE SOETE G, CARROLL J D and DESARBO W S (1986), 'The wandering ideal point model: a probabilistic model multidimensional unfolding model for paired comparisons data', *Journal of Mathematical Psychology*, **32**, 449–465.

EARTHY P J, MACFIE H J H and HEDDERLEY D (1997), 'Effect of question order on sensory perception and preference in central location trials', *Journal of Sensory Studies*, **12**, 215–237.

ELMORE J R, HEYMANN H, JOHNSON J and HEWETT J E (1999), 'Preference mapping: relating acceptance of 'creaminess' to a descriptive sensory map of a semi-solid', *Food Quality and Preference*, **10**, 465–475.

ENNIS D M and JOHNSON N L (1994), 'A general model for preferential and triadic choice in terms of central F distributions', *Psychometrika*, **59**, 91–96.

FABER N, MOJET J and POELMAN A A M (2003), 'Simple improvement of consumer fit in external preference mapping', *Food Quality and Preference*, **14**, 455–461.

FAYE P, BRÉMAUD D, TEILLET E, COURCOUX P, GIBOREAU A and NICOD H (2006), 'An alternative to external preference mapping based on consumer perceptive mapping', *Food Quality and Preference*, **17**, 604–614.

GEEL L, KINNEAR M and DE KOCK H L (2005), 'Relating consumer preferences to sensory attributes of instant coffee', *Food Quality and Preference*, **16**, 237–244.

GREENHOFF K and MACFIE H (1994), 'Preference mapping in practice', in H.J.H. MacFie and D.M.H. Thomson (eds), *Measurement of Food Preferences*. Blackie Academic & Professional, London (1994), pp. 137–166.

GUINARD J-X, UOTANI B and SCHLICH P (2001), 'Internal and external mapping of preferences for commercial lager beers: comparison of hedonic ratings by consumers blind versus with knowledge of brand and price', *Food Quality and Preference*, **12**, 243–255.

HEDDERLEY D and WAKELING I (1995), 'A comparison of imputation techniques for internal preference mapping, using Monte Carlo simulation', *Food Quality and Preference*, **6**, 281–297.

HELGESEN H, SOLHEIM R and NAES T (1997), 'Consumer preference mapping of dry fermented lamb sausages', *Food Quality and Preference*, **8**, 97–109.

HERSLETH M, BERGGREN R, WESTAD F and MARTENS M (2005), 'Perception of bread: a comparison of consumers and trained assessors' *Journal of Food Science*, **70**(2), S95–S101.

HEYD B and DANZART M (1998), 'Modelling consumers' preferences of coffees: evaluation of different methods', *Lebensmittel-Wissenschaft und -Technologie*, **31**, 607–611.

HORSFIELD S and TAYLOR L J (1976), 'Exploring the relationship between sensory data and acceptability of meat', *Journal of Science and Food Agriculture*, **27**, 1044–1056.

HOUGH G and SÁNCHEZ R (1998), 'Descriptive analysis and external preference mapping of powdered chocolate milk', *Food Quality and Preference*, **9**, 197–204.

HOUGH G, WAKELING I, MUCCI A, CHAMBERS E, MENDEZ-GALLARDO I and RANGEL ALVES L (2006), 'Number of consumers necessary for sensory acceptability tests', *Food Quality and Preference*, **17**, 522–526.

HUON DE KERMADEC F H, DURAND J F and SABATIER R (1997), 'Comparison between linear and nonlinear PLS methods to explain overall liking from sensory characteristics', *Food Quality and Preference*, **8**, 395–402.

JAEGER S R, ANDANI Z, WAKELING I N and MACFIE H J H (1998), 'Consumer preferences for fresh and aged apples: a cross-cultural comparison', *Food Quality and Preference*, **9**, 355–366.

JAEGER S R, WAKELING I N and MACFIE H J H (2000), 'Behavioural extensions to preference mapping: the role of synthesis', *Food Quality and Preference*, **11**, 349–359.

JAEGER S R, ROSSITER K L, WISMER W V and HARKER F R (2003), 'Consumer-driven product development in the kiwifruit industry', *Food Quality and Preference*, **14**, 187–198.

JAEGER S R, LUND C M, LAU K and HARKER F R (2003), 'In search of the 'Ideal' pear (*Pyrus* spp.): results of a multidisciplinary exploration', *Journal of Food Science*, **68**(3), 1108–1117.

JÖRESKOG K and SÖRBOM D (1989), *LISREL 7 – A Guide to the Program and its Applications*, 2nd edn, SPSS Publications, Chicago.

KÄLVIÄINEN N, SALOVAARA H and TUORILA H (2002), 'Sensory attributes and preference mapping of muesli oat flakes', *Journal of Food Science*, **67**(1), 455–460.

LAWLESS H and HEYMANN H (1998), *Sensory Evaluation of Food: Principles and Practice*. Chapman and Hall, New York.

LÊ S and LEDAUPHIN S (2006), 'You like tomato, I like tomato: segmentation of consumers with missing values', *Food Quality and Preference*, **17**, 228–233.

MACFIE H J H, BRATCHELL N, GREENHOFF K and VALLIS L V (1989), 'Designs to balance the effect of order of presentation and first-order carry-over effects in hall tests', *Journal of Sensory Studies*, **4**, 129–148.

MACKAY D (2005), 'Probabilistic scaling analyses of sensory profile, instrumental and hedonic data', *Journal of Chemometrics*, **19**, 180–190.

MACKAY D (2006), 'Chemometrics, econometrics, psychometrics – how best to handle hedonics?', *Food Quality and Preference*, **17**, 529–535.

MACKAY D and ZINNES J (1995), 'Probabilistic multidimensional unfolding: an anisotropic model for preference ratio judgments', *Journal of Mathematical Psychology*, **39**, 99–111.

MARTENS H, ANDERSSEN E, FLATBERG A, GIDSKEHAUG L, HØY M and WESTAD F (2005) 'Regression of a data matrix on descriptors of both its rows and of its columns via latent variables: L-PLSR', *Computational Statistics and Data Analysis*, **48**, 103–123.

MARTÍNEZ C, SANTA CRUZ J M, HOUGH G and VEGA M J (2002), 'Preference mapping of cracker type biscuits', *Food Quality and Preference*, **13**, 535–544.

MATTILA V (2003), 'Semantic analysis of speech quality in mobile communications: descriptive language development and mapping to acceptability', *Food Quality and Preference*, **14**, 441–453.

MCEWEN J A (1996), 'Preference mapping for product optimization', in Næs, T. and Risvik, E. (eds), *Multivariate Analysis of Data in Sensory Science*. Elsevier Science, Amsterdam.

MCEWAN J A and THOMSON D M H, (1989), 'The repertory grid method and preference mapping in market research: a case study on chocolate confectionery', *Food Quality and Preference*, **1**, 59–68.

MEULLENET J-F, XIONG R, MONSOOR M A, BELLMAN-HOMER T, DIAS P, ZIVANOVIC S, FROMM H and LIU Z (2002), 'Preference mapping of commercial toasted white corn tortilla chips', *Journal of Food Science*, **67**, 1950–1957.

MONTELEONE E, FREWER L, WAKELING I and MELA D J (1998), 'Individual differences in starchy food consumption: the application of preference mapping', *Food Quality and Preference*, **9**, 211–219.

OKAYASUAND H and NAITO S (2001), 'Sensory characteristics of apple juice evaluated by consumer and trained panels', *Journal of Food Science*, **66**(7), 1025–1029.

PAGLIARINI E, MONTELEONE E and WAKELING I (1997), 'Sensory profile description of mozzarella cheese and its relationship with consumer preference', *Journal of Sensory Studies*, **12**, 285–301.

PLAEHN D and LUNDAHL D S (2005), 'An L-PLS preference cluster analysis on French consumer hedonics to fresh tomatoes', *Food Quality and Preference*, **17**, 243–256.

POPPER R, ROSENSTOCK W, SCHRAIDT M and KROLL B J (2004), 'The effect of attribute questions on overall liking ratings', *Food Quality and Preference*, **15**, 853–858.

POSRI W (2001), 'Improving the predictability of consumer preferences from Central Location Test (CLT) in tea', The University of Reading, UK, Ph.D. thesis.

PRESCOTT J, YOUNG O and O'NEILL L (2001), 'The impact of variations in flavour compounds on meat acceptability: a comparison of Japanese and New Zealand consumers', *Food Quality and Preference*, **12**, 257–264.

RICHARDSON-HARMAN N, STEVENS R, WALKER S, GAMBLE J, MILLER M, WONG M and McPHERSON A (2000) 'Mapping consumer perceptions of creaminess and liking for liquid dairy products', *Food Quality and Preference*, **11**, 239–246.

RIVIÈRE P, MONROZIER R, ROGEAUX M, PAGÈS J and SAPORTA G (2006), 'Adaptive preference target: contribution of Kano's model of satisfaction for an optimized preference analysis using a sequential consumer test', *Food Quality and Preference*, **17**, 572–581.

SCHIFFMAN S S, REYNOLDS M L and YOUNG F W (1981), *Introduction to Multidimensional Scaling. Theory, Methods, and Applications*. Academic Press, New York.

SERRANO-MEGÍAS M, PÉREZ-LÉPEZ A J, NÓÑEZ-DELICADO E, BELTRÁN F and LÓPEZ-NICOLÁS J M (2005), 'Optimization of tropical juice composition for the Spanish market', *Journal of Food Science*, **70**(1), 28–33.

SUWANSRI S, MEULLENET J-F, HANKINS J A and GRIFFIN K (2002), 'Preference mapping of domestic/imported jasmine rice for US-Asian consumers', *Journal of Food Science*, **67**(6), 2420–2431.

TENENHAUS M, PAGÈS J, AMBROISINE L and GUINOT C (2005), 'PLS methodology to study relationships between hedonic judgements and product characteristics', *Food Quality and Preference*, **16**, 315–325.

THOMPSON J L, DRAKE M A, LOPETCHARAT K and YATES M D (2004) 'Preference mapping of commercial chocolate milks', *Journal of Food Science*, **69**, 406–413.

THYBO A K, KÜHN B F and MARTENS H (2004), 'Explaining Danish children's preferences for apples using instrumental, sensory and demographic/behavioural data', *Food Quality and Preference*, **15**, 53–63.

VAN KLEEF E, VAN TRIJP H C M and LUNING P (2005), 'Consumer research in the early stages of new product development: a critical review of method and techniques', *Food Quality and Preference*, **16**(3), 181–201.

VIGNEAU E and QANNARI E M (2002), 'Segmentation of consumers taking account of external data. A clustering of variables approach', *Food Quality and Preference*, **13**, 515–521.

VIGNEAU E, QANNARI E M, PUNTER P H and KNOOPS S (2001), 'Segmentation of a panel of consumers using clustering of variables around latent directions of preference', *Food Quality and Preference*, **12**, 359–363.

WAKELING I N and MACFIE H J H (1995), 'Designing consumer trials balanced for first and higher orders of carry-over effect when only a subset of k samples from t may be tested', *Food Quality and Preference*, **6**, 299–308.

WESTAD F, HERSLETH M and LEA P (2004), 'Strategies for consumer segmentation with applications on preference data', *Food Quality and Preference*, **15**, 681–687.

WILLIAMS E J (1949), 'Experimental designs balanced for the estimation of residual effects of treatments', *Australian Journal of Science Research A*, **2**, 149–168.

WOLD S, KETTANEH N and TJESSEM K (1996), 'Hierarchical multi-block PLS and PC models for easier interpretation and as an alternative to variable selection', *Journal of Chemometrics*, **10**, 463–482.

XIONG R and MEULLENET, J F (2004), 'Application of multivariate adaptive regression splines (MARS) to the preference mapping of cheese sticks', *Journal of Food Science*, **69**, 131–139.

YACKINOUS C, WEE C and GUINARD J-X (1999), 'Internal preference mapping of hedonic ratings for ranch salad dressings varying in fat and garlic flavour', *Food Quality and Preference*, **10**, 401–409.

YOUNG N D, DRAKE M, LOPETCHARAT K and McDANIEL M (2004), 'Preference mapping of cheddar cheese with varying maturity levels', *Journal of Dairy Science*, **87**, 11–19.

Index

UNIVERSITY OF STRATHCLYDE

2 8 JUN 2007

UNIVERSITY LIBRARY